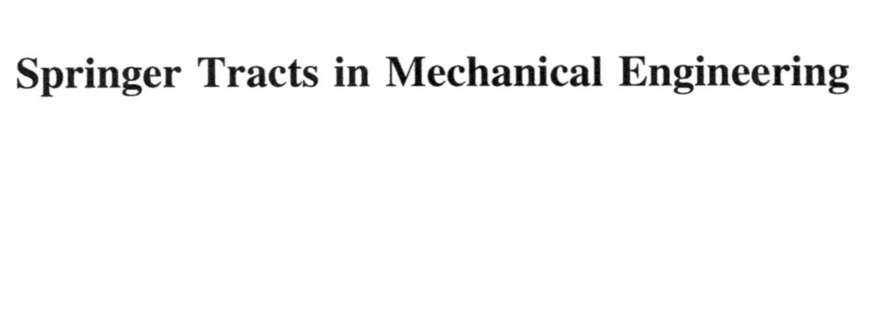

Springer Tracts in Mechanical Engineering

For further volumes:
http://www.springer.com/series/11693

K. J. Vinoy · G. K. Ananthasuresh
Rudra Pratap · S. B. Krupanidhi
Editors

Micro and Smart Devices and Systems

Editors
K. J. Vinoy
Electrical Communication Engineering
Indian Institute of Science
Bangalore
Karnataka
India

Rudra Pratap
Centre for Nano Science and Engineering
Indian Institute of Science
Bangalore
Karnataka
India

G. K. Ananthasuresh
Mechanical Engineering
Indian Institute of Science
Bangalore
Karnataka
India

S. B. Krupanidhi
Materials Research Centre
Indian Institute of Science
Bangalore
Karnataka
India

ISSN 2195-9862 ISSN 2195-9870 (electronic)
ISBN 978-81-322-1912-5 ISBN 978-81-322-1913-2 (eBook)
DOI 10.1007/978-81-322-1913-2
Springer New Delhi Heidelberg New York Dordrecht London

Library of Congress Control Number: 2014938988

Printed on acid-free paper

Springer is part of Springer Science+Business Media (www.springer.com)

Dedicated to

Prof. Vasudev K. Aatre

for his inspiring vision, unwavering conviction, and tireless effort that have resulted in creating and nurturing a vibrant multidisciplinary research field of micro and smart systems in India

Prof. Vasudev Kalkunte Aatre was born in 1939 in Bangalore where he also spent most of his childhood and formative years. He obtained his B.E. from UVCE (then under Mysore University) in 1961, M.E. from the Indian Institute of Science (IISc), Bangalore in 1963, and Ph.D. from the University of Waterloo, Canada, in 1967, all in Electrical Engineering. He worked as Professor of Electrical Engineering at Technical University of Nova Scotia, Canada, from 1968 to 1980. He was also a Visiting Professor at IISc in 1977. In 1980 he joined the Defence Research and Development Organisation (DRDO) of India.

Prof. Aatre worked in India's Ministry of Defence in various capacities for 24 years. He started his career in DRDO as a Principal Scientific Officer (1980–1984). Subsequently, he became the Director of the Naval Physical Oceanographic Laboratory (1984–1991), the Chief Controller (1991–1999), and finally, led the organization as the Director General and Scientific Advisor to Defence Minister (1999–2004). During this long period of dedicated service, he designed and developed sonar suites for surface ships, submarines, and the air arm of the Indian Navy. He was also instrumental in the development of integrated electronic warfare systems for the Indian Army, Navy, and Air Force, and he established GaAs MMIC fabrication facility and VLSI design centers for the Ministry of Defence. Prof. Aatre is also the founding president of the Institute of Smart Structures and Systems (ISSS) and has led the national programs on smart materials and micro and smart systems.

He has published over 60 papers in the fields of active filters, digital signal processing, and defense electronics, and has two books entitled Network Theory and Filter Design and Micro and Smart Systems, both published by John Wiley & Sons. He is a Fellow of the IEEE (USA) and the National Academy of Engineering (India), a Distinguished Fellow of IETE (India) and several other societies. Dr. Aatre is the recipient of the prestigious Padma Bhushan Award of the Government of India.

Foreword

Since the dawn of civilization, Nature has been man's greatest teacher. We have learned by observing and mimicking Nature and natural phenomena with the ultimate goal of building systems as complex, efficient, and optimal as biological systems created by Nature. Such systems, if they have to mimic biological systems, need to continuously sense the environment and respond, to a degree, optimally to achieve certain objectives or perform certain tasks. Although over the centuries, especially for the last century and a half, man has developed materials, devices, and systems, which have found application in myriad fields, competing with natural systems is a dream yet to be fulfilled. The recent advances in smart, micro, and nano systems have opened up a possibility of achieving this goal.

Institute of Smart Structures and Systems (ISSS) was started by a group of scientists, technologists, and engineers in India from academic institutions, space and defense departments in 1998 to trigger research and development in potentially highly application-oriented areas of micro and smart systems. ISSS actively participated in formulating two National Programs—National Program on Smart Materials (NPSM) followed by the National Program on Micro and Smart Systems (NPMASS), both sponsored by the five Scientific Departments of the Government of India and funded by the Department of Defence, India.

While setting up infrastructural facilities such as MEMS foundries, LTCC packaging facility, and developing sensors, actuators and subsystems for aeronautical, automobile, and biomedical applications were the principal goals of NPSM and NPMASS, supporting research projects in materials, sensors, and actuators and developing human resources in this area were equally important goals of the two programs. Towards this, the two national programs sponsored several R&D projects to academic institutions and national laboratories besides establishing 65 National MEMS Design Centers (NMDCs) in institutions across the country. These institutions and centers have conducted research, trained undergraduate and postgraduate students, thus creating a large body of human resources capable of pursuing developments in the general area of micro and nano systems. This special edition gives a glimpse of the R&D work carried out in these institutions and centers.

The contents of this special edition clearly bring out two facts. The first and foremost is the large number of institutions involved in such R&D work and their geographical spread in India. This augurs well for the future development of this field and for the development of the required human resources thereof. The second is that the R&D activities cover the entire spectrum of the field from materials to systems and applications.

The founding members of ISSS were guided by one conviction that "India had missed the microelectronic revolution but should not miss the micro-machine and nano revolution." The happenings of the last decade and a half give great hope. I wish special editions like this were brought out once in three years to coincide with the triennial International Conference organized by ISSS.

Bangalore, March 2014 V. K. Aatre

Preface

This book covers multiple facets of micro and smart systems technologies. Miniaturization of sensors and actuators through effective use of smart materials forms the core of the book. Related aspects of material processing and characterization; modeling and simulation; and applications are also given due importance. Twenty nine chapters written by competent research teams from academia and government research labs comprise a valuable resource that gives a bird's-eye view of the state of the art of the field in India. While the technological details of the work described in this book are self-explanatory, it is pertinent to introspect on how it all happened in India not too long after the miniaturization revolution transpired elsewhere in the world.

Generous financial support and guidance from the government, vision and driving force of a leader, and a professional society that can enthuse an able workforce are perhaps three necessary factors to initiate and establish a new research area in a country. India has had all of these in the last 15 years to lay a firm foundation for micro and smart systems technologies. First, the Defence Research and Development Organisation (DRDO) and four other science and technology departments of the Government of India, initiated and ran two large research programs, namely, National Programme on Smart Materials (NPSM, 2000–2006) and National Programme on Micro and Smart Materials and Systems (NPMASS, 2007–2014), with a combined budget of nearly Rs. 270 crores ($45 M today). Second, Prof. V. K. Aatre, gave unstinted leadership and support to numerous researchers and research administrators whom he inspired and nurtured. Third, a professional society, ambitiously christened, the *Institute of Smart Structures and Systems* (ISSS) was founded in 1998 to bring together experts from multiple disciplines to create a research community in micro and smart systems in India. As a result of these efforts, India today is proud to claim its presence in the field. This edited monograph, with the exception of one chapter, is a record of the work done in India and thus it stands as a testimony to the success of a well-conceived and ably executed endeavor.

Many researchers from the academia and government research laboratories contributed to NPSM and NPMASS, which were admirably administered by the Aeronautical Development Agency (ADA) under the guidance of the Board for Smart materials Research and Technology (B-SMART). Constant support from the past and present Heads of DRDO and its higher management has helped run these programs well. Program offices of NPSM and NPMASS, which operated out of ADA, Bangalore, since 2000, did exemplary work in liaising with various arms of the programs and grantees, bringing synergy and effective program management. The chairs and members of Programme Assessment and Recommendation Committees (PARCs) looked after the technical details of the funded projects. The result of the untiring efforts of all these and many more individuals—too many to mention here—is widespread awareness of micro and smart systems technologies and engagement into research and development activities in almost all parts of India. NPSM and NPMASS have paid particular attention to human resource development by establishing more than 65 National MEMS Design Centres (NMDCs) in many states of India covering the length and breadth of the country. Hundreds of researchers have been involved in more than 150 projects funded by NPSM and NPMASS.

The most significant outcome of this concerted effort is that the spirit of multidisciplinary research in micro and smart technologies now pervades all parts of India. Most researchers began with modeling and design. Not too long ago in India, possessing a license of a microsystems simulation software meant being engaged in research in this area. But today it has changed; with the establishment of state-of-the-art well-equipped cleanrooms and characterization facilities, researchers in India are able to fabricate and even package devices and systems. All aspects of the field, development of microsensors and microactuators; material processing and characterization; fabrication; advanced modeling, design, and simulation; and systems design have all begun. Packaging and transfer of technology have also commenced. The chapters in this book are indeed organized accordingly.

The final link in this chain of events is commercialization of the developed technology. This step needs conscious effort and copious resources, perhaps an order of magnitude more than what went into creating the able research community. The time is now ripe to involve the established industries and to nurture entrepreneurship. One hopes that the same level of commitment and financial support will be given to incubating companies in micro and smart technologies in order to create a thriving industry in these areas in India.

Bangalore, April 2014

K. J. Vinoy
G. K. Ananthasuresh
Rudra Pratap
S. B. Krupanidhi

Acknowledgments

We thank all the contributors for their timely response. We also thank all the reviewers who read the chapters and made valuable suggestions for improvement. Ms. Meera Rao copy-edited early versions of the chapters. Dr. Santosh D. B. Bhargav helped with formatting and Mr. T. S. Bharath scanned numerous corrected sheets. Dr. K. Vijayaraju and Mr. V. Sudhakar of Aeronautical Development Agency (ADA), Bangalore, gave valuable suggestions during the preparation of this book. This book owes much to the Institute of Smart Structures and Systems (ISSS) for its support. Finally, we thank the Springer team comprising Mr. Aninda Bose, Ms. Kamiya Khatter, Ms. P. Kavitha, and Ms. Nathalie Jacobs for helping us throughout.

Bangalore, April 2014

K. J. Vinoy
G. K. Ananthasuresh
Rudra Pratap
S. B. Krupanidhi

Contents

Contributors

G. K. Ananthasuresh Computational Nanoengineering (CoNe) group and Mechanical Engineering, Indian Institute of Science, Bangalore, India

Rajeev R. Ashokan Naval Physical and Oceanographic Laboratory, Thrikkakara, Kochi, India

Sreenath Balakrishnan Computational Nanoengineering (CoNe) group and Mechanical Engineering, Indian Institute of Science, Bangalore, India

K. N. Bhat Centre for Nano Science and Engineering, Indian Institute of Science, Bangalore, India

Navakanta Bhat Centre for Nano Science and Engineering, Indian Institute of Science, Bangalore, India

Tarun Kanti Bhattacharyya Department of Electronics and Electrical Communication Engineering and Advanced Technology Development Centre, Indian Institute of Technology, Kharagpur, India

S. K. Bhaumik Materials Science Division, CSIR-National Aerospace Laboratories, Bangalore, India

Dhananjay Bodas Centre of Nanobioscience, Agharkar Research Institute, Pune, India

Jeevanjyoti Chakraborty Advanced Technology Development Centre, Indian Institute of Technology Kharagpur, Kharagpur, West Bengal, India; Mathematical Institute, University of Oxford, Oxford, UK

Suman Chakraborty Advanced Technology Development Centre, Indian Institute of Technology Kharagpur, Kharagpur, West Bengal, India; Mechanical Engineering Department, Indian Institute of Technology Kharagpur, Kharagpur, West Bengal, India

Deepali Chandratre IIT Bombay, Mumbai, India

Dhiman Chatterjee Department of Mechanical Engineering, Indian Institute of Technology Madras, Chennai, India

Varsha Chaware Centre for Materials for Electronics Technology (C-MET), Panchawati, Pune, India

Amitava DasGupta Department of Electrical Engineering, Indian Institute of Technology Madras, Chennai, India

S. DattaGupta Department of Electrical Engineering, Indian Institute of Technology, Mumbai, India

B. Dattaguru Institute of Aerospace Engineering and Management, Jain University, Bangalore, India; TechMahindra, Bangalore, India

Sukomal Dey Centre for Applied Research in Electronics, Indian Institute of Technology Delhi, Hauz Khas, New Delhi, India

Soma Dutta Materials Science Division, CSIR-National Aerospace Laboratories, Bangalore, India

Paul Braineard Eladi Department of Electrical Engineering, Indian Institute of Technology Madras, Chennai, India

S. A. Gangal Department of Electronic Science, University of Pune, Ganeshkhind Road, Pune, India

R. Ganguli Department of Aerospace Engineering, Indian Institute of Science, Bangalore, India

Shyam Gaurav Centre for Nano Science and Engineering, Indian Institute of Science, Bangalore, India

Neena Gilda Department of Electrical Engineering, Indian Institute of Technology Bombay, Mumbai, India

S. Gopalakrishnan Department of Aerospace Engineering, Indian Institute of Science, Bangalore, India

Gurudat Centre for Nano Science and Engineering, Indian Institute of Science, Bangalore, India

Robert E. Harbaugh Department of Neurosurgery, Penn State University Medical School, Hershey, PA, USA

Munshi Imran Hossain IIT Bombay, Mumbai, India

Jeyabal Centre for Nano Science and Engineering, Indian Institute of Science, Bangalore, India

C. S. Jog Computational Nanoengineering (CoNe) group and Mechanical Engineering, Indian Institute of Science, Bangalore, India

Abhay Joshi Department of Electronic Science, University of Pune, Ganeshkhind Road, Pune, India

Manoj Kandpal Department of Electrical Engineering, Indian Institute of Technology Bombay, Mumbai, India

M. Kathiresan Naval Physical and Oceanographic Laboratory, Thrikkakara, Kochi, India

P. K. Khanna CSIR-Central Electronics Engineering Research Institute, Pilani, Rajasthan, India; Academy of Scientific and Innovative Research (AcSIR), New Delhi, India

V. K. Khanna CSIR-Central Electronics Engineering Research Institute, Pilani, Rajasthan, India; Academy of Scientific and Innovative Research (AcSIR), New Delhi, India

D. K. Kharbanda CSIR-Central Electronics Engineering Research Institute, Pilani, Rajasthan, India; Academy of Scientific and Innovative Research (AcSIR), New Delhi, India

A. H. Kiranmayee CSIR-Central Electronics Engineering Research Institute, Pilani, Rajasthan, India

Swathi Korrapati Bigtec Private Limited, Bengaluru, India

Shiban K. Koul Centre for Applied Research in Electronics, Indian Institute of Technology Delhi, Hauz Khas, New Delhi, India

S. B. Krupanidhi Materials Research Centre, Indian Institute of Science, Bangalore, India

Abhijeet Kshirsagar Department of Electronic Science, University of Pune, Ganeshkhind Road, Pune 411007, India; Department of Electrical Engineering, Indian Institute of Technology, Mumbai, India

Shrikant Kulkarni Centre for Materials for Electronics Technology (C-MET), Panchawati, Pune, India

Anil Kumar Indian Institute of Technology Bombay, Mumbai, India

Prashanth Kumar Department of Electrical Engineering, University of Arkansas, Fayetteville, AR, USA

S. Kumar CSIR-Central Electronics Engineering Research Institute, Pilani, Rajasthan, India

Vijay Kumar Department of Mechanical Engineering, Nanomaterials and National MEMS Design Centre, Birla Institute of Technology and Science, Pilani, India

T. Kundu IIT Bombay, Mumbai, India

Bhoopesh Mahale Department of Electronic Science, University of Pune, Ganeshkhind Road, Pune, India

Ashok K. Majji Indian Institute of Technology Bombay, Mumbai, India

S. Manish Centre for Nano Science and Engineering, Indian Institute of Science, Bangalore, India

S. Mohan Indian Institute of Science, Bengaluru, India

S. Mukherji IIT Bombay, Mumbai, India

R. Mukhiya CSIR-Central Electronics Engineering Research Institute, Pilani, Rajasthan, India; Academy of Scientific and Innovative Research (AcSIR), New Delhi, India

Chandrashekhar B. Nair Bigtec Private Limited, Bengaluru, India

Shiny Nair Naval Physical and Oceanographic Laboratory, Thrikkakara, Kochi, India

K. Natarajan Department of Telecommunication Engineering, MS Ramaiah Institute of Technology, Bangalore, India

V. Natarajan Naval Physical and Oceanographic Laboratory, Thrikkakara, Kochi, India

M. M. Nayak Centre for Nano Science and Engineering, Indian Institute of Science, Bangalore, India

Sechang Oh Department of Electrical Engineering, University of Arkansas, Fayetteville, AR, USA

P. C. Panchariya CSIR-Central Electronics Engineering Research Institute, Pilani, Rajasthan, India; Academy of Scientific and Innovative Research (AcSIR), New Delhi, India

P. K. Panda Materials Science Division, CSIR-National Aerospace Laboratories, Kodihalli, Bangalore, India

Pandian Centre for Nano Science and Engineering, Indian Institute of Science, Bangalore, India

Jayanta Parui Materials Research Centre, Indian Institute of Science, Bangalore, India

Kunal D. Patil Computational Nanoengineering (CoNe) group and Mechanical Engineering, Indian Institute of Science, Bangalore, India

Rajul S. Patkar Department of Electrical Engineering, Indian Institute of Technology Bombay, Mumbai, India

Justin K. Paul Bigtec Private Limited, Bengaluru, India

Girish Phatak Centre for Materials for Electronics Technology (C-MET), Panchawati, Pune, India

T. V. Prabhakar Department of Electronic Systems Engineering, Indian Institute of Science, Bangalore, India

Rudra Pratap Center for Nano Science and Engineering and Department of Mechanical Engineering, Indian Institute of Science, Bangalore, India

Pratyush Rai Department of Electrical Engineering, University of Arkansas, Fayetteville, AR, USA

K. V. Ramaiah Materials Science Division, CSIR-National Aerospace Laboratories, Bangalore, India

Vivek Rane G. M. Vedak College of Science, Raigad, Maharashtra, India

V. Ramgopal Rao Department of Electrical Engineering, Indian Institute of Technology Bombay, Mumbai, India

J. Raviprakash Bigtec Private Limited, Bengaluru, India

Prasenjit Ray Department of Electrical Engineering, Indian Institute of Technology Bombay, Mumbai, India

Anindya Lal Roy Advanced Technology Development Centre, Indian Institute of Technology, Kharagpur, India

Anish Roychowdhury Center for Nano Science and Engineering and Department of Mechanical Engineering, Indian Institute of Science, Bangalore, India

B. Sahoo Materials Science Division, CSIR-National Aerospace Laboratories, Kodihalli, Bangalore, India

C. N. Saikrishna Materials Science Division, CSIR-National Aerospace Laboratories, Bangalore, India

Shilpa K. Sanjeeva Bigtec Private Limited, Bengaluru, India

D. Saranya Materials Research Centre, Indian Institute of Science, Bangalore, India

N. N. Sharma Department of Mechanical Engineering, Nanomaterials and National MEMS Design Centre, Birla Institute of Technology and Science, Pilani, India

R. Sharma CSIR-Central Electronics Engineering Research Institute, Pilani, Rajasthan, India; Academy of Scientific and Innovative Research (AcSIR), New Delhi, India

Sudhir Kumar Sharma Indian Institute of Science, Bengaluru, India

S. A. Shivashankar Centre for Nano Science and Engineering, Indian Institute of Science, Bangalore, India

Jasmine Sinha Johns Hopkins University, Baltimore, USA

G. Suresh Naval Physical and Oceanographic Laboratory, Thrikkakara, Kochi, India

K. Suresh Department of Instrumentation and Control Engineering, National Institute of Technology, Tiruchirappalli, India

K. A. Thomas Naval Physical and Oceanographic Laboratory, Thrikkakara, Kochi, India

Linet Thomas Centre for Nano Science and Engineering, Indian Institute of Science, Bangalore, India

Vijay Thyagarajan Centre for Nano Science and Engineering, Indian Institute of Science, Bangalore, India

G. Uma Department of Instrumentation and Control Engineering, National Institute of Technology, Tiruchirappalli, India

M. Umapathy Department of Instrumentation and Control Engineering, National Institute of Technology, Tiruchirappalli, India

Vijay K. Varadan Department of Electrical Engineering, University of Arkansas, Fayetteville, AR, USA; Department of Biomedical Engineering, University of Arkansas, Fayetteville, AR, USA; Department of Neurosurgery, Penn State University Medical School, Hershey, PA, USA; Global Institute of Nanotechnology in Engineering and Medicine, Fayetteville, AR, USA

E. Varadarajan Naval Physical and Oceanographic Laboratory, Thrikkakara, Kochi, India

Vijay Kumar Centre for Nano Science and Engineering, Indian Institute of Science, Bangalore, India

S. Vijay Mohan Bigtec Private Limited, Bangalore, India

K. J. Vinoy Department of Electrical Communication Engineering, Indian Institute of Science, Bangalore, India

S. R. Viswamurthy National Aerospace Laboratories, Council of Scientific and Industrial Research, Bangalore, India

Sathyadeep Viswanathan Bigtec Private Limited, Bengaluru, India

Abbreviations

ADC	Analog-to-digital converter
AFE	Antiferroelectric
AFE-FE	Antiferroelectric to ferroelectric switching
AFP	Amplifying fluorescent polymer
A-IgG	Anti-immunoglobulin G
APA	Amplified piezo actuator
ASME	American society for mechanical Engineers
ASSURED	Affordable, sensitive, specific, user-friendly, rapid and robust, equipment-free, and delivered to those who need it
ASTM	American society for testing and materials
at.%	Atom percentage
ATS	Alkyltricholrosilane
Bo	Bond number
CA	Contact angle
CBP	Cardiac biopotentials
CFRP	Carbon fiber reinforced plastics
CMOS	Complementary metal oxide semiconductor
CNT	Carbon nanotubes
CSIR	Council of scientific and industrial research, India
CVD	Cardiovascular diseases
CVI	C for virtual instrumentation
DAC	Digital-to-analog converter
DAQ	Data acquisition
DC	Direct current
DI	De-ionized
DNA	Deoxy-ribonucleic acid
DNT	2,4-Dinitrotoluene
DoS	Department of Space, India
DRDO	Defence Research and Development Organization, India
DRIE	Deep reactive ion etching
DSC	Differential scanning calorimeter
DST	Department of Science and Technology, India
EC	Electrocaloric
ECG	Electrocardiograph

EDC	*N*-(3-Dimethylaminopropyl)-*N*′ Ethyl-carbodiimide Hydrochloride
EESSs	Electrical energy storage systems
EGFET	Extended-gate field-effect transistor
EMI	Electro magnetic interference
FE	Ferroelectric
FEA	Finite element analysis
FE-AFE	Ferroelectric to antiferroelectric switching
FEM	Finite element method
FGA	Forming gas anneal
FITC	Fluorescein Isothiocyanate
FSO	Full scale output
FWHM	Full width at half maxima
GFRP	Glass fiber reinforced plastics
GPa	Giga Pascal
GPRS	General packet radio service
HMX	High melting explosive (Octahydro-1,3,5,7-tetranitro-1,3,5,7-tetrazocine)
ICAF	International Committee for Aeronautical Fatigue
ICCMS	International Congress on Computational mechanics and Simulation
IDE	Inter Digited electrode
IgG	Immunoglobulin G
IPA	Iso-propylalcohol
ISFET	Ion-sensitive field-effect transistor
KMF	Kotagiri Mission Fellowship Hospital
KOH	Potassium hydroxide
LCD	Liquid crystal display
LED	Light emitting diode
LPCVD	Low-pressure chemical vapor deposition
LSCO	Lanthanum strontium cobaltate
LTCC	Low-temperature co-fired ceramic
LUMO	Lowest unoccupied molecular orbital
MEMS	Micro electro mechanical systems
ML	Multi-layered
MOSFET	Metal-oxide semiconductor field-effect-transistor
MuA	11-Mercapto-undecanoic acid
MWCNT	Multi-wall carbon nanotubes
NDE	Non destructive evaluation
NHS	*N*-Hydroxysuccimide
Ni	Nickel
NiTi	Nickel Titanium
NPMASS	National Programme on Micro and Smart Systems
NPSM	National Programme on Smart Materials
Op-Amp	Operational amplifier
OTS	Octadecyltrichlorosilane
PANP	Polyaniline nanoparticles

PBS	Phosphate buffer saline
PCB	Printed circuit board
PDE	Partial differential equation
PDMS	Polydimethylsiloxane
PDMS	Poly-dimethyl Siloxane
PE	Paraelectric
PECVD	Plasma enhanced chemical vapor deposition
PEDOT	PSS:poly (3,4-ethylenedioxythiophene) poly(styrenesulfonate)
PETN	Penta erythritol tetra nitrate
pHpzc	pH at the point of zero charge
PLD	Pulsed laser deposition
PLZT	Lead lanthanum zirconate titanate
PMN	Lead magnesium niobate
pMUTs	Piezoelectric micromachined ultrasonic transducers
POC	Point-of-care
PT	Lead tiatanate
PTFE	Poly-tetrafluoro ethylene
PVDF	Polyvinylidene fluoride
PZ	Lead zirconate
PZT	Lead zirconate titanate
RC	Resistance capacitance network
RCA	Radio Corporation of America
RDX	Research Department Explosive (1,3,5-Trinitro-1,3,5-triazacyclohexane)
REFET	Reference field-effect transistor
RF	Radio Frequency
RIE	Reactive ion etching
RIU	Refractive index unit
SAMs	Self assembled monolayers
SEM	Scanning electron microscopy
SHM	Structural health monitoring
SIF	Stress intensity factor
SMA	Shape memory alloys
SOI	Silicon on insulator
SPR	Surface plasmon resonance
SPW	Surface plasmon wave
SWCNT	Single walled carbon nanotubes
TCAD	Technology computer-aided design
TEM	Transmission electron microscopy
Ti	Titanium
TM	Transverse magnetic
TMAH	Tetra-methyl ammonium hydroxide
TNT	2,4,6-Trinitrotoluene
T-NT	Titanium-rich nickel titanium
UART	Universal asynchronous receiver/transmitter

UTM	Universal testing machine
VCG	Vectorcardiograph
WCA	Water contact angle
XRD	X-ray diffraction

Part I
Microsensors

Design, Development, Fabrication, Packaging, and Testing of MEMS Pressure Sensors for Aerospace Applications

K. N. Bhat, M. M. Nayak, Vijay Kumar, Linet Thomas, S. Manish, Vijay Thyagarajan, Pandian, Jeyabal, Shyam Gaurav, Gurudat, Navakanta Bhat and Rudra Pratap

Abstract In this chapter we present the design, fabrication, packaging, and calibration of silicon micro machined piezo-resistive pressure sensors for operation in the pressure range of 1.2–400 bar. Based on the detailed Finite Element Analysis (FEA), the diaphragm dimensions and the optimized locations for the piezo-resistors are designed, to achieve the best performance parameters over a wide range of pressures, with minimum nonlinearity and adequate burst pressure. The process parameters are optimized and the pressure sensors fabricated in the Centre for Nano Science and Engineering (CeNSE) at IISc. The wafers are diced, the devices mounted on headers, wire bonded and packaged suitably, tested, and calibrated at the cell level to determine the adequacy of the performance parameters of the sensors for different pressure ranges. The results achieved on the pressure transducer assembly with Active Temperature Compensation and the offset compensation using electronics and EMI filters in a single package are presented. Excellent linearity within 0.5 % in the output voltage versus pressure is demonstrated, over the specified pressure ranges (i) 0−1.2 bar and (ii) 0−400 bar, and over the temperature range of −40°C to +80°C.

Keywords Pressure sensors · Microfabrication · Packaging · Aerospace applications · Finite element analysis · Pressure sensor calibration

1 Introduction

Pressure sensors cater to about 60 % of the MEMS market. Among the various types of pressure sensors, piezo-resistive pressure sensors are easy to design to suit a wide range of pressure ranges. They are simple to fabricate with a suitably

K. N. Bhat (✉) · M. M. Nayak · V. Kumar · L. Thomas · S. Manish · V. Thyagarajan
Pandian · Jeyabal · S. Gaurav · Gurudat · N. Bhat · R. Pratap
Centre for Nano Science and Engineering, Indian Institute of Science, Bangalore, India
e-mail: knbhat@gmail.com

K. J. Vinoy et al. (eds.), *Micro and Smart Devices and Systems*,
Springer Tracts in Mechanical Engineering, DOI: 10.1007/978-81-322-1913-2_1,

designed diaphragm acting as the sensing element and piezo-resistors serving as transducers. Miniaturization and batch processing of this device is achieved with great precision using the silicon micromachining technique for realizing the diaphragm [1–3]. The piezo-resistors and the interconnections are achieved using photolithography, diffusion, ion implantation, and thin film deposition, which are well established in the microelectronics technology. All the finer aspects of design, fabrication, packaging, characterization, and calibration of silicon micro machined piezo-resistive pressure sensors ranging from 1.2 (0.12) to 400 bar (40 MPa) fabricated at the National Nano Fabrication Centre (NNFC) of the Centre for Nano Science and Engineering (CeNSE) at the Indian Institute of Science, Bangalore in India are presented in the following sections of this chapter. The analysis and the experimental results on the devices fabricated have shown that the best results in terms of sensitivity and accuracy can be achieved by appropriately laying out the resistors either inside or outside the diaphragm, depending on the diaphragm dimensions and aspect ratios, which are decided by the maximum pressure range of operation. The chapter also brings out the various critical issues encountered during packaging and calibrating the pressure sensors.

2 FEM Analysis and Diaphragm Design Considerations

Based on the theory of plates [4], which assumes that the diaphragm is anchored all around its edges (as shown in Fig. 1a), the location of the maximum stress is usually identified to be at the center of the edge of the diaphragm and hence the piezo-resistors are conventionally embedded on the inner side of the diaphragm edge. However, for the silicon micro machined diaphragms, realized either by Deep Reactive Ion Etching (DRIE) or wet chemical anisotropic etching, the physical location of the anchor position is on the backside of the chip as shown in Fig. 1b and c. Hence, the maximum stress position with respect to the diaphragm and the positions of piezoresitors on the chip need to be assessed.

A rigorous 3D FEM-based COMSOL [5, 6] simulation of these bulk micro machined diaphragms has indeed shown that the position of the longitudinal peak stress component, estimated along the line XX' (marked in Fig. 1d) lies outside the diaphragm edge at a distance *Xp* (marked in Fig. 1b and c), when the aspect ratio, length/height (L/h), of the diaphragm is low. In Fig. 2 we show the typical result obtained for the case of a DRIE etched diaphragm having thickness $h = 200$ μm and lateral dimension $L = 750$ μm, for different magnitudes of pressure applied on this DRIE diaphragm as shown in the inset. For this case, the L/h ratio is 3.75 and the location of the peak stress lies outside the diaphragm edge at a distance $X_p = 75$ μm. It is also interesting to note that the position of the peak stress remains the same irrespective of the magnitude of the pressure applied.

The effects of aspect ratio (L/h) on the position X_p of the peak longitudinal stress, for an applied pressure $P = 100$ bar is studied using the FEM simulation tool, considering L/h ratios ranging from 2.5 to 8.5 by varying L from 500 to

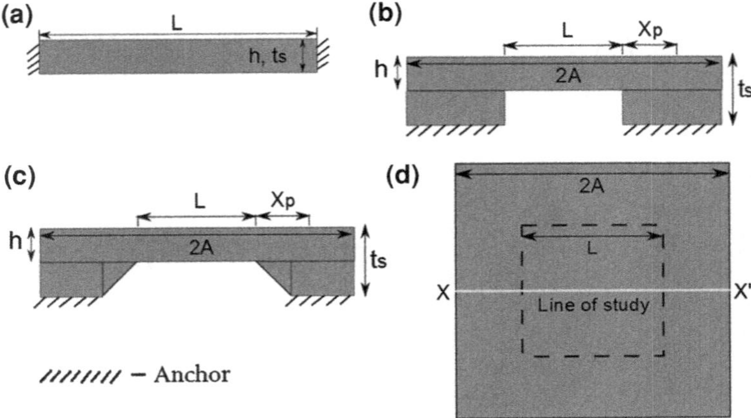

Fig. 1 Cross-section of the diaphragm **a** anchored on all *sides* (as in the theory of plates) **b** and **c** anchored on the *backside* (as obtained in the DRIE and KOH etched silicon). **d** *Top view* of the chip showing the *square* diaphragm of side length L (= 2a) by *dashed line*. The side length of the *square* chip is shown as 2A

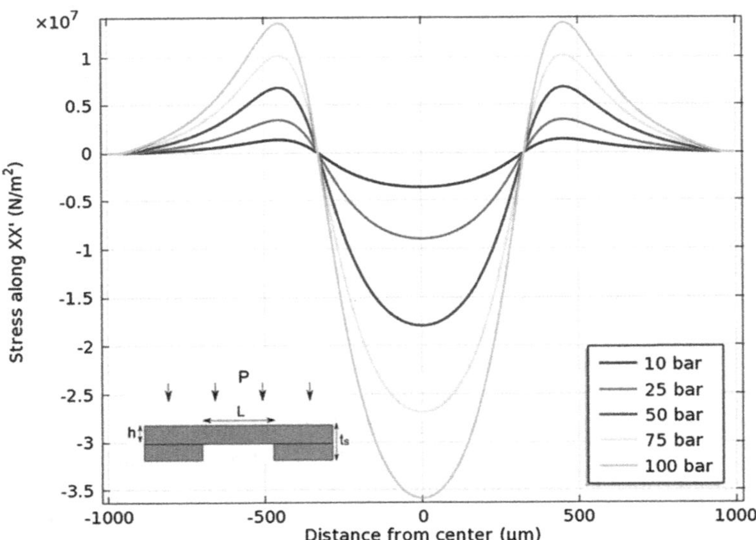

Fig. 2 FEM analysis results showing the stress distribution along XX' for different magnitudes of pressure P (in bars) = 10, 25, 50, 75, and 100 applied on a DRIE diaphragm having $h = 200$ μm and $L = 750$ μm, and a *square* chip having the side length 2A = 2 mm, $t_s = 400$ μm

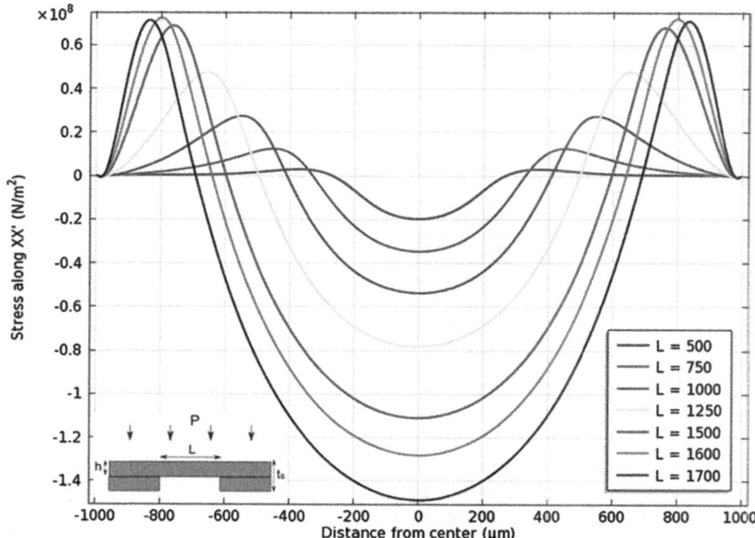

Fig. 3 FEM analysis results showing the variation of stress along XX' for a 200 μm thick DRIE etched diaphragm, for different diaphragm side lengths (*L*) as running parameter, when a pressure of 100 bar is applied

1700 μm and keeping the DRIE etched diaphragm thickness fixed at $h = 200$ μm. The results are shown in Fig. 3. The distance (*Xp*) between the edge of the diaphragm and the position of peak as a function of *L*/*h* ratio determined for the cases having $h = 200$ μm and $h = 20$ μm are shown in Table 1a and b.

We also determined the effect of the diaphragm lateral length *L* on the position of the peak stress (*Xp*), with a fixed diaphragm thickness, $h = 100$ μm, considering both DRIE etched and KOH etched diaphragms with crosssection shown in Fig. 1b, c. The results obtained from all the simulation studies involving various thicknesses and *L*/*h* ratios are shown in Fig. 4, where *Xp* is plotted versus *L*/*h* ratio. It can be seen that in general, *Xp* is positive when the aspect ratio is below eight indicating that the peak stress lies outside the diaphragm portion for these cases. As it can be seen from Fig. 4, in the case of thick diaphragms with $h = 200$ μm that are used for high-pressure applications, the location of peak stress lies at $Xp = +50$ μm outside the diaphragm edge when *L*/*h* ratio is equal to 5, corresponding to $L = 1000$ μm. On the other hand when $L/h > 8$, corresponding to the cases $L > 1600$ μm, with $h = 200$ μm, *Xp* becomes negative showing that the location of the peak stress shifts onto the diaphragm surface. In this situation as the diaphragm is too large the pressure range of operation decreases. In the case of thin diaphragms of the order of 20 μm thickness, $Xp = 0$ or very small ($Xp = +1$ to $+3$ μm). Thus, in situations where the diaphragm is thin, the theory of plates holds good with the maximum stress occurring at the diaphragm edge.

Table 1 Variation of peak position relative to the diaphragm edge (*Xp*) and the maximum total stress as the *L/h* ratio is changed for a DRIE etched diaphragm of thickness **a** h = 200 μm. *P* = 100 bar and **b** *h* = 20 μm, *P* = 1 bar

L (μm)	*L/h*	*Xp* (μm)	Max total stress (MPa)
(*a*) h = 200 μm, P = 100 bar			
500	2.5	120	43.78
750	3.75	75	63.563
1000	5.0	50	98.735
1250	6.25	30	147.37
1500	7.5	10	222.55
1600	8.0	0	245.36
1700	8.5	-15	313.67
(*b*) h = 20 μm, P = 1 bar			
60	3.0	5	0.228
100	5.0	4	0.396
130	6.5	3	0.873
160	8.0	2	1.290
200	10	0	1.976

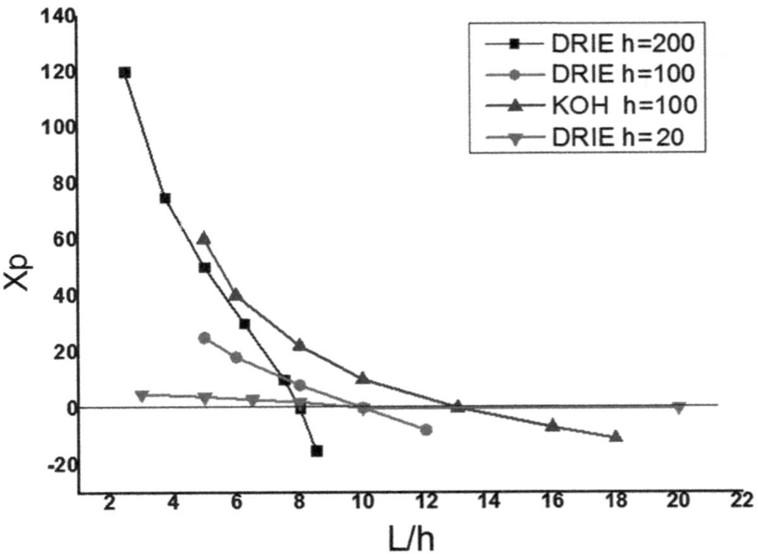

Fig. 4 Variation of the Location *Xp* (of peak stress with respect to the diaphragm edge) versus the diaphragm aspect ratio (*L/h*) for the DRIE etched diaphragms having, respectively, *h* = 20, 100, 200 μm, and for the KOH etched diaphragms with *h* = 100 μm

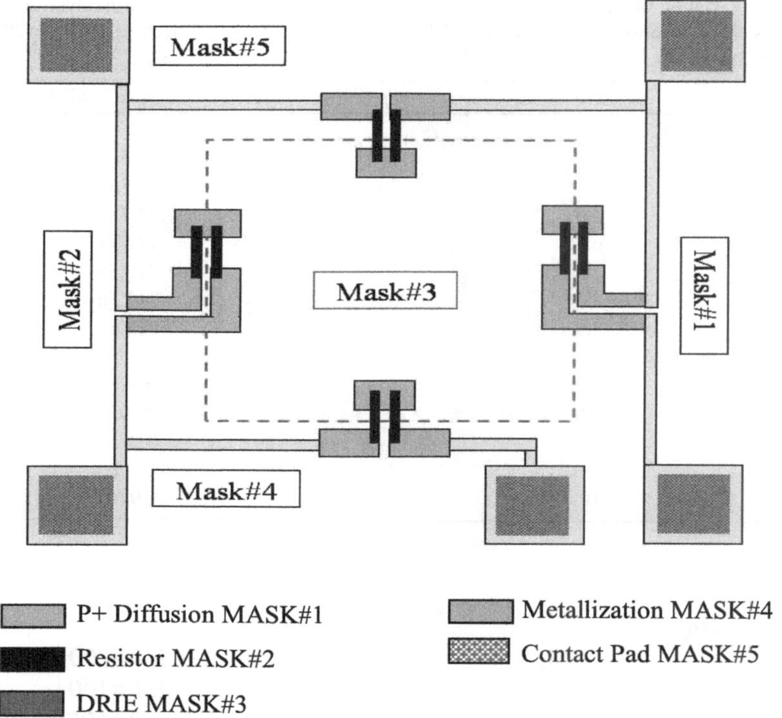

Fig. 5 Composite mask layout for the 400 bar Pressure sensor

3 Pressure Sensor Design and Fabrication

3.1 Pressure Sensor for Operation up to 400 bar

The lateral dimension of chip was chosen to be 2 × 2 mm, and based on the FEM simulations a reasonable diaphragm dimension of 750 × 750 μm and a diaphragm thickness of 210 μm was chosen to ensure that the maximum stress on chip is only 0.3 GPa at 400 bar, so that the maximum stress is well below 1 Gpa even at 1,000 bar. Once again the maximum stress versus pressure was determined to be linear over the pressure range 0–400 bar. The resistors of the Wheatstone bridge of the sensors were placed outside the diaphragm at a distance of 77 μm from the diaphragm edge. Considering a sheet resistance of 200 ohm/square, individual resistor dimensions were decided to be $L/W = 20$. A mask layout for a five-mask process was designed and the mask plates were prepared at the NNFC of CeNSE at IISc. Figure 5 shows the composite mask layout of the five-mask process.

Each of the process steps are marked alongside the individual masks. The starting wafer is an SOI (100) wafer with N-type device layer thickness 210 μm,

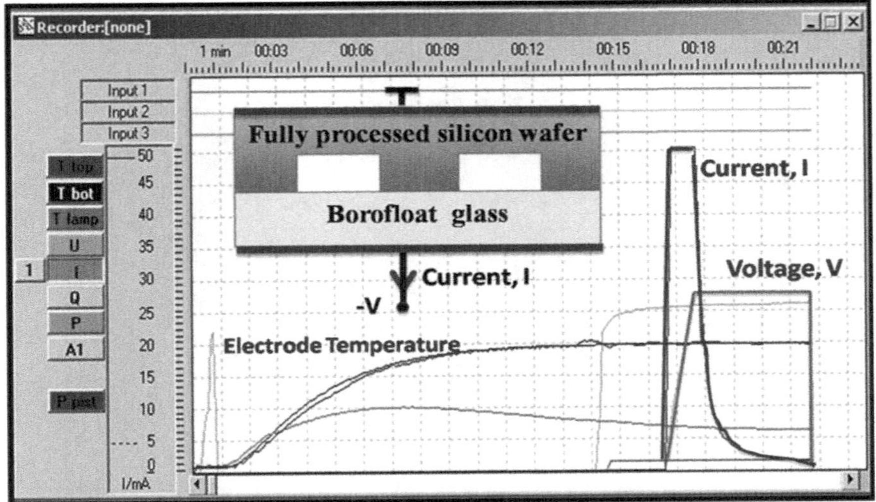

Fig. 6 Screenshot taken during anodic bonding of silicon wafer to glass. The high current pulse indicates the successful completion of bonding process

BOX layer of 1 μm, and P type handle wafer thickness 400 μm. This wafer is thermally oxidized and P+ diffusion window is opened using Mask#1 in the oxide and boron diffusion is carried out in this region at 1,100 °C for 30 min to achieve a sheet resistance in the range 5–6 ohm/square. This is followed by ion implantation at 80 KeV with a dose of 3×10^{14} Boron ions/cm^2 into the region opened in the oxide using Mask#2. Thermal annealing is done at 1,000 °C in Nitrogen ambient for 30 min to achieve a sheet resistance of 180–190 ohms/square. Following this step, backside alignment lithography is carried out aligned with respect to the patterns already created, using the Mask#3 to open window in the backside. DRIE is carried out using Bosch process with etch rates of 10 μm/min initially and 4 μm/min during the last 10 min, using the photoresist as the masking layer. After stripping of the PR in oxygen plasma aluminum metallization is carried out and patterned using Mask#4. Next, the Forming Gas Anneal (FGA) is carried out for 30 min at 450 °C. This is followed by deposition of silicon dioxide by PECVD at 350 °C to a thickness of 0.1 μm. In the final step, lithography is carried out using Mask#5 and the pad oxide and scribe lines are etched. The backside oxide is removed protecting the front surface and borofloat glass is anodically bonded on to the backside in an evacuated bonding system. During the bonding process the temperature is maintained at 390 °C and the glass is biased at a negative voltage of 1,000 V for 30 min as shown in Fig. 6.

A photograph of the completed pressure sensor after the anodic bonding is shown in Fig. 7. The wafer is next diced using a diamond saw and then dies mounted on a header and wire bonded. A photograph of the diced sensor and the die

Fig. 7 Photograph of processed and anodic bonded 100 mm silicon wafer showing about 700 pressure sensors. The devices occupy only the 75 mm diameter area

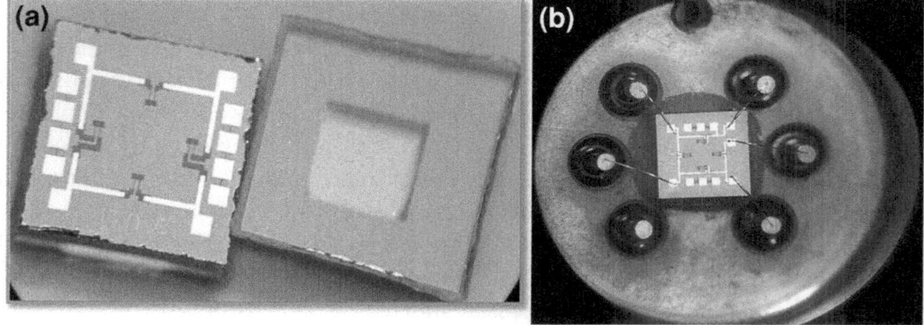

Fig. 8 **a** Front side view and backside view of the pressure sensor die after dicing. **b** Pressure sensor die mounted on the header and wire bonded

mounted and wire bonded sensor is shown in Fig. 8a and b respectively. Figure 9 shows the packaged 400 bar pressure sensor with compensating electronics.

3.2 Pressure Sensor for Operation up to 1.2 bar

Based on the FEM analysis similar to that presented in Sect. 2, a thin square diaphragm of $h = 25$ µm and $L = 1$ mm is chosen for this pressure range (0–1, 2 bar). The FEM-based COMSOL simulated stress distribution across the

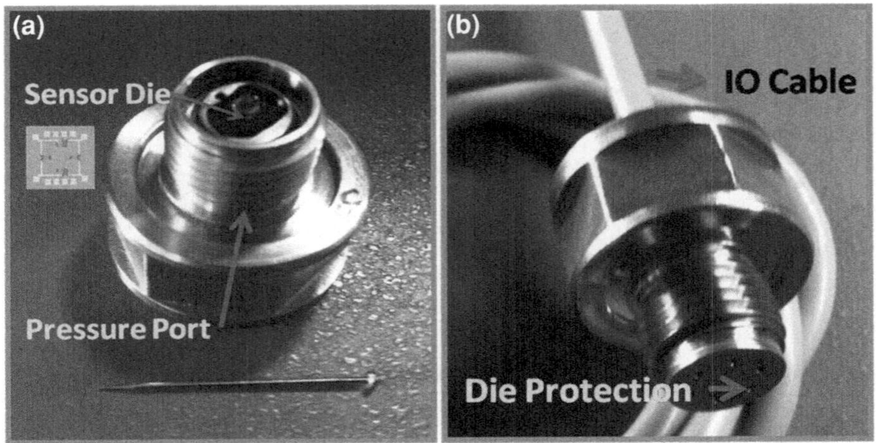

Fig. 9 Photograph of the packaged 400 bar pressure sensor along with electronics, **a** Without the connector and pressure port. **b** With the connector and the welded pressure port

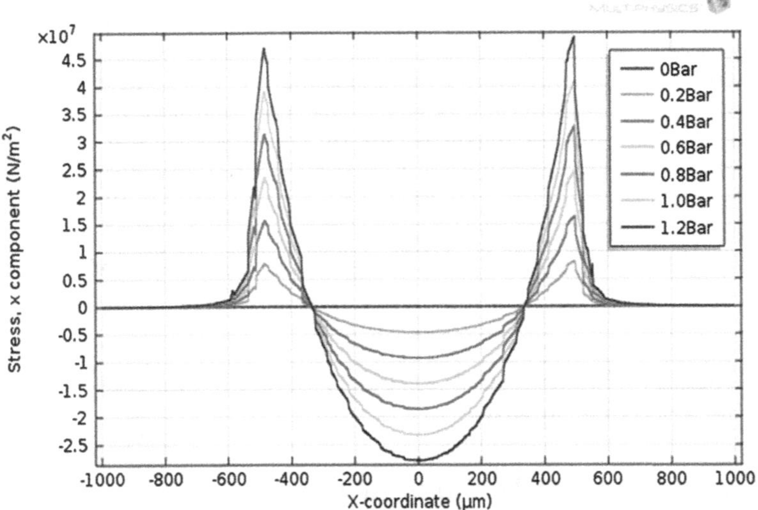

Fig. 10 Stress distribution across the *top* surface of a *square* diaphragm with its thickness $h = 25$ μm and side length $L = 1$ mm for pressure $P = 0{-}1.2$ bar

diaphragm along the XX' line (as in Fig. 1d) is shown in Fig. 10. It can be seen that the maximum stress occurs at the edge of the diaphragm. For this situation of high L/h ratio $= 40$, the theory of plates is applicable as discussed in Sect. 2 and hence the maximum longitudinal stress is estimated to be equal to 48 MPa at the center of the diaphragm edge, for the applied pressure $P = 0.12$ MPa (1.2 bar) using the

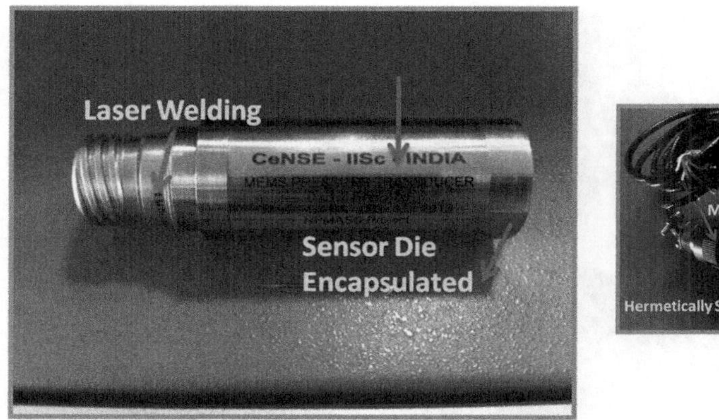

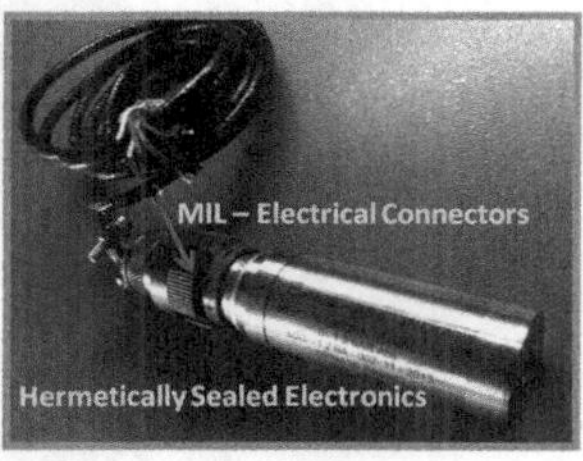

Fig. 11 Photographs of 1.2 bar pressure sensors packed and hermetically sealed with electronics

analytical expression $\sigma_L = P(L/2h)^2$. This value is very close to the value of 46 MPa determined using the FEM-based analysis with COMSOL (see Fig. 10).

As in the case of 400 bar sensor, in this case also the starting wafer is SOI wafer, the only difference being in the SOI layer thickness, which in this case is $h = 25$ µm. The mask layout, the number of masks, and process parameters are exactly the same as the 400 bar sensor. However, as the aspect ratio is $L/h = 40$ in this case, the position of the piezo-resistors is at the edge on the inner side of the diaphragm. At the end of the five-mask process, anodic bonding of the wafer to glass is carried out and the individual dies are separated by dicing and followed by die mounting wire bonding and other packaging steps. The packaged devices are calibrated and electronically compensated for temperature drift. Figure 11 shows the photographs of the packaged 1.2 bar pressure sensors alongwith electronics and EMI filters.

4 Testing, Calibration and Temperature Compensation of the Pressure Sensors

4.1 Setup for Testing Packaged Sensors

The photograph of a Hydraulic Pressure Calibrator, procured from M/s. DHB, UK. Model DHB 42 H-542, capable of operation in the range of 0–1,200 Bar is shown in Fig. 12. The transducer to be tested is mounted on the calibrator adapter and electrical connections are made to stabilized power supply, digital voltmeter (DVM), and decade box. This setup is used to calibrate the 0–400 bar MEMS sensor in five equal ascending and descending steps. In addition, proof pressure application, burst pressure trials, and testing of glass to metal seals are also carried out up to 800 Bars.

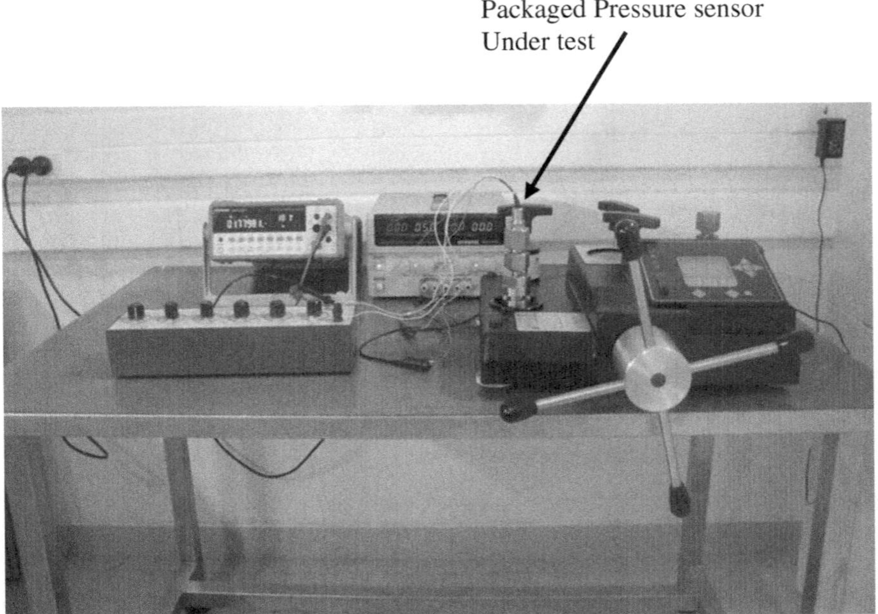

Fig. 12 Hydraulic calibration set-up to calibrate 0−400 bar pressure transducers at CeNSE, IISc

4.2 Testing and Temperature Compensation of 1200 milli Bar Pressure Sensor

The 0−1,200 milli bar MEMS-based pressure transducers were wire bonded, encapsulated, and then tested through a wide range of temperatures (−40 °C, 0 °C, +25 °C, +50 °C, and +80 °C). Figure 13 shows the raw milli volt output at various temperatures with respect to input pressure.

From the above results, the drift in the range of $1-13\times10^{-3}$/FSO/°C may be noted in the Zero Offset and Full Scale Output (FSO) with temperature. In terms of actual values, the drift in Zero Offset is as high as 50.34 mV between 25 °C and −40 °C and the drift in the full scale output voltage is 56.95 mV at 1,200 milli bars. Similarly, the drift in Zero Offset and FSO between 25 °C and 80 °C was found to be 15.83 mV and 24.12 mV respectively. To compensate for these drifts, we utilize active temperature compensation technique using advanced differential sensor signal conditioning CMOS IC ZMD 31050. This IC contains programmable gain amplifier, multiplexer, built-in temperature sensor, ADC, DAC, PWM, ROM, EEPROM, analog output, I2C, digital interface and other blocks. After programming the compensation circuit and retesting the overall drift is reduced to $1\times10^{-4}-1\times10^{-5}$/FSO/°C and thus the Zero Offset and FSO can be adjusted to suit many requirements. Figure 14 shows the linear and repeatable output at

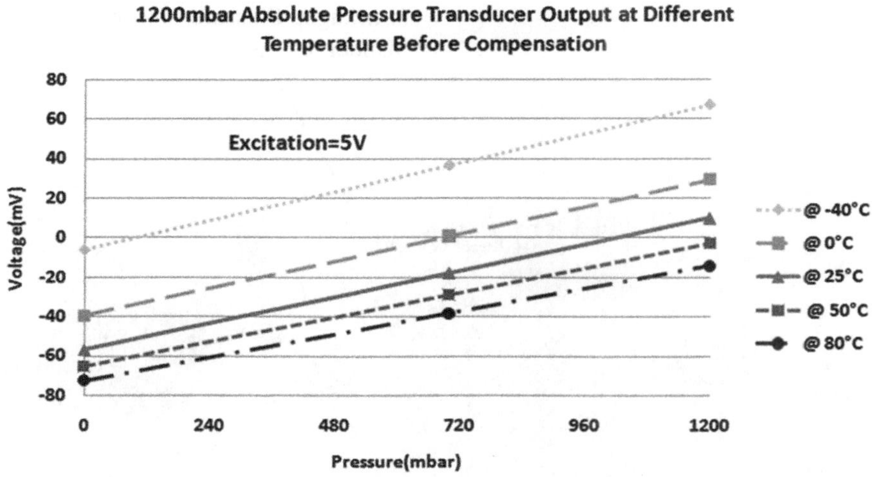

Fig. 13 Output voltage versus pressure at various temperatures with the 1.2 bar pressure sensor before compensation

+25 °C, −33.5 °C and +65 °C, randomly chosen temperatures for the validation of active compensation. The maximum nonlinearity and hysteresis was also found to be within 0.5 % FSO. To utilize these transducers for aerospace application, specific EMI filters and necessary protection diodes were also incorporated in the circuit and tested from 16 to 36 V DC excitation.

The transducer was subjected to proof pressure of 2.4 bar, twice the nominal pressure and random vibration tests for their survival under harsh environment. The drift after these tests is found to be within the allowable limits.

4.3 Testing and Calibration of 400 Bar (Gauge) Pressure Transducer Output at Different Temperatures

The 0−400 bar absolute MEMS pressure transducer was temperature compensated. Figure 15 shows a linear output from −40 to +80 °C and the overall temperature drift was found to be within 1×10^{-4}− 6×10^{-5}/FSO/°C. The sensor was also calibrated at +80 °C, +50 °C, +25 °C, −20 °C, and −40 °C for five ascending and descending equal pressure intervals (80 bar). The inset shows the photograph of a 0−400 bar Pressure sensor with built-in electronics having a voltage regulator and ZMD 31050 signal conditioning IC for active temperature compensation along with sensor protection filter at the pressure end. The input excitation can be from 16 to 36 V DC. The nominal excitation is 28 V DC. The Maximum nonlinearity and hysteresis of this transducer was found to be within 0.855 % FSO, as calculated using least squares best fit straight line method. The pressure end connection has M14×1.5 thread for 10 mm length with an "O"

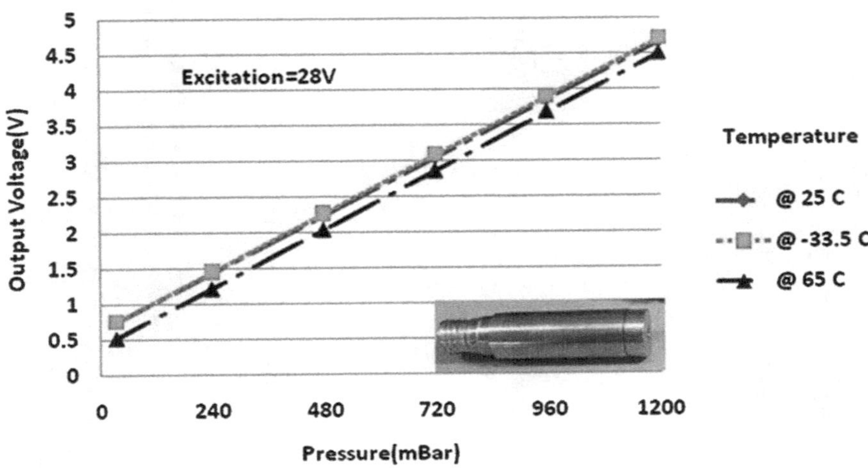

Fig. 14 Temperature compensated output versus pressure at different temperatures. for the 1.2 bar pressure (Gauge) sensor

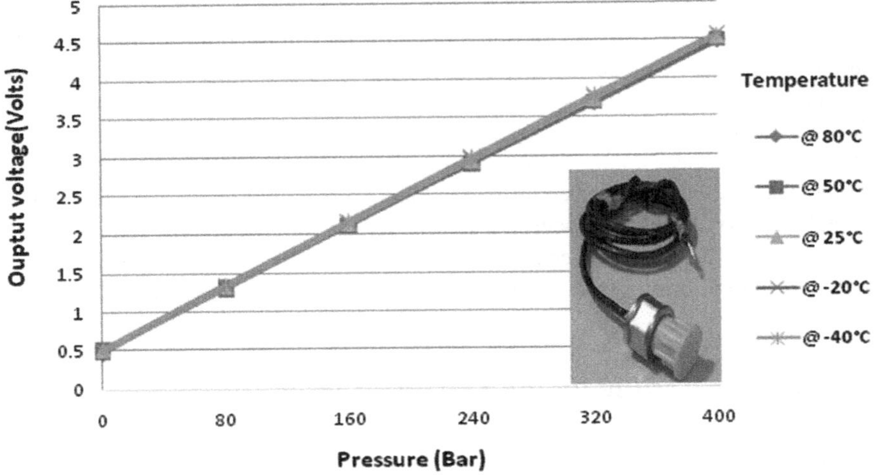

Fig. 15 Temperature compensated output voltage versus pressure from the 400 bar pressure transducer at different temperatures

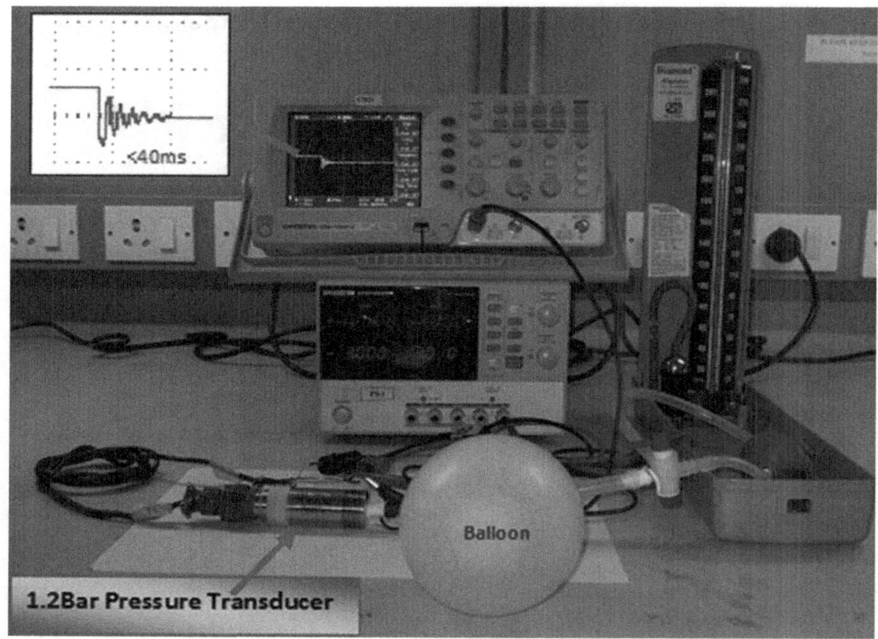

Fig. 16 Dynamic pressure testing on 1,200 milli bar absolute MEMS-based pressure transducer

ring groove for arresting the leak during mechanical interfacing. We employ laser or electron beam welding, AISI 304L stainless steel assembly for corrosion resistance and robust mechanical integrity.

The MEMS piezo-resistive sensor is mounted on a glass to metal seal (normally called as "Header") made out of Kovar metal with matching glass to withstand a pressure of above 800 bars. The insulation resistance between all the leads of this glass to metal seal with the body of the transducer is above 100 mega ohms at 50 V DC for proper isolation as per aerospace requirements. This assembly is also helium leak tested for a leak rate better than 10^{-8} sccm/sec for long-term operational stability. This transducer is tested under random vibration conditions in X and Y axes at 13.5 g rms, over a frequency range of 20 Hz–2 KHz for 3 minutes. After, all the calibration and testing, the drift was within allowable limits.

4.4 Dynamic Testing on the Pressure Sensor

To experimentally verify the output response of 1,200 milli bar pressure transducer, a simple dynamic test setup was rigged up as shown in Fig. 16 consisting of a low pressure source, 3-way valve, required tubing, clamp, power supply, digital voltmeter, storage oscilloscope, and related accessories. A balloon was pressurized so that the considerable output can be recorded through the transducer under test before, during, and after the balloon burst using either a sharp needle or lighted matchstick. The balloon was pressurized through a 3-way valve. While the output waveform was being monitored through the oscilloscope, the balloon was burst using a sharp needle. It is well known that, the dynamic response varies with inlet tube diameter, length, dead volume, and type of fluid in it. With 10 cm length, 3 mm internal diameter silicone tube having ambient pressure, the balloon burst output wave form is recorded and the response is as shown in the inset of Fig. 16. We observed a transient response of less than 40 milli sec. This can be further reduced by utilizing proper interface and by direct coupling.

5 Conclusions

In summary, the design, fabrication, packaging, calibration, and compensation details of micro machined piezo-resistive pressure sensors for operation in the pressure ranges 0–1.2 and 0–400 bar have been presented in this chapter. Based on the FEM simulation, it is shown that the location for placing the piezo-resistors of the Wheatstone bridge need to be carefully chosen depending on the ratio of the square diaphragm length (L) to the thickness (h) ratio. It is shown that when the L/h ratio is greater than eight the peak stress and hence the piezo resistor placement position lies inside the diaphragm at its edge center. For high- pressure sensors where h is in the range 100–200 μm, the piezo-resistors need to be placed outside the diaphragm edge at an optimum distance that depends on the L/h ratio to achieve optimized sensitivity. The details of the process steps for fabricating the pressure sensors for operation in the range 0–1.2 and 0–400 bar are presented. These devices have been suitably packaged, calibrated, and temperature compensated in the temperature range −40 to +80 °C.

Acknowledgments The authors thank the NPMASS for funding and supporting the project entitled "Design, Development, Fabrication, Packaging and Testing of MEMS Pressure Sensors for Aerospace applications" under PARC#3 Project No. NPMA-30. We are grateful to Dr. Aatre Chairman BSMART for his unconditional support and advice on various aspects of this project

References

1. Anthasuresh GK, Vinoy KJ, Gopalakrishnan S, Bhat KN, Aatre VK (1993) Micro and smart systems. Wiley, New Jersey
2. Kovacs Gregory TA (1998) Micromachined transducers source book. WCB McGraw-Hill, New York
3. Bhat KN (2007) Silicon micromachined pressure sensors. J Ind Inst Sci 87:11
4. Timoshenko S, Woinosky-Krieger S (1987) Theory of plates and shells, 2nd edn. McGraw-Hill, New York
5. Thyagarajan V, Bhat, KN (2013) In: ISSS national conference on MEMS Pune
6. Bhat KN, Nayak MM (2013) MEMS pressure sensors - an overview of challenges in technology and packaging. J ISSS 2(1): 39–71

MEMS Piezoresistive Accelerometers

Tarun Kanti Bhattacharyya and Anindya Lal Roy

Abstract MEMS piezoresistive accelerometers are inertial sensors which measure acceleration of the reference frame to which they are attached. These devices provide extremely localized acceleration-induced stress sensing with low noise outputs and have been the subject of academic as well as commercial research for quite a few years. This chapter deals with the basics micromachined piezoresistive accelerometers, tracing their evolution and typical analyses to sensor fabrication and characterization.

Keywords MEMS · Piezoresistive accelerometer · Wheatstone bridge · SEM · LDV · Wafer probe station · Sensitivity · Cross-axis sensitivity · Offset

1 Introduction

Much of our understanding of and interaction with nature occurs through various forms of electromechanical signal transduction. As a result, engineering systems that replicate such transduction techniques have come into being for myriad applications ranging from our daily lives to industrial usage to strategic deployment. Miniaturization of such conventional electromechanical systems has become imperative today as the need for high sensitivity devices with smaller footprints grows, along with the prospect of integration with signal conditioning electronics in order to execute high precision operations.

The rapid evolution of the state of the art in integrated circuit and microfabrication technology over the past few decades has played a major role in the development and maturing of the integrated microelectromechanical systems

T. K. Bhattacharyya (✉) · A. L. Roy
Advanced Technology Development Centre, Indian Institute of Technology, Kharagpur, India
e-mail: tkb@ece.iitkgp.ernet.in

K. J. Vinoy et al. (eds.), *Micro and Smart Devices and Systems*,
Springer Tracts in Mechanical Engineering, DOI: 10.1007/978-81-322-1913-2_2,

(MEMS) industry. MEMS employ a wide selection of sensing and actuation schemes to carry out a varied range of specialized functions at microscale regimes, thereby effectively replacing power-hungry, area-inefficient macroscale bulk sensor, and actuator systems. Figure 1 shows a couple of MEMS accelerometer structures from both commercial as well as academic research and development laboratories that have been developed using different micromachining processes and operated through different signal transduction mechanisms.

2 An Overview of the Piezoresistive Effect

Piezoresistance in semiconductors (silicon, germanium) was discovered in 1954 [1] as a phenomenon whereby the application of stress induced a proportional alteration of material resistivity. This principle is schematically illustrated in Fig. 2 where the dependence of material resistance on all the terms found in its mathematical expression is shown. Dominance of resistivity changes in comparison to dimensional changes as observed in semiconductor materials have led to its usage in applications involving stress sensitivity.

Semiconductors like silicon exhibits anisotropic **E–J** relationship dependent on its crystallographic axes. The matrix equation governing semiconductor resistivity is given by

$$E = \begin{bmatrix} E_x \\ E_y \\ E_z \end{bmatrix} = \rho_{\text{Anisotropic}} \begin{bmatrix} J_x \\ J_y \\ J_z \end{bmatrix} = \begin{bmatrix} \rho_{xx} & \rho_{xy} & \rho_{xz} \\ \rho_{yx} & \rho_{yy} & \rho_{yz} \\ \rho_{zx} & \rho_{zy} & \rho_{zz} \end{bmatrix} \begin{bmatrix} J_x \\ J_y \\ J_z \end{bmatrix} = \rho_{\text{Anisotropic}} J \tag{1}$$

where $\rho_{\text{Isotropic}}$ is the resistivity of metallic materials and $\rho_{\text{Anisotropic}}$ is the resistivity of semiconducting materials comprising ρ_{ij}, the components of the anisotropic resistivity matrix, and E_i and J_i denote the components of the electric field and current density vectors in Cartesian coordinates. The symmetry of this resistivity matrix (Onsager's theorem) [2] ensures six unique components: $\rho_1 \equiv \rho_{xx}$; $\rho_2 \equiv \rho_{yy}$; $\rho_3 \equiv \rho_{zz}$; $\rho_4 \equiv \rho_{yz} \equiv \rho_{zy}$; $\rho_5 \equiv \rho_{zx} \equiv \rho_{xz}$; $\rho_6 \equiv \rho_{xy} \equiv \rho_{yx}$

$$\rho_{\text{Anisotropic}} = \begin{bmatrix} \rho_1 & \rho_6 & \rho_5 \\ \rho_6 & \rho_2 & \rho_4 \\ \rho_5 & \rho_4 & \rho_3 \end{bmatrix} \tag{2}$$

Applied stress on such anisotropically affected materials may be expressed as a matrix with six independent stress components corresponding to appropriate cause and effect directions similar to (3) which are given by

$$\sigma = \begin{bmatrix} \sigma_1 & \sigma_6 & \sigma_5 \\ \sigma_6 & \sigma_2 & \sigma_4 \\ \sigma_5 & \sigma_4 & \sigma_3 \end{bmatrix}$$

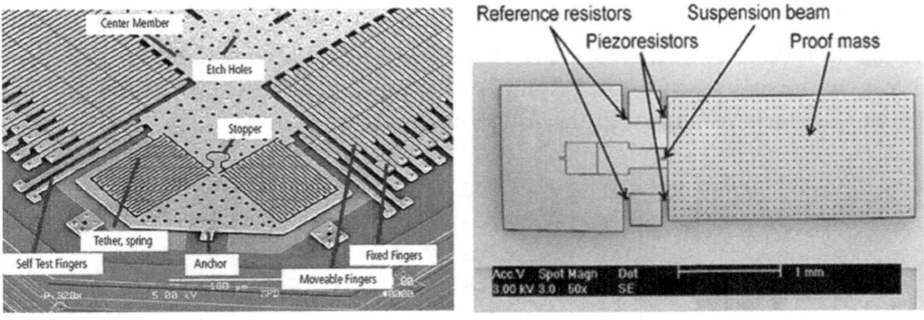

Fig. 1 (*left*) Analog Devices Inc. MEMS capacitive accelerometer. (*right*) MEMS piezoresistive accelerometer developed at the University of California, Irvine

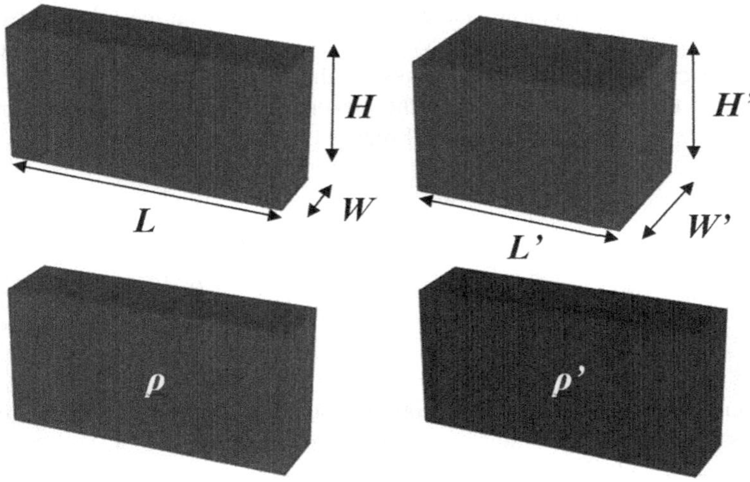

Fig. 2 (*top left*) Resistor to be subjected to stress-induced dimensional changes. (*top right*) Resistor after stress-induced deformation. (*bottom left*) Resistor to be subjected to stress-induced resistivity change. (*bottom right*) Resistor after stress-induced resistivity change

$$\left(\sigma_1 = \sigma_{xx}; \sigma_2 = \sigma_{yy}; \sigma_3 = \sigma_{zz}; \sigma_4 = \sigma_{yz} = \sigma_{zy}; \sigma_5 = \sigma_{zx} = \sigma_{xz}; \sigma_6 = \sigma_{xy} = \sigma_{yx}\right) \tag{3}$$

and the resistivity matrix of the material can be expressed as the superposition of unstressed (first matrix term) and stressed (second matrix term) conditions

$$\begin{bmatrix} \rho_1 & \rho_6 & \rho_5 \\ \rho_6 & \rho_2 & \rho_4 \\ \rho_5 & \rho_4 & \rho_3 \end{bmatrix} = \rho_0 \begin{bmatrix} 1 & 0 & 0 \\ 0 & 1 & 0 \\ 0 & 0 & 1 \end{bmatrix} + \rho_0 \begin{bmatrix} \delta_1 & \delta_6 & \delta_5 \\ \delta_6 & \delta_2 & \delta_4 \\ \delta_5 & \delta_4 & \delta_3 \end{bmatrix} \tag{4}$$

where, $\delta_i = \frac{\rho_i - \rho_0}{\rho_0}$ for $i = 1, 2, \ldots, 6$ and ρ_0 is the unstressed resistivity. The relative change in resistivity (with respect to the unstressed material resistivity) as defined above can be expressed as the product of the piezoresistive coefficient matrix and the stress matrix as shown in the following expression:

$$\begin{bmatrix} \delta_1 \\ \delta_2 \\ \delta_3 \\ \delta_4 \\ \delta_5 \\ \delta_6 \end{bmatrix} = \begin{bmatrix} \pi_{11} & \pi_{12} & \pi_{13} & \pi_{14} & \pi_{15} & \pi_{16} \\ \pi_{21} & \pi_{22} & \pi_{23} & \pi_{24} & \pi_{25} & \pi_{26} \\ \pi_{31} & \pi_{32} & \pi_{33} & \pi_{34} & \pi_{35} & \pi_{36} \\ \pi_{41} & \pi_{42} & \pi_{43} & \pi_{44} & \pi_{45} & \pi_{46} \\ \pi_{51} & \pi_{52} & \pi_{53} & \pi_{54} & \pi_{55} & \pi_{56} \\ \pi_{61} & \pi_{62} & \pi_{63} & \pi_{64} & \pi_{65} & \pi_{66} \end{bmatrix} \begin{bmatrix} \sigma_1 \\ \sigma_2 \\ \sigma_3 \\ \sigma_4 \\ \sigma_5 \\ \sigma_6 \end{bmatrix} \tag{5}$$

The relative change of resistivity with applied stress is formulated in [5] where the piezoresistive coefficient matrix plays a crucial role. By virtue of its octahedral diamond structure with a basic face-centered cubic (FCC) lattice, silicon possesses the symmetry of the O_h group [3]. Symmetric operations on such groups suggest that in the crystallographic coordinates of the material, there exist only three independent nonzero components of the piezoresistive coefficient matrix and the piezoresistive coefficient matrix of [5] reduces to

$$[\pi] = \begin{bmatrix} \pi_{11} & \pi_{12} & \pi_{12} & 0 & 0 & 0 \\ \pi_{12} & \pi_{11} & \pi_{12} & 0 & 0 & 0 \\ \pi_{12} & \pi_{12} & \pi_{11} & 0 & 0 & 0 \\ 0 & 0 & 0 & \pi_{44} & 0 & 0 \\ 0 & 0 & 0 & 0 & \pi_{44} & 0 \\ 0 & 0 & 0 & 0 & 0 & \pi_{44} \end{bmatrix} \tag{6}$$

We can draw a couple of assumptions about the coefficients of the resistor elements based on the values of the matrix components [3]: for *p*-type, $\pi_{11} = \pi_{12} = 0$ and for *n*-type, $\pi_{44} = 0$. These allow us to calculate the relative change in material resistivity due to the magnitude of the applied stress. The relative change in resistivity (equivalently, relative change in resistance) for a generalized applied acceleration stress is given by the expression

$$\frac{\Delta\rho}{\rho} = \pi_l\sigma_l + \pi_t\sigma_t + \pi_s\sigma_s \tag{7}$$

where π_l, π_t, π_s are the longitudinal, transverse, and shear coefficients of piezoresistance and σ_l, σ_t, σ_s are the corresponding longitudinal, transverse, and shear stresses applied to the sensing element.

From the perspective of extrinsic (doping and temperature) dependence of piezoresistance, the piezoresistive coefficient Π for a doping concentration N at a temperature T is given by

$$\Pi(N,T) = P(N,T)\Pi(300\,\mathrm{K}) \tag{8}$$

where P is the (N,T)-dependent piezoresistance factor and Π(300 K) is the lightly doped piezoresistive coefficient at 300 K. The variation of the piezoresistance factor P of both n-type and p-type silicon with doping concentration and temperature is illustrated in Fig. 3. It can be clearly seen that the piezoresistance factor decreases in magnitude with both temperature and concentration. Temperature dependence can be attributed to band structure properties while concentration dependence can be due to increased metallic nature of the material, both of which leads to a degradation in the piezoresistive property.

3 MEMS Piezoresistive Accelerometers

Physical sensing systems such as pressure sensors, accelerometers, and gyroscopes comprise the largest fraction of the industrial implementation of MEMS technology and among these, the MEMS accelerometer has been in the forefront in terms of its widespread utility and successful system realization. Research on piezoresistive accelerometers began through simple beam-mass structures [4, 5] with piezoresistive elements thermally diffused at the stress maxima of the compliant mechanical element in a Wheatstone bridge topology [6] for sensing the inertial forces experienced by the seismic mass. This fundamental sensor evolved into a wide array of diffused piezoresistive element multiple beam-mass systems [7, 8].

Advancement of semiconductor doping technology has led to the replacement of thermal diffusion with ion implantation [9, 10] techniques which were preferred for their accurate doping capabilities, precision profiling of the doped elements along with reduced thermal spread of resistors. Later, anisotropic bulk micromachining methodologies [11, 12] gave way to precision high aspect-ratio dry etch techniques [13] which resulted in greater refinement of the mechanical structure and added flexibility to the types of geometries explored [14, 15]. Piezoresistive sensors provide good sensitivity but are also extremely susceptible to ambient temperature conditions [16] as well as self-heating [17] which has led to a research decline recently. With the level of progress in modern microfabrication technology, a very robust sensor with high sensitivity can be realized for automotive and aerospace applications. Commercial piezoresistive accelerometers [18, 19] have maintained simple geometries with practicality in implementation and its resulting sensitivity being the primary figures of merit.

4 Design Analyses of MEMS Piezoresistive Accelerometers

Sensor specification requirements drive the design phase of the MEMS piezoresistive accelerometer structure which comprises one of the most critical steps in any microsystem development and involves the establishment of the

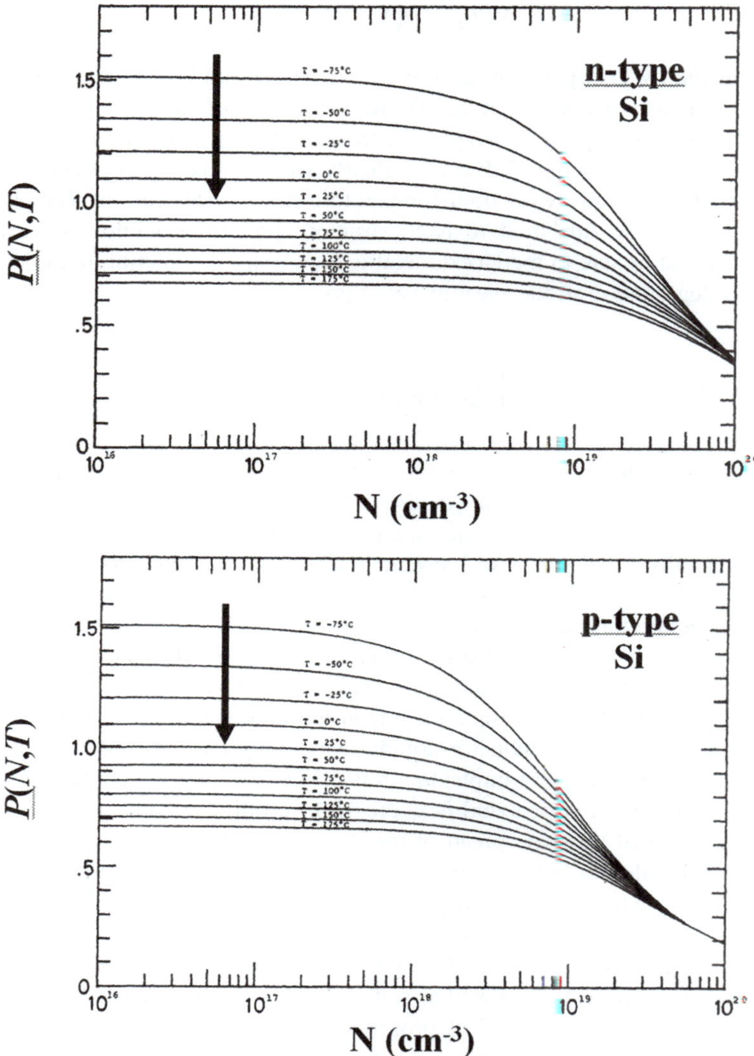

Fig. 3 Variation of the piezoresistance factor P for n-type (*top*) and p-type (*bottom*) silicon with dopant concentration and temperature

micromechanical structure and electrical transducer element design specifications through analysis and simulations. The basic design of the MEMS accelerometer comprises generic flexure beam/seismic mass structure attached to a rigid frame with acceleration-induced stress sensing piezoresistor elements positioned at the stress maxima of the beam flexures and connected in the form of a Wheatstone bridge as shown in the variant of Fig. 4. Wheatstone bridges form the backbone of

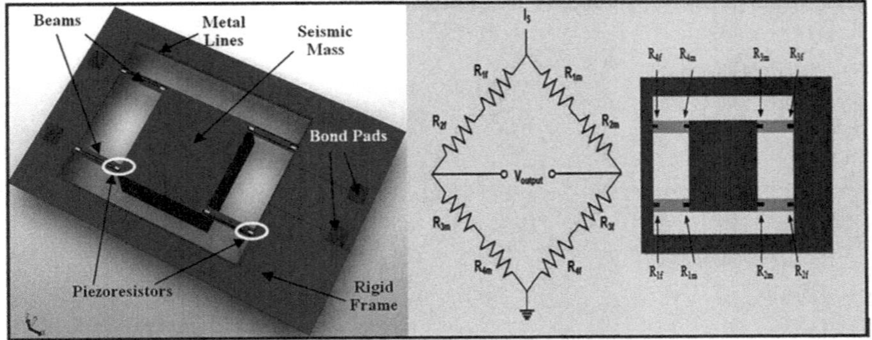

Fig. 4 (*left*) MEMS piezoresistive accelerometer variant discussed in this test. (*right*) Wheatstone bridge arrangement of stress sensing piezoresistors and their mapping on the MEMS accelerometer

resistive sensing as their balanced and unbalanced conditions correspond to different perturbation inputs along different axes of operation.

Piezoresistive accelerometers may be designed for single-axis as well as multiaxis inertial sensing. In this chapter, we primarily deal with single-axis accelerometers with minimal cross-axis sensitivity. As seen from Fig. 4, mass edge-alignment of the beams imparts maximum immunity to off-axis perturbations for the geometry in question while maintaining sensitivity along the desired axis. In theory, the Wheatstone bridge implementation of the piezoresistive sensing elements has two very important consequences:

(i) Enhancement of desired axis signals and auto-cancelation of the off-axis perturbations
(ii) Auto-compensation of temperature variation of individual piezoresistor values.

The basic operation of the accelerometer along its axis under external sinusoidal mechanical excitation can be approximated as a single degree of freedom (SDOF) mass-spring-damper (MSD) system which is defined by

$$m_{\mathrm{eff}}\frac{d^2z}{dt^2} + c_{\mathrm{eff}}\frac{dz}{dt} + k_{\mathrm{eff}}z = F_M = mA\sin(\Omega t) \tag{9}$$

$$c_{\mathrm{eff}} = 2\zeta\omega_0 m_{\mathrm{eff}} = 2\frac{(0.42\mu LB^3)}{t^3} \tag{10}$$

$$k_{\mathrm{eff}} = m_{\mathrm{eff}}\omega_0^2 = \frac{4Ebh^3}{l^3} \tag{11}$$

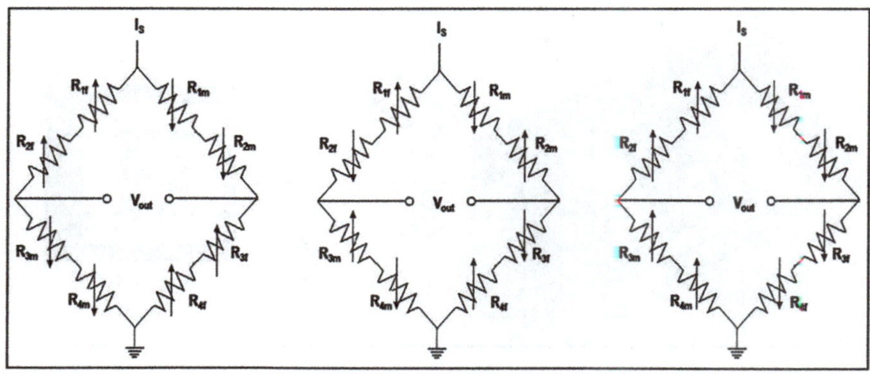

Fig. 5 Wheatstone bridge configuration changes due to acceleration along (*left*) out-of-plane *z*-axis (*middle*) in-plane *x*-axis and (*right*) in-plane *y*-axis

where m_{eff} is the seismic mass, c_{eff} is the damping coefficient, t is the air gap, k_{eff} is the equivalent spring constant of the quad-beam structure, ω_0 is the natural frequency (angular), A denotes the amplitude of excitation and Ω is the input excitation frequency. Here, F_M is considered sinusoidal because of its relevance to vibration shake table characterization of the system where the MSD sensor oscillates relative to the rigid frame. From beam mechanics and fluid (squeezed-film) damping [3], we can extract the effective spring constant, the natural frequency of the structure and the damping coefficient, thereby enabling us to tune the structural dimensions according to requirement. Eight piezoresistors occupy the stress maxima regions of the four beam flexures supporting the mass and are connected by metal lines in the form of a current biased Wheatstone bridge as shown in Fig. 4. The sensor is wafer-encapsulated using etched glass caps on either surface for over-range protection and fluidic squeezed-film damping. The separation between the sensor surface and the glass cap can be evaluated from the damping coefficient calculated. The Wheatstone bridge configuration serves to nullify the effect of temperature drifts in individual piezoresistors as well as canceling out off-axis acceleration inputs as illustrated in Fig. 5.

From the changes in the bridge arm resistances, the corresponding bridge output voltage is given by

$$V_{\text{output}} = I_{\text{bias}} R_{\text{input}} \frac{\pi_{44}\sigma_l}{2} \tag{12}$$

considering the assumptions of purely longitudinal stress (σ_l) and the sole influence of π_{44}. As far as the bridge offset and sensitivity [20] are concerned, their origins and dependence on temperature have been explored in theory and are presented as subsections below.

4.1 Wheatstone Bridge Offsets

Bridge offsets are a result of mismatch in the resistance values of the piezoresistive sensing elements comprising the acceleration sensor and can be attributed to the following reasons:

- Process variations resulting from the tolerances of the nonideal MEMS fabrication techniques employed result in resistor nominal value mismatch.
- Post-release stress relaxation mechanisms occurring in compliant MEMS structures results in the generation of residual stresses which in turn causes mismatch in the nominal piezoresistance values.
- Bias current flow induced localized heating within the wafer-level device causes the piezoresistive sensing elements to be subjected to different temperatures due to asymmetric crossing of the isotherms over the resistors creating mismatch in the nominal values.

Of the above, the thermal effect is shown to dominate this variation in resistance values and thus, symmetry of the sensing element layout is essential in the design as well as the allowance of sufficient stabilizing time before carrying out measurements at different temperatures. From the above discussion, it is clear that the bridge offset voltage is quite a complex quantity dependent on individual resistances (with their foundry tolerances) and their individual variations (due to the reasons outlined above) and as such, theoretical approximations have not been attempted.

4.2 Wheatstone Bridge Sensitivity

Bridge sensitivity, on the other hand, is a highly nonlinear function of temperature and occurs despite the equality of nominal resistance values (at 300 K). An individual piezoresistor and bridge output at temperature T and stress σ (neglecting bridge offset) can be defined as

$$R(\sigma, \Gamma) = R_0(1 + \alpha_{P1}\Gamma + \alpha_{P2}\Gamma^2 + \cdots)[1 \pm S_0(1 + \alpha_{N1}\Gamma + \alpha_{N2}\Gamma^2 + \cdots)\sigma] \quad (13)$$

$$V_{\text{output}}(\Gamma) = V_{\text{bias}}S_0\sigma(1 + \alpha_{P1}\Gamma + \alpha_{P2}\Gamma^2 + \cdots)(1 + \alpha_{N1}\Gamma + \alpha_{N2}\Gamma^2 + \cdots) \quad (14)$$

where R_0 and S_0 are the nominal values of resistance (at 300 K) and the coefficient of stress sensitivity (+ for tensile, − for compressive), respectively, α_P and α_N are the resistance and stress sensitivity temperature coefficients (in practice, $|\alpha_P| < |\alpha_N|$ which results in the second bracketed term dominating the first bracketed term), σ is the stress due to the applied acceleration, I_{bias} is the bridge biasing current and Γ represents the difference between T and the reference temperature T_0 (300 K). The bridge sensitivity is defined as the derivative of the bridge output with respect to the applied stress (due to acceleration) as:

$$\frac{\partial V_{\text{output}}}{\partial \sigma} = I_{\text{bias}} R_0 S_0 (1 + \alpha_{P1}\Gamma + \alpha_{P2}\Gamma^2 + \cdots)(1 + \alpha_{N1}\Gamma + \alpha_{N2}\Gamma^2 + \cdots) \quad (15)$$

implying that sensitivity estimation requires accurate curve fitting of experimental data. In case of piezoresistance in semiconductors, the temperature coefficient of resistance is negligible in comparison to the temperature coefficient of stress sensitivity and as a result, calculations are often made on the basis of the assumption that the first bracketed term is approximately unity.

5 Sensor Fabrication Processes

The MEMS accelerometer design presented earlier can be implemented on silicon substrates for integration with interfacing circuits to create a complete monolithic or multichip MEMS-CMOS sensor system. The basic outline of the fabrication process followed for realizing the MEMS piezoresistive accelerometer is presented below:

(a) High resistivity and high purity silicon substrates are generally used such that substrate resistivity does not interfere with the functioning of the Wheatstone bridge. In addition, defect-free substrate surface layers are preferred since the electrical sensing element is laid out on the wafer surface.

(b) Piezoresistors can either be diffused or ion implanted at the surface with implantation providing greater precision in the definition of resistor profile and junction depth. As a result, the piezoresistors can be accurately localized to the stress maxima positions for maximum sensitivity.

(c) Wheatstone bridge metallization may be carried out using any metal depending upon the requirements. In general, metals with high melting points and nonsusceptibility to oxide formation are preferred due to structural release issues and wafer-level packaging constraints.

(d) Structural release can be achieved using either wet or dry etching. Wet etching is relatively cost-effective when compared to dry etching techniques but raises issues such as reduction of the seismic mass due to trapezoidal etch profiles, thereby reducing mechanical sensitivity. In addition, certain metals must be avoided as the etchant often attacks the metal layer as well. The advantages of dry etching include accurate profile definition of the mechanical elements with no such decrease in the seismic mass due to vertical sidewall generation as well as lack of etch affinity toward metals.

(e) Wafer-level encapsulation is a technique whereby the device wafer is capped on either surface by an appropriately patterned wafer for the provision of air damping. In case of bulk micromachined structures such as the one being explored in this text, wafer bonding is the only viable option for encapsulation. Wafer bonding can be of many types based on the wafer materials. In this case, glass wafers are patterned to form cavities which are then bonded on to the device wafer on either side. Hermetic or air-tight seals can be obtained using anodic or eutectic bonding techniques while quasi-hermetic seals can be realized using simpler adhesive bonding processes.

A wide range of fabrication processes may be employed to realize MEMS piezoresistive accelerometers, as is apparent from the literature reviewed in this chapter, that are tailored to meet specific goals and as such, there is no optimum process technology that can be used to implement all piezoresistive variants. The above process flow merely serves as a possible guideline for fabricating structures similar to the one being developed in this chapter. Quality of sensor fabrication directly determines operational performance of the accelerometer and can be estimated using standard tests developed over time to characterize such electro-mechanical systems. Subsequent tuning of process tolerances based on characterization data serves as a critical "feedback" loop in the quest to realize even better devices. The next section deals with the various characterization tools and techniques used for micromechanical inertial sensor testing in general.

6 Characterization of MEMS Accelerometers

MEMS accelerometers and most other micromechanical sensors and actuators are subjected to detailed testing using different testing equipment and measurement setups. Such tests are conducted at various levels, as defined below, in order to obtain the most accurate estimate of device performance. Testing level, as is apparent from the name, is defined by the "state" of the subject being tested and may be classified broadly into three types:

(i) *Wafer-level tests:* These tests are conducted on whole or diced wafers which provide an estimate of characteristics such as sheet resistance, film thicknesses, and feature dimensions to name a few. This level of testing is mostly concerned with the bulk substrate properties and the proper handling of the same so as to enable the realization of devices as close to ideal as possible.

(ii) *Device-level tests:* Placed higher in the testing hierarchy, these tests are carried out on individual devices that may or may not be wafer-encapsulated. Perhaps the most critical stage of testing, the device-level experiments reveal information about the natural frequency of the device, signal transduction reliability, electrical characteristics under bias such as offsets and output signal levels and the like. This level deals with individual devices without any external support and is useful in predicting integrated behavior to some extent.

(iii) *System-level tests:* The most complete level of characterization, these tests are done by interfacing the MEMS with all the accessories such as signal conditioning front-end and back-end electronics and subjecting the entire system to the perturbations that emulate real-life conditions to which the

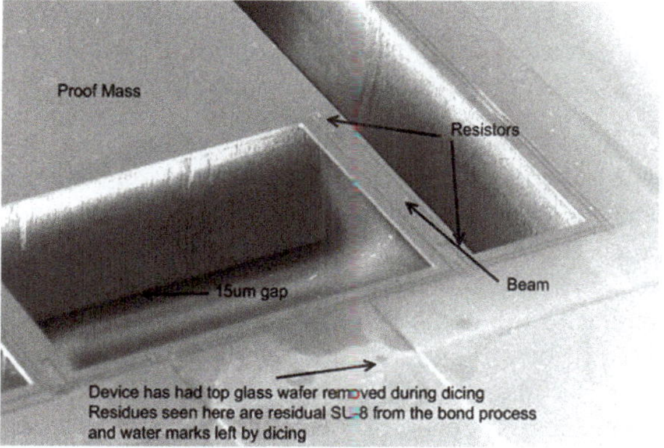

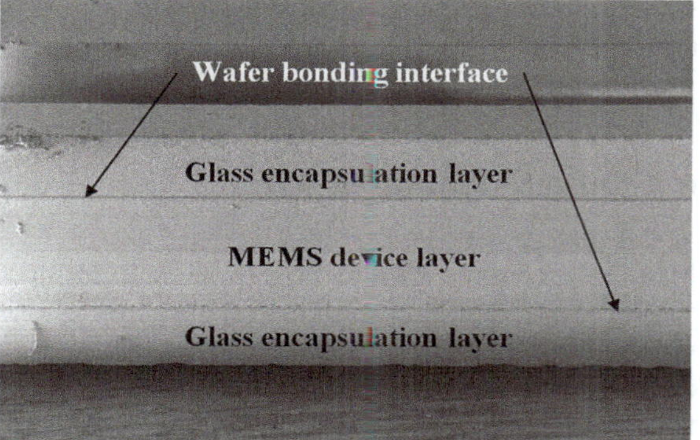

Fig. 6 (*top*) Scanning electron micrograph of the beam–mass configuration. (*bottom*) Wafer-encapsulated accelerometer system (encapsulated wafer details not visible) (image courtesy [21])

system is to be exposed in practical applications. This advanced level of testing gives an estimate of how the system behaves under standard operating conditions and is strongly dependent on the wafer-level and device-level performance of the MEMS.

These different levels of tests generally require corresponding characterization equipment although there are multipurpose tools which can be effectively employed at more than one level. A few most important of these are explored in the rest of this section starting with visual characterization using advanced imaging techniques.

Figure 6 shows a scanning electron microscope image of the beam-mass system of the MEMS piezoresistive accelerometer that has been used as reference in this chapter as well as its complete wafer-encapsulated implementation. Proper feature realization can be inferred from these images which corroborates the validity of the design realization processes. Scanning electron microscope (SEM) imaging is a wafer-level visual characterization technique which uses the reflection of a focused beam of electrons to assess substrate features and topography. In case of encapsulated MEMS accelerometers, observation of device features requires the removal

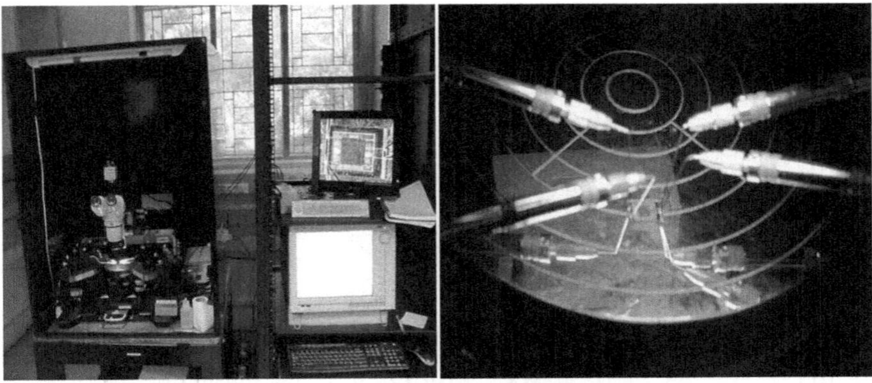

Fig. 7 (*left*) Wafer probe station interfaced with device analyzer. (*right*) Encapsulated MEMS piezoresistive accelerometer probing on thermal stage

of the capping layers as the impinging electrons cannot penetrate the capping surface and provide imaging of underlying layers as can be seen in Fig. 6. SEM images can also be used to study interface integrity, possible delamination defects and dimensional aberrations resulting from process tolerances and wafer handling.

The wafer probe station is an essential electrical characterization instrument which is used to measure the electrical and electronic properties of the MEMS device under test (DUT). The probe station is not stand-alone equipment and it is necessary to interface instruments like device analyzers to evaluate various performance metrics. An open stage wafer probe station interfaced with a device analyzer is shown in Fig. 7 and is used to study the current-voltage (I-V) characteristics of MEMS accelerometers. It is to be noted that the wafer mounting stage of many probe station variants is temperature regulated such that studies on temperature dependence of certain parameters may be carried out. In addition, there are vacuum chamber probe stations which have the stage enclosed in vacuum for precision measurements.

One of the most important device-level characterization tool for vibratory MEMS devices is the laser Doppler vibrometer (LDV) shown in Fig. 8, which estimates vibration amplitudes and frequencies of MEMS device surfaces using frequency heterodyning and laser interferometric techniques. LDV measurements yield precise vibration frequencies and their corresponding mode shapes using Fourier transform methods and device surface sampling. Frequency response of inertial sensors is extremely critical for optimal performance as operating frequencies and device natural frequencies should often be isolated so as to prevent amplitude resonance and consequent device damage. Additionally, damping characteristics of MEMS accelerometers need to be studied carefully for the design of precisely damped systems in the future due to the fact that most accelerometers are critically damped and deviations from specified values may lead to unwanted oscillations and response delays.

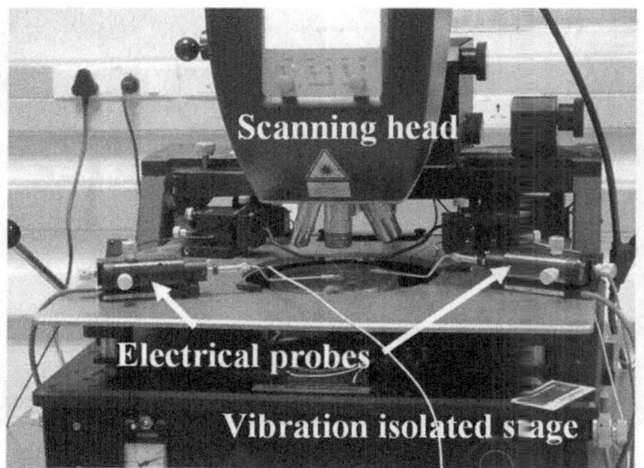

Fig. 8 Laser Doppler vibrometer used for vibration analysis of MEMS components

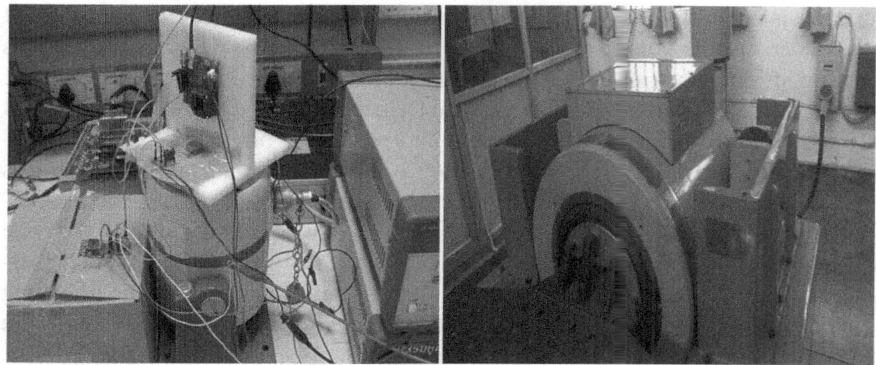

Fig. 9 (*left*) A table-top electromagnetic shaker. (*right*) A large-scale hydraulic shaker

System-level characterization forms the final level of testing the functionality of MEMS accelerometers. Generally, the MEMS sensors are interfaced with their signal conditioning and readout circuits and mounted on a vibratory stage. These vibratory stages are designed so as to provide precision vibrations like sinusoidal and shock impulse for evaluating the dynamic performance of the device. Such setups are called "shakers" and may of different types. Two versions are shown in Fig. 9 with one being an electromagnetic table-top shaker targeted toward testing miniaturized assemblies while the other is a hydraulic shaker meant for vibration analysis of large structures.

7 Conclusions

MEMS piezoresistive accelerometers have been covered in this chapter with their usage being overviewed. The basic principle behind the sensing mechanism is discussed and the subsequent evolution of the MEMS piezoresistive accelerometer is covered ranging from academia to industry. A sample device structure is chosen for description purposes and its design is analyzed for mechanical functionality and subsequent electrical signal transduction issues. Various forms of characterization of such MEMS sensors are explored at different levels with the utility of each being presented with the hope that the overall content of this chapter shall enable the understanding of such sensors at the basic level in the very least.

Acknowledgments The authors would like to thank the National Programme on Micro Smart Systems (NPMASS) for their financial support without which this work would not have materialized.

References

1. Smith CS (1954) Piezoresistance effect in germanium and silicon. Phys Rev 94:42–49
2. Smith CS (1958) Macroscopic symmetry and properties of crystals. Solid State Phys Adv Res Appl 6:175–249
3. Roylance LM, Angell JB (1979) Batch-fabricated silicon accelerometer. IEEE Trans Electron Devices 26:1911–1917
4. Bao M (2008) Micro mechanical transducers elsevier handbook of sensors and actuators, vol 8. Elsevier, Oxford, pp 215–217
5. Crazzolara H, Flach G, von Miinch W (1993) Piezoresistive accelerometer with overload protection and low cross-sensitivity. Sens Actuators, A 39:201–207
6. Kal S et al (2006) CMOS compatible bulk micromachined silicon piezoresistive accelerometer with low off-axis sensitivity. Microelectron J 37:22–30
7. Shen S, Chen J, Bao M (1992) Analysis on twin-mass structure for a piezoresistive accelerometer. Sens Actuators, A 34:101–107
8. Sim JH et al (1998) Eight-beam piezoresistive accelerometer fabricated by using a porous-silicon etching method. Sens Actuators, A 66:273–278
9. Partridge A et al (2000) A high performance planar piezoresistive accelerometer. IEEE/ASME J Microelectromech Syst 9:58–66
10. Park WT et al (2000) Encapsulated submillimeter piezoresistive accelerometers. IEEE/ASME J Microelectromech Syst 16:507–514
11. Kwon K, Park S (1998) A bulk-micromachined three-axis accelerometer using silicon direct bonding technology and polysilicon layer. Sens Actuators, A 66:250–255
12. Ravi Sankar A, Lahiri SK, Das S (2009) Performance enhancement of a silicon MEMS piezoresistive single axis accelerometer with electroplated gold on a proof mass IOP. J Micromech Microeng 19:11
13. Amarasinghe R et al (2005) Design and fabrication of a miniaturized six-degree-of-freedom piezoresistive accelerometer IOP. J Micromech Microeng 15:1745–1753
14. Atwell AR et al (2003) Simulation, fabrication and testing of bulk-micromachined 6H-SiC high-g piezoresistive accelerometers. Sens Actuators, A 104:11–18
15. Lin Y (2011) Micromachined piezoresistive accelerometers based on an asymmetrically gapped cantilever. IEEE/ASME J Microelectromech Syst 20:83–93

16. Kim SC, Wise KD (1983) Temperature sensitivity in silicon piezoresistive pressure transducers. IEEE Trans Electron Devices 30:802–810
17. Doll JC, et al (2011) Self-heating in piezoresistive cantilevers. Appl Phys Lett 98:3
18. Endevco Meggit Sensing Systems https://www.endevco.com
19. Honeywell Sensing and Control https://measurementsensors.honeywell.com
20. Dutta A (2012) Design and implementation of high precision variable-gain amplifier with on-chip temperature compensation scheme for MEMS sensors (MS Dissertation), IIT Kharagpur
21. Roy AL, Bhattacharyya TK (2013) Design, fabrication and characterization of high performance SOI MEMS piezoresistive accelerometers Microsystem Technologies (in press)

A Handheld Explosives Detector Based on Amplifying Fluorescent Polymer

Anil Kumar, Jasmine Sinha, Ashok K. Majji, J. Raviprakash, Sathyadeep Viswanathan, Justin K. Paul, S. Vijay Mohan, Shilpa K. Sanjeeva, Swathi Korrapati and Chandrashekhar B. Nair

Abstract Explosive detection has become more relevant today due to increased threats of terrorist activities and chemical warfare. In this context, amplifying fluorescent polymers, AFPs, provide an attractive platform as they are easy to synthesize and exhibit a high fluorescence quantum yield in solid state which is beneficial for a handheld detector. Furthermore, the easy functionalizing of AFPs allows one to introduce various receptors to broaden the scope of detection. The introduction of a "click"able pendant group in AFP enables one to design and develop various biomedical and chemical sensors based on the guest–host chemistry.

Keywords Explosive detection · Chemical sensor · Fluorescent polymers · Optical detection

1 Introduction

With the growing need for public security, it has become paramount to develop an indigenous system for the detection of various analytes such as explosives, chemical and biological warfare agents, narcotic substances, etc. The detection of explosives is further complicated by the camouflaging of the analytes. The development criteria for explosive detectors are that they should be rugged, easy-

A. Kumar (✉) · A. K. Majji
Indian Institute of Technology Bombay, Mumbai, India
e-mail: anilkumar@iitb.ac.in

J. Sinha
Johns Hopkins University, Baltimore, USA

J. Raviprakash · S. Viswanathan · J. K. Paul · S. V. Mohan · S. K. Sanjeeva · S. Korrapati
C. B. Nair
Bigtec Private Limited, Bengaluru, India

K. J. Vinoy et al. (eds.), *Micro and Smart Devices and Systems*,
Springer Tracts in Mechanical Engineering, DOI: 10.1007/978-81-322-1913-2_3,

to-use by a semiskilled person, minimum false positives and be fully automatic, battery-operated, and handheld with a rapid sensing response. Various detection methods based on fluorescence [1–6], surface acoustics [7], and photo-fragmentation [8] are known for the vapor-phase detection of explosives. Detection selectivity is principally determined by transduction methods (e.g., absorption, fluorescence, conductivity, etc.), and the design of a new material has been focused on the selectivity in it, toward the analyte [9−11]. In this direction, AFPs provide an attractive platform and have been used for the successful design and fabrication of handheld detectors for various explosives, chemical and biological warfare agents as well as narcotic substances. Their ease of syntheses along with the ability to fine-tune their properties makes them ideal candidates for the development of various detectors. Of particular relevance to the work discussed here is the use of AFPs whose emission radiation could be changed by an oxidative quenching mechanism, when contact is made with an electron-depleting explosive compound. Traditional fluorescent-based sensors utilize quenching as the mechanism of sensing, wherein the fluorescent polymers are excited with an independent light source resulting in steady-state fluorescence. In this chapter, we present various AFP-based approaches as well as our efforts toward the development of a handheld explosives detector and the scope for future work in this important area.

2 Fluorescence-Based Detectors

In general, a detector consists of three major components: (a) a receptor (based on either a chemical or physical interaction); (b) a transducer, and (c) a readout circuit. The success of such devices depends upon the fidelity of the receptor–analyte complex formation. Signal transduction is the readable signal change that is directly related to the measurable amount of energy change when the receptor interacts with the analyte. The readout of a fluorescent chemosensor is usually measured as a change in intensity of fluorescence, decay lifetime, or shift in the emission wavelength [12−18]. The fluorescence technique has the advantage of capturing the signal of analyte binding in a very short time, which makes possible the real-time and real-space detection of the analyte and imaging associated with analyte distribution [19, 20]. In general, measuring the change in fluorescent intensity is the most commonly employed method for detection, wherein the detector can work either as "turn on" (intensity increases in the presence of an analyte) or as "turn off" (intensity decreases in the presence of an analyte). Furthermore, the change in fluorescent intensity could result either from the photo-induced electron transfer or from an energy transfer process [21–23].

AFPs based on conjugated polymers offer an advantage in fluorescence sensing due to the amplification from energy migration. The amplification of the signal in AFPs can be explained using the analogy of the lamps in a series versus lamps in a parallel arrangement. In parallel arrangement, one needs to turn on/off many lamps before a significant change in signal can be detected. However, in case of AFPs,

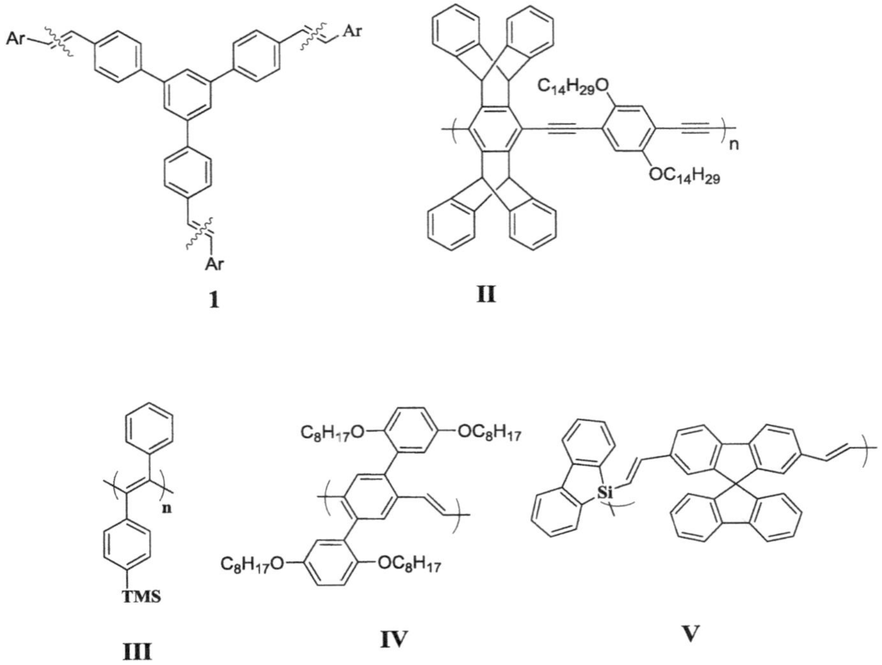

Fig. 1 Structure of various AFPs used for explosive detectors

lamps are in a series arrangement and turning on/off a single lamp results in a drastic change in the signal, and hence, amplification. The combination of the two, amplification and sensitivity, leads to the production of new systems of unparalleled sensitivity. These utilities of AFPs for fluorescence-based sensing were first demonstrated by Zhou and Swager [24, 25]. AFPs (Fig. 1) based on poly(*tris*-phenyleneethynylenes) (TPV) network (**I**) [26], poly(*p*-aryleneethynylene)s (**II**) [12, 13], polyacetylenes (**III**) [27], poly(pheneylenevinylene)s (**IV**) [28], and polysilanes (**V**) [29] have been explored for the detection of various explosives. Currently, AFP-based explosive detectors are available commercially under the brand name "**FIDO**".

3 Development of Explosive Detector Beagle-Z

IIT Bombay in collaboration with Bigtec Labs, Bangalore, undertook a project sponsored by the National Program on Micro and Smart Systems (NPMASS) of Aeronautical Development Agency (ADA) Bangalore for the development of a fully automatic handheld explosive detector based on AFPs. After going through

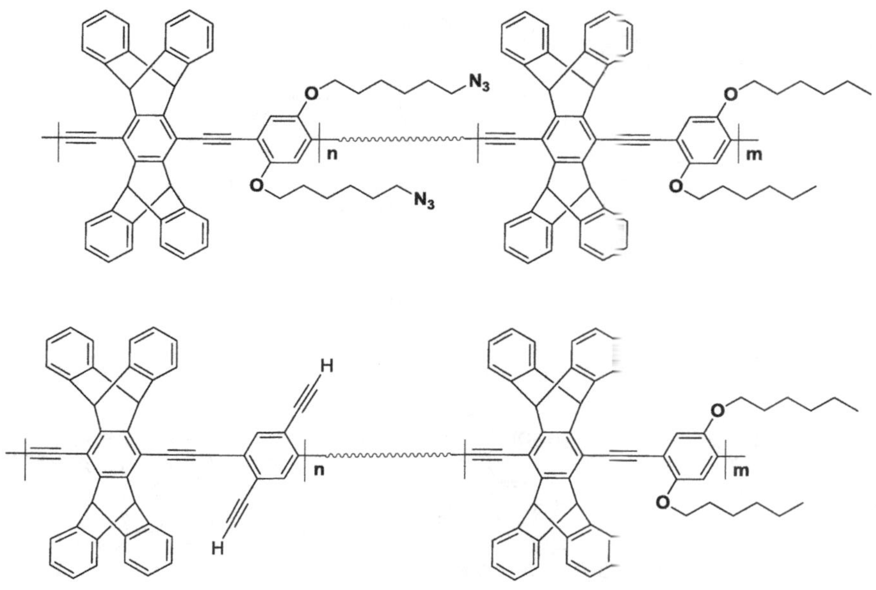

Fig. 2 "Click"able AFPs

various reported structures and detectors, it became clear that poly(aryleneethynylenes) (**II**) are the best candidates for explosive detection. In the case of fluorescence-quenching sensing in AFPs, the quenching happens due to the transfer of the photo-excited electron from AFP to the LUMO level of the analyte. In order to design a system with successful electron transfer from AFP to the analyte, apart from exhibiting fluorescence in the solid state, there are three criteria which need to be satisfied. These are: (a) the LUMO of the analyte should be at lower energy than the excited state of AFP; (b) the analyte should bind with AFP; and (c) the film should be porous so that the analyte can diffuse into the polymer matrix. However, the number of synthetic variants which can be explored to fulfill these criteria tend to be very large and hence synthesis becomes a laborious process. Therefore, we developed a novel synthetic route wherein we designed and synthesized "Click"able AFPs [30] based on poly(aryleneethynylenes) and the structure of these polymers is shown in Fig. 2.

The advantage of the "click"able AFP was that it is much easier now to generate various AFPs starting from a single AFP. However, these AFPs still exhibited limited sensitivity as they could only detect trinitrotoluene (TNT). In order to further broaden the scope of detecting other explosives, molecular "kinks" were introduced in the backbone as shown in Fig. 3. Furthermore, in order to improve the binding of nonaromatic nitro explosives, substituted silyl groups were introduced (Fig. 3) as it is known that the nitro group forms an Lewis acid–base complex with silanes. These AFPs were now found to be suitable for the detection of various aliphatic nitro-based explosives. All these AFPs were tested in

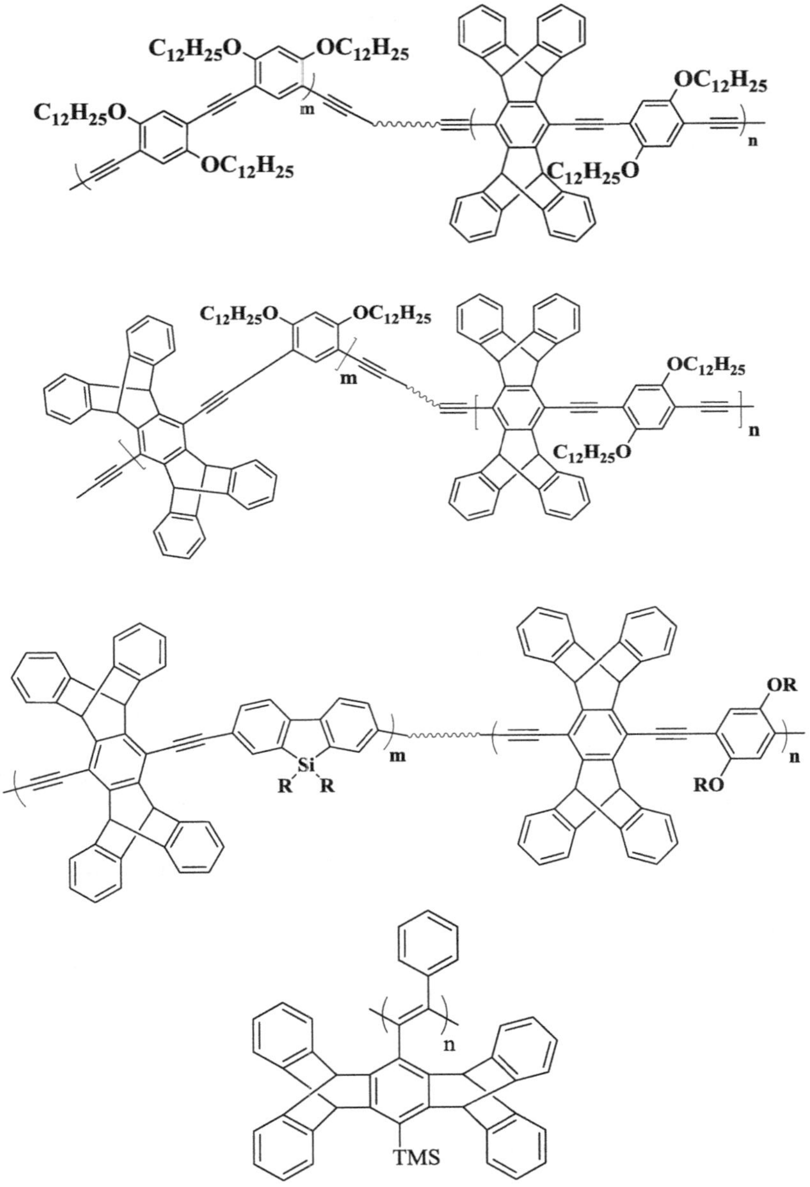

Fig. 3 AFPs with molecular "kinks" and silanes

laboratory conditions for their sensing ability for various explosives using fluorescence quenching studies, both in solution and in the solid state. A typical solid state quenching profile is shown in Fig. 4.

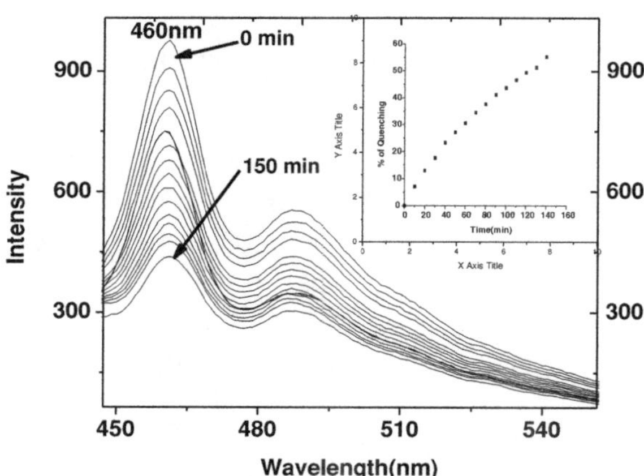

Fig. 4 The time-dependent fluorescence intensity in a 25-Å thick film upon exposure to TNT vapor (room temperature) at 0, 10, 20, 30, 40, 50, 60, 70, and 150 min (*top* to *bottom*), and the fluorescence quenching (%) as a function of time (inset)

3.1 Technical Details of Beagle-Z

3.1.1 Overview

The explosive detection device is used for the trace-level detection of organic explosives like TNT and RDX. The device is portable and battery operated with a RS-232 interface that enables real-time sensing. It uses an amplifying fluorescent polymer (AFP) as the active sensing material. When the AFP is excited by ultraviolet light radiation of a particular wavelength (360–400 nm), the material emits photons. When TNT or RDX is present, the molecule binds to the AFP surface and quenches the photon emission. A recorded drop in the emitted signal indicates the presence of the specific explosive.

The device is used for the detection of trace levels of organic explosives in the field. The possible scenarios could be active warfare, homeland surveillance and security, and crowded public places like railway stations and airports. It is designed to be used by security personnel with minimal training. Therefore, the device is designed to be portable and easy to operate.

3.1.2 Device Design

The device consists of a two-part silicone molded (polyurethane + ABS) casing that houses the electronics and the mechanical parts (Fig. 5). The device has three subsystems:

1. A sensing element coated with a sensing polymer (amplified fluorescent polymer—AFP) that binds specifically to aromatic nitro compounds and fluoresces at a characteristic wavelength.
2. A sampling and optical detection module that detects a change in fluorescence.

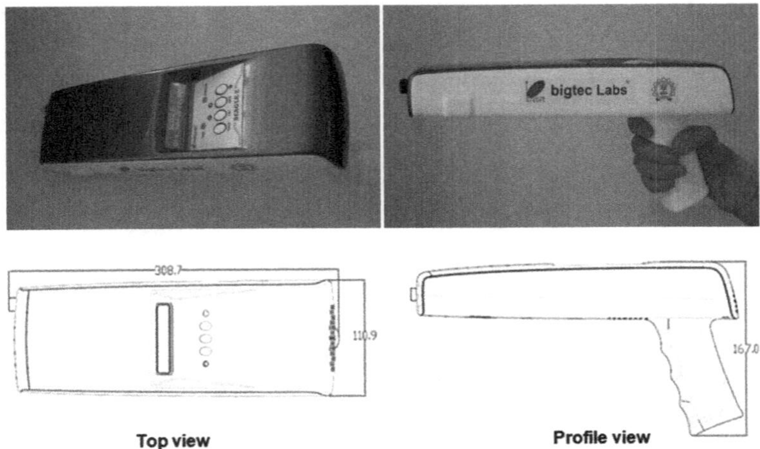

Fig. 5 Picture and outline of Beagle-Z with dimensions

3. Electronics that displays the response of the sensing element in a readable form and controls all other aspects of the unit.

The details of these subsystems are as follows:

Sensing Element

A 65-mm-long borosilicate glass tube, dip-coated with AFP serves as the active sensing material that responds to the presence of explosive vapors in ambient air. The AFP consists of fluorescing chromophores linked together in polymer chains. When the AFP is exposed to ultraviolet light of a particular wavelength, the material undergoes a stimulated emission with a stream of photons being emitted by the polymer film. TNT, DNT, and/or RDX molecules bind to the AFP surface and quench the emission (Fig. 6). The inherent structure of the chromophore chain in an AFP dramatically amplifies the quenching effect, resulting in ultralow detection limits. In conventional fluorescence quenching, the binding of a single explosive molecule quenches only the chromophore to which the explosive molecule binds. However, with AFP, binding of a single molecule of explosive quenches the fluorescence of multiple chromophores linked in a chain.

Sampling/Optical Detection Module

This module consists of an excitation source, focusing optics, a detector, a flow sensor, and a pump. The excitation source is an ultraviolet (UV) light-emitting diode (LED) with a peak emission at 380 nm. The focusing optics consists of a collimating lens followed by a band pass filter (380 nm, 10 nm FWHM) and a plano-concave

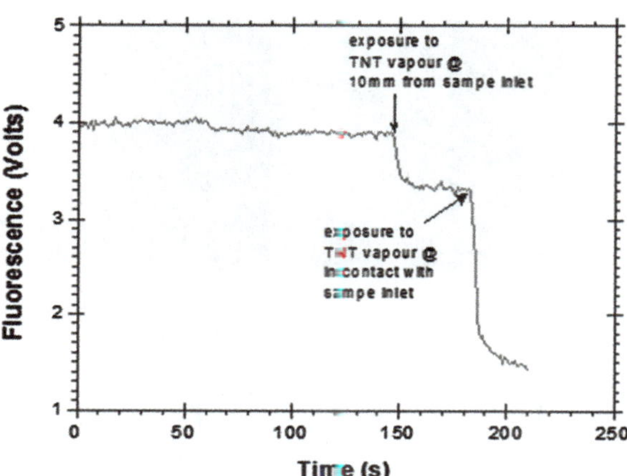

Fig. 6 Typical response of the device to TNT vapors

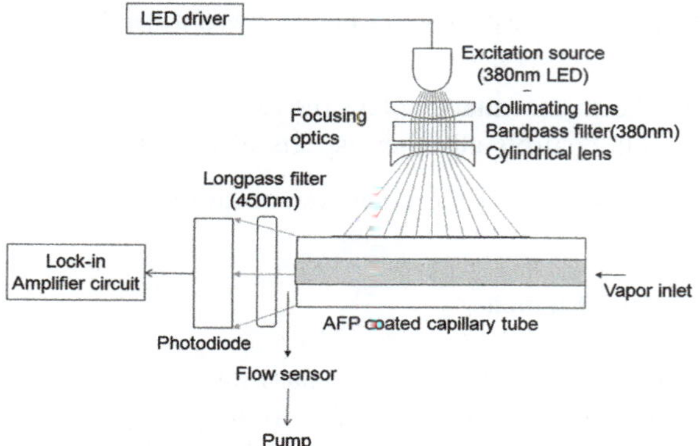

Fig. 7 Block diagram of sampling/optical detection module

cylindrical lens to laterally spread the collimated and filtered light beam to illuminate a predefined segment of the glass tube coated with AFP. A fluorescence signal at wavelengths greater than 450 nm is detected by a silicon photodiode and amplified. This is achieved by filtering light from the glass tube through a long pass 450-nm filter. The details of the optical unit are depicted in Fig. 7. The vapor is drawn through the glass tube by a diaphragm pump. The flow is monitored by a flow sensor that triggers a warning if the flow drops due to a blockage in the air-flow path.

The optical components and the sensing element are housed in a custom assembly that is machined from Aluminium and is anodized. Figure 8 illustrates a 3D solid model of the mechanical assembly housing the optics, the sensing tube, and the detection electronics (LED and detector).

Electronics

The device is powered by a 7.5-V Li-ion battery and is controlled by a microcontroller-based circuit. The principal parts of the electronic circuit are as follows:

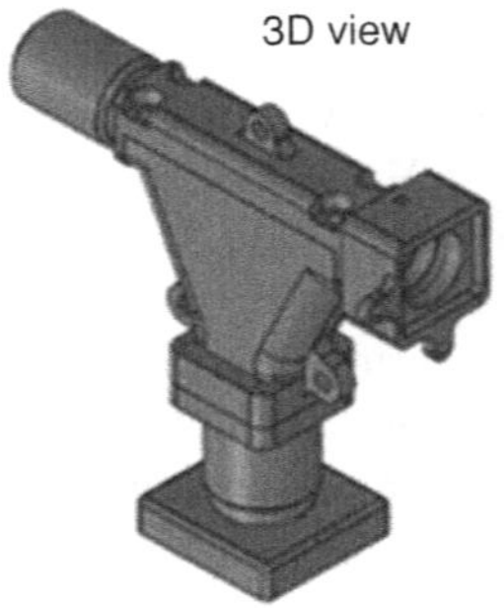

Fig. 8 Mechanical assembly for the optical detection module

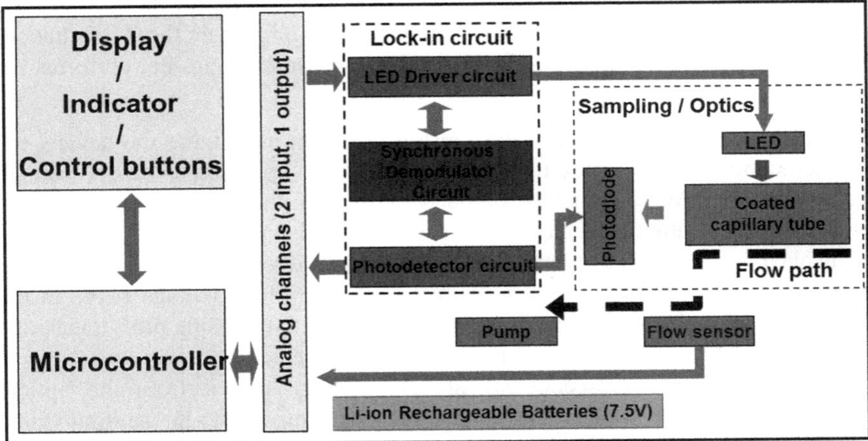

Fig. 9 Block diagram of vapor detection device

- Power management circuit: It generates regulated voltages and manages the battery charging.
- Data acquisition and control circuit: A microcontroller-based circuit that interfaces and controls analog/digital parts of the circuit.
- Optical detection circuit: LED driver circuits and amplifier circuits for detector.
- Flow sensor circuit: It monitors the air flow.
    - Peripheral electronics: It consists of display, indicators (LED) and user control.

Various parts of the electronic circuit responsible for the functioning of the device are illustrated in Fig. 9.

Power Management Circuits

These circuits provide regulated power supplies to all the subassemblies of the device. The circuit generates +5 V, −5 V, and an additional +6 V exclusively for the diaphragm pump. It also includes the battery charging and the low-battery indicator circuits. It also seamlessly switches between battery power (without an external adaptor) and external power (when the adaptor is connected). The external power is a 12 V dc adaptor. When the adaptor is connected, the load is taken off the battery as it is charged and the device draws power from the adaptor.

Data Acquisition and Control Circuit

The data acquisition circuit comprises a microcontroller that is serially (SPI) interfaced to an analog to digital convertor (ADC) and digital to analog convertors (DAC). All the analog signals are routed through the ADC and the DAC to the microcontroller. The microcontroller averages 20 readings from the ADC that are taken in 150 ms duration. The embedded code in the microcontroller performs the following functions:

- Calibration and validation of the sensing tube: On initializing the device, the microcontroller calibrates the tube by giving an appropriate off-set voltage through the DAC to adjust the output voltage from the detector circuit to 4 V. The detector output is displayed on the screen and is updated every second. The displayed data are an average of three readings. The sensing tube is validated to ensure that the optical signal is within a predetermined voltage range in the absence of explosive vapor. On identifying a defective sensing tube, a message is displayed on the screen and an alarm is triggered.
- Monitoring the fluorescence signal: The voltage output from the optical detection circuit is measured every second and compared to the starting value, obtained at the onset of the scanning operation. A predetermined threshold value can be set to trigger an alarm and an LED, on the front panel, indicating a positive detection.
- Monitor the flow sensor output: If the flow sensor output voltage goes outside a preset voltage range, the microcontroller triggers an alarm and displays an error message on the LCD screen.

Optical Detection Circuit

The circuit comprises an LED driver and a lock-in amplifier. The lock-in amplifier consists of a trans-impedance amplifier, voltage preamplifier, synchronous demodulator, low pass filter, and postamplifiers. The circuit drives the LED (excitation source 380 nm) at a fixed frequency and measures the fluorescence at the same frequency thus improving the signal to noise ratio. The details of the various parts of the circuit are as follows:

- Oscillator and LED driver circuits: A function generator generates a reference square wave of 1 KHz frequency. This reference is fed to the LED driver circuit which consists of an MOSFET switch in series with the LED and a current-limiting resistor. This signal is also fed to the lock-in amplifier reference input. The LED current is 20mA (pulsed).
- Trans-impedance amplifier: This makes up the first stage of the lock-in circuitry and converts the photodiode current into a voltage with a very high gain of 10^6 V/Amp. A current signal of the order of pico-amps is converted into micro-volts.
- Preamplifiers: The trans-impedance amplifier output goes to a chain of three voltage amplifies. The amplifiers have a total gain of about 250 V/V and have DC filters to eliminate DC and low frequency (line voltage) interference. The output of this stage contains both a signal and noise.
- Synchronous demodulator: This stage comes after the preamplifiers and forms the heart of the lock-in amplifier as it separates the signal from the noise. It receives the reference signal form the function generator and rectifies the amplified signal from the preamplifier synchronously with the reference. Synchronous rectification gives an output, the average value of which depends only on the magnitude of the signal (as it is in sync with the reference) and not that of the background noise (noise, being of random frequency, is out of sync).
- Low-pass filter: This circuit extracts the average value form the demodulator output and gives a DC voltage which is proportional to the fluorescence intensity.
- Postamplifiers: The low-pass filter output is amplified and scaled by postamplifiers and we get a final output scaled in a 0–5 V range.

Flow Sensor Circuit

It consists of the flow sensor and the associated signal conditioning circuit for providing a DC voltage output which represents the flow rate. The output signal is calibrated to about 2.25 V for normal flow and decreases when there is a block in the air-flow path. A change in the signal triggers a warning message that is displayed on the screen.

Peripheral Electronics

This collectively refers to the LCD display, indicator lights (LEDs), and a buzzer in the device. A 16-character, 2-line display is used in the device. The microcontroller directly controls the indicator LEDs and the buzzer. The LCD display is interfaced to the microcontroller through a parallel interface.

4 Summary and Future Directions

We have shown that these devices, "Beagle-Z," are capable of detecting various explosives at room temperature such as TNT, RDX, PETN, etc. The device is portable and battery operated with a RS-232 interface for data transfer to PC. The device uses an amplifying fluorescent polymer (AFP) as the active sensing material. It should be noted that while TNT could be detected within a few seconds of exposure to TNT vapors, RDX, HMX, and PETN took many seconds for detection. The slow response of Beagle-Z could be due to either the low vapor pressure of these explosives as compared to TNT, or could be inherently slow detection by the AFP. Future directions will aim toward the development of new polymers wherein detection toward aliphatic nitro-based explosives could be accelerated along with the detection of a broad spectrum of explosives. Furthermore, "click"able AFPs provide an attractive platform wherein suitable biomolecular guest–host interactions could be explored for the design and fabrication of the next generation of portable biomedical and chemical sensors.

Acknowledgment We gratefully acknowledge the funding received from the NPMASS program of ADA Bangalore, India (Project Ref No. PARC# 4.9) for the development of Beagle-Z. Ashok Kumar Majji acknowledges UGC India for senior research fellowship.

References

1. Goldman ER, Anderson G, Lebedev N et al (2003) Analysis of aqueous 2,4,6-trinitrotoluene (TNT) using a fluorescent displacement immunoassay. Anal Bioanal Chem 375:471–475
2. Charles PT, Rangasammy JG, Anderson GP et al (2004) Microcapillary reversed-displacement immunosensor for trace level detection of TNT in seawater. Anal Chim Acta 525:199–204
3. Wilson R, Clavering C, Hutchinson A (2003) Electrochemiluminescence enzyme immunoassays for TNT and pentaerythritoltetranitrate. Anal Chem 75:4244–4249
4. Bromberg A, Mathies RA (2003) Homogeneous immunoassay for detection of TNT and its analogues on a microfabricated capillary electrophoresis chip. Anal Chem 75:1188–1195
5. Gauger PR, Holt DB, Patterson CH et al (2001) Explosives detection in soil using a field-portable continuous flow immunosensor. J Hazard Mater 83:51–63
6. Bakaltcheva IB, Ligler FS, Patterson CH et al (1999) Multi-analyte explosive detection using a fiber optic biosensor. Anal Chim Acta 399:13–20
7. Yang X, Du X-X, Shi J et al (2001) Molecular recognition and self-assembled polymer films for vapour phase detection of explosives. Talanta 54:439–445
8. Heflinger D, Arusi-Parpar T, Ron Y et al (2002) Application of a unique scheme for remote detection of explosives. Opt Commun 204:327–331
9. Rose AZZ, Madigan C, Swager TM et al (2005) Sensitivity gains in chemosensing by lasing action in organic polymer. Nature 434:876–879
10. Yang JS, Swager TM (1998) Fluorescent porous polymer films as TNT chemosensors: electronic and structural effects. J Am Chem Soc 120:11864–11873
11. Burroughts JH, Bradley DDC, Brown AR et al (1990) Light-emitting diodes based on conjugated polymers. Nature 347:539–541

12. Juan Z, Swager TM (2006) Poly(aryleneethynylene)s in chemosensing and biosensing. Adv Polym Sci 177:151–179
13. Sarah JT, Trogler WC (2006) Polymer sensors for nitroaromatic explosives detection. J Mater Chem 16:2871–2883
14. Thomas SW, Joly GD, Swager TM (2007) Chemical sensors based on amplifying conjugated polymers. Chem Rev 107:1339
15. Czarnik AW (1993) Fluorescent chemosensors for ion and molecule recognition. American Chemical Society, Washington
16. De Silva AP, Gunaratne HQN, Gunnlaugsson T et al (1997) Signaling recognition events with fluorescent sensors and switches. Chem Rev 97:1515–1566
17. Martinez-Manez R, Sancenon F (2003) Fluorogenic and chromogenic chemosensors and reagents for anion. Chem Rev 103:4419–4476
18. Valeur B, Leray I (2003) Design principles of fluorescent molecular sensors for cation recognition. Coord Chem Rev 205:3–40
19. Gunnlaugsson T, Glynn M, Tocci GM et al (2006) Anion recognition and sensing in organic and aqueous media using luminescent and colorimeter sensors. Coord Chem Rev 250:3094–3117
20. Jiang P, Guo Z (2004) Fluorescent detection of zinc in biological systems: recent development on the design of chemosensors and biosensors. Coord Chem Rev 248:205–229
21. Lakowicz JR (1999) Principles of fluorescence spectroscopy. Kluwer Academic/Plenum, New York
22. Turro NJ (1991) Modern molecular photochemistry. University Science Books, Sauslito
23. Guillet J (1985) Polymer photophysics and photochemistry. Cambridge University Press, Cambridge
24. Zhou Q, Swager TM (1995) Method for enhancing the sensitivity of fluorescent chemosensors: energy migration in conjugated polymers. J Am Chem Soc 117:7017–7018
25. Zhou Q, Swager TM (1995) Fluorescent chemosensors based on energy migration in conjugated polymers: The molecular wire approach to increased sensitivity. J Am Chem Soc 117:12593–12602
26. Gopalakrishnan D, Dichtel WR (2013) Direct detection of RDX vapour using a conjugated polymer network. J Am Chem Soc 135:8357–8362
27. Liu Y, Mills RC, Boncella JM et al (2001) Fluorescent polyacetylene thin film sensor for nitroaromatics. Langmuir 17:7453–7455
28. RoseA ZhuZ, Madigan CF et al (2005) Sensitivity gains in chemosensing by lasing action in organic polymers. Nature 434:876–879
29. Sanchez JC, Trogler WC (2008) Efficient blue-emitting silafluorene–fluorene-conjugated copolymers: selective turn-off/turn-on detection of explosives. J Mater Chem 18:3143–3156
30. Sinha J, Kumar A, Pullela PK (2008) Indian Patent File No. 2319/MUM/2008

[illegible] inserted on [illegible]

[illegible] (2003) [illegible] Sci [illegible]

[illegible] W [illegible] (1983) [illegible] 2853

[illegible] (1997) [illegible] Chem Rev 97:1515

[illegible] Washington DC [illegible]

[illegible] Gunnlaugsson [illegible] sensors and switches. Chem Rev [illegible]

[illegible] (2000) [illegible] Chem Rev 100:2477 [illegible]

[illegible] Chem Rev 105:[illegible]

[illegible]

[illegible] development in the design of [illegible]

[illegible] (1999) Principles of fluorescence spectroscopy. [illegible] Academic [illegible] New York

[illegible] (1990) [illegible]

[illegible] (1995) [illegible]

[illegible] (2003) [illegible]

[illegible] Science [illegible]

[illegible] (2008) [illegible] Nature [illegible]

[illegible] (2008) [illegible]

Development of a Surface Plasmon Resonance-Based Biosensing System

S. Mukherji, Munshi Imran Hossain, T. Kundu and Deepali Chandratre

Abstract Surface Plasmon Resonance (SPR)-based biosensing systems have become very popular among biologists and biochemists due to their ability to sense biomolecular interaction kinetics. It can be easily surmised that such systems may be of great use in diagnostic applications as well. However, the principal impediment to wide-scale deployment of these devices is the cost of the instrument. This effort to develop a low cost, widely deployable instrument was born out of a desire to initially popularize this instrument and technique among colleges and universities with budget constraints and move toward developing diagnostic applications on this platform. A low-cost device with significant sensitivity was developed, and this chapter describes the development on this front.

Keywords Surface plasmon resonance · Biosensor · Biomolecular interaction · Immobilization of antibodies · Optical detection

1 Introduction

This chapter discusses the development of a Surface Plasmon Resonance (SPR)-based biosensor. This is a kind of optical biosensor that makes use of SPR, a phenomenon that occurs due to the interaction of light with a metal–dielectric interface.

Biosensors may be defined as devices that use specific biochemical reactions mediated by isolated enzymes, immunosystems, tissues, organelles, or whole cells to detect chemical compounds usually by electrical, thermal, or optical signals [1]. Optical biosensing has some advantages over other sensing techniques. One

S. Mukherji (✉) · M. I. Hossain · T. Kundu · D. Chandratre
IIT Bombay, Mumbai, India
e-mail: mukherji@iitb.ac.in

K. J. Vinoy et al. (eds.), *Micro and Smart Devices and Systems*,
Springer Tracts in Mechanical Engineering, DOI: 10.1007/978-81-322-1913-2_4,

important advantage of this class of sensors is that they are unaffected by electrical noise. This, in turn, makes the associated instrumentation system easy to design. Another advantage is that these sensors are easily given to microfabrication which allows for the possibility of miniaturization of the sensor and therefore smaller, portable instruments which can be used on fields.

Among the label-free, real-time optical biosensors, SPR is emerging as a very impactful phenomenon for making biosensors because of its high sensitivity and specificity.

1.1 Organization of the Chapter

The contents of this chapter can be broadly distinguished into the following parts that appear in order. The physics behind SPR, and how the phenomenon is exploited for biosensing, is explained. This is followed by a discussion of the setup that was used to perform experiments along with associated procedures. Thereafter, the results obtained have been discussed along with methods that were used to analyze the experimental data. The chapter concludes with a brief outline of the scope of future work in the development of this instrument.

2 Fundamentals of SPR and SPR-Based Sensing

The phenomenon of SPR occurs when light, more specifically *p*-polarized light, is incident on a metal–dielectric interface. At this interface, the light excites free electrons in the metal causing charge density oscillations. There are certain criteria that need to be fulfilled for the phenomenon to occur. These have been stated later in the chapter.

SPR was first reported by R. W. Wood in 1902 in his seminal paper [2]. When Wood allowed polarized light to be incident on a mirror with a diffraction grating on its surface, he observed a pattern of light and dark bands in the reflected light. The phenomenon was, however, explained later by Otto [3] and Kretschmann [4] independently in 1968. This phenomenon was first used for biosensing in 1983 by Liedberg et al. [5]. They used SPR for immunosensing and could detect the presence of a-IgG (anti-immunoglobulin G) down to 0.2 μg/ml.

When *p*-polarized (transverse magnetic) light passing through a prism is incident on a thin metal film, at an angle greater than the critical angle for the metal-prism interface, the light is reflected back due to total internal reflection. When the angle of incidence is changed, it is seen that the intensity of reflected light is at a minimum for a particular angle of incidence. At this angle of incidence, the light excites free electrons of the metal layer causing them to oscillate. This sets up oscillations in the charge density of the metal film. These charge density oscillations are known as plasmons. The field vector of this wave has maxima at the interface of the metal and dielectric and decays exponentially into both the media [6]. This is schematically shown in Fig. 1. The generated surface plasmon wave (SPW) can travel up to a few microns along the surface.

The absorption of energy by the free electrons causes a reduction in the intensity of the reflected light. The angle at which the intensity of reflected light is minimum is the SPR angle.

In order to generate SPW, light that is polarized parallel to the plane of incidence is required. Such a light ray is said to be TM (transverse magnetic) polarized. The momentum vector k of TM polarized light may be resolved into two orthogonal components, one of which is parallel to the metal surface. This component of momentum, k_x excites plasmons at the metal–dielectric interface [2].

$$k_x = \left(\frac{\omega}{c}\right)\sqrt{\varepsilon}\sin\theta$$

In the equation, c is the speed of light in a vacuum, ω is the angular frequency of light in a vacuum, and ε is the dielectric function of the coupling medium.

The necessary condition for the existence of the SPW is that the dielectric constants of the two media, at whose interface the SPW is generated, are of opposite signs. The real part of the permittivity for dielectrics is usually positive, and that for metals is negative. In fact, metals such as gold, silver, and aluminum have a negative real part of permittivity in the visible and near infrared region of the spectrum [7]. The propagation constant, k_{sp} of the SPW is given by the following equation.

$$k_{\text{sp}} = \left(\frac{\omega}{c}\right)\sqrt{\frac{\varepsilon_M \varepsilon_D}{\varepsilon_M + \varepsilon_D}}$$

Here ε_M and ε_D are the dielectric functions of metal and dielectric, respectively. The phenomenon of SPR occurs when

$$k_x = k_{\text{sp}}$$

For the phenomenon of SPR to occur, the real part of ε_M should be negative and its absolute value must be smaller than ε_D. Metals like gold, silver, aluminum, etc., have a negative real part of permittivity in the visible and near infrared region of the spectrum. A major field of the SPW is concentrated in the dielectric, so the propagation constant of the SPW is sensitive to refractive index changes of the dielectric [6].

Direct excitation of the metal–dielectric interface cannot cause SPR. Another condition that should be satisfied is that the wave vector propagating along the interface between metal and dielectric should be equal. The wave vector of light in free space (k_o) is usually smaller than the wave vectors of surface plasmons propagating along the interface (k_{sp}). This warrants a higher refractive index material such as prisms, gratings, and optical waveguides that help to match the propagation constant of the light wave to that of the SPW.

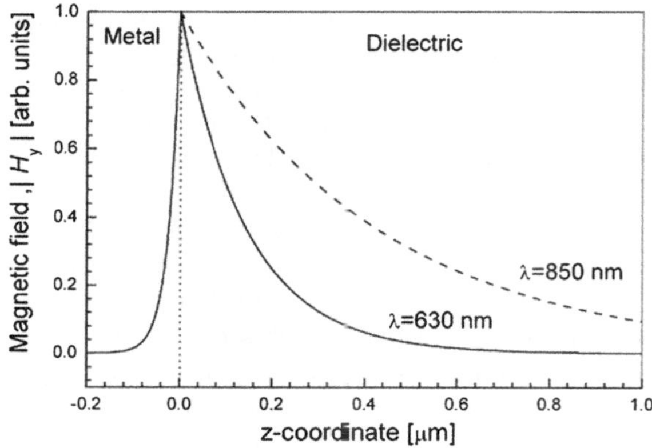

Fig. 1 Distribution of magnetic field intensity for a SPW at the interface between gold and dielectric (r.i. = 1.32) in the direction perpendicular to the interface, wavelength = 630 nm and 850 nm [21]

There are three main methods of coupling *p*-polarized light into the metal–dielectric interface. These are:

1. Prism-coupled system
2. Optical waveguide coupling-based system
3. Grating coupler-based system.

2.1 Prism-Coupled System

The most common configuration in which prisms are used is the Kretschmann configuration. A light wave is passed through a prism of a high refractive index and suffers total internal reflection from the base of the prism. The evanescent field thus generated penetrates through the thin metal layer at the base of the prism and excites plasmons in the metal film. By adjusting the angle of incidence of the light, the propagation constant of the evanescent wave can be matched with that of the SPW, i.e.,

$$\left(\frac{\omega}{c}\right)\sqrt{\varepsilon}\sin\theta = \mathrm{Re}\left\{k_{\mathrm{sp}}\right\}$$

This configuration is schematically shown in Fig. 2.

In this project, an SF11 (refractive index (r.i.) = 1.778) prism is used in the Kretschmann configuration. The prism is hemicylindrical in shape which ensures that the incident light is always normal to the curved surface of the prism and, hence, prevents refraction of the light at this face.

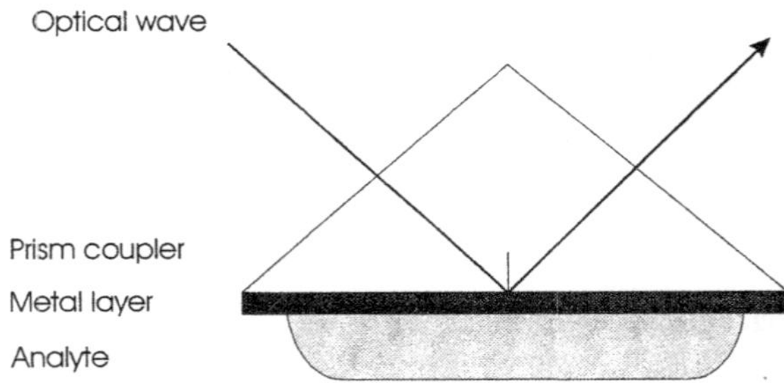

Fig. 2 Prism-coupled SPR configuration [6]

2.2 Optical Waveguide Coupling-Based System

Optical waveguides are structures that guide light waves. These are also called optical fibers. When light passes through an optical waveguide it generates an evanescent wave at the core–cladding interface. If a part of the fiber is decladded and coated with a metal film, surface plasmons can be excited in this region by the evanescent wave. The metal film can be suitably processed for biosensing applications. Figure 3 is a schematic of the optical waveguide coupling system.

2.3 Grating Coupler-Based System

Diffraction gratings are used to couple the light to the metal film. Sinusoidal gratings are used most often. The component of the wave vector of the diffracted waves parallel to the interface is diffraction-increased by an amount which is inversely proportional to the period of the grating and can be matched to that of an SPW [8]. This method of coupling is illustrated in Fig. 4.

Unlike in the other two configurations, in grating-coupled systems, the light passes through the sample solution resulting in a decreased signal from the sensor [8].

2.4 Metals for SPR: Their Thickness and Reflectance

For SPR, metals must have a band of electrons capable of resonating at a particular wavelength. The metals for SPR must have their resonating wavelength within the electromagnetic spectrum. They should also be highly pure for good resonance because the formation of oxide, sulfides, etc., on the metal surface can make performance of SPR deteriorate.

Metals that can be used for SPR are silver, gold, copper, aluminum, sodium, and indium. Indium is very expensive, sodium too reactive, copper and aluminum are too broad in their SPR response, and silver too susceptible to oxidation though it has a longer propagation length of SPW resulting in a very fine and sharp dip, as

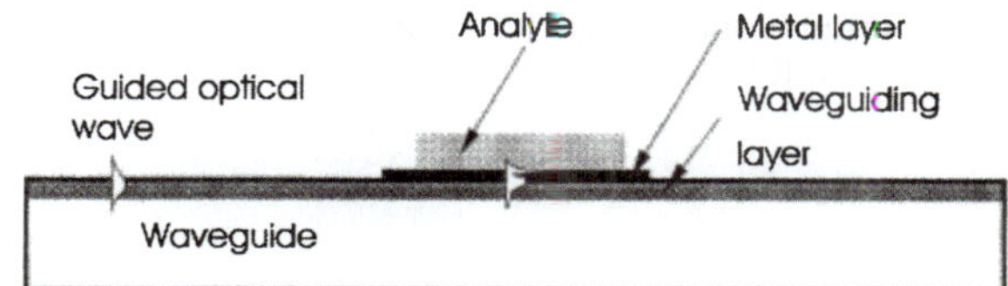

Fig. 3 Optical waveguide being used to couple light to a metal layer [6]

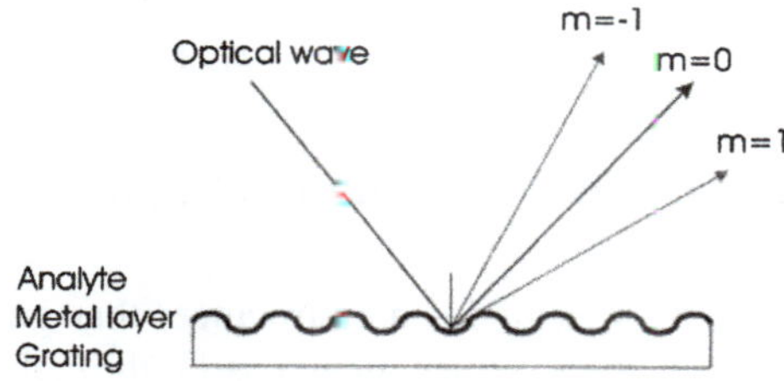

Fig. 4 Diffraction grating used for light coupling to a metal film [6]

shown in Fig. 5 [9]. Gold is very resistant to oxidation, and is compatible with many biochemical interactions, but has a shorter SPR propagation length resulting in a broad dip, as shown in Fig. 5. A silver–gold bimetallic layer is used, in which the silver sustains the SPW and the gold layer provides a biochemically stable surface. The total thickness of the metal layer is about 50 nm for proper resonance. The effect of different metals on the sharpness of the reflectance minimum is shown in Fig. 5.

The thickness of the metal films also affects the reflectance of light. The best contrast is obtained around 55 nm and it progressively deteriorates on both the higher and lower side, as shown in Fig. 6 [10]. The adhesion layer material like chromium, titanium, etc., and its thickness can also affect the contrast of dip.

2.5 SPR Used in Biosensing

The conditions for SPR are very sensitive to changes in the optical properties of the dielectric medium adjacent to the metal film. Therefore, changes in this dielectric medium can be monitored by the interaction of the SPW and the polarized light. A change in the refractive index of the dielectric on the metal film, changes its permittivity. This causes a change in the k_{sp} of the surface plasmon. As a result, the condition for resonance is satisfied for a different incident angle (if the wavelength is constant)—angular interrogation; or at a different wavelength (if the angle of incidence is constant)—wavelength interrogation.

SPR-based sensors are used for studying antigen-antibody binding, protein–protein, and protein–DNA (de-oxy ribonucleic acid) binding. In the biosensors, a biorecognition element is immobilized on the surface of the sensor. A binding reaction causes a change in the refractive index of the dielectric layer and, therefore, the condition for SPR. The change in the refractive index produced by

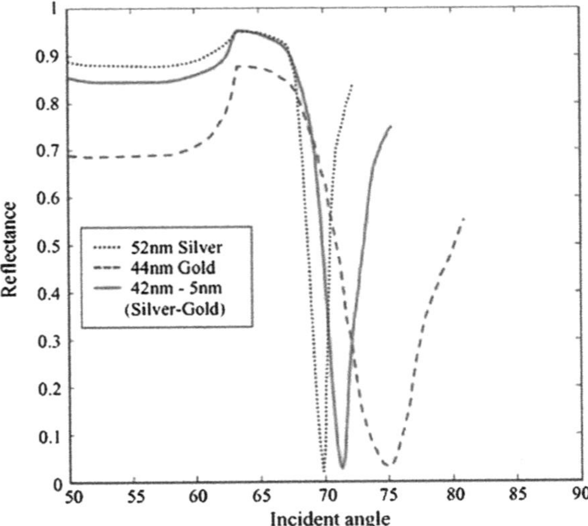

Fig. 5 Experimental SPR curves showing effect of metal on dip broadness with BK7prism and glucose solution [22]

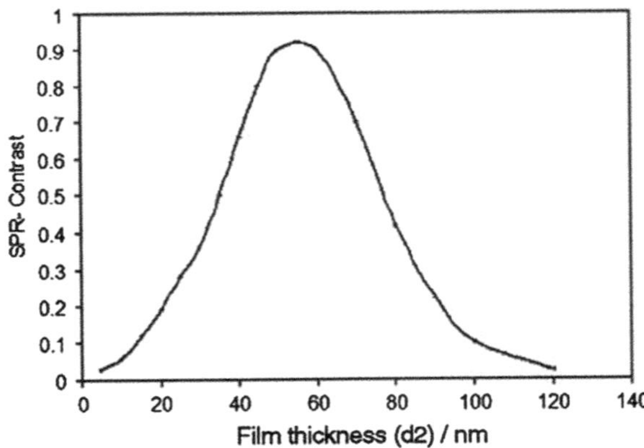

Fig. 6 SPR dip contrast versus metal thin film thickness [10]

the capture of analyte molecules depends on the concentration of the analyte molecules at the sensor surface. Figure 7 shows the schematic of the principle of biosensing in SPR.

3 The Experimental Setup for SPR

A substantial amount of work has been done on prism-based SPR biosensors at the Bioinstrumentation group, in IIT Bombay. Initial angular experiments were carried out using a He–Ne laser (632.8 nm), polarizer, focusing optics, rotary stage, and an

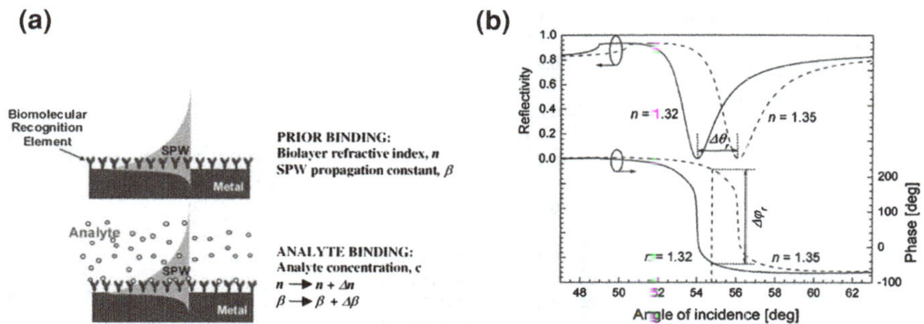

Fig. 7 **a** Schematic illustration of the concept of biosensing by SPR and **b** shift in reflectance minima because of change in dielectric medium [21]

optical power meter. Initial studies have reported characterization of the metal film thickness and the effect of different refractive indices on the dip characteristics [11, 12].

The metal layer was directly sputtered on the surface of the SF10 (r.i. = 1.728) triangular prism or BK7 (r.i. = 1.515) hemicylindrical prism. Later, BK7 cover-slips were sputtered with the metal layer and coupled to the BK7 prism using index matching oil (r.i. = 1.513, Invitrogen, USA). The cysteamine-glutraldehyde protocol and MuA (11-Mercapto-undecanoic acid)-EDC (N-(3-Dimethylaminopropyl)-N' ethyl-carbodiimide hydrochloride)/NHS (N-Hydroxysuccimide) protocol for immobilizing antibodies were also experimented with [13, 14].

Figure 8 shows the angular interrogation setup that was initially used. It used a 512 pixel linear CMOS (complementary metal oxide semiconductor) detector (S9227, Hamamatsu Photonics, Japan) to detect the dip in reflected light. The detector has a pixel pitch of 12.5 microns and a height of 500 microns. A microcontroller MSP430-based circuit was used to control the CMOS detector, and was serially interfaced to PC [15]. A Lab windows CVI-based program was developed to display and store data.

This setup was subsequently modified. The source of light was changed to a red laser diode (635 nm) DL-3148-025(Sanyo, Japan). The laser diode has a vertical beam divergence of 30° on both sides of the lasing layer and a horizontal beam divergence of 10° on both sides of the lasing layer. This laser diode oscillates in the TM mode. The polarization direction is perpendicular to the junction plane. Thus, the light emitted by the laser diode contains *p*-polarized components. By properly orienting the laser diode, the need for a polarizer was eliminated. It is followed by a biconvex lens with a focal length of 50 mm and a diameter of 12.7 mm held in a plastic lens holder. The metal-sputtered coverslip coupled to the BK7 prism using index matching oil is placed on a rotary stage such that the metal side of the coverslip faces the Teflon flow cell. The flow cell has an inlet and outlet tubing. An analyte is inserted into the flow cell with the help of a syringe. The prism-flow cell holder is mounted on a rotary stage which, in turn, is mounted on a platform with railings for *x* and *y* translations. The *z*-axis translation was achieved

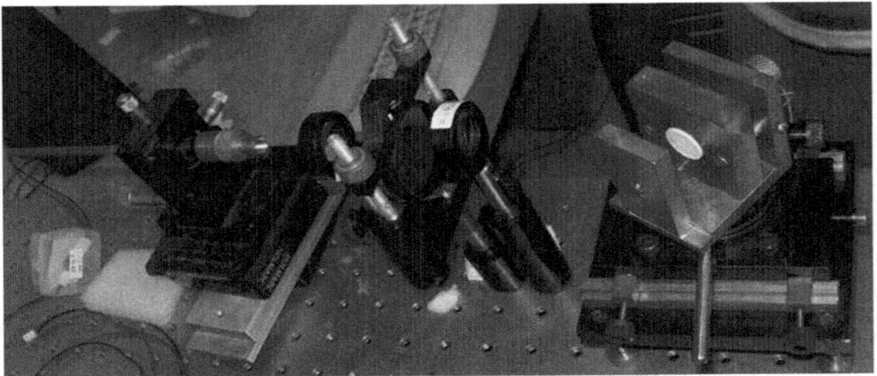

Fig. 8 Angular interrogation setup for SPR experiments [17]

using a support screw system. CMOS linear image sensor-S9227 was used to capture the SPR dip, the same one used in the previous setup.

In the horizontal setup mentioned above, the flow cell and the analyte in the flow cell remain vertical during the experiments. This position is mechanically unfavorable for liquid analytes. So, there was a need for a setup in which the flow cell remains horizontal and the setup remains portable. The setup was therefore modified to a vertical angular interrogation setup as shown in Fig. 9.

It consists of a modular prism holder and a flow cell. The prism with a metal-sputtered coverslip coupled to it was tightened on the prism holder such that the metal side of coverslip faced the Teflon flow cell. The laser diode and the biconvex lens were mounted on the railing-mounted rotary stage. With this type of arrangement, the focal point does not shift when the source is rotated. The holder for the 512 pixel CMOS detector array was a right-angled bracket on which the PCB (printed circuit board) for the detector was mounted. The detector had to be manually moved and positioned to detect the dip.

All the previous SPR angular interrogation setups were portable, but a need was felt to develop a small, light-weighted, rugged, user friendly, semi-automated, and modular setup. The new portable setup was designed to be used as a vertical or a horizontal setup, with motion of the source and detector in synchrony in opposite directions, to track light and the simple assembly and fabrication of modules. The setup is shown in Fig. 10 and is described below.

The framework and supports for various components such as—a beam restrictor, prism platform, and coverslip platform were all made of acrylic sheets. This drastically reduced the weight of the setup, and also the size, to an extent.

The detector PCB was mounted on the vertical acrylic sheet. A polarizer in the p-polarization direction was placed over the CMOS detector. The analog video signal from the detector was digitized with an internal analog to digital converter and was serially transmitted. A Lab Windows-based program was used to receive serially transmitted data from the microcontroller and display it. A regulated

Fig. 9 Vertical angular interrogation setup for SPR experiments [17]

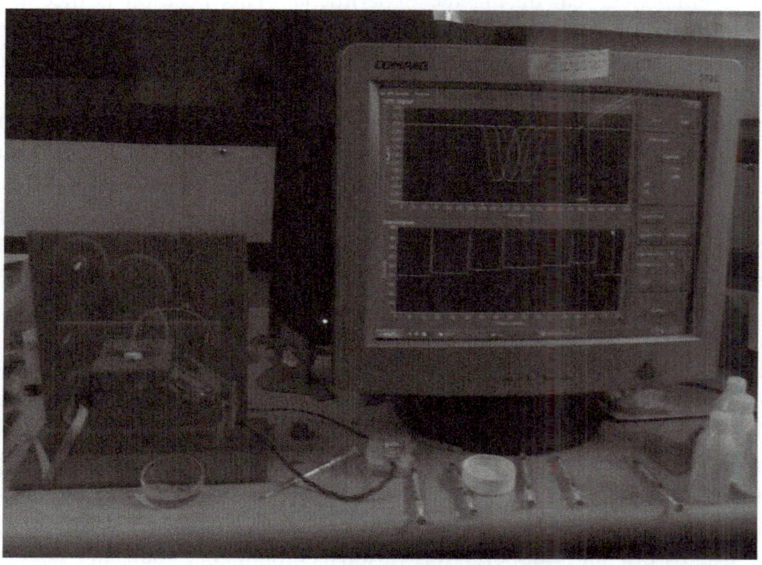

Fig. 10 Acrylic-based portable angular interrogation setup used in tests [16]

power supply (Tektronix PS280) was used to supply DC voltages to the MSP430 board and the laser diode driver.

However, it was not the kind that could be handheld and, hence, was not deployable in the field. This setup was subsequently modified to make it handheld and more modular than it was, previously. The electronics part of the instrument was housed in a separate compartment, thereby isolating the areas (optical and fluidic) where there might be an ingress of water. The cover slip sputtered with metal, and the flow cell, are components that need to be frequently replaced in the system. Therefore, a provision was made for these in the upper exposed part of the instrument. The remaining part of the instrument below these has been covered, as shown in Fig. 11, and is not accessible to the user.

4 Experiments and Results Using Sucrose Solutions

The new setup was used for sucrose calibration tests. This required the preparation of the active sensor surface. This active surface is the metal-sputtered cover slip.

SF11 glass prisms (r.i. = 1.778) and cover slips (Lensel Optics, Pune) were used for the experiments. The cover slips were of size 18 × 18 × 0.5 mm. The refractive index matching fluid (r.i. = 1.785, Cargille, USA) was used to couple the cover slip to the prism.

Sucrose solutions of different concentrations were used for the calibration of the sensor. The different concentrations of Sucrose GR (Merk Chemicals) were prepared in deionized water (r.i. = 1.33299). The different concentrations of sucrose were approximately 2 % (r.i. = 1.33587), 5 % (r.i. = 1.34026), 8 % (r.i. = 1.34477), 10 % (r.i. = 1.34783), and 12 % (r.i. = 1.35093) wt/wt.

4.1 Preparation of the Active Sensor Surface

The cover slip surface is sputtered with Ti–Au. However, before the sputtering, this surface needs to be cleaned appropriately so as to remove contaminants. The following procedure is followed to clean the cover slips [16].

4.1.1 Cleaning of Cover Slips

1. Sulfochromic acid cleaning: The cover slips, after antigen–antibody or sucrose calibration experiments, need to be recycled. The cover slips were kept in sulfochromic acid for 24 h which removes gold and other impurities from the surface. Later, these cover slips were rinsed in deionized water.
2. Etching of metal: Sometimes, sulfochromic acid is unable to remove gold and the metal layer beneath the gold sensing layer completely. Hence, a gold etching solution (potassium iodide and iodine) was used to etch the gold and followed by rinsing in concentrated HCl (to remove titanium).
3. Sonication in acetone: The cover slips were then immersed in acetone in a Petri dish and sonicated for 15 min. The cover slips were rinsed in deionized water and dried in a nitrogen jet.

Fig. 11 The portable SPR setup [23]

4. Piranha cleaning: A Piranha solution is a mixture of 30 vol. % H_2O_2 and 70 vol. % H_2SO_4. The Piranha solution, being highly reactive, removes organic and inorganic contaminants. Then, the cover slips were rinsed in deionized water and dried in a nitrogen jet.

4.1.2 Metal Sputtering

The cleaned SF11 cover slips were then sputtered with a bimetal layer in a metal sputtering system (NORDIKO Metal Sputter) at Electrical Engineering, IIT Bombay. The sputter chamber is vacuumised and then filled with Argon gas. The metal to be deposited is the cathode and the cover slips (substrate) are the anode. Argon plasma is used to deposit metals on the samples.

The sputtering time influences the thickness of metal deposition. Previous work has optimized the sputtering time and temperature for the Ti–Au bilayer [17]. The cover slips were first sputtered with titanium for 35 s at 150 W and gold for 50 s at 80 W.

4.2 Experimental Procedure

The cover slip is cleaned with isopropyl alcohol prior to use. A drop of index matching fluid is introduced on the flat surface of the prism, and the metal-sputtered cover slip is placed on the prism with the metal side facing out. The flow cell is then placed on the cover slip and tightened with the help of acrylic pieces on the sides of the module using nuts.

The laser is powered on, and the arm on which it is supported is rotated to obtain a dip in the reflection spectra. The spectrum is viewed on a computer with the help of a DAQ (data acquisition) card and software developed for this system. An appropriate filter is chosen to get a smooth output. The dip with air as dielectric is obtained at around 37°.

Using a syringe, analytes (different concentrations of sucrose solutions in water) are injected into the flow cell. The source arm is moved to appropriate angles to obtain a dip. By setting the proper pixel limits for finding the minima, a sensorgram is started so as to track the minima with respect to time. During the SPR experiments, a background reading is established in deionized water. With deionized water set as the background, a series of sucrose solutions are injected into the flow cell and recorded using a CMOS detector.

When deionized water is injected back after a 12 % sucrose solution, the minima is found to revert back to nearly the same pixels as at the beginning of the experiment.

A refractive index change from deionized water (r.i. = 1.33299) to 12 % sucrose (r.i. = 1.35093) was used for sensitivity calculations. The minima shifted by 74 pixels giving a sensitivity of 0.242 mRIU/pixel.

The results of the sensor calibration experiments using different concentrations of the sucrose solution are shown below in Figs. 12 and 13.

4.3 Data Analysis

An algorithm is implemented with the aim of finding the pixel at which resonance occurs which can consequently give a measure of the SPR angle and, therefore, the r.i. of the analyte. The algorithm was proposed by Thirstrup and Zong [18]. The algorithm essentially finds the centroid of the SPR profile. However, it uses a dynamic baseline that can compensate for fluctuations in the intensity of the light source and any dark signal from the detector. A diagrammatic representation of the concept is shown in Fig. 14. The algorithm is described in the following steps:

1. A dynamic baseline P_B is chosen according to the following equation [18],

$$\gamma A_0 = \gamma \int_{\theta_0}^{\theta_1} (P_B - P(\theta))d\theta = A_1 = \int_{\theta_1}^{\theta_2} (P_B - P(\theta))d\theta$$

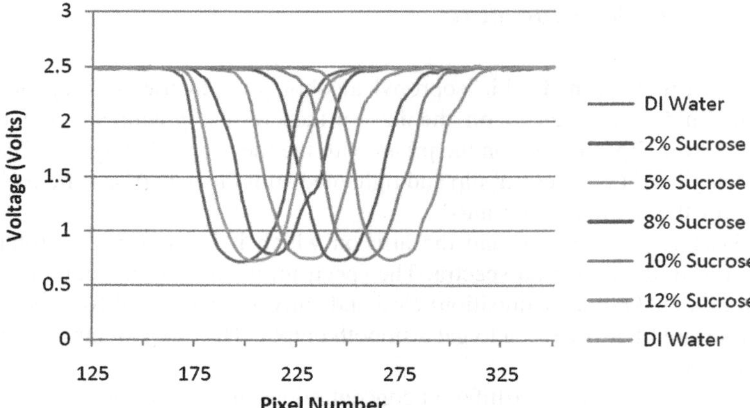

Fig. 12 Results of sensor calibration using sucrose on Ti–Au cover slip that is continuously heated at 125 °C by sputtering

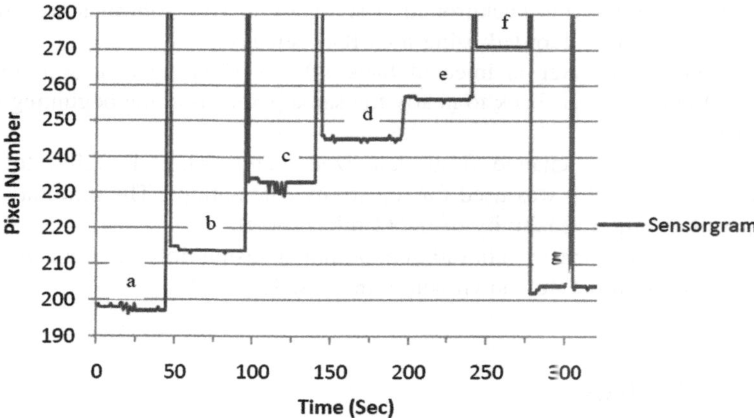

Fig. 13 Sensorgram for the sensor calibration experiment with Ti–Au-sputtered cover slip continuously heated at 125 °C (detector distance: 50 mm). (*a, b, c, d, e, f,* and *g* are minima positions when deionized water, 2, 5, 8, 10, 12 % sucrose solutions and deionized water are injected, respectively)

In the discrete domain, this equation may be written as,

$$\gamma\left(\sum_{\theta=\theta_0}^{\theta_1} P_B - P(\theta)(\theta_1 - \theta_0)\right) = \left(\sum_{\theta=\theta_1}^{\theta_2} P_B - P(\theta)\right)(\theta_2 - \theta_1)$$

The criterion for choosing γ is that there should be a sufficient number of data points in A_0 and A_1.

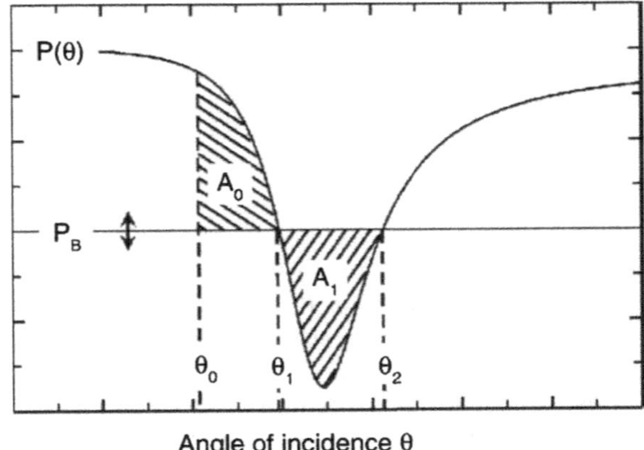

Fig. 14 SPR profile with the dynamic baseline P_B and the areas A_0 and A_1 of the curve *above* and *below* it, respectively [18]

Therefore, $P_B = \frac{\gamma \sum_{p=p_0}^{p_1} P(p) - \sum_{p=p_1}^{p_2} P(p)}{\gamma(p_1 - p_0 + 1) - (p_2 - p_1 + 1)}$ where p is the pixel number.

2. The algorithm suggested in [18] uses an SPR profile that is plotted against the angle of incidence. However, this may also be implemented using an SPR profile that is plotted against a pixel number, since the pixel number has a one-to-one correspondence with the angle of incidence. Accordingly, the equation that is used to compute the centroid is,

$$\text{Pixel}_{\text{resonance}} = \frac{\sum_{i=i_1}^{i_2} \left(P_{i,b} - P_i\right) i}{\sum_{i=i_1}^{i_2} \left(P_{i,b} - P_i\right)}$$

3. The centroid may be taken as a reasonable estimate of the pixel at which resonance occurs. The SPR angle may be estimated by a linear transformation from the pixel number domain to the angle of incidence domain.

The centroid calculation from the raw data, however, is not very reliable. This is because the noise in the photo detector and source intensity fluctuations can affect the output of the algorithm. In order to overcome this shortcoming, a local similarity matching algorithm is used on the raw data.

This algorithm [19] computes a moving dot product on the SPR profiles obtained for air and liquid (analyte in solution). The regions where the profiles are very similar give a high value for the dot product and a low value (valleys in the curve) for regions where the greatest dissimilarity occurs. The position of the valleys is where the resonance occurs. This algorithm is robust in the sense that it is better immune to noise in the data. The centroid algorithm can now be applied to this data to obtain an estimate of the minima. The governing equation for this algorithm is,

$$\cos\theta_{j,s} = \frac{L_{j,s}A_{j,s}}{|L_{j,s}||A_{j,s}|} = \frac{\sum_{i=j-s/2}^{j+s/2} l_i a_i}{\sqrt{\sum_{i=j-s/2}^{j+s/2} l_i^2 \sum_{i=j-s/2}^{j+s/2} a_i^2}}$$

where L and A are vectors of lengths of elements l and a, respectively, from SPR profiles of analyte and air, respectively.

The following curves, shown in Fig. 15, were obtained for the air and liquid (analyte in solution).

The voltage levels at each pixel comprise elements of a vector, one each for air and analyte. These vectors are used for the computation of a moving dot product.

The values of $\cos\theta$ so obtained can be interpreted as follows: In regions where the value is close to 1, the profiles are nearly similar. At places, where there is a wide disparity in the profiles, the value of $\cos\theta$ is low. At these positions, valleys are seen to occur. The SPR angle for air is usually higher than that for analytes. Therefore, valleys at the lower end are indicative of resonance due to analytes. These can now be inspected using a dynamic baseline centroid algorithm to obtain an estimate of the pixel at which resonance occurs.

The local similarity matching gives multiple valleys in the $\cos\theta$ profile. This makes it difficult to choose the correct valley to estimate the minima.

The dynamic baseline algorithm was then implemented using the SPR profiles that were obtained from the experiments. The output of the algorithm is an estimate of the pixel number at which resonance occurs. This is, then, transformed into the angle domain using the following equation.

$$\theta = \left[\left(9.642 \times 10^{-3}\right) \times p\right] + 46.893$$

This equation is obtained by a linear transformation using the pixels at which resonance occurs and the standard SPR angles for the sucrose solutions. This equation is, however, specific to the 512 pixel linear CMOS detector and the SF 11 prism used in the system. The estimated r.i. is then calculated using the following equation [18]

$$n_s = n_p \times \sin\theta$$

where n_s is the r.i. of the analyte and n_p is the r.i. of the SF 11 prism (=1.778).

The results are depicted in the graph shown in Fig. 16.

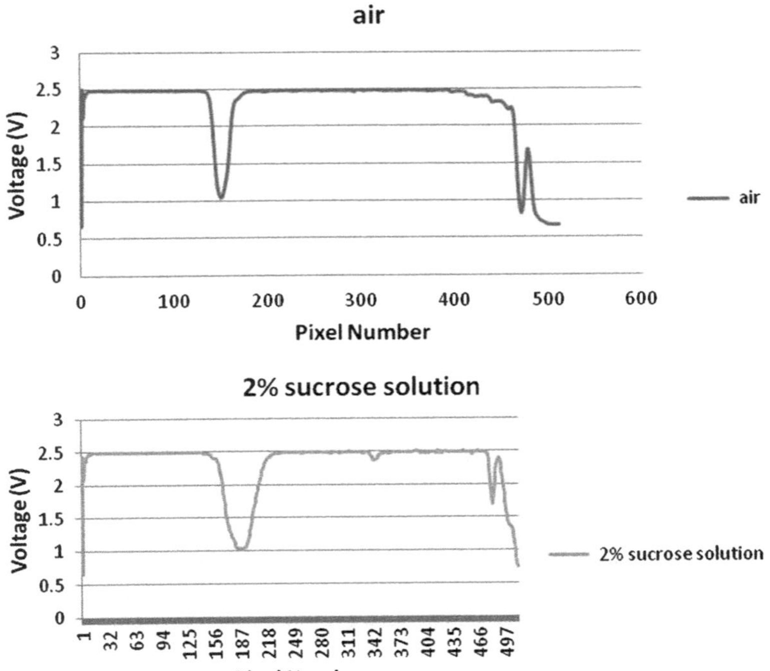

Fig. 15 SPR profiles obtained for air and 2 % sucrose solution

5 The SPR System in Biosensing

The primary purpose of an SPR system is to assess the kinetics of biomolecular interactions using the SPR phenomenon. The tests using sucrose solutions of varying concentrations revealed that the system is capable of resolving refractive index variations in the microenvironment of the metal surface down to a few milli-RIU (refractive index units) or even micro-RIU. In order to use the system for biosensing, the metal layer had to be modified for immobilization of biological receptor molecules on the surface. The most common forms of biorecognition events used in biosensing are:

(1) Antibody–antigen interactions,
(2) Nucleic acid interactions,
(3) Enzymatic interactions,
(4) Cellular interactions (i.e., microorganisms, proteins), and
(5) Interactions using biomimetic materials (i.e., synthetic bioreceptors).

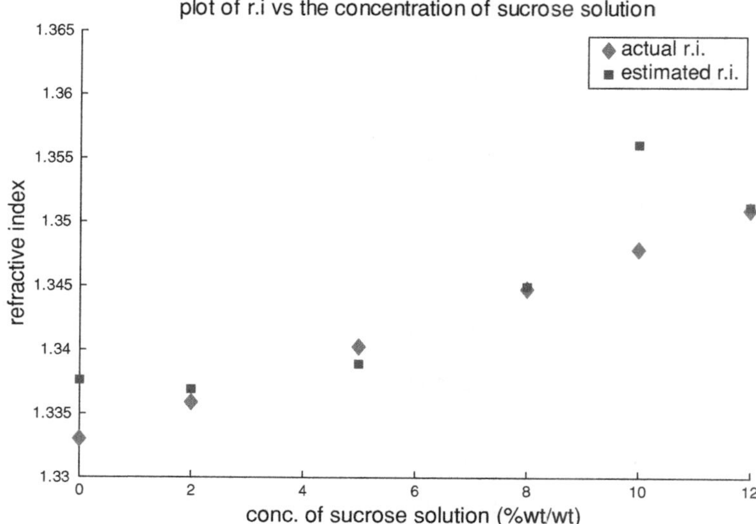

Fig. 16 Plot showing the actual and the estimated r.i. values versus the concentration of sucrose solution. The Euclidean distance or 2-norm error was calculated for the obtained estimates. This is equal to 0.0096. This figure suggests that most of the estimated values are within an error margin of 1 %

5.1 Immobilization of Antibodies

In our experiment, we used antigen–antibody interactions as biological recognition events that could be sensed using the SPR system. In this case, the antibodies were immobilized on the gold surface. For covalently binding antibodies on gold, two types of approaches were used.

5.1.1 Cystamine–Glutaraldehyde

- **Preparation of Buffer**: (PBS) Phosphate buffer saline (Ph: 7.4; 10 mM) was made by dissolving a PBS buffer tablet in 1000 ml water.
- **Cleaning of gold surface**: Coverslips coated with gold were first soaked in a 1.2 M NaOH solution for 5 min, then washed in DI (deionized) water followed by incubation in 1.2 M HCl for 5 mins, and again washed with DI water.
- **Cystamine layer on substrate**: The surface was incubated in cystamine solution (10 mM cystamine in 50 mM PBS, Ph: 7.4) for 1 h at room temperature. The surface was then washed with a buffer 2 or 3 times and dried using a nitrogen air jet.
- **Glutaraldehyde binding**: 10 % of Glutaraldehyde was added and incubated for 30 min. The surface was washed 2 or 3 times with a buffer. The surface was then ready for immobilizing antibodies.

- **Antibody immobilization**: Substrates were incubated in a 0.02 mg/ml of Goat anti-Human IgG antibody solution prepared in phosphate buffer saline (PBS pH 7.4). After an overnight incubation at 4 °C, the substrates were washed in PBS thoroughly to remove any loosely adsorbed molecules.
- **Antigen detection**: The immobilization of antibodies on the surface was tested by subjecting the surface to different concentrations of FITC (Fluorescein isothiocyanate)-tagged Human IgG antibodies for 10 min.

5.1.2 MuA: EDC–NHS Protocol

- **Cleaning of gold surface**: Coverslips coated with gold were first soaked in a 1.2 M NaOH solution for 5 min then washed in DI water followed by incubation in 1.2 M HCl for 5 min and again washed with DI water.
- **MuA layer on substrate**: The surface was incubated in a MuA solution (0.1 M MuA in absolute ethanol) for 30 min at room temperature. The surface was then washed with absolute ethanol 2 or 3 times followed by DI water and dried using a nitrogen air jet.
- **EDC–NHS Crosslinking**: The MuA-coated coverslips were then incubated in a solution of NHS (60 mM in DI water) and EDC (30 mM in DI water) for 2 h. Following this, the surface was washed 2 or 3 times with DI water making it ready for immobilizing antibodies.

Antibody preparation: Substrates were incubated in a 0.02 mg/ml of Goat anti-Human IgG antibody solution prepared in phosphate buffer saline (PBS pH 7.4). After an incubation of 40 min at room temperature, substrates were washed in PBS thoroughly to remove any loosely ordered molecules. The rest of the procedure is the same as earlier (Fig. 17).

5.2 Antigen–Antibody Interactions

5.2.1 Detection Using Cystamine Protocol

A BK7 prism was coupled with a coverslip sputtered with a silver–gold bilayer. An antibody (Goat anti-human IgG—GaHIgG) was immobilized on the metal bilayer using cystamine protocol. The SPR response was recorded after completion of the whole treatment of immobilizing with water as dielectric. This set of reading acts as a reference. Following this various concentrations of Human IgG antibodies were injected into the flow cell and the results were captured (Fig. 18).

From the graph (Fig. 19) and table (Table 1), we can see that there is minimum variation at the tip, whereas the variation was highest at two-thirds of the first half of the slope (15.6 mV/1 nM) and second highest at the middle of the second half of slope (8.7 mV/1 nM). In MSP430, a 12-bit ADC (analog to digital converter)

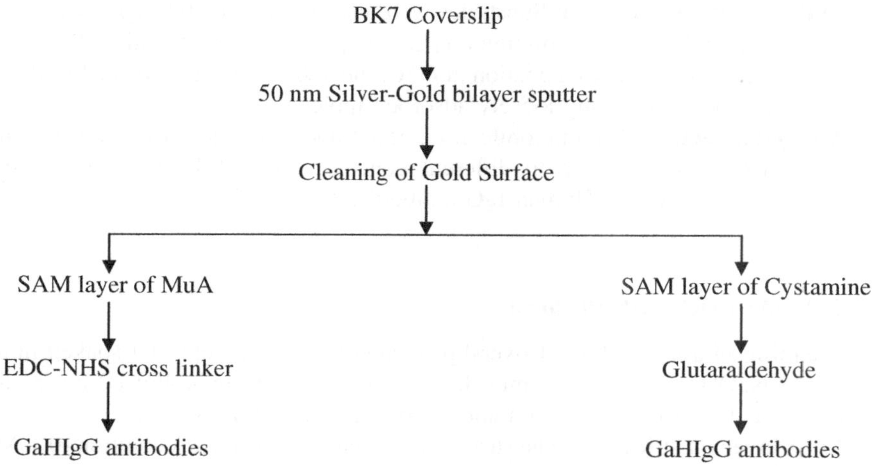

Fig. 17 Flow diagram summarizing the sample preparation procedure

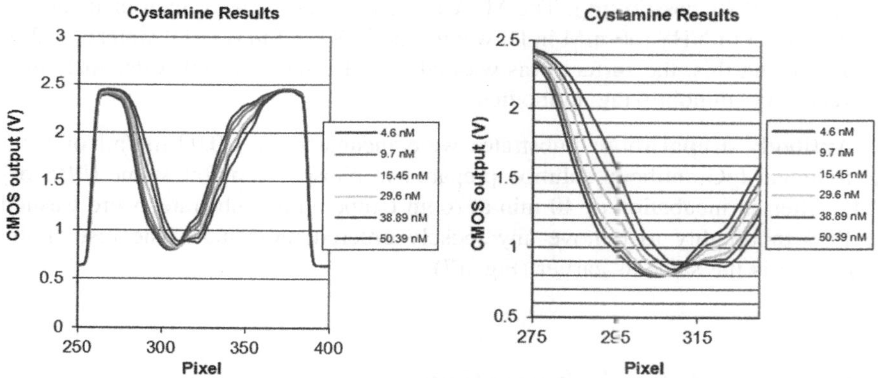

Fig. 18 Output on CMOS detector array using varying concentrations of HIgG antibodies as analytes after immobilizing GaHIgG antibodies using cystamine protocol to test for immobilization

(2.5 V full range) is used. Thus, we have 2.5 V distributed over 4096 bits. This leaves us with a resolution of 0.6 mV/bit. Even if we allow noise within 5 bits, we have a resolution of 3 mV/5 bits.

The first half of the slope is steeper than the second half of the slope. This also means that a span of analog data represented by one pixel is greater in the case of the first half slope than in the second half slope. Therefore, we can observe that despite having a better deflection on the first slope, the results are more clubbed together as compared to the second half. In the second half slope, the difference due to changing concentrations can be easily made out. Thus, it can be said that

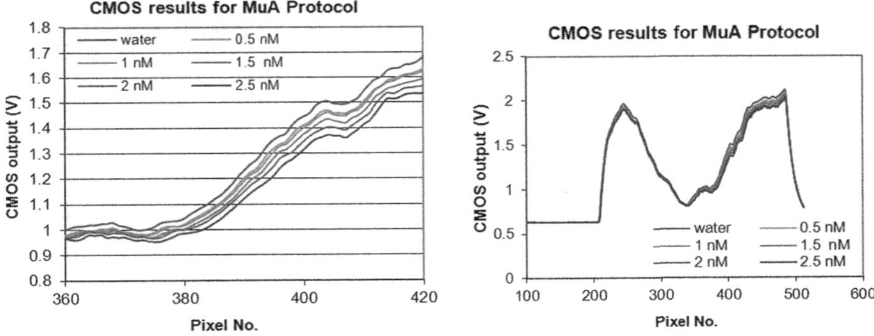

Fig. 19 CMOS sensor output for varying concentrations of HIgG antibodies in water for GaHIgG antibodies immobilized using MuA–EDC–NHS on silver–gold bilayer

Table 1 Comparison between concentration and reflectivity at various pixel positions for cystamine-coated slides

Concentration	Concentration	CMOS output (V)			
Nm	(ug/ml)	Pixel: 306 (tip)	Pixel: 291	Pixel: 345	Pixel: 328
4.6	0.66	0.797	1.063	2.232	1.538
9.7	1.393	0.805	1.135	2.177	1.446
15.45	2.218	0.819	1.219	2.116	1.376
29.6	4.26	0.863	1.400	2.003	1.253
38.89	5.58	0.916	1.584	1.945	1.207
50.39	7.23	0.998	1.788	1.820	1.115
Slope		**0.004**	**0.0156**	**−0.0086**	**−0.0087**

calculations on the second half slope would give the optimum bargain between resolution constraints due to ADC and sensitivity of detection.

Another important thing that should be noticed in the graph is that all 512 pixels do not participate in detection. Only that part of the CMOS sensor on which the laser output falls is analyzed, and is of interest. As we increase the distance between the reflecting surface and the sensor, the intensity of light decreases and the width of the dip broadens. Thus, if we place the sensor at an optimum position such that the laser output has sufficient measurable intensity, the dip would get broadened up. This would reduce the span of the analog data represented by one pixel which would, in turn, increase sensitivity.

5.2.2 Detection Using MuA–EDC–NHS Protocol

This prism had a refractive index of 1.515. In this case, first silver was sputtered for 20 s and then gold for 20 s, on a BK7 coverslip which was later coupled with the prism using an index matching fluid. Antibodies (Goat anti-human IgG

Table 2 Comparison between concentration and reflectivity at various pixel positions for MuA–EDC–NHS-coated slides

Concentration (nM)	Pixel No 390	Pixel No 400
0	1.232	1.445
0.5	1.202	1.411
1	1.191	1.400
1.5	1.169	1.376
2	1.1503	1.347
2.5	1.121	1.314
Slope	**−0.04192**	**−0.04978**

antibody) were immobilized on the metal bilayer by MuA SAMs (self-assembled monolayers) activation and EDC–NHS crosslinking. After completion of the whole treatment of immobilizing, the SPR response was recorded with water as dielectric, and at varying concentrations of Human IgG antibodies. The CMOS sensor output for varying concentrations of antibodies and water was seen as in Fig. 19. It is possible that the relationship between the resonating angle and refractive index is Gaussian. However, in this narrow range, it was found to be close to linear.

As can be seen, 73 ng (i.e., 500 pM) of antibodies were injected into the flow cell and were successfully detected. Table 2 shows that the sensitivity of the system was approximately 50 mV for 1 nM. Thus, comparing with ADC resolution and noise constraints, the minimum concentration that could be detected would be about 60 pM.

6 Conclusions

The setup that is used currently is reasonably light in weight, modular, and portable. However, there are some directions in which more development may be done.

Other methods may be explored to improve the results that are obtained. These may involve suitable modeling of the system as a function of various parameters of interest such as the r.i. of the analyte, thickness of the analyte layer, etc.

The data analysis algorithms have been implemented in MATLAB. However, these can be easily converted to the C language and written into the MSP430 microcontroller that is used in the system. This will, then, make the system capable of being a standalone system without the need for an external computer for data analysis.

More work may be done to improve the mechanical design to make the setup user friendly. The feasibility of using Ag films may be explored as suggested in [20] because Ag films give a sharper dip in the SPR profile than Au films.

Acknowledgments The development of the SPR system was partially supported by the NPMASS project "A lab on chip of Cardiac Diagnostics". The authors also acknowledge the contribution of a large number of students toward this development viz. Kanak Mhatre, Joaquim Ignetious Monteiro, Rahul Bharadwaj, Vamsi Ravali, Gauri Shukla, Tanneru Kumaraswami, and Dhananjay Patil.

References

1. IUPAC (1997) Compendium of chemical terminology. In: McNaught AD, Wilkinson A (eds) The "Gold Book", 2nd edn. Blackwell Scientific Publications, Oxford (1997)
2. Wood RW (1902) On a remarkable case of uneven distribution of light in a diffraction grating spectrum. Philos Mag 4:396–402
3. Otto A (1968) Excitation of non radiative surface plasma waves in silver by the method of frustrated total reflection. Z Phys 216:398–410
4. Kretschmann E, Raether H (1968) Radiative decay of non radiative surface plasmons excited by light. Z Naturforsch 23:2135–2136
5. Liedberg B, Nylander C, LunDstrOm L (1983) Surface plasmon resonance for gas detection and biosensing. Sens Actuators 4:299–304
6. Homola J, Yee SS, Gauglitz G (1999) Surface plasmon resonance sensors: review. Sens Actuators B 54:3–15
7. Wolfbeis OS (2006) Springer series on chemical sensors and biosensors—methods and applications. Springer, Berlin
8. Homola J (1997) On the sensitivity of surface plasmon resonance sensors with spectral interrogation. Sens Actuators B 41:207–211
9. SPR pages (2011) Website http://www.sprpages.nl/SprTheory/SprTheory.htm. Accessed 15 May 2011
10. Neff H, Zong W, Lima AMN, Borre M, Holzhuter G (2006) Optical properties and instrumental performance of thin gold films near the surface plasmon resonance. Thin Solid Films 496:688–697
11. Dhananjay VP (2006) Surface plasmon resonance based biosensor for detection of waterborne pathogens. Master's thesis, IIT Bombay
12. Bharadwaj R (2007) Surface plasmon resonance based biosensor for detection of waterborne pathogens. Master's thesis, IIT Bombay
13. Chandratre D (2008) Development of a surface plasmon resonance based biosensor. Master's thesis, IIT Bombay
14. Tanneru K (2009) Design and development of hemi cylindrical prism based surface plasmon resonance biosensor. Master's thesis, IIT Bombay
15. Dhawangle A (2009) MSP430 applications in biomedical system. Master's thesis, IIT Bombay
16. Mhatre AK (2011) Development of a portable surface plasmon resonance system. Master's thesis, IIT Bombay
17. Monteiro J (2010) Development of a portable surface plasmon resonance biosensor. Master's thesis, IIT Bombay
18. Thirstrup C, Zong W (2005) Data Analysis for surface plasmon resonance sensors using dynamic baseline algorithm. Sens Actuators B 106:796–802
19. Owega S, Poitras D (2007) Local similarity matching algorithm for determining SPR angle in surface plasmon resonance sensors. Sens Actuators B 123:35–41
20. Abdelghani A et al (1997) Surface plasmon resonance fibre-optic sensor for gas detection. Sens Actuators B 38–39:407–410
21. Homola J (2003) Present and future of surface plasmon resonance biosensors. Anal Bioanal Chem 377:528–539

22. Ong BH, Yuan X, Tjin SC, Zhang J, Ng HM (2006) Optimised film thickness for maximum evanescent field enhancement of a bimetallic film surface plasmon resonance biosensor. Sens Actuators B 114:1028–1034
23. Hossain MI (2012) Development of a surface plasmon resonance based biosensing system. Master's thesis, IIT Bombay

Design and Development of Ion-Sensitive Field-Effect Transistor and Extended-Gate Field-Effect Transistor Platforms for Chemical and Biological Sensors

V. K. Khanna, R. Mukhiya, R. Sharma, P. K. Khanna, S. Kumar, D. K. Kharbanda, P. C. Panchariya and A. H. Kiranmayee

Abstract This chapter deals with the design, fabrication, packaging, and instrumentation of ISFET and EGFET devices as pH-sensing platforms for chemical and biological sensors. The ISFETs have been developed in two geometrical layouts: a twin-FET configuration on a single chip comprising ISFET and a reference field-effect transistor (REFET); and a single ISFET with linear gate geometry. Packaging challenges have been successfully overcome by thick film alumina and low-temperature cofired ceramic (LTCC) technologies. Characterization studies have been performed on ISFETs. An ISFET-based pH meter, consisting of a sensor probe, a signal conditioning circuit, microcontroller, and display has been developed. The study has been carried out on a pH-sensitive extended gate electrode with a Ta_2O_5 sensing film by connecting it to a standard MOSFET gate to develop an EGFET device.

Keywords ISFET · EGFET · pH sensor · Thick-film packaging · LTCC packaging · pH meter

1 Introduction

To achieve wide recognition of credible point-of-care (POC) devices, the World Health Organization (WHO) guidelines for chemical/biochemical sensors are being followed for the development of new devices/technologies [1–4]. POC

V. K. Khanna (✉) · R. Mukhiya · R. Sharma · P. K. Khanna · S. Kumar
D. K. Kharbanda · P. C. Panchariya · A. H. Kiranmayee
CSIR-Central Electronics Engineering Research Institute, Pilani 333031, Rajasthan, India
e-mail: vkk@ceeri.ernet.in; vkkhanna.ceeri@gmail.com

V. K. Khanna · R. Mukhiya · R. Sharma · P. K. Khanna · D. K. Kharbanda
P. C. Panchariya
Academy of Scientific and Innovative Research (AcSIR), New Delhi 110001, India

K. J. Vinoy et al. (eds.), *Micro and Smart Devices and Systems*,
Springer Tracts in Mechanical Engineering, DOI: 10.1007/978-81-322-1913-2_5,

devices are needed for early and quick detection at remote locations. To achieve these boundaries, mature and well-understood microfabrication/MEMS technologies are used to develop such devices. ISFETs/EGFETs are devices which have the potential to fulfill these requirements. This chapter presents the research carried out at the authors' laboratory on ISFET and EGFET devices as pH-sensing platforms to be used for chemical and biochemical sensors.

ISFET acts like a classical MOSFET, in which the gate electrode is replaced by the electrolytic solution and a reference electrode. For ISFET, the threshold voltage V_T is given by Eq. (1) [5]

$$V_{\mathrm{T}} = E_{\mathrm{REF}} - \psi + \chi^{\mathrm{sol}} - \frac{\Phi_{\mathrm{Si}}}{q} - \left(\frac{Q_{\mathrm{IN}} + Q_{\mathrm{SS}} + Q_{\mathrm{B}}}{C_{\mathrm{IN}}}\right) + 2\phi_{\mathrm{f}} \tag{1}$$

where E_{REF} is the constant potential of the reference electrode , $(-\psi + \chi^{sol})$ is the interfacial potential at the electrolyte/insulator interface of which ψ is the chemical input parameter depending on the solution pH, χ^{sol} is the surface dipole potential of the solvent, Φ_{Si} is the work function of silicon, q is the elementary electronic charge, Q_{IN}, Q_{SS} and Q_B are the charges located in the insulator, the insulator-silicon interface states and the depletion charge in silicon, respectively, C_{IN} is the insulator capacitance per unit area. ϕ_f determines the onset of inversion depending on the doping level of silicon given by Eq. (2), where k is the Boltzmann constant, N_A is the acceptor concentration, and n_i is the intrinsic carrier concentration of the substrate.

$$\phi_{\mathrm{f}} = \left(\frac{kT}{q}\right) \ln\left(\frac{N_{\mathrm{A}}}{n_{\mathrm{i}}}\right). \tag{2}$$

A *Site-binding model* connects the interfacial potential ψ to the concentration of hydrogen ions in the electrolytic solution, expressed by Eq. (3) [6–8]. For SiO_2, the amphoteric oxide surface contains three types of sites Si–O$^-$, Si–OH, and Si–OH$_2{}^+$. The pH value at which the surface is neutral, is called the point of zero charge pH_{PZC} ($pH_{PZC} = 2.2$ for SiO_2) [8]. At a pH $< pH_{PZC}$, the oxide surface is positively charged and at a pH $> pH_{PZC}$, the surface is negatively charged. Hence depending upon the pH, the surface potential (ψ) changes, which in turn modulates V_T, and thus the drain current also. The sensitivity parameter β, given by Eq. (4), is expressed in terms of acidic and basic equilibrium constants, K_a and K_b, respectively, of the electrolytic surface reaction, the double-layer capacitance, C_{DL}, as per the Gouy-Chapman-Stern model [9], and the number of surface sites, N_s, per unit area.

$$\psi = 2.3\left(\frac{kT}{q}\right)\left(\frac{\beta}{\beta+1}\right)(\mathrm{pH}_{\mathrm{PZC}} - \mathrm{pH}) \tag{3}$$

$$\beta = \frac{2q^2 N_s (K_b/K_a)^{1/2}}{kTC_{DL}} \tag{4}$$

Equation (1) differs from the standard MOSFET equation by the incorporation of parameters ψ and χ^{sol}. As χ^{sol} is fixed for a given solvent, the only variable is ψ, which is obtained from the *site-binding model*. Various gate dielectric materials used in ISFET (SiO_2, Si_3N_4, Ta_2O_5 and Al_2O_3) differ in the ψ values, giving pH sensitivity from 37 to 58 mV/pH [10, 11]. For the Si_3N_4 dielectric used for ISFET fabrication, the pH sensitivity is $\sim$50 mV/pH.

EGFET is a low cost and simpler version of ISFET, having an extended gate arm fabricated separately and later connected electrically to the gate terminal of the commercial MOSFET device [12]. Its design methodology considers a MOSFET whose gate potential varies according to the potentials generated on the dielectric surface of the extended arm and transferred to the MOSFET gate.

2 ISFET/EGFET Design, Simulation, and Fabrication

ISFET design is based on the well-known MOSFET design equations along with the ISFET threshold voltage Eq. (1), and interfacial potential Eq. (3) based on the *site-binding model.* For a given process technology, the ISFET design parameters are similar to the MOSFET. They are operating and breakdown voltages, and the channel aspect ratio, defined as the ratio of channel width (*W*) to the channel length (*L*), controlling the transconductance of the device [13]. Operating voltage requirements impose constraints on the resistivity of the silicon wafer. Threshold voltage is determined by material resistivity in addition to gate dielectric thickness. Transconductance is mainly dictated by the channel aspect ratio for a given technology. Apart from the design equations and analysis given in Sect. 1, another important layout design consideration is the positioning of the source/drain metal contacts. For reliable device packaging, the source/drain metal contacts must be located far from the gate region.

In the present case, two different ISFET design configurations are considered, as shown in Fig. 1 [13, 14]. The twin transistor configuration was designed for differential measurements using one transistor as a reference FET (REFET). For REFET fabrication, a polyurethane membrane was deposited on ISFET giving pH sensitivity $\leq$4.2 mV/pH [15]. Further experiments are necessary to confirm the reliability of this REFET. Here, all the measurements were performed using one transistor with an Ag/AgCl reference electrode. The details about the design parameters of the two ISFET structures are summarized in Table 1.

A cross-sectional view of the ISFET and extended electrode of EGFET is shown in Fig. 2. Process and device simulations were carried out using the SILVACO® TCAD tool on the AthenaTM and AtlasTM modules for the Al-gate *N*-MOSFET device to predict the characteristics of the ISFET with respect to change in the gate voltage for a step change of 50 mV, which is approximately

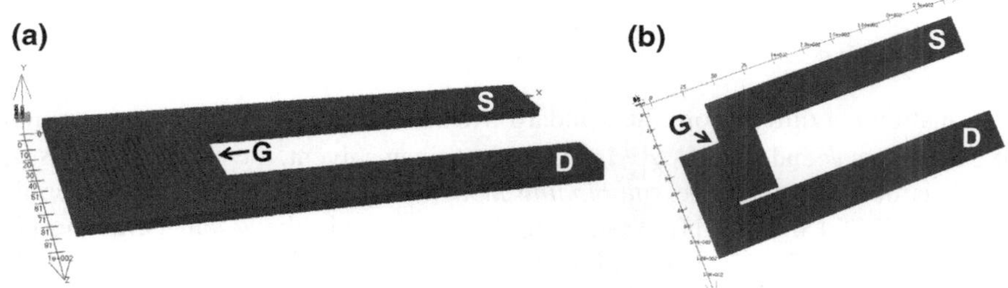

Fig. 1 ISFET structures: **a** Interdigitated; and **b** Linear (not to scale)

Table 1 ISFET design parameters

Parameter	Design 1	Design 2
Geometry	Interdigitated source-drain structure	Linear source-drain structure
Type	*N*-channel	*N*-channel
Gate dielectric	Thermal SiO_2 (500 Å) + LPCVD silicon nitride (750 Å)	Thermal SiO_2 (500 Å) + LPCVD silicon nitride (750 Å)
Channel length	12 μm	20 μm
Channel width	4800 μm	500 μm
Aspect ratio	400	25
Transconductance	5.6 mS	0.35 mS

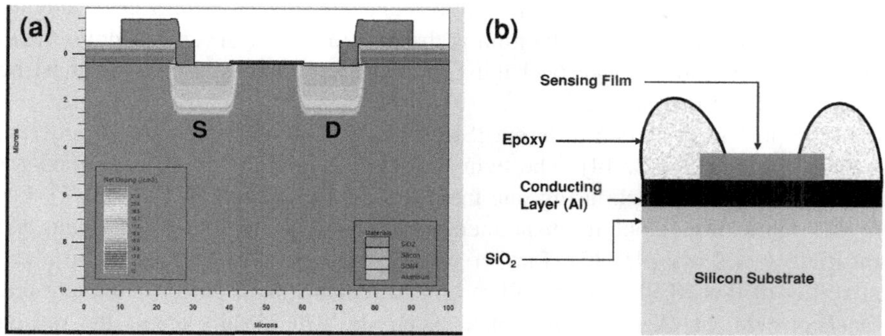

Fig. 2 Cross-sectional view: **a** ISFET with diffusion profile; and **b** Extended gate electrode with sensing film of EGFET

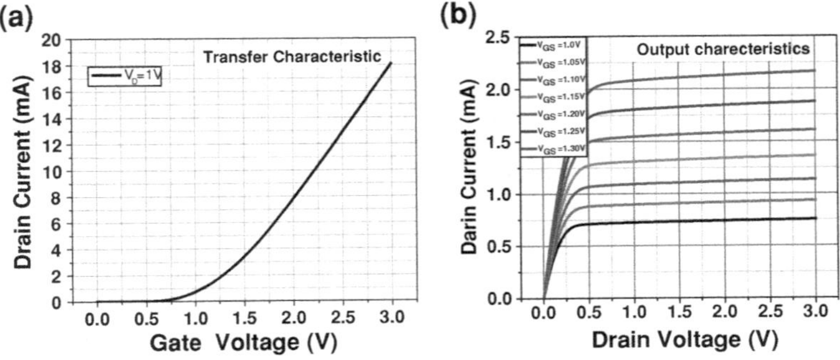

Fig. 3 Simulated **a** transfer and **b** output characteristics of the ISFET design 1

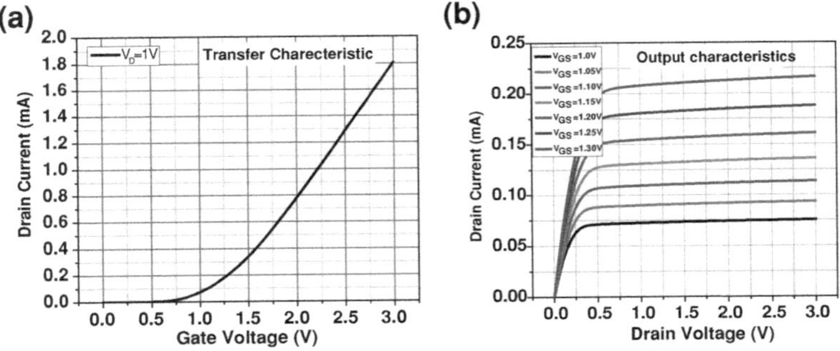

Fig. 4 Simulated **a** transfer and **b** output characteristics of the ISFET design 2

equivalent to a unit pH change for the gate dielectric used [15]. Simulated transfer and output characteristics for the two designs of Al-gate *N*-channel MOSFET devices are shown in Figs. 3 and 4, respectively.

The process steps of the *N*-channel self-aligned silicon nitride-based ISFET are briefly presented in Table 2. The ISFET device is fabricated by MOS technology using four masks: (i) Active area definition mask; (ii) Source/drain diffusion mask; (iii) Contact window opening mask; and (iv) Metal pattern definition mask.

The process steps of the metal oxide-based extended arm of EGFET are briefly presented in Table 3.

3 Packaging and Characterization of ISFET/EGFET

The materials used in packaging were tested in pH solutions 3–11 to ensure that they were chemically resistant to these solutions. The working of these devices namely, ISFET and EGFET requires the sensing film to be exposed for pH

Table 2 ISFET fabrication process steps [13, 14]

S. No	Process step	Process parameters
1	Substrate	Silicon CZ (100), 4″ diameter, *p*-type, boron doped, resistivity 10–20 Ω-cm, thickness 525 μm
2	Cleaning	RCA-1 and RCA-2
3	Field oxidation	~1 μm, @ 1,100 °C for dry-wet-dry cycle of 20-120-20 min
4	Photolithography#1	Active area definition using mask#1
5	Gate oxidation	~500 Å, Trichloroethane oxidation @ 1,000 °C for 60 min
6	LPCVD nitride	Low-pressure-chemical-vapor silicon nitride deposition using ammonia (80 sccm) and dichlorosilane (20 sccm) gases @ 800 °C for 25 min at a pressure of 270 mTorr
7	Photolithography#2	Source-drain definition for phosphorus diffusion using mask#2
8	Reactive ion etching	RIE for nitride and oxide using CF_4/SF_6 (40 sccm) and O_2 (4 sccm) gases for 4 min and 30 s at 600 W power
9	Phosphorous diffusion	Pre-deposition: @ 1,050 °C for 30 min using $POCl_3$ source Drive-in: @ 1,100 °C for 30 min in oxygen ambient; Sheet resistance: 1.64 Ω/Sq
10	Photolithography#3	Contact window opening for source and drain regions using mask#3
11	Metallization	Aluminum metallization using sputtering at 400 W for 35 min at a pressure of 5×10^{-3} mbar on both sides of the wafer
12	Photolithography#4	Front side photolithography for source and drain contact area definition and aluminum etching, protecting the rear aluminum using mask#4
13	Sintering	@ 450 °C for 30 min in forming gas ambient
14	Dicing	To separate the individual devices from the wafer

Table 3 Fabrication process steps for extended electrode of EGFET

S. No	Process step	Process parameters
1	Substrate	As in Table 2
2	Cleaning	RCA-1 and RCA-2
3	Oxidation	As in Table 2
4	Metallization	As in Table 2
5	Photolithography#1	Extended gate electrode definition
6	Sintering	As in Table 2
7	Photolithography#2	Pattern definition for sensing film deposition for lift-off process
8	Sputtering	Sensing film Ta_2O_5 deposition: Target—Ta_2O_5, RF power—150 W, Pressure—2.8×10^{-2} mbar, Time—50 min, Thickness—1,200 Å
9	Lift-off	Removal of photoresist and sensing film from undesired locations on the wafer
10	Dicing	As in Table 2

Table 4 ISFET/EGFET packaging steps for thick-film technology

S. No	Process step	Description/Process parameters
1	Conceptualization of design strategy	After obtaining the necessary parameters like physical dimensions of the device, area of device to be exposed, etc., a packaging strategy was finalized, as shown in Fig. 5
2	Design layout	Graffy HYDE software was used to make the design layout for the packaging of the devices. Using this layout, masks were fabricated to make the screens for the screen-printing process
3	Screen preparation	Using stainless steel mesh of mesh count 325 openings per linear inch
4	Base fabrication	Screen printing was carried out on alumina substrates of size 2″ × 2″ using Pd–Ag paste followed by firing and dicing
5	Die bonding	Using conductive adhesive compound for ISFET and non-conductive adhesive compound for EGFET device, followed by curing
6	Wire attachment	Using 1 mil diameter Au wire and conductive adhesive compound. Curing was done for 10 min at 120 °C
7	Encapsulation	Using the *Dam-and-Fill* technique

detection while protecting the rest of the device. For this, encapsulation of the remaining area is done using the *Dam-and-Fill* technique [16, 17]. This technique utilizes a dam around the device for the purpose of preventing the flow of the epoxy to the adjacent area, i.e., the sensing part, which has to be exposed. The dam is formed by a high-viscosity potting compound surrounded by a low-viscosity compound. The packaging of ISFET and EGFET devices has been done using thick film alumina and LTCC technologies. The process steps used for the packaging are given in Tables 4 and 5.

A design strategy for the packaging of the ISFET/EGFET microsensor using thick film technology was conceptualized in the manner illustrated in Fig. 5. The packaging of the EGFET device involves the printing of a single pad for wire bonding because the electrical connections are taken from one terminal rather than three terminals as in the ISFET devices for source, drain and ground connections. This, in turn, reduces the Pd–Ag conductor tracks from three to one. The packaged ISFET and EGFET microsensor photographs are shown in Fig. 6.

Similarly, a design strategy for the packaging of the ISFET/EGFET microsensor using LTCC technology was executed, as shown in Fig. 7. In contrast to the packaging of an EGFET device, three conductor tracks and two wire attachment pads of Pd–Ag were used for the ISFET packaging. The LTCC-packaged EGFET device is shown in Fig. 8.

The packaged ISFET/EGFET devices were characterized based on the constant drain-source voltage (V_{DS}) and constant drain-source current (I_{DS}) measurement technique [13]. The device and the reference electrode were immersed in different pH solutions (pH 4, 7, and 10), and measurements were taken. The measurement circuit details are given in Sect. 4. Figure 9 depicts a graph between I_{DS} and V_{REF} (voltage applied between reference electrode and source) of fabricated silicon

Table 5 ISFET/EGFET packaging steps for LTCC technology

S. No	Process step	Description/Process parameters
1	Conceptualization of design strategy	Packaging strategy was finalized as shown in Fig. 7
2	Design layout	Masks were fabricated to make the screens for screen printing process
3	Screen preparation	Stainless steel mesh of mesh count 325 was used to prepare the screens
4	Preparation of base, frame and lid using LTCC tapes	Base (Fig. 7a) fabrication followed by screen printing using Pd–Ag for die bonding, wire attachment and lead attachment pads Similarly, a frame with cavity (Fig. 7b), and a lid (Fig. 7c) were fabricated using LTCC tapes. The length of the frame and lid were kept smaller than the base so that leads for electrical interconnections could be attached to the base after capping. The frame covers the base and provides encapsulation using the *Dam-and-Fill* technique. The lid was used for capping the frame and also to provide access to a pH solution
5	Die Bonding	The devices were die bonded individually over the die bonding pad of the base using a non-conductive adhesive compound. Curing of potting compound was done for 20 min at 100 °C
6	Wire attachment	An Au wire of 1 mil diameter was used for attachment. Curing of conductive adhesive compound was done for 10 min at 120 °C. To avoid failure of the device due to the breakage of gold wire, two wires were attached to the two different pads of the device
7	Encapsulation	Using the *Dam-and-Fill* technique
8	Capping of lid	Lid having a small window like opening was capped on the top of the frame

nitride ISFET for different pH solutions. The sensitivity calculated for the ISFET device is 50.0 mV/pH.

In the case of the EGFET, the extended gate arm with a Ta_2O_5 sensing film was immersed in the electrolytic solutions and connected with the gate of the commercial MOSFET (MC14007UB). The applied voltage V_{REF} through the reference electrode was varied and I_{DS} was measured for a constant V_{DS} of 5 V. The measurement circuit is shown in Fig. 10. The sensitivity calculated for the Ta_2O_5-based EGFET device is 55.0 mV/pH.

4 ISFET/EGFET Instrumentation

At a fixed V_{DS}, changes in the gate potential were compensated for by the modulation of V_{GS}. This adjustment was carried out in such a way that the changes in V_{GS} applied to the reference electrode were exactly opposite to those in the gate

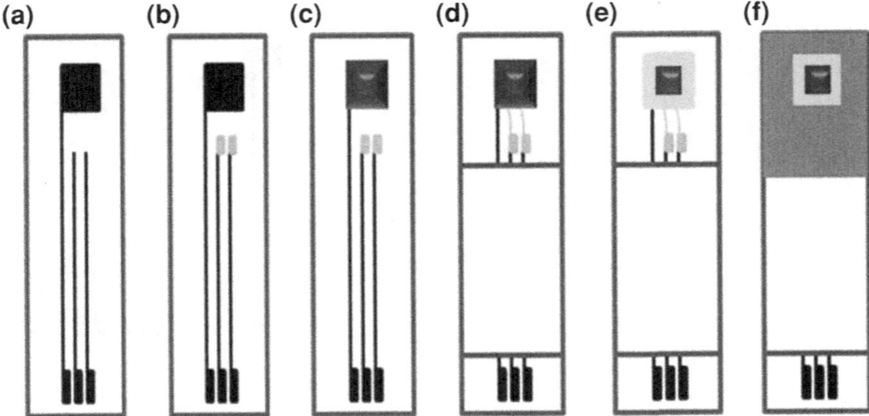

Fig. 5 Design strategy for packaging of ISFET microsensor using thick-film technology: **a** Pd–Ag screen printing for die bonding pad, interconnecting lines and soldering pads; **b** Screen printing of Au for wire attachment pads; **c** Die bonding stage; **d** Wire attachment stage and capping; **e** Dam formation; and **f** Filling around the dam

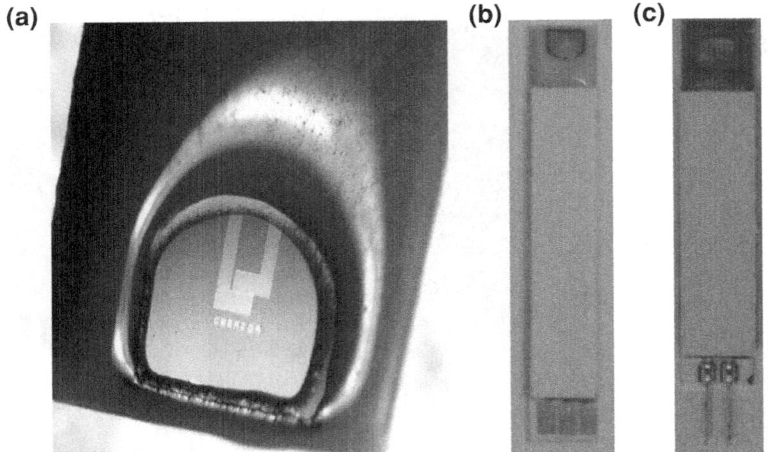

Fig. 6 Packaged devices: **a** Enlarged view of packaged ISFET device showing the sensing gate area; **b** Packaged ISFET microsensor; and **c** Packaged extended arm of EGFET

insulator potential. This was automatically performed by an ISFET amplifier with feedback which allows obtaining a constant I_{DS}. The block diagram of the pH meter using the ISFET sensor is presented in Fig. 11. The developed system consists of a sensor unit, excitation circuit, and signal conditioning unit with a microcontroller.

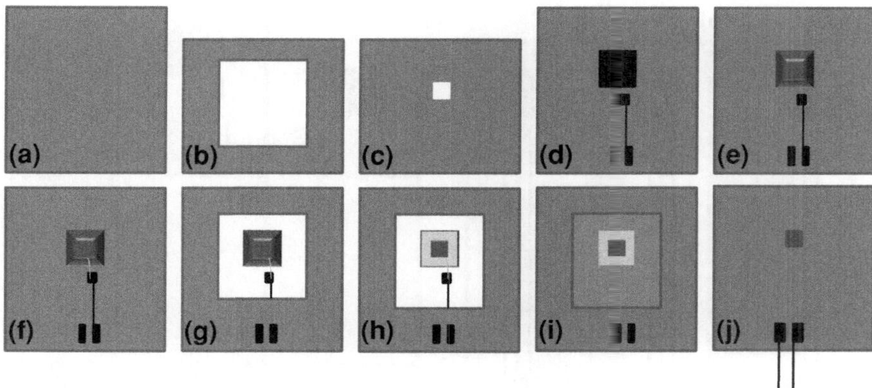

Fig. 7 Design strategy for packaging of EGFET microsensor using LTCC technology: **a** Fabricated base; **b** Frame for capping over base; **c** Lid for exposure to chemical solution; **d** Pd–Ag screen printing for die bonding and electrical connections; **e** Die bonding stage; **f** Wire attachment stage; **g** Capping of frame over base; **h** Dam formation; **i** Filling around the dam; and **j** Capping of lid over frame and lead interconnection

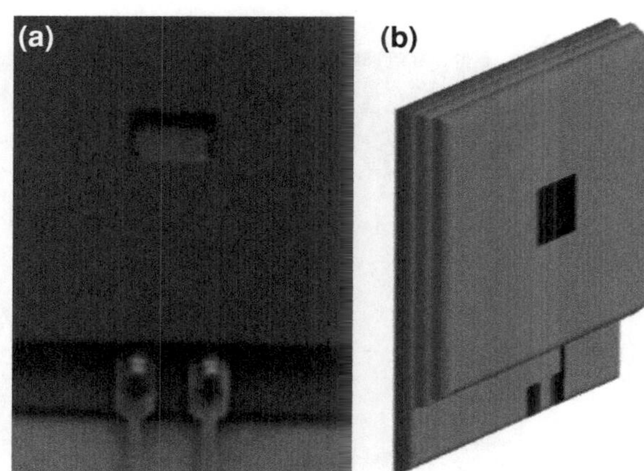

Fig. 8 LTCC packaged EGFET device: **a** Top-view; and **b** Side-view

System Description:

(a) *Sensor unit*. It consists of ISFET/EGFET and the Ag/AgCl reference electrode.

(b) *Excitation circuit and signal conditioning unit*. The ISFET sensor was operated under constant V_{DS} and constant I_{DS} conditions, i.e., the I_{DS} of the ISFET was maintained constant at constant V_{DS}. As a result, V_{GS} varies in accordance with the ionic concentration on the exposed gate surface under measurement. This was achieved by the use of an amplifier with feedback. The excitation circuitry is shown in Fig. 12. Any change in the pH of the solution changes the gate potential. The change is compensated for by the

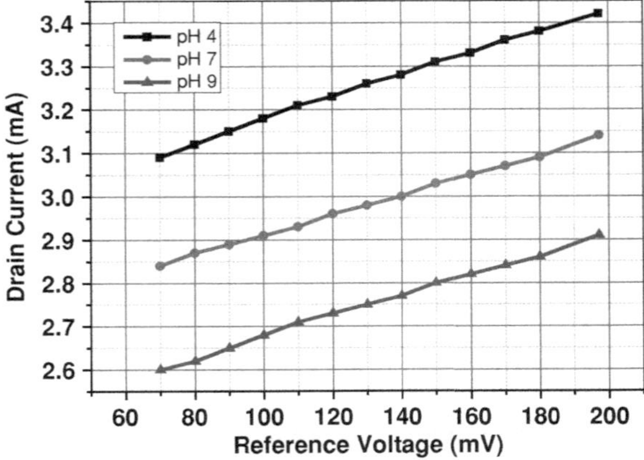

Fig. 9 Reference voltage versus drain current for ISFET

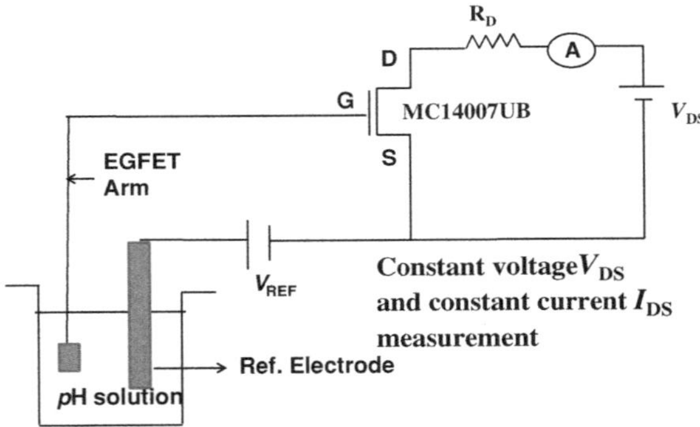

Fig. 10 EGFET (Extended electrode + MOSFET)

modulation of V_{GS} so as to counteract the changes produced in the gate insulator potential, thereby maintaining constant I_{DS} through the device. The change in the gate potential due to the variation in ionic concentrations, therefore, reflects at the reference electrode where V_{GS} is measured by keeping V_{DS} and I_{DS} constant. Depending on the design of the sensor, and the material used for the gate (sensing films/immobilized enzymes), affecting the sensitivity and selectivity, the typical values of V_{DS} and I_{DS} were chosen. Based on these specifications of V_{DS} and I_{DS} at which the sensor was to be operated, the values of V_A, V_B and R_D were selected. In the present case, an INA128 instrumentation amplifier has been used for signal conditioning to set the required gain by the use of a single resistor R_G.

The signal conditioning circuit was designed according to the characteristics of the ISFET sensor and the sensor output over the entire range of pH (1−14). Since the input range of the Analog-to-Digital Converter (ADC) of the microcontroller is 0 to +5 V, the signal conditioning circuitry is designed so as to get an output of 0 V at 1 pH and +5 V at 14 pH. The resultant transfer function of the signal conditioning unit is given by

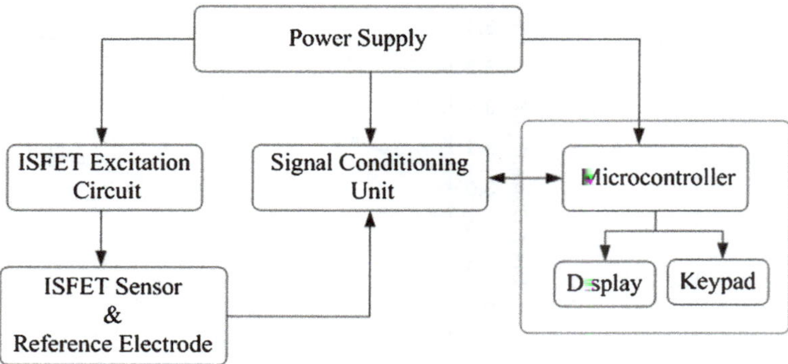

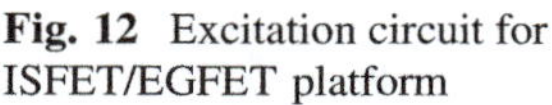
Fig. 11 Block diagram of the developed pH meter

Fig. 12 Excitation circuit for ISFET/EGFET platform

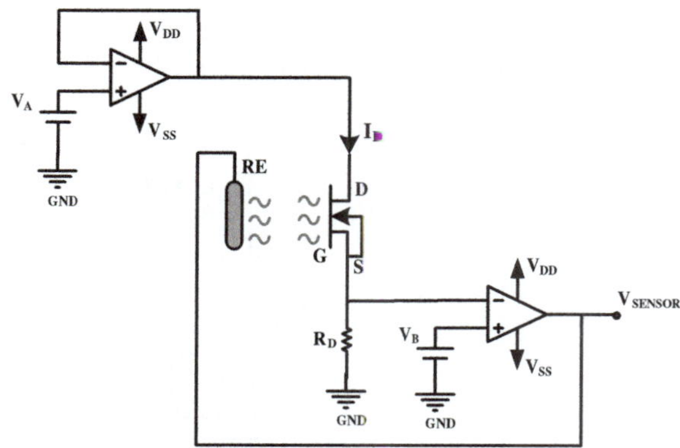

Fig. 13 Typical ISFET response with different pH values

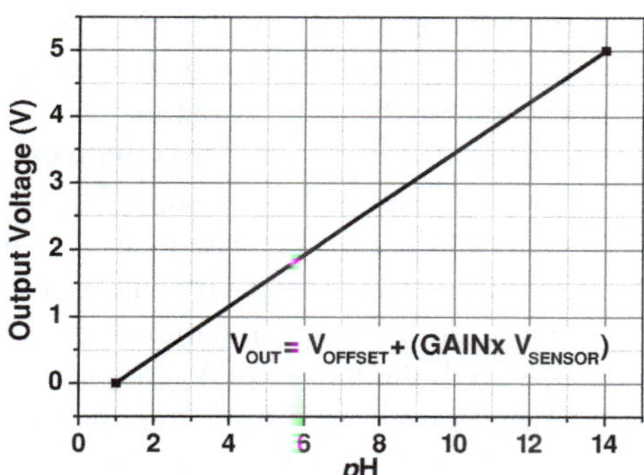

$$V_{\mathrm{OUT}} = V_{\mathrm{OFFSET}} + (\mathrm{GAIN} \times V_{\mathrm{SENSOR}}) \tag{5}$$

Figure 13 shows a typical plot of the characteristics of the signal conditioning unit. The output of the sensor is also affected by extraneous parameters like light and temperature. Therefore, care was taken to minimize the light effect

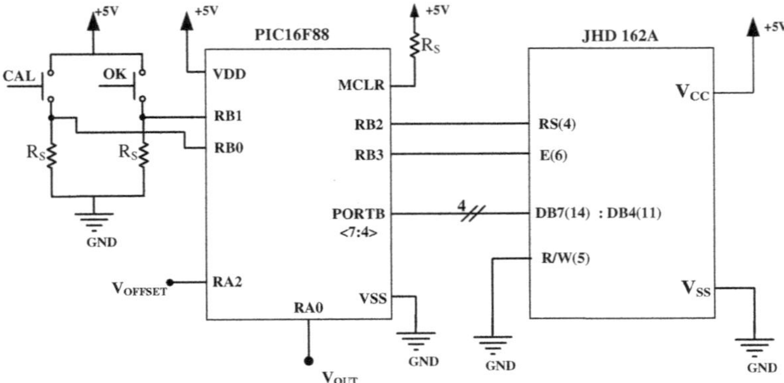

Fig. 14 Interfacing of LCD with microcontroller unit

by placing the sample under test in a closed darkened holder. The signal V_{OFFSET} is provided by the Digital-to-Analog Converter (DAC) of the microcontroller. The output of the signal conditioning circuit is given as the input to the ADC of the microcontroller.

(c) *Microcontroller unit.* It comprises a microcontroller, a liquid crystal display (LCD), and a keyboard. A (PIC16F88) microcontroller belonging to the mid-range family of the PICmicro^{TM} devices was selected to develop the system. The main peripherals of the microcontroller used were ADC and DAC. Both the LCD and the keyboard were interfaced to the I/O (Input/Output) port pins of the microcontroller. The pH value of a solution was measured at the output voltage of the signal conditioning circuit. An equation which derives the value of pH based on the output voltage of the signal conditioning unit was modeled. The final modeled equation for calculating the pH of the electrolyte solution is generically written as a function of V_{OUT}:

$$\text{pH} = f(V_{\text{OUT}}) \tag{6}$$

The microcontroller acquires the voltage value indicative of the pH of the solution using the built-in ADC, converts this voltage into digital values, calculates the value of pH of the solution using the modeled equation and the processed values, and then displays the result on the LCD. The keyboard, on the other hand, is for the online calibration of the system, whenever needed. The PIC16F88 has been interfaced to the JHD162A, a 16 $\times$ 2 character LCD display. The LCD has been operated in a 4-bit data interface mode. When operating in the 4-bit mode, two characters/command are required. The LCD module has three control signals: Enable (E), Read/Write (R_W), and Register Select (RS). The selection of a 4-bit or 8-bit data transfer mode is strictly a program memory size vs. an I/O resource trade-off. A driver for this standard 2 line by 16-character LCD has been written. The driver sets the LCD up for the 4-bit mode, and is write only. Hence only six I/O pins of the microcontroller have been used. Data lines 4–7 of the LCD displays were connected to the pins RB0-RB3 of PORTB, and the Register Select and Enable lines have been connected to pins RB4 and RB5 of PORTB, respectively, as shown in Fig. 14. The keypad consisting of two switches "CAL" and "OK" was interfaced via pins RB0 and RB1, respectively. Pressing the switch "CAL," starts the calibration process and the sensor is calibrated. The system software comprises two parts: calibration and measurement. An interrupt is generated when the "calibration" key on the keypad is pressed and the system is in calibration mode, otherwise the system runs in the measurement mode.

5 Conclusions

Chemical and biological sensors have attracted attention due to the growing demand of analytical and diagnostic tools providing label-free detection. ISFETs and EGFETs are capable of qualifying under "ASSURED" criteria established by WHO as a benchmark to decide if sensors are: Affordable, Sensitive, Specific, User friendly, Rapid and robust, Equipment-free and Deliverable to end-users [4]. In line with these requirements, this chapter has presented the research done towards the development from the device (ISFET/EGFET sensor) level to the system level. The developed ISFET has a silicon nitride sensing film with the sensitivity of 50 mV/pH. Silicon nitride ISFETs are prone to temporal drift. Other sensing films such as Ta_2O_5 and Al_2O_3 are being optimized for better sensitivity and reliability, with low drift. From the simulations of the two designed structures of ISFET, it is observed that in the high aspect-ratio design 1, the I_{DS} is one order higher than design 2 because of high transconductance, and imposes less challenges to the circuit designer.

EGFET is a low-cost version of the ISFET and is more suitable for disposable biological applications, where the devices are restricted to single use. The Ta_2O_5-based extended arm of the EGFET showed a higher sensitivity of 55 mV/pH. The structure being a conducting electrode with sensing film, the EGFET extended arm is insensitive to light and temperature, which also makes the system design easy for EGFET.

ISFET/EGFET demonstrated as pH sensors have the capability to be used for chemical and biological sensors. To fabricate these sensors, special functionalization techniques/coatings are done over the sensing region [13].

Further research is in progress to lower the detection limit (resolution) of the sensor for early disease detection using nanowire-based silicon ISFET sensors. ISFET/EGFET microsensors do not integrate well with conventional large reference electrodes. To overcome this limitation, polymer-based on-chip reference electrodes are also being developed. Reliable REFFET fabrication is an interesting research area [15]. Its realization will help in replacing the conventional reference electrode.

Acknowledgments Authors wish to thank the Director, CSIR-CEERI, Pilani for his generous support and encouragement. Financial support from the following agencies is gratefully acknowledged: 1. NPMASS-ADA, Bangalore, "Development of Microsensors for Biomedical,

Food, and Environmental Applications," PARC: 4, No.: 3 (2009–2013). 2. NPMASS-ADA, Bangalore, "Augmentation of Processing Equipment under 6′ Silicon Wafer Fabrication Facility for R & D Projects", PARC: 2, No.: 15 (2010-2014). 3. DST, New Delhi, under ITPAR/Phase-II/ Framework: Microsystems, "Development of Biosensor based on Micro- and Nano-Fabrication Technologies" (2010–2013). 4. CSIR Network Project, "Developing Capabilities and Facilities for Microelectromechanical Systems (MEMS) and Sensors," Code No.: CMM-0011 (2002–2007).

References

1. Urdea M, Penny LA, Olmsted SS et al (2006) Requirements for high impact diagnostics in the developing world. Nature 444–1:73–79
2. Yager P, Domingo GJ, Gerdes J (2008) Point-of-care diagnostics for global health. Annu Rev Biomed Eng 10:107–144
3. Mabey D, Peeling RW, Ustianowski A et al (2004) Diagnostics for the developing world. Nat Rev Microbiol 2:231–240
4. Lee WG, Kim Y-G, Chung BG et al (2010) Nano/microfluidics for diagnosis of infectious diseases in developing countries. Adv Drug Deliv Rev 62:449–457
5. Bergveld P (2003) Thirty years of ISFETOLOGY What happened in the past 30 years and what may happen in the next 30 years. Sens Actuators B 88:1–20
6. Olthuis W (2005) Chemical and physical FET-based sensors or variations on an equation. Sens Actuators B 105:96–103
7. Yates DE, Levine S, Healy TW (1974) Site-binding model of the electrical double layer at the oxide/water interface. J Chem Soc Faraday Trans 170:1807
8. Bousse L (1982) Single electrode potentials related to flat-band voltage measurements on EOS and MOS structures. J Chem Phys 76:5128–5133
9. Oldham KB (2007) A Gouy-Chapman-Stern model of the double layer at a (metal)/(ionic) interface. J Electro Chem 613:131–138
10. Khanna VK (2007) Advances in chemical sensors, biosensors and microsystems based on ion-sensitive field-effect transistor. Ind J Pure Appl Phys 45:345–353
11. van den Berg A, Bergveld P, Reinhoudt DN et al (1985) Sensitivity control of ISFETs by chemical surface modification. Sens Actuators 8:129–148
12. Batista PD, Mulato M (2005) ZnO extended-gate field-effect transistors as pH sensors. Appl Phys Lett 87:143508-1–143508-3
13. Khanna VK, Kumar A, Jain YK et al (2006) Design and development of a novel high-transconductance pH-ISFET (ion-sensitive field-effect transistor)-based glucose sensor. Int J Electron 93–2:81–96
14. Khanna VK (2012) Fabrication of ISFET microsensor by diffusion-based Al gate NMOS process and determination of its pH sensitivity from transfer characteristics. Ind J Pure Appl Phys 50:199–207
15. Khanna VK, Oelbner W, Guth U (2009) Interfacial and adhesional aspects in polyurethane (PUR) membrane coating on Si_3N_4 surface of ISFET gate for REFET fabrication. Appl Surf Sci 255:7798–7804
16. Bowman AC (2002) A selective encapsulation solution for packaging an optical microelectromechanical system. MS Thesis, Department of Material Engineering, Worcester Polytechnic Institute, U.S
17. Kharbanda D, Khanna PK (2013) Packaging of EGFET devices for pH monitoring applications. In: Abstracts of the sixth ISSS (Institute of smart structures and systems) national conference on MEMS, smart materials, structures and systems, R&DE (Engrs), Pune, India, 06–07 September 2013

Part II
Microactuators

RF MEMS Single-Pole-Multi-Throw Switching Circuits

Shiban K. Koul and Sukomal Dey

Abstract Radio frequency micro electromechanical system (RF MEMS) based switches are widely used in microwave and millimeter wave communication systems for their low loss, excellent linearity, low power consumption and compact size compared to other contemporary solid state devices [1]. Surface micromachining technology is typically used to develop different types of single pole multi-throw (SPMT) switches like SPST, SPDT and SP4T. MEMS series and shunt switches are extensively used in SPMT RF switches for multiband selectors, filter banks, reconfigurable antennas and phase shifter applications in transmit/receive modules.

Keywords Micro-electro-mechanical switches · Coplanar waveguide · Single pole double through switch · High isolation switch · Single pole four throw switch

This chapter mainly concentrates on in-line ohmic contact switches. A gold based surface micromachining process is used to develop different kinds of MEMS switches using an alumina ($\varepsilon_r = 9.8$) substrate. All switches are implemented using coplanar waveguide (CPW) transmission lines and actuated by electrostatic actuations. Furthermore, the switch profile analysis, mechanical response, electrical response, transient behavior, power handling capability, temperature stability, intermodulation distortion (IMD) and loss performances of SPST MEMS switches are discussed. This chapter also includes a brief design overview of different SPMT switch configurations with a metal to metal contact SPST switch.

S. K. Koul (✉) · S. Dey
Centre for Applied Research in Electronics, Indian Institute of Technology Delhi, Hauz Khas, New Delhi, India
e-mail: shiban_koul@hotmail.com

K. J. Vinoy et al. (eds.), *Micro and Smart Devices and Systems*,
Springer Tracts in Mechanical Engineering, DOI: 10.1007/978-81-322-1913-2_6,

1 MEMS DC-Contact SPST Switch

In this section, design and electromechanical modelling of a DC contact in-line MEMS switch are discussed. The main purpose is to study the effect of switch design parameters on switch electromechanical parameters like, mechanical resonance frequency, pull-in and release voltage and switching and release time.

1.1 Design of the MEMS DC Contact In-Line Series Switch

MEMS in-line series switches are implemented on 50 Ω CPW configurations and with an electrostatic actuator, as shown in Fig. 1. A cantilever beam is fixed at the source and the free end comes into contact with a drain upon electrostatic actuation at a gate and it also serves as a signal line. When sufficient DC voltage is applied between the cantilever beam and the fixed electrode, the beam snaps down under the electrostatic force and makes a contact with the drain, resulting in an on-state condition. When the DC bias is gradually reduced, the mechanical stress overcomes the stiction force and the cantilever beam moves up to its initial position, resulting in an off-state condition. Simple cantilever beam structures are used in this study on which all performances are evaluated with different testing environments.

1.2 Electromechanical Modelling of MEMS Switch

The electromechanical modelling of a DC-contact MEMS switch can be carried out using a Coventor ware saber platform, as shown in Fig. 2 [2]. Switch mechanical resonance frequency, stress, pull-in voltage and a transient analysis are carried out for the DC-contact cantilever switch. All results are discussed with a measured response, validated with theoretical view point up to a reasonable extent and presented in the subsequent sections. The next section deals with the fabrication process followed by different levels of characterizations.

2 Fabrication

The process starts with a 0.025″ thick alumina substrate ($\varepsilon_r = 9.8$) polished on both sides. After the RCA cleaning of the wafer, the first resistive layer of Titanium Tungsten (TiW) is deposited and patterned using the lift-off technique. This layer is used to make an electrical bias circuit by mask 1 (Fig. 3a). A 0.7 μm thick SiO_2 is deposited at 250 °C by plasma enhanced chemical vapor deposition (PECVD) and patterned using reactive ion etching (RIE) (Fig. 3b). RIE is

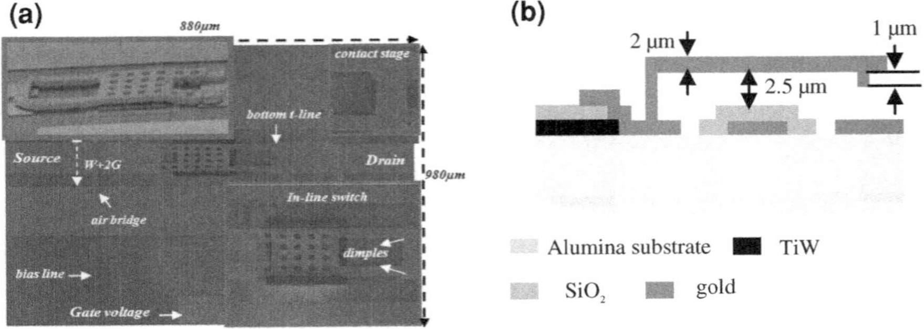

Fig. 1 **a** SEM image and **b** cross-sectional view of MEMS in-line series switch

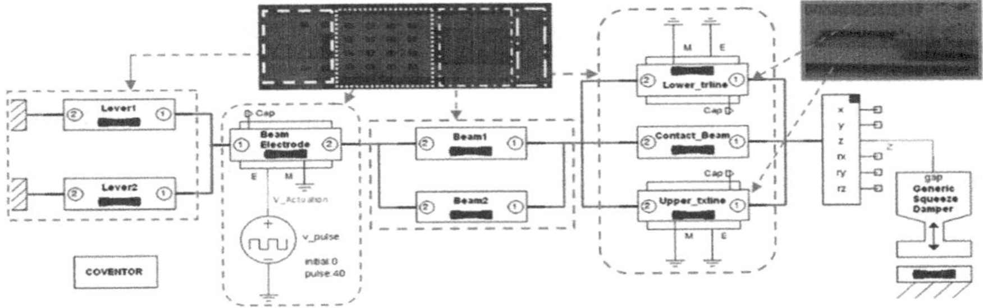

Fig. 2 Lumped model representations of MEMS switch in saber architect

employed to pattern the oxide and remove the patterning photoresist (PR). This is a passivation layer and deposited on the last layer (TiW) by mask 2. An evaporated 400 Å chromium/70 nm gold bilayer is deposited as a seed layer.

A photo-resist mould is formed in the third lithographic step and 2 μm gold is electroplated inside the mould (Fig. 3c). The mould and seed layers will be removed afterwards. Chromium is applied as an adhesion layer for the gold. This layer is used to pattern the CPW line, fixed electrodes and bias pad by mask 3. A 300 Å of TiW film is sputtered followed by the deposition of 0.5 μm silicon oxide using PECVD at 250 °C (Fig. 3d). The dielectric and TiW layers are dry etched in RIE in order to pattern them using a mask layer. The TiW layer serves as an adhesion layer for the silicon oxide to the gold. This layer is patterned by mask 4 to make an insulation layer on the CPW line and on electrodes. Spin coated Polyimide (PI) is used as the sacrificial layer for this process. Initially, it is coated to a thickness of 2.5 μm (Fig. 3e). Next, it is patterned by mask 5 (anchor mask) in an RIE step to etch the PI and fully clear the anchor holes. Furthermore, the dimple openings are performed in polyimide using an RIE etching step using the pattern of mask 6. The depth of the etching is set to be 1 μm. The top gold layer consists of a sputtered gold seed layer and an electroplated gold. The total thickness of this layer is 2 μm, and it is used as the structural layer for the devices (Fig. 3f).

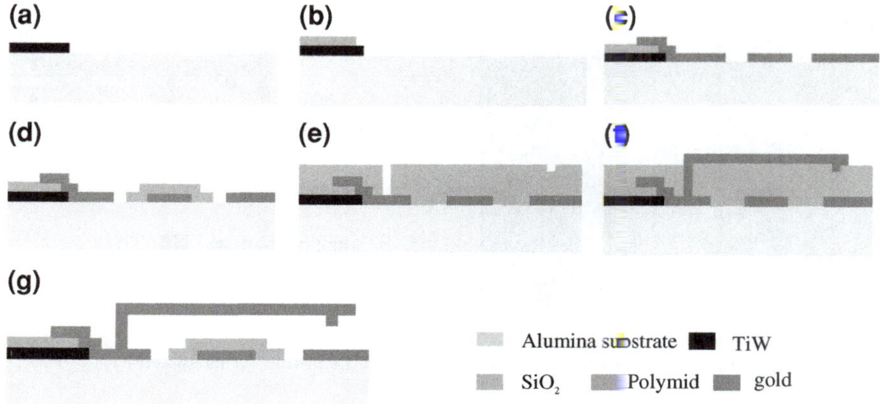

Fig. 3 Schematic view of micro-fabrication process steps of MEMS DC contact in-line switch

The negative PR moulding method is used to define this layer. Next, the layer is patterned to make a mobile electrode by mask 7. Last step is the release process. In this process, the sacrificial layer is removed in an oxygen plasma dry etch process in RIE (Fig. 3g).The fabrication process steps are depicted in Fig. 3.

3 MEMS Switch Design, Simulation and Characterization

The switch profile is measured using an Optical Profilometer. It predicts the beam tip deflection and surface roughness of the DC-contact MEMS switch. A mechanical analysis is carried out using a Laser Doppler Vibrometer and switch pull-in voltage is extracted using electrical measurement through a LCR meter. Switching and release time, RF response, power handling, and a linearity analysis are performed on the MEMS switch at standard room temperature without any device package. All measured responses are validated with simulated responses and theoretical viewpoints.

3.1 Switch Profile Analysis

After the successful release of the cantilever beam, non-uniform in-plane residual stress is induced on the beam. As a result, the cantilever beam deflects either upward or downward. A positive stress gradient deflects the beam in the upward direction, whereas a negative gradient bends the beam in the downward direction [3]. In the process mentioned above, the induced residual stress is tensile in nature and varies from 110–150 MPa. The deflection profile of the switch (Fig. 4a) shows

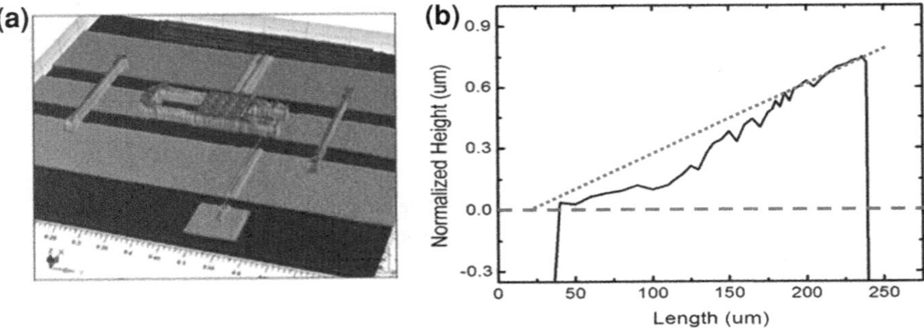

Fig. 4 **a** 3D view of switch using optical profilometer and **b** Normalized gap height along the length of the cantilever beam

an extra 0.76 μm tip deflection due to +4 MPa/μm stress gradient, as shown in Fig. 4b.

If the stress of the beam varies linearly across the film thickness, then the stress gradient and tip deflection of the cantilever beam can be expressed as [3]

$$\Delta\sigma = \frac{\mathrm{d}\sigma}{\mathrm{dh}} = \frac{2E_e\delta}{l^2} \tag{1a}$$

$$\Delta Z_{\max} = \frac{\Delta\sigma l^2}{2E_e} \tag{1b}$$

where, l is the length of the cantilever beam and $E_e = E/(l - \upsilon)$ is the biaxial Young's modulus of electroplated gold (E = 45 GPa, υ = 0.4). All the design parameters of the DC contact MEMS switch are tabulated in Table 1.

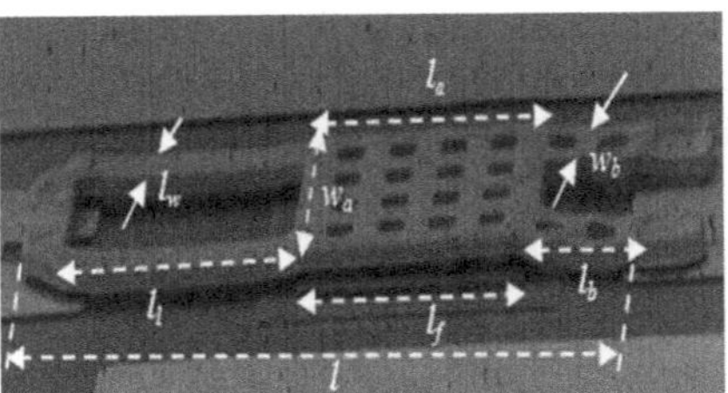

3.2 Mechanical Analysis

The mechanical resonance frequency (f_c) of the cantilever beam depends on beam dimensions and material properties, such as Young's modulus, Poisson's ratio, residual stress and material density. The fundamental resonance frequency of the cantilever switch is given by [3]

Table 1 Designed structural dimensions of the MEMS switch

Parameter	Value (μm)
Fixed electrode length, l_f	100
Beam length, l_b	46
Beam width, w_b	30
Lever length, l_l	125
Lever width, w_l	20
Length of movable electrode, l_e	100
Width of movable electrode, w_e	110
Total length of the beam, l	275
Width of the beam, w	100
Thickness of beam, t	2
Dimple area	10 × 10

$$f_c = 0.16154\frac{h}{l^2}\sqrt{\frac{E_e}{\rho}} \tag{2}$$

where h is the thickness of the cantilever beam and ρ is the material density of electroplated gold and its value is 19,300 kgm^{-3}.

The measured residual stress and average deformation of the cantilever beam are included in the saber model. The simulation shows resonance frequency of 10.2 KHz, as shown in Fig. 5a.

The switch mechanical response is extracted using Laser Doppler Vibrometer (LDV) where an electrical connection is established between the beam and the fixed electrode by a wafer prober [2]. A chirp voltage with a frequency sweep of 0–170 KHz has been applied at the cantilever. The switch shows measured f_c at 9.7 KHz with 3 V actuation voltage which is much less than its pull-in voltage, as shown in Fig. 5b and it is closely matched with simulated frequency of 10.2 KHz (Fig. 5a). The measured Q factor of the switch is 1.2.

Initially, device design parameters are found out from the static solutions followed by a dynamic response of the bridge. The functional parameters are extracted from experimental results and with the parametric extraction method considered in the system level model in the Saber Architect platform (Fig. 1). Finally, these are verified with Harmonic EM simulation in Coventor ware.

3.3 Electromechanical Analysis

The pull-in and release behaviour of the MEMS switch are defined by applied electrostatic force and zero external force, respectively and it can be expressed as

$$m_a\frac{\mathrm{d}^2\mathrm{x}}{\mathrm{dt}^2} + b\frac{\mathrm{dx}}{\mathrm{dt}} + k_a x = \frac{1}{2}\frac{\varepsilon_0 A V_s^2}{\left(g_0 + \frac{t_d}{\varepsilon_r} - x\right)}(\mathrm{pull-in}) \tag{3a}$$

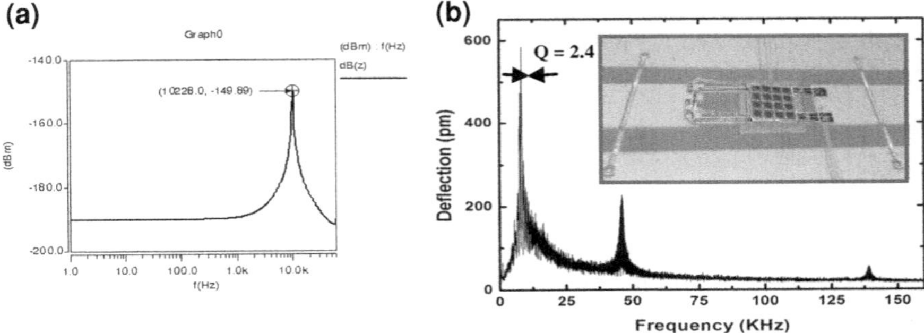

Fig. 5 **a** Simulated and **b** measured resonance frequency of the MEMS switch, inset in figure shows shape of the fundamental mode of vibration under excitation. Copyright/used with permission of/courtesy of Institute of Physics and IOP publishing limited

$$m_n \frac{\mathrm{d}^2 x}{\mathrm{d}t^2} + b\frac{\mathrm{d}x}{\mathrm{d}t} + k_n x = 0 \text{ (release)} \tag{3b}$$

This results in two different quality factors; natural spring constant Q_n ($=k_n/(\omega_0 b)$) and by actuation spring constant Q_a ($=k_a/(\omega_0 b)$), respectively. The effective mass of the cantilever beam can be calculated by equating and equal to ω_o. The switch pull-in voltage can be found analytically by [2]

$$V_{\mathrm{pi}} = \left(g_0 - \mathrm{d}_{\mathrm{pi}}\right)\sqrt{\frac{2(k_1 + k_2)\mathrm{d}_{\mathrm{pi}}}{\varepsilon_0 A}} \tag{4}$$

where V_{pi} is the pull-in voltage ($g_O - d_{\mathrm{pi}}$) the gap as the switch closes, ε_O is the permittivity of free space, A is the electrostatic overlap area and d_{pi} is the pull-in distance. In this equation, k_1 is the structural stiffness which is determined by Young's modulus and device geometry, and k_2 is the stiffness due to biaxial residual stress which is very less in cantilever type structure.

When the dimple of the switch touches the bottom transmission line, it gives a contact resistance (R_c) which is the summation of the constriction resistance (R_1) and contamination film resistance (R_2). The Maxwellian spreading resistance theory [4] defines the R_1 from contacting the surface topography and Abbott and Firestone's material deformation model [5] revises the constriction resistance (Eq. 10) under plastic deformation condition using Holm's contact resistance equation. These can be expressed as

$$R_1 = \frac{\rho}{2r_{\mathrm{eff}}} \tag{5a}$$

$$R_2 = 0.866\rho\sqrt{\frac{H}{F_c}} \tag{5b}$$

where, r_{eff} is the effective radius of a circular contact area. The hardness and resistivity values used for gold–gold contacts are 2–2.3 GPa and 2.5–2.8 μΩ cm, respectively. F_c, the contact force is given by

$$F_c = \frac{\varepsilon_0 A_e V^2 a^2}{4l^3 \left(g_0 - d_{\text{pi}}\right)^2}(3l - a) \tag{6}$$

where a is the applied load position on the cantilever beam.

The pull-in voltage of the MEMS switch is extracted by an Agilent 4284A LCR meter with a probe station [2]. The C–V characteristics of MEMS switches are found with a voltage sweep of 0–40 V. The voltage is applied to the top suspended beam using the probes, whereas the bottom fixed electrodes are grounded. To measure the capacitance value, a small ac signal of 5 MHz is given on the DC actuation voltage and open circuit offset measurement corrections were ensured before recording the capacitance values. The beam gives a measured pull-in voltage at 30 V as shown in Fig. 6.

To simulate the MEMS switch, all structural and mechanical parameters such as switch initial deformation (Z_{max}), damping (b), and quality factor (Q) are used in the saber model. As a result, the simulated pull-in voltage is found to be 36 V using the model defined in Fig 2. The fringing field effect is ignored during the simulation due to mathematical complexity, and intrinsic residual stress is assumed to be zero. Figure 7 shows that a residual stress of 160–190 MPa gives 0.7–0.85 μm tip deflection and pull-in voltage shifts to 28–30 V. This observation is closely validated with a 5.8 % deviation from the measured result (30 V) and tip deflection under a process-induced residual stress of 170 MPa.

3.4 Switching and Release Time Analysis

The dynamics of the MEMS switch is based on mechanical movement between the open and closed states, and the actuation speed is limited by the mechanical response time. The switching time can be expressed by a dynamic 1D equation [1]

$$m_a \frac{\mathrm{d}^2 \mathrm{x}}{\mathrm{dt}^2} + b\frac{\mathrm{dx}}{\mathrm{dt}} + k_a x = F_e + F_{\text{LJ}} \tag{7a}$$

$$F_{\text{LJ}} = \frac{c_1 A}{\left(g_0 - z_{\max}\right)^3} - \frac{c_2 A}{\left(g_0 - z_{\max}\right)^9} \tag{7b}$$

$$b = \frac{k_1 + k_2}{2\pi f_c Q_e} \tag{7c}$$

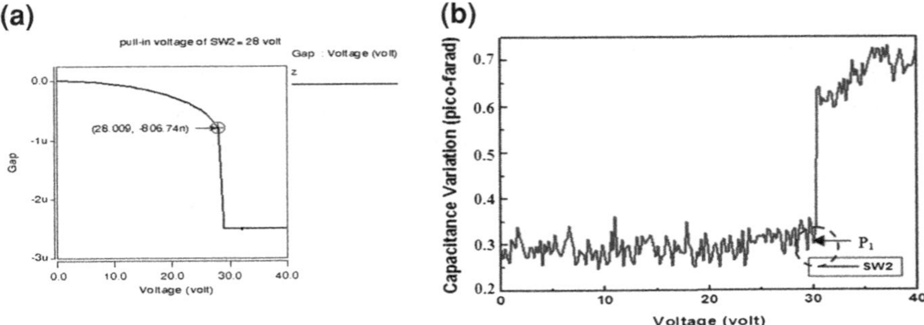

Fig. 6 **a** Simulated pull-in voltage of SW2 is 28 V, and **b** C–V measurement of MEMS switch shows pull-in at 30 V. Copyright/used with permission of/courtesy of Institute of Physics and IOP publishing limited

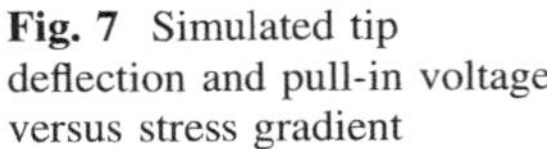
Fig. 7 Simulated tip deflection and pull-in voltage versus stress gradient

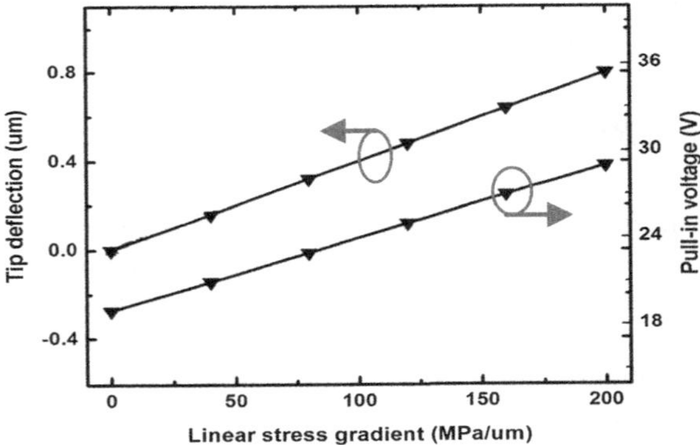

In Eq. (7b), the first term is the van der Waals component and the second term is the repulsive component. The values of c_1 and c_2 are 10^{-80} N m and 10^{-75} N m^8, respectively. The switching time (t_s) values are strongly affected by chemical properties and morphology at the contact surface (surface roughness). The holes on the beam reduce the damping coefficient and increase the Q of the structure. The displacement of the beam is a function of time and it spends the maximum time to reach up to the pull-in limit; afterwards the beam suddenly snaps down. The above 1-D equation can be solved for the switching time versus voltage for $x = 0$ and $x = 3.2$ µm (considering an extra 0.8 µm deflection). The minimum t_s can be found by

$$t_s = 3.67\frac{V_p}{V_s\omega_0} \quad \text{where } V_s = 1.5V_p \tag{8}$$

The switching time measurement is carried out using Agilent infiniium DSO-X 92504 25 GHz high frequency digital storage oscilloscope, as shown in Fig. 8a. A square wave of 0–40 V with a frequency of 1 KHz is given to the actuation pad and the corresponding effects on switching (78 μs) and release time (123 μs) are recorded from Agilent Digital Storage Oscilloscope (DSO) as depicted in Fig. 8b. The release time is higher than the switching time due to the switch contact bouncing effect which takes a few more cycles to settle down at its initial position. The switch is lifted up at 86 μs and additional 37 μs takes to complete settle down the switch at its initial position. The pull-in and release voltage of the switch are 30 and 16 V, respectively. The release time can be analytically formulated by [2]

$$\begin{aligned} t_r &= \frac{\pi}{2\omega_n} \text{ for } Q_m < 1 \\ &= \frac{1.6}{\omega_n Q_m} \text{ for } Q_m > 1 \end{aligned} \tag{9}$$

Switch damping limited switching time and release time (t_r) both depend on the mechanical Q factor (Q_m). For lower quality factors, damping significantly increases the switching time.

3.5 RF Power Handling Capabilities

The RF power handling capability of a MEMS switch depends on the thermal conductivity of the metal beam and non uniform heating along the beam. Low thermal resistance results in a low temperature rise on the beam. The medium value of the spring constant and high release voltage improve the power handling capability of a MEMS switch. The equivalent rms RF voltage across the switch V_{RF}^{rms} is given by (10)

$$V_{sw} = \frac{2V_{pk}}{\sqrt{1 + 4\omega^2 C_{RF}^2 Z_0^2}} \approx \frac{V_{pk}}{2\omega C_{RF} Z_0} \text{ for } \omega\, C_{RF} Z_0 \gg 1 \tag{10}$$

The input power (P_{in}), power required to actuate the switch (P_{act}) and the maximum RF power to hold the switch down (P_h) can be expressed by [6]

$$P_{in} = \frac{V^2}{2Z_0} \tag{11a}$$

$$P_{act} = \frac{kz(g_0 - z_{max})^2}{2\varepsilon_0 A Z_0} \tag{11b}$$

$$P_h = \frac{4\omega C_{RF} k (g_0 - z_{max})\left(g_0 + \frac{t_d}{\varepsilon_r}\right)^2}{\varepsilon_0 A} \tag{11c}$$

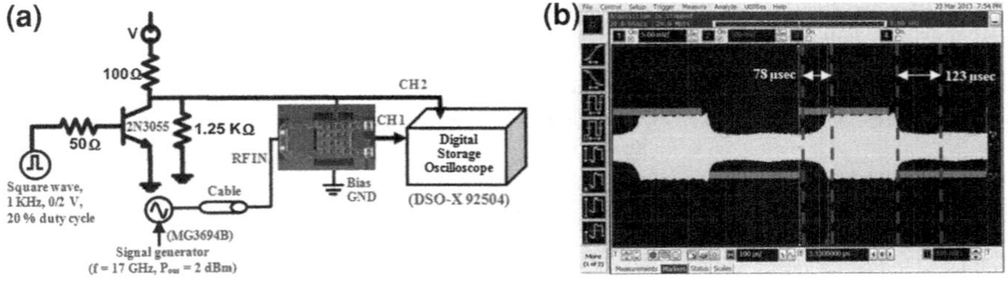

Fig. 8 **a** Measurement set up of switching and release time and **b** Measured switching and release time of DC contact MEMS switch. Copyright/used with permission of/courtesy of Institute of Physics and IOP publishing limited

The power handling capability of the MEMS switch is measured using the setup shown in Fig. 9. The measurement shows that the switch can handle 2.5 W of RF power at 17 GHz without any self actuation ($V_{RFrms} > V_{PI}$) or latching ($V_{RFrms} > V_r$). The switch has the ability to handle up to 34 dBm ($\approx$2.5 W) of input power and the measured pull-in voltage changes from 30 to 26 V with an R_c variation of 2.5–2.8 Ω as shown in Fig. 9.

3.6 S-Parameter Analysis

A cross section of the DC contact MEMS switches is shown in Fig. 1. The dimple is used to minimize the dielectric charging problems and to create a better metal contact between the beam and the bottom *t*-line. When the MEMS switch is actuated (*on-state*), it gives a very small contact resistance to provide an RF through path. When the beam is at the up-state, it exhibits a very small capacitance which will block the RF path (*off-state*). The equivalent circuit model for an in-line DC-contact MEMS switch in the two states is shown in Fig. 10.

As shown in Fig. 10, in the *off-state*, the switch is modeled as an inductance (L) of 113 pH in a series with a small transmission line resistance (R_t) of 0.5 Ω and a capacitance (C_{off}) of 12 fF. The parallel capacitance to ground (C_p) is 39 fF. The capacitance C_{off} is between the beam and the bottom contact metal connecting to the output transmission line, while the C_p is the parallel plate capacitance between the membrane and the bottom fixed electrode. The L_s and C_s are the *t*-line inductance and capacitance which contribute 50 Ω to the RF input and output transmission lines.

When the bias voltage is applied between the membrane and the bottom fixed electrode, the switch exhibits a contact resistance (R_{ON}) of 2.9 Ω. The inductance (115 pH) is in series with a resistance R_{ON} and is connected in parallel with a capacitance (C_p) of 35 fF. The C_g ($\approx$4–6.6 fF) refers to the signal-line coupling capacitance due to the gap between each broken signal line. This can be found from the fundamental capacitance formula given in [1],

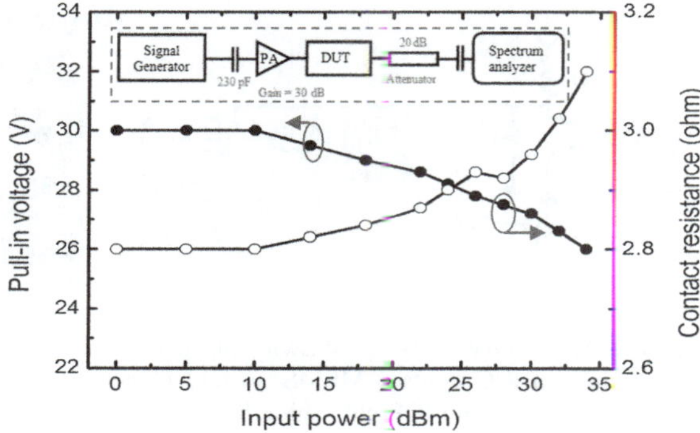

Fig. 9 Measured variation of switch pull-in voltage and contact resistance with applied input power of up to 34 dBm

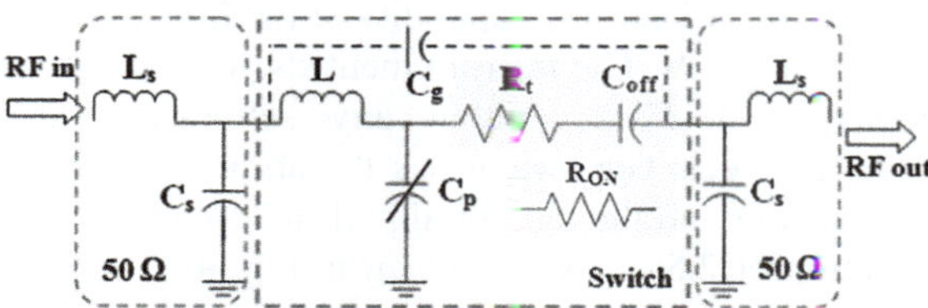

Fig. 10 Equivalent circuit model of the MEMS DC-contact series switch

$$C_p = \frac{\varepsilon_0 A_e}{g_0 + \frac{t_\varepsilon}{\varepsilon_r}} + C_f \tag{12}$$

where C_f is the fringing field capacitance which is 20–60 % of C_p, depending on the switch dimensions and height (g_O).

The switch inductance can be expressed as

$$L = \frac{Z_h l \sqrt{\varepsilon_{\text{eff}}}}{c} \quad \text{for } l < \frac{\lambda_g}{12} \tag{13}$$

where, c is the velocity of light in free space, Z_h is the high impedance line, λ_g is the guide wavelength and ε_{eff} is the effective permittivity. This inductance (L) can also be found from switch return loss (S_{11}), as given in [1]

$$|S_{11}|^2 = \left(\frac{\omega L}{2Z_0}\right)^2 \tag{14}$$

The switch OFF-state capacitance (C_{off}) can be found from the isolation response (S_{21}) of the switch [1]

$$|S_{21}|^2 \approx 4\omega^2 C_{\text{off}}^2 Z_0^2 \tag{15}$$

The ON-state resistance is the summation of transmission line resistance and contact resistance (R_c). The contact resistance can be obtained from switch on-state return loss (S_{11}), as given in [1]

$$|S_{11}|^2 = \left(\frac{R_{\mathrm{ON}}}{2Z_0}\right)^2 \tag{16}$$

where ω is the operating frequency and Z_0 is the characteristic impedance (50 Ω).

The switch on-state and off-state performances are simulated using a 3D full wave electromagnetic based simulation tool ANSYS High Frequency Structure Simulator (HFSS) V.13. The measurements on the switch are performed using an Agilent 8510C vector network analyzer with GSG RF probes, and a Cascade DC probe. The short-open-line-through (SOLT) method was used for the calibration. The measured loss performance of the switch is validated by HFSS and with the circuit model in the Agilent ADS platform (using parameter extraction method from measured S-parameter response) (Fig. 11). The switch gives measured return loss of a better than 24 dB, a worst case insertion loss of 0.38 dB and isolation better than 20 dB. The measured results closely validate the simulation results within a 6 % tolerance limit which is mostly due to signal leakage via the bias line.

3.7 Linearity Analysis

In contact type micro switches, the capacitance in the *off-state* is very small (11–12 fF) so it is not the major source of intermodulation distortion (IMD). In the *on-state*, the switch resistance is a weak function of contact force; the contact force typically changes only a few percent, even when RF voltage is applied. The switch resistance (R_{ON}) is expected to change a little less because of the increased contact force caused by the applied RF voltage. Moreover, the modulation of switch resistance by ohmic heating of switch materials appears to dominate in the in-line series switch. The mechanical motion of the beam also takes part in IMD generation [7].

To find the IMD, two input frequencies (ω_1 and ω_2) which are close together with a spacing of $\Delta\omega = (\omega_1 - \omega_2)$ have been applied to the switch Δ where $\omega \ll \omega_n$. Two incident signals on the *t*-line ($V_i = V_0\,[\sin(\omega_1 t) + \sin(\omega_2 t)]$) whose output signal is given as (17)

$$V_0 \approx V_{\mathrm{bias}} + V_0[\sin(\omega_1 t + \varphi) + \sin(\omega_2 t + \varphi)] \tag{17}$$

With an output phase and magnitude of

$$\varphi = -2\omega C_{\mathrm{off}} Z_0, \; |S_{11}| \approx 1 \tag{18}$$

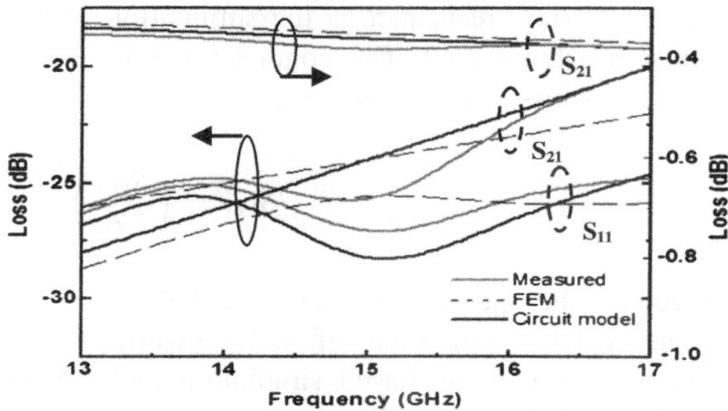

Fig. 11 Measured versus simulated S-parameters response of the switch

The input signal can also be expressed by

$$V_{\text{in}} = 2V_0 \cos(\delta t) \sin(\omega t) \tag{19}$$

where $\delta = \frac{\omega_1 - \omega_2}{2}$ and $\omega = \frac{\omega_1 + \omega_2}{2}$

The magnitude of the signal is $2V_0$ with an average frequency ω being modulated at a frequency δ.

Therefore, the third order intercept point (IIP3) is the fictitious input power for which the power of the sideband would be equivalent to the power of the input signal if all the input power is transmitted to the output and it can be given by [8]

$$\text{IIP3} = \frac{2}{R}\left[\left(\frac{2\left(2R + R_0'\right)^2}{CRR_0'}\right)\left(\frac{R}{2R + R_0'}\right)^{1+\alpha}\right]^{2/\alpha} \tag{20}$$

where C is a constant that is independent of power but depends on signal frequency and α is equal to 2.11. The load and source resistance are taken to be R which is 50 Ω and R_0' is the average resistance.

IIP3 is experimentally obtained using two signal generators with carriers frequencies $f_1 = 16.999998$ GHz and $f_2 = 17.0000013$ GHz respectively with a spacing of 3.3 KHz which is much less than the switch mechanical resonance frequency (8 KHz). The changes in modulation levels during zero and applied bias have been obtained experimentally and are shown in Fig. 12. The measured IIP3 of the switch is found to be 70 dBc.

4 MEMS High Isolation SPST Switch

MEMS SPST switch isolation can be improved using three identical MEMS DC-contact switches with series-shunt configuration [9]. The micro-photograph of the SPST switch is shown in Fig. 13. All dimensions of the MEMS switch are scaled down to make mechanically robust switch with high reliability. The total length of

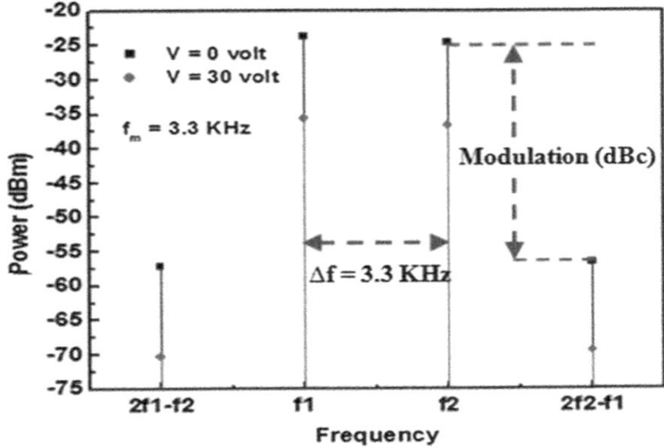

Fig. 12 Measured IIP3 of the switch with input power of 0 dBm. Copyright/used with permission of/courtesy of Institute of Physics and IOP publishing limited

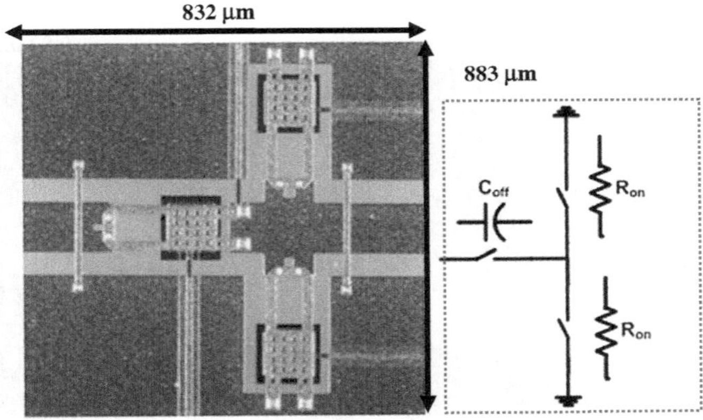

Fig. 13 Micro-fabricated image of series shunt SPST switch. Schematic shows high-isolation stage

the switch is 200 μm. One switch is placed in-line and the other two are connected to the ground. This topology gives a measured *on-state* (series switch is actuated) return loss of better than 24.5 dB and a worst case insertion loss of 0.96 dB within 13–17 GHz as shown in Fig. 13. The *off-state* (both shunt switches are actuated) isolation is better than 49.5 dB up to 17 GHz as depicted in Fig. 14. This configuration avoids the fabrication complexity and biasing problem where three identical metal-contact MEMS switches are actuated with 40–50 V.

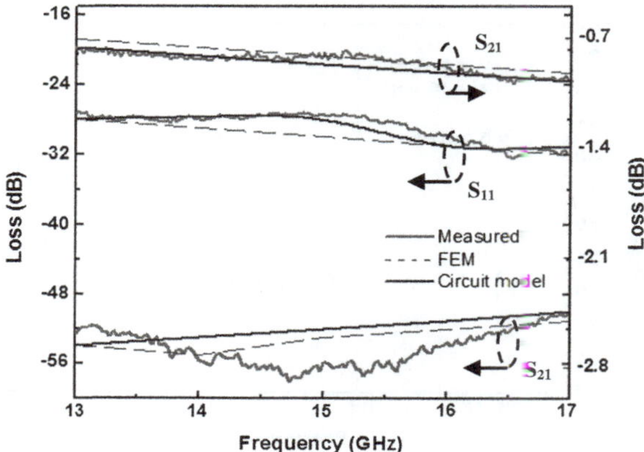

Fig. 14 Measured versus simulated *off-state* and *on-state* S-parameters of series-shunt SPST switch

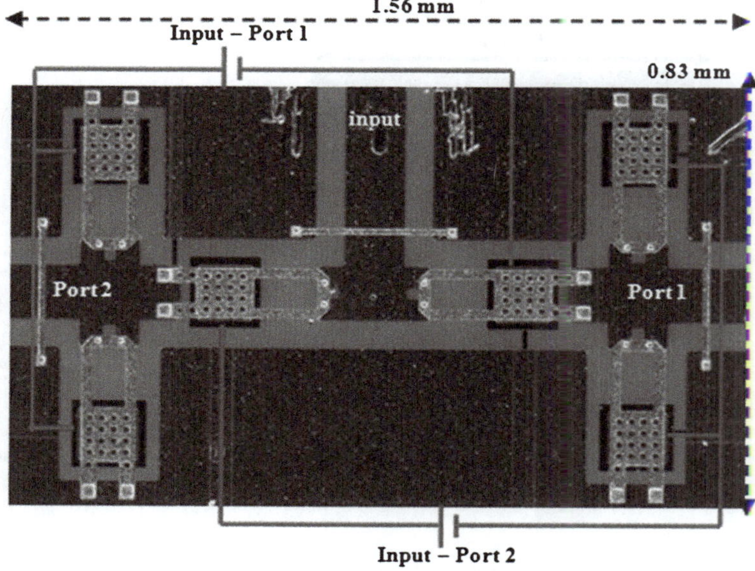

Fig. 15 Microscopic image of MEMS SPDT switch. The total chip area is 1.25 mm^2

5 MEMS SPDT Switch

The MEMS SPDT switch is made by combining two series-shunt SPST cells as shown in Fig. 15 where the actuation scheme is shown with a blue line [9]. A measured return loss of better than 22 dB, a worst case insertion loss of 1.5 dB and

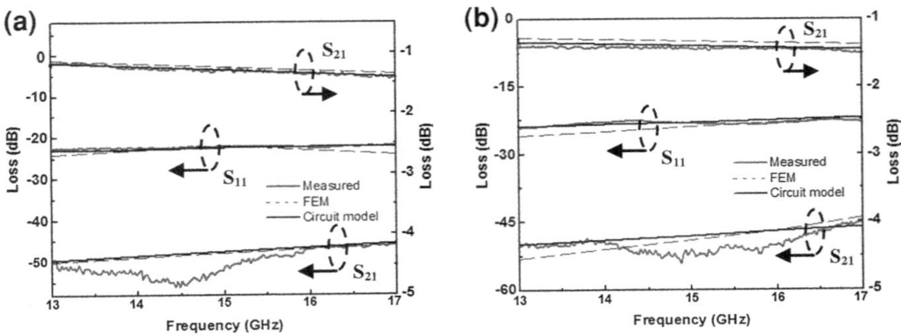

Fig. 16 Measured versus simulated *on-off* state S-parameters **a** input to port 1 and **b** input to port 2 of the SPDT switch

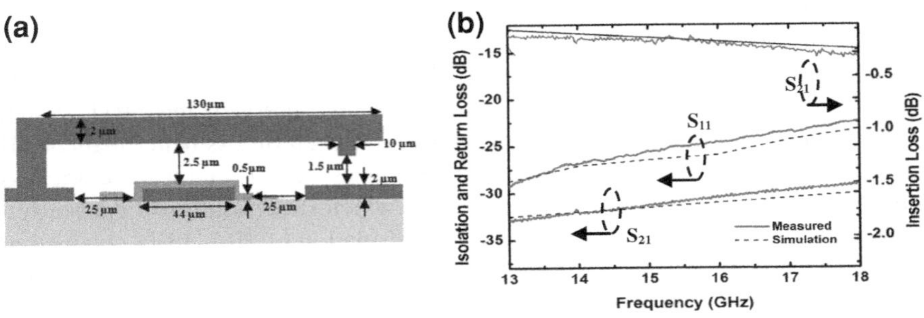

Fig. 17 **a** Design parameters of cantilever switch and **b** measured versus simulated loss of a dc-contact MEMS switch

isolation better than 45 dB are obtained from the SPDT switch as shown in Fig. 16. Measured port to port isolation is 29 dB. A maximum deviation of 8% was observed between measured and simulated loss performance. This is attributed to the overall gap height non-uniformities in the in-line MEMS switches which leads to an asymmetric distribution of capacitances (C_{off}, C_{p}) and resistances (R_{ON}) throughout the signal line and signal leakage through the TiW bias line.

6 MEMS SP4T Switch

The MEMS SP4T switch is made with four DC-contact cantilever switches that are placed 72° apart with respect to each other. A single MEMS switch gives a measured return loss of 25 dB and a worst case insertion loss of 0.27 dB (Fig. 17) with 53 V actuation. The measured characteristics from 13 to 18 GHz indicate that

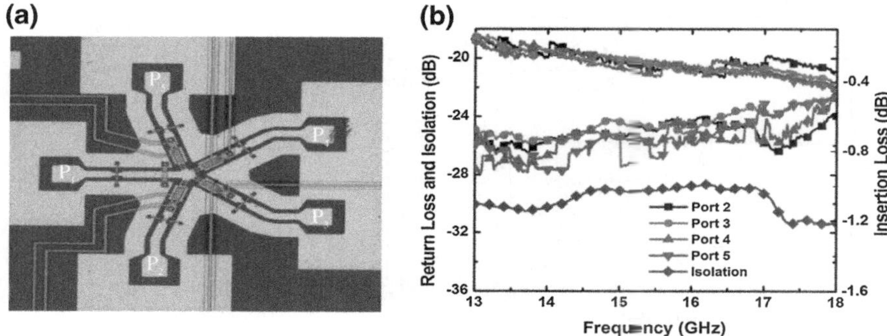

Fig. 18 **a** Microphotograph of MEMS SP4T switch and **b** S-parameter response of the switch

the fabricated MEMS switch offers good isolation characteristics (between 33.0 and 29.5 dB) as shown in Fig. 17.

The microphotograph image of the SP4T switch is shown in Fig. 18a and its measured characteristics are shown in Fig. 18b. As observed, the SP4T switch demonstrates less than a 0.38 dB insertion loss and a return loss better than 24 dB for all ports within the 13–18 GHz band. The measured deviation of return loss between the near port to the far port is 3.6 dB which successfully validates a full wave simulated return loss deviation of 3 dB. Isolation of 29.7 dB is obtained at one on-port condition (Fig. 18b) which is 1.6 dB worse (31.3 dB) than all ports off-condition. The total area of the MEMS SP4T switch is 0.75 mm^2.

7 Conclusions

The MEMS DC contact switch design followed by different types of characterization is presented in this chapter. These include a switch profile analysis, mechanical and electrical response, switching and release time, loss, power handling and linearity. A high isolation of 50 dB is achieved from the SPST switch using series-shunt topology from 13 to 17 GHz. The measurement of the SPDT configuration shows a low insertion loss (<1.5 dB), high return loss (>20 dB) and excellent isolation characteristics (>45 dB) from 13 to 18 GHz within the smallest possible area of 1.25 mm^2. The SP4T switch configuration uses 4 simple cantilever MEMS switches and exhibits excellent performance from 13 to 18 GHz within a 0.75 mm^2 area. The electromechanical modeling and circuit modeling of the MEMS switches are presented. Comprehensive modeling and the design of different MEMS switching circuits are verified up to a reasonable extent.

Acknowledgments The authors would like to express their profound gratitude to the National Program on Micro and Smart Systems (NPMASS), Govt. of India for setting up a MEMS design lab and RF characterization facilities at CARE, Indian Institute of Technology, Delhi, India. The work reported here was carried out under the project titled; "Design and Development of MEMS phase shifter and SPDT switches".

References

1. Rebeiz GM (2003) RF MEMS theory, design, and technology. Wiley, Hoboken
2. Dey S, Koul SK (2014) Design and development of a CPW-based 5-bit switched-line phase shifter using inline metal contact MEMS series switches for 17.25 GHz transmit/receive module application. J Micromech Microeng 24:1–24
3. Baek CW, Kim YK, Ahn Y, Kim YH (2005) Measurement of the mechanical properties of electroplated gold thin films using micromachined beam structures. Sens Actuat A Phys 117:17–27
4. Holm R (1969) Electric contacts: theory and applications. Springer, Berlin
5. Abbot E, Firestone F (1933) Specifying surface quantity—a method based on the accurate measurement and comparison. ASME Mech Eng 55:569
6. Peroulis D, Pacheco SP, Katehi LPB (2004) RF MEMS switches with enhanced power-handling capabilities. IEEE Trans Microw Theor Tech 52: 1881–1890
7. Laurent D, Rebeiz Gabriel M (2003) Intermodulation distortion and power handling in RF MEMS switches, varactors and tunable filters. IEEE Trans Microw Theor Tech 51:1247–1256
8. Jeffrey J, Adams George G, McGruer Nicol E (2005) Determination of intermodulation distortion in a contact-type MEMS microswitch. IEEE Trans Microw Theor Tech 53:3615–3620
9. Dey S, Koul SK (2013) Design and development of miniaturized high isolation MEMS SPDT switch for Ku-band T/R module application. In: IEEE MTT-S international microwave and RF conference (IMaRC 2013), 14–16th Dec 2013

Piezoelectric Actuators in Helicopter Active Vibration Control

R. Ganguli and S. R. Viswamurthy

Abstract Vibration is a major problem for helicopters due to flexible rotating blades, an airspeed varying radially and temporally at each rotor blade section, and a highly unsteady aerodynamic environment due to nonuniform wake and dynamic stall. The complexity of the forcing means that efforts to reduce vibration through vibration absorbers and isolators yield only meager vibration reduction and lead to a large weight penalty. The performance of such passive devices also degenerates quickly away from the tuned flight condition. Piezoelectric stack actuators offer the possibility of active vibration control in helicopter rotors, through the actuation of judiciously placed trailing edge flaps at appropriate higher harmonic multiples of the main rotor speed. Such actuators need to be coupled with amplification mechanisms to generate sufficient rotary motion of about 2 degrees which is needed to dramatically reduce vibration by 70–90 %. However, hysteresis is inherent in stack actuators, and this complicates their use in helicopter vibration control. In this chapter, we look at hysteresis compensation methods which can be used along with harmonic optimal control for vibration reduction in periodic rotating systems. The danger of ignoring hysteresis effects on the controller is also highlighted.

Keywords Piezostack · Hysteresis · Helicopter · Vibration · Control · Nonlinear

R. Ganguli (✉)
Department of Aerospace Engineering, Indian Institute of Science, Bangalore, India
e-mail: ganguli@aero.iisc.ernet.in

S. R. Viswamurthy
National Aerospace Laboratories, Council of Scientific and Industrial Research, Bangalore, India

K. J. Vinoy et al. (eds.), *Micro and Smart Devices and Systems*,
Springer Tracts in Mechanical Engineering, DOI: 10.1007/978-81-322-1913-2_7,

1 Introduction

Helicopters endure severe vibration levels compared to fixed wing aircraft. Most active vibration control systems in rotorcraft are designed to cancel the vibratory loads before they reach the airframe [1]. Individual blade control (IBC) can be implemented in several ways in a helicopter to reduce vibration. In the conventional IBC, the entire blade is excited in the pitch direction at its root using servo-actuators [2]. The shortcomings of this conventional IBC approach are high actuation power and high pitch link loads.

A modern implementation of IBC involves twisting the entire rotor blade by actuating embedded or surface mounted piezoelectric materials [3–7]. On-blade partial span trailing edge flaps can also be actuated at higher harmonics of the rotor rotational speed to generate new unsteady airloads. Trailing edge flap actuation needs much less actuator authority than twisting the full rotor blade. When phased properly, these unsteady airloads can lead to a considerable reduction in hub vibration [8, 9]. Each flap is individually controlled in the rotating frame. This approach appears as a strong contender for full-scale implementation due to its simplicity and enhanced airworthiness, since the flap control is typically independent of the primary control system [10].

Piezoceramic stack actuators are ideal for actuation of on-blade trailing edge flaps. They have high energy density and high bandwidth. Hall and Prechtl developed a servo-flap actuator and successfully conducted tests on a 1/6 Mach scaled model rotor system [11]. Lee and Chopra demonstrated the performance of a piezostack-based actuation device along with an amplification mechanism in an open-jet wind tunnel [12]. In other studies, bi-directional piezoelectric stack actuator designs were developed for actuation of on-blade trailing edge flaps [13, 14]. These bi-directional actuators can be operated in both push and pull directions and have higher free stroke, blocked force, and energy efficiency as compared to parallel-configured single-stack actuators [14]. However, a major limitation of piezoceramic actuators is the presence of nonlinearity and hysteresis. Hysteresis causes the piezoceramic expansion to depend not only on the current voltage excitation but also on the history of excitation. In addition, hysteresis in the amplification mechanism due to friction and freeplay can couple strongly with the piezoceramic material hysteresis to yield significant hysteresis at the structural level. Unmodeled hysteresis in the actuator can generate amplitude-dependent phase shifts in the output leading to inaccuracy in open-loop control and probably even loss of stability in closed-loop control [15].

Macroscopic models such as the classical Preisach model (CPM) are often used to characterize piezoceramic hysteresis. Lee et al. used a discretized Ishlinskii model (sometimes referred to as The Maxwell resistive capacitor model) to model the hysteresis in piezoceramic wafers [16]. They also proved that the Ishlinskii model and its inverse are a particular subset of CPM. In another paper, Lee et al. investigated the effect of piezoceramic hysteresis on structural vibration control of a simply supported beam with a surface bonded PZT wafer [17]. They used the discretized Ishlinskii hysteresis model in the PZT wafer and found that the piezoceramic hysteresis adversely affected the performance of vibration control schemes. A disadvantage of phenomenological models is the difficulty in updating the model parameters due to variations in operating conditions. Serpico and Visone used feed-forward neural networks to model ferromagnetic hysteresis [18]. The neural network was trained by back-propagation. Some other researchers have also used neutral networks for hysteresis modeling [19, 20].

Control strategies developed for systems with hysteresis often compensate for the hysteresis, for example, using an inverse hysteresis operator. Such a compensator can be a part of an open-loop filter, which yields a linear relationship between the input and output. However, developing an accurate hysteresis model of the system whose inverse can be determined is difficult. Galinaitis and Rogers derived closed-form expressions for the inverse operator for some bivariate Preisach models [21]. In another work, they used the Krasnoselskii-Pokrovskii (KP) operator to model hysteresis in a piezoelectric stack actuator. They constructed an approximate hysteresis model and the unknown model parameters were determined using major hysteresis loop data. Finally, they developed the inverse operator for the approximate hysteresis model and verified it using computer simulation [22].

In this chapter, we emphasize the effect of piezoceramic actuator hysteresis on the performance of the trailing edge flap system while also presenting an approach to minimize its undesirable effects. This chapter is organized in the following way: Sect. 2 describes a computational model of the helicopter rotor in forward flight. Section 3 briefly discusses the control algorithm used to obtain the flap motion. Section 4 describes the Preisach model and an approximate compensator. Section 5 presents the performance of the compensator and the deleterious effect of partial or inaccurate hysteresis compensation. A modified control algorithm that considers the actuator voltage as the control input (instead of the flap motion itself) is outlined in Sect. 6, the authors summarize this chapter. The material in this chapter is derived from past publications by the authors [23, 24].

2 Helicopter Mathematical Model

A mathematical model of the helicopter is needed to study the effect of hysteresis and trailing edge flaps on vibration reduction [24]. The helicopter is represented by a nonlinear aeroelastic model of several elastic rotor blades dynamically coupled to a rigid fuselage. Each blade experiences out-of-plane bending, in-plane bending, and torsion and axial displacement. Small strains and moderate deflections are assumed. The aerodynamic loads acting on the blade section and the flap are calculated for the forward flight condition.

The model is based on a generalized Hamilton's principle applicable to non-conservative systems:

$$\int_{\psi_1}^{\psi_2} (\delta U - \delta T - \delta W) d\psi = 0 \tag{1}$$

The δU, δT, and δW are the virtual strain energy, kinetic energy, and work, respectively. Beam finite elements are used to discretize the governing equations of motion. Each beam finite element has 15 degrees of freedom. The spatial functionality is eliminated by using finite element discretization, and partial differential equations are converted to ordinary differential equations. The finite element equations are transformed into modal space. Four out-of-plane, four in-plane bending, and two torsion modes are used in this chapter. The blade response is solved in modal space using a finite element in time. Eight time elements are used and fifth-order polynomials are used as shape functions. A convergence study is used to obtain the number of elements and modes. The blade nonlinear response and vehicle equilibrium equations are solved simultaneously. Steady and vibratory components of the rotating frame blade loads are calculated by integrating blade inertia and aerodynamic forces over the length of the blade. Fixed frame hub loads are calculated by summing the contributions of individual blades at the root.

3 Control Law for Optimal Flap Motion

Johnson's frequency domain algorithm is used to obtain the optimal flap motion to minimize hub vibrations [25]. The periodic nature of the helicopter blade response in steady forward flight permits the transformation of the control problem from the time domain to the frequency domain. For an N_b-bladed helicopter rotor, the trailing edge flap is most effective when deflected at $(N_b - 1)$, N_b, and $(N_b + 1)$ harmonics of the rotor rotational speed. The flap can be moved at any one of the above frequencies or by using a combination of all three harmonics using a harmonic control law. For a 4-bladed helicopter rotor considered in this chapter, the trailing edge flap is actuated according to the following *harmonic control law*:

$$\begin{aligned} \delta(\psi) = \delta_{3c}\cos(3\psi) + \delta_{3s}\sin(3\psi) + \delta_{4c}\cos(4\psi) \\ + \delta_{4s}\sin(4\psi) + \delta_{5c}\cos(5\psi) + \delta_{5s}\sin(5\psi) \end{aligned} \tag{2}$$

An optimal control algorithm is used to find the unknown flap harmonics in the above equation. The algorithm minimizes a scalar objective function based on the six 4/rev (4Ω) hub vibratory loads (Z) and six flap control harmonics (u).

$$\begin{aligned} J &= Z^T W_Z Z + u^T W_u u \\ Z &= \begin{bmatrix} F_{xH} & F_{yH} & F_{zH} & M_{xH} & M_{yH} & M_{zH} \end{bmatrix}_{4p}^T \\ u &= \begin{bmatrix} \delta_{3c} & \delta_{3s} & \delta_{4c} & \delta_{4s} & \delta_{5c} & \delta_{5s} \end{bmatrix}^T \end{aligned} \tag{3}$$

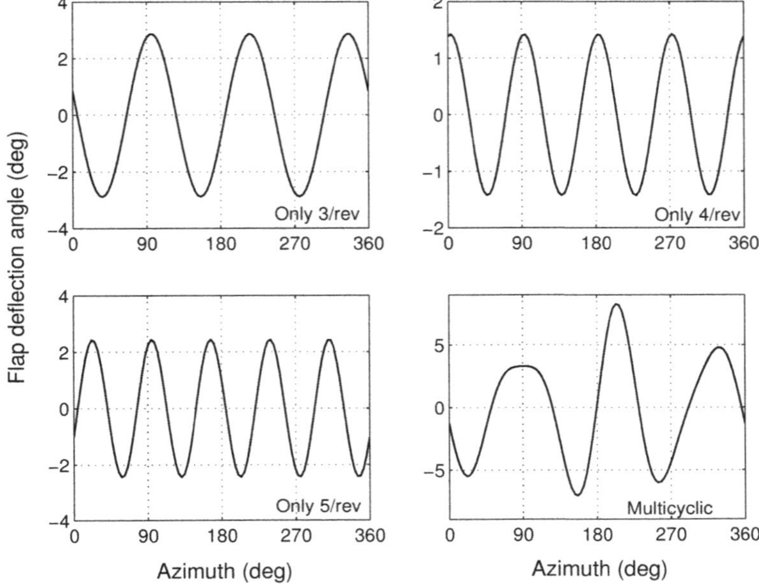

Fig. 1 Desired flap motion for individual and multicyclic inputs obtained from optimal controller

Here W_Z and W_u are weighting matrices for the hub vibratory loads and flap control harmonics, respectively. They are adjusted such that the flap deflection angles are within practically achievable limits for piezoelectric actuators. A *global controller* is used to determine the optimal flap motion.

The optimal motion of a trailing edge flap located at a 70 % blade span on a 4-bladed hingeless rotor is shown in Fig. 1. The properties of the rotor are given in Table 1. Figure 1 shows the desirable flap motion when 3/rev (δ_{3c}, δ_{3s}), 4/rev (δ_{4c}, δ_{4s}), and 5/rev (δ_{5c}, δ_{5s}) control input are applied independently and when applied together (δ_{3c}, δ_{3s}, δ_{4c}, δ_{4s}, δ_{5c}, δ_{5s}, multicyclic input) at a forward speed of 60 m/s.

4 Hysteresis Model and Compensation

Hysteresis in the actuator is modeled using the CPM. The Preisach model was originally developed to model hysteresis in ferromagnetic materials [26]. Krasnoselskii disembodied Preisach's model from its physical roots and morphed it into a pure mathematical form [27]. As a result, this model is now used for the mathematical description of hysteresis from any system. The CPM has several attractive features such as its ability to model complex hysteresis types, an identification algorithm, and a numerical simulation form [28]. A brief explanation of the model is given below.

Table 1 Baseline blade and flap properties

Blade properties	
N_b	4
c/R	0.055
Solidity, σ	0.07
Lock number, γ	5.2
C_T/σ	0.07
$EI_y/m_0\Omega^2R^4$	0.0108
$EI_z/m_0\Omega^2R^4$	0.02686
$GJ/m_0\Omega^2R^4$	0.00615
m_0 (kg/m)	6.46
Ω (rpm)	383
R (m)	4.94
Trailing edge flap properties	
L_f/R	0.12
c_f/c	0.2
m_f/m_0	0.1
X_g^f/c_f	0.2

4.1 Classical Preisach Model

The CPM can be represented in the following mathematical form:

$$\delta(\psi) = \Gamma[u(\psi)] = \iint_{\alpha \geq \beta} \mu(\alpha, \beta)\hat{\gamma}_{\alpha\beta}[u(\psi)]\mathrm{d}\alpha\mathrm{d}\beta \tag{4}$$

where Γ is the Preisach operator, $\hat{\gamma}_{\alpha\beta}$ is the elementary hysteresis operator (also called hysteron), δ, i.e., the output of the actuator (flap deflection), u is the voltage input to the actuator, and ψ is the nondimensional time measure (rotor azimuth). The output of each elemental hysteresis operator is a rectangular loop on the input-output diagram switching from -1 to $+1$ when the input increases above the threshold α. The output switches from $+1$ to -1 when the input decreases below the value of β as shown in Fig. 2. The weighting function, $\mu(\alpha, \beta)$ is a characteristic of the hysteresis transducer. Equation (4) can be solved directly by integration if the weighting function is known.

The function $\mu(\alpha, \beta)$ may not be known explicitly. Then, experimental data are used to estimate the weighting function through geometric interpretation and numerical implementation [28]. As a first step, the unknown weighting function $\mu(\alpha, \beta)$ is estimated using experimental data from the X-frame actuator developed by Hall and Prechtl [11]. Figure 3a shows only the piezostack material hysteresis of the X-frame actuator. Figure 3b shows the combined effect of both the material and mechanical hysteresis of the actuator, which is significantly greater. Piezoelectric stack actuators clearly show considerable hysteresis.

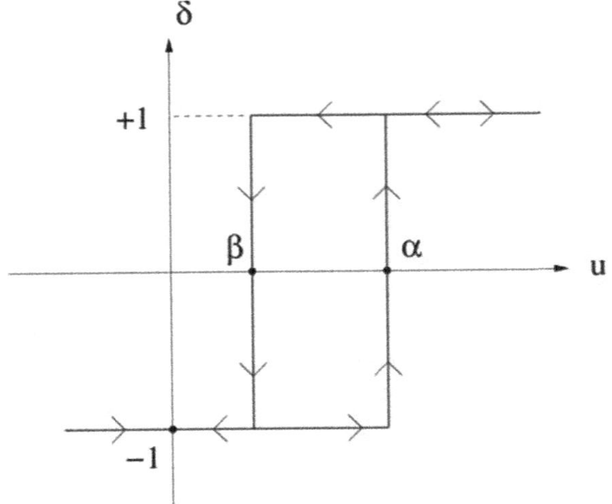

Fig. 2 Elementary hysteresis operator, also called hysteron

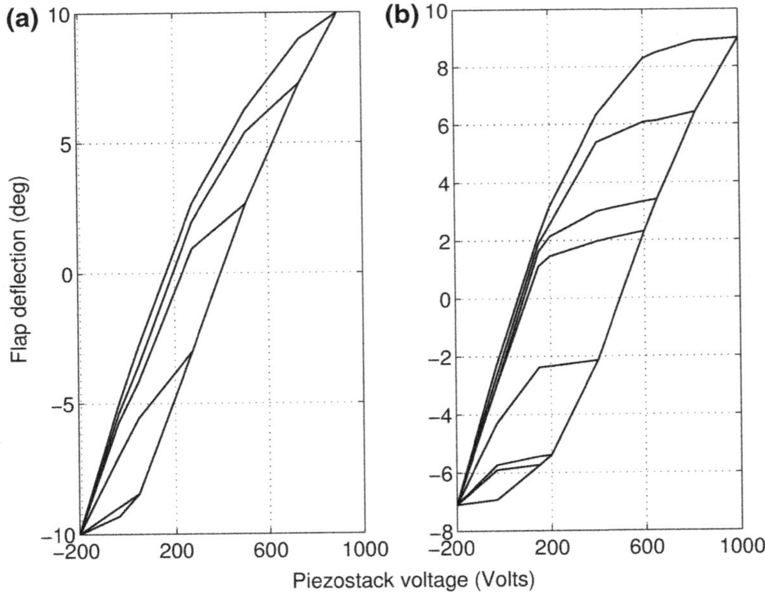

Fig. 3 Trailing edge flap actuator hysteresis from Hall and Prechtl [11]. **a** Only material hysteresis and **b** both material and mechanical hysteresis

Equation 4 can be written as:

$$\delta(\psi) = \iint\limits_{S+(\psi)} \mu(\alpha,\beta)\mathrm{d}\alpha\mathrm{d}\beta - \iint\limits_{S-(\psi)} \mu(\alpha,\beta)\mathrm{d}\alpha\mathrm{d}\beta \tag{5}$$

where $S^{+}(\psi)$ is the set of all points (α, β) in the half-plane $\alpha \geq \beta$ for which $\hat{\gamma}_{\alpha\beta}[u(\psi)] = 1$and where $S^{-}(\psi)$ is the set of all points (α, β) in the half-plane $\alpha \geq \beta$ for which $\hat{\gamma}_{\alpha\beta}[u(\psi)] = -1$. The interface between $S^{+}(\psi)$ and $S^{-}(\psi)$ is a staircase line whose vertices have α and β coordinates coinciding with the local maxima and minima of inputs at previous instants of time. This interface ϕ is a function of time and is also known as the *memory curve*, since it contains the information about the state of any hysteron. A detailed description of the model is given in Ref. [28].

4.2 Hysteresis Compensation

We seek the appropriate piezostack voltage using the compensation algorithm that yields the desired flap motion shown in Fig. 1. An "ideal" compensator placed in series with a hysteretic system makes the overall system "exactly" linear. In such a case, linear control theory techniques are used to control the actuator. However, it is difficult to develop an exact compensation scheme. So while the system cannot be made exactly linear, it can be made approximately linear. In this chapter, an approximate inverse of the Preisach operator is used to compensate for actuator hysteresis. This nonlinear inversion algorithm called the *closest match algorithm* was proposed by Tan et al. [29]. They discretized the Preisach plane ($\alpha - \beta$ plane) into a finite number of grids. The algorithm determines the input whose output matches the desired output most closely among all possible inputs. Since the Preisach plane is discretized, the input can only take values from a finite set $U = \{u_l, 1 \leq l \leq L\}$ with each $u_l, 1 \leq l \leq L$, representing an input level. We define,

$$\Delta u = \frac{u_{\max} - u_{\min}}{L - 1} \tag{6}$$

Then $u_l = u_{\min} + (l - 1)\Delta u$. Given the initial memory curve $\phi^{(0)}$ (from which the initial input $u^{(0)}$ and the output $\delta^{(0)}$ can be derived) and a desired output δ^{*}, the inverse operator seeks to determine the input $u^{*} \in U$, such that:

$$\left|\Gamma\left(u^{*};\varphi^{(0)}\right) - \delta^{*}\right| = \min_{u\in U}\left|\Gamma\left(u;\varphi^{(0)}\right) - \delta^{*}\right| \tag{7}$$

The algorithm should also return the resulting memory curve ϕ^{*}. The algorithm is as follows:

- Step 0. Set n = 0. Compare $\delta^{(0)}$ and δ^* : if $\delta^{(0)} = \delta^*$, let $u^* = u^{(0)}$, $\phi^* \approx \phi^{(0)}$, go to step 3; if $\delta^* \approx \delta^{(0)}$, go to step 1.1; otherwise go to step 2.1;
- Step 1.
  - Step 1.1: If $u^n = u_L$, let $u^* = u^n$, $\phi^* = \phi^n$, go to step 3; otherwise $u^{(n+1)} = u^{(n)} + \Delta u$, $\bar{\phi} = \phi^{(n)}$ [backup the memory curve], $n = n + 1$, go to step 1.2;
  - Step 1.2: Evaluate $\delta^{(n)} = \Gamma\left(u^n; \phi^{(n-1)}\right)$ and (at the same time) update the memory curve to $\phi^{(n)}$. Compare δ^n with δ^*: if $\delta^{(n)} = \delta^*$, let $u^* = u^n$, $\phi^* = \phi^n$, go to step 3; if $\delta^{(n)} < \delta^*$, go to step 1.1; otherwise go to step 1.3;
  - Step 1.3: If $\left|\delta^{(n)} - \delta^*\right| \leq \left|\delta^{(n-1)} - \delta^*\right|$, let $u^* = u^n$, $\phi^* = \phi^n$, go to step 3; otherwise $u^* = u^{(n-1)}$, $\phi^* = \tilde{\phi}$ [restore the memory curve], go to step 3;
- Step 2.
  - Step 2.1: If $u^n = u_1$, let $u^* = u^n$, $\phi^* = \phi^n$, go to step 3; otherwise $u^{(n+1)} = u^{(n)} - \Delta u$, $\tilde{\phi} = \phi^{(n)}$ [backup the memory curve], $n = n + 1$, go to step 2.2;
  - Step 2.2: Evaluate $\delta^{(n)} = \Gamma\left(u^n; \phi^{(n-1)}\right)$ and (at the same time) update the memory curve to $\phi^{(\mathrm{n})}$. Compare δ^n with δ^*: if $\delta^{(n)} = \delta^*$, let $u^* = u^n$, $\phi^* = \phi^n$, go to step 3; if $\delta^n < \delta^*$, go to step 2.1; otherwise go to step 2.3;
  - Step 2.3: If $\left|\delta^{(n)} - \delta^*\right| \leq \left|\delta^{(n-1)} - \delta^*\right|$, let $u^* = u^n$, $\phi^* = \phi^n$, go to step 3; otherwise $u^* = u^{(n-1)}$, $\phi^* = \bar{\phi}$ [restore the memory curve], go to step 3;
- Step 3. Exit.

  The above algorithm yields the best input u^* within L iterations.

5 Effect of Incomplete/Inaccurate Compensation

Three different cases are considered to study the effect of material and mechanical hysteresis on the performance of the helicopter vibration control system (Fig. 4). They represent different levels of modeling and compensation of X-frame actuator hysteresis:

1. Case (1): The compensator is derived based on complete actuator nonlinearity (Fig. 3b). This model is more accurate than the other two cases.
2. Case (2): The compensator is derived based only on material nonlinearity in the actuator (Fig. 3a).
3. Case (3): No compensation is made for the actuator hysteresis and, hence, this is the least accurate of the three models. However, it is simple to implement

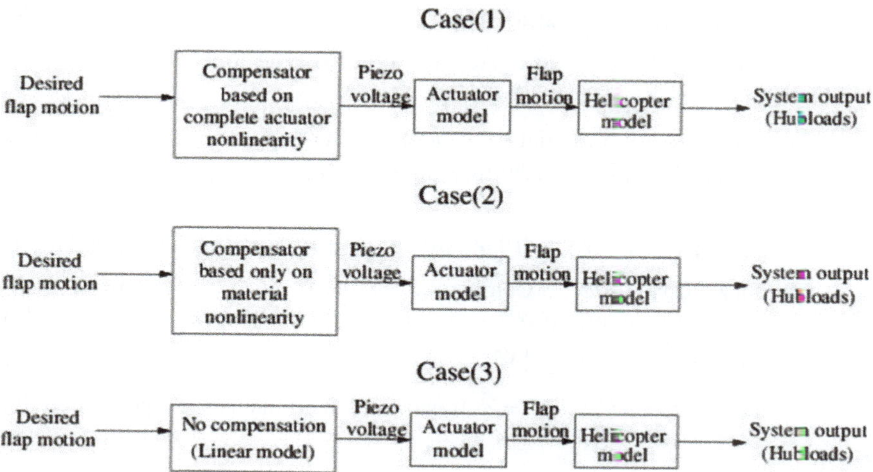

Fig. 4 Three different hysteresis compensation levels

since it does not involve an inversion of the hysteresis model. Case (3) highlights the pitfalls of ignoring hysteresis.

5.1 3Ω Control Input

Initially, the desired 3/rev flap motion from Fig. 1 is given as input to the three different compensators mentioned above. Figure 5 shows the piezostack voltage obtained from the compensators. The voltage history is different in all three cases. The actuator response to the voltage inputs are shown in Fig. 6. The desired flap motion history is also shown in this figure. The result from case (1) is almost identical to the desired flap motion, since the compensator includes complete actuator nonlinearity. The helicopter aeroelastic model predicts a reduction of 79 % in hub vibration levels due to this flap motion. The actuator responses in cases (2) and (3) deviate substantially from the desired flap motion due to the use of inaccurate compensation. The helicopter aeroelastic model predicts a 42 and 36 % reduction in hub vibration in case (2) and case (3), respectively. Figure 7 shows the reduction in the 4/rev component of individual hub loads.

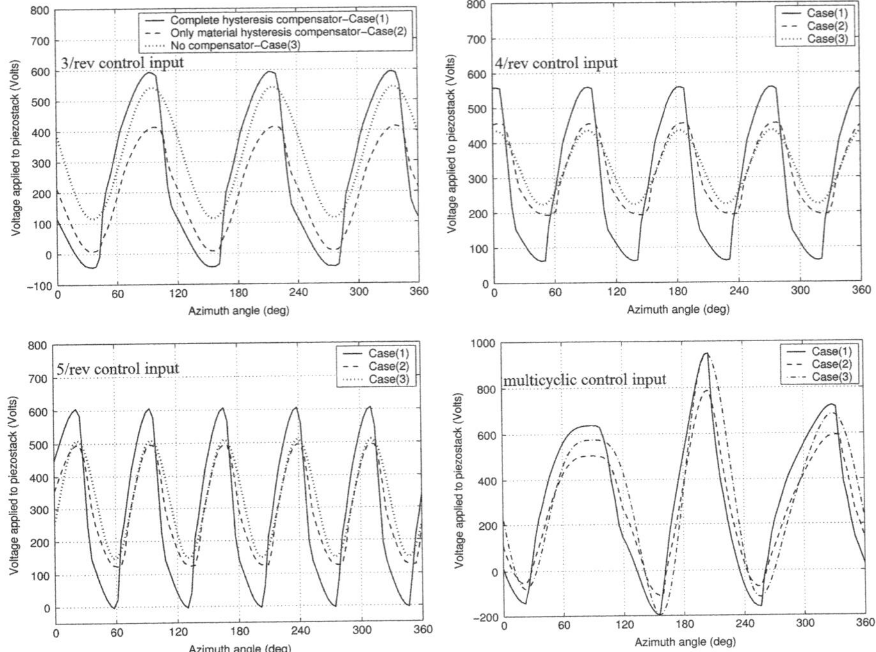

Fig. 5 Piezostack voltage computed from the compensators

5.2 4Ω Control Input

Figure 5 shows the time history of the voltage input to the piezostack. Again, there is a notable difference in the magnitude and phase of the voltage input obtained for the three cases. The flap motion obtained in case (1) is almost identical to the desired flap motion, since the compensator in this case was based on complete hysteresis inversion (Fig. 6). The helicopter aeroelastic model predicts about a 70 % reduction in 4/rev hub loads due to this flap motion. However, the flap motions in the other two cases are entirely different from the desired flap motion, thereby leading to a very poor performance in terms of helicopter hub vibration reduction (Fig. 7). In case (2), the overall hub vibration was reduced by only 14 %, and in case (3), it was even lower at 8 %.

5.3 5Ω Control Input

The voltage input and flap motion for the 5/rev control input are also shown in Figs. 5 and 6, respectively. Again, it is seen that the flap motion due to complete

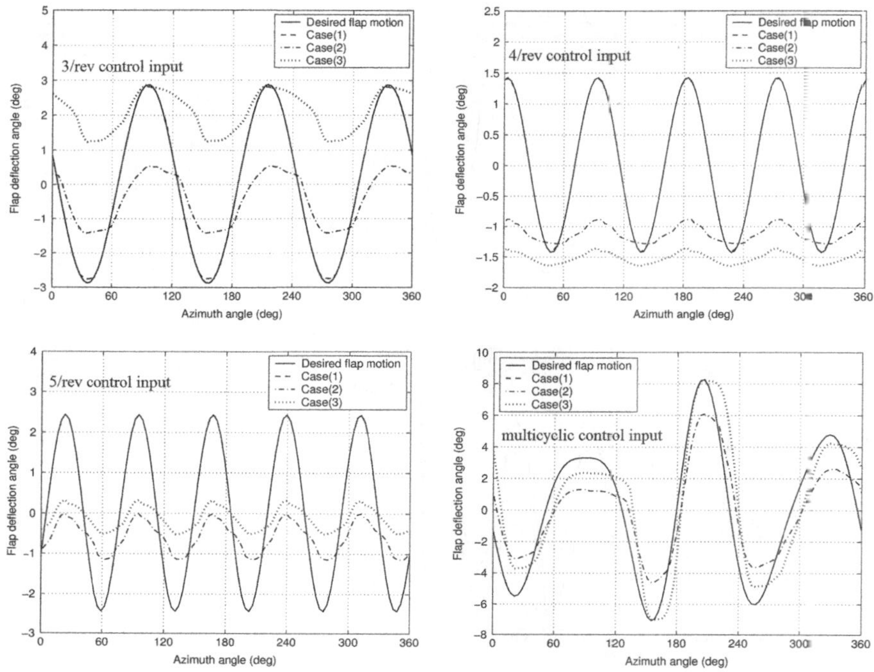

Fig. 6 Flap motion history for the three cases shown in Fig. 4

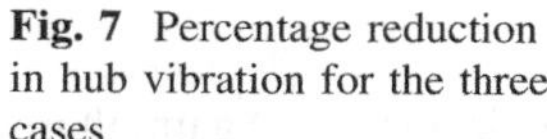
Fig. 7 Percentage reduction in hub vibration for the three cases

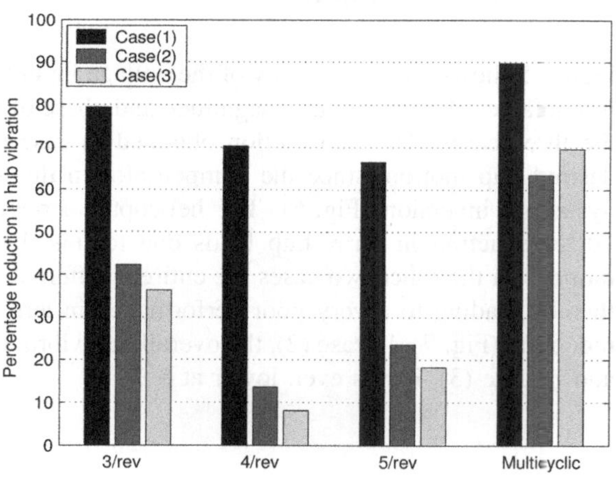

hysteresis compensation is identical to the desired flap motion. In this case, the hub vibration is reduced by about 66 % from baseline conditions. Hub vibration levels were reduced by about 24 and 18 % from baseline conditions in the case of material hysteresis compensation and that of no compensation, respectively. This is due to the large difference between the actual flap motion in these two cases and the desired flap motion. Figure 7 shows the reductions in individual hub vibratory forces and moments.

5.4 *Multicyclic Control Input (3Ω, 4Ω, and 5Ω)*

In this section, 3, 4, and 5/rev harmonics of the control input are given simultaneously (Eq. 2). Figure 5 shows the piezostack voltage obtained from the compensators. The actuator response along with the desired flap motion is shown in Fig. 6. In case (1), where the compensator is based on complete actuator hysteresis inversion, the hub vibration is reduced by about 90 % from baseline conditions.

Figure 7 summarizes the numerical results obtained in terms of percentage reduction in hub vibration ($Z^T Z$) for all three cases. It is evident that partial or no compensation of the actuator hysteresis (case 2, case 3) leads to deterioration in the performance of the vibration control system. The compensator in case (2) performed poorly due to significant mechanical hysteresis in the X-frame actuator compared to the piezostack material hysteresis (Fig. 3). The hysteresis in the amplification mechanism due to friction and freeplay can be ameliorated by reducing the number of joints or by using compliant mechanisms. However, since it is difficult to obliterate mechanical hysteresis, both material and mechanical hysteresis must be addressed when using piezostack actuators for vibration control

6 Summary

Hysteresis and nonlinearity with memory, are key traits of piezoceramic stack actuators. Such actuators are the leading contenders for use in helicopter vibration control for actuating trailing edge flaps. In this chapter, the effect of hysteresis of piezoceramic stack actuators on helicopter vibration control is investigated. Data are obtained from the published literature and a classical Preisach hysteresis model is used. The hysteresis models are integrated with a mathematical model of a helicopter rotor based on finite elements in space and time. An optimal control algorithm is then used to study the effect of hysteresis on helicopter vibration control using trailing edge flaps. The possibility of compensating for the hysteresis effect is studied. Numerical results are obtained with single and multicyclic harmonic inputs at low and high speed flight. The importance of the hysteresis phenomenon on helicopter vibration control is quantified and clearly brought out. The research discussed in this chapter is useful for the design and development of smart rotor systems.

Acknowledgments The authors are grateful for the NPMASS project titled *Adaptive Trailing Edge Flaps for Active Flow Control* (PARC #3.1) for funding part of this research.

References

1. Nguyen K, Chopra I (1990) Application of higher harmonic control to rotor operating at high speed and thrust. J Am Helicopter Soc 35:336–342
2. Friedmann PP, Millott TA (1995) Vibration reduction in rotorcraft using active control: a comparison of various approaches. J Guid Control Dyn 18:664–673
3. Thakkar D, Ganguli R (2004) Dynamic response of rotating beams with piezoceramic actuation. J Sound Vib 270:729–753
4. Thakkar D, Ganguli R (2006) Use of single crystal and soft piezoceramics for alleviation of flow separation induced vibration in a smart helicopter rotor. Smart Mater Struct 15:331
5. Thakkar D, Ganguli R (2006) Single-crystal piezoceramic actuation for dynamic stall suppression. Sens Actuators 128:151–157
6. Thakkar D, Ganguli R (2007) Induced shear actuation of helicopter rotor blade for active twist control. Thin-walled structures 45:111–121
7. Cesnik ECS, Shin S, Wilbur ML (2001) Dynamic response of active twist rotor blades. Smart Mater Struct 10:62
8. Ravichandran K, Chopra I et al (2013) Trailing-edge flaps for rotor performance enhancement and vibration reduction. J Am Helicopter Soc 58:1–13
9. Viswamurthy S, Ganguli R (2004) An optimization approach to vibration reduction in helicopter rotors with multiple active trailing edge flaps. Aerosp Sci Technol 8:185–194
10. Maurice J-B, King FA, Fichter W (2013) Derivation and validation of a helicopter rotor model with trailing-edge flaps. J Guid Control Dyn 36:1375–1387
11. Hall SR, Prechtl EF (1999) Preliminary testing of a mach-scaled active rotor blade with a trailing edge servo-flap. In: Proceedings of SPIE conference on smart structures and materials, Newport Beach, p 14–21
12. Lee T, Chopra I (2001) Design of piezostack-driven trailing edge flap actuator for helicopter rotors. Smart Mater Struct 10:15–24
13. Straub FK, Charles BD (2001) Aeroelastic analysis of rotors with trailing edge flaps using comprehensive codes. J Am Helicopter Soc 46:192–199
14. Heverly DE, Wang KW, Smith EC (2004) Dual-stack piezoelectric device with bidirectional actuation and improved performance. J Intell Mater Syst Struct 15:565–574
15. Kurdila AJ, Li J, Strganac T et al (2003) Nonlinear control methodologies for hysteresis in PZT actuated on-blade elevons. J Aerospace Eng 16:167–176
16. Lee SH, Royston TJ, Friedman G (2000) Modeling and compensation of hysteresis in piezoceramic transducers for vibration control. J Intell Mater Syst Struct 11:781–790
17. Lee SH, Ozer MB, Royston TJ (2002) Piezoceramic hysteresis in the adaptive structural vibration control problem. J Intell Mater Syst Struct 13:117–124
18. Serpico C, Visone C (1998) Magnetic hysteresis modeling via feed-forward neural networks. IEEE Trans Magn 34:623–628
19. Sunny Mohammed R, Kapania RK (2014) Artificial neural network based identification of a modified dynamic preisach model. Int J Comput Methods Eng Sci Mech 15:45–53
20. Wang Xiangjiang, Alici G et al (2014) Modeling and inverse feedback control for conducting polymer actuators with hysteresis. Smart Mater Struct 23:25015–25023
21. Galinaitis WS, Rogers RC (1997) Compensation for hysteresis using bivariate preisach models. Proc SPIE—Smart Struct Mater 3039:538−547
22. Galinaitis WS, Rogers RC (1998) Control of a hysteretic actuator using inverse hysteresis compensation. Proc SPIE—Smart Struct Mater 3323:267−277
23. Viswamurthy S, Ganguli R (2007) Modeling and analysis of piezoceramic actuator hysteresis for helicopter vibration control. Sens Actuators 35:810–810
24. Viswamurthy S, Ganguli R (2006) Effect of piezoelectric nonlinearity on helicopter vibration control using trailing edge flaps. AIAA J Guidance Control Dyn 29:1201–1209
25. Johnson W (1982) Self-tuning regulators for multicyclic control of helicopter vibration. NASA technical paper

26. Preisach F (1935) On magnetic aftereffect. Zeitschrift fur Physiks 94:277–302
27. Krasnoselskii MA, Pokrovskii AV (1989) Systems with hysteresis. Springer, New york, p 1−57
28. Doong T, Mayergoyz ID (1985) On numerical implementation of hysteresis models. IEEE Trans Magn 21:1853–1855
29. Tan X, Venkataraman R, Krishnaprasad PS (2001) Control of hysteresis: theory and experimental results. Proc SPIE—Smart Struct Mater: Model Signal Process Control Smart Struct 4326:101−112

Design and Development of a Piezoelectrically Actuated Micropump for Drug Delivery Application

Paul Braineard Eladi, Dhiman Chatterjee and Amitava DasGupta

Abstract Micropumps form the heart of several microfluidic systems like micro total analysis system (μTAS) and drug delivery devices, which have resulted from the advancement of silicon micromachining technology. Among the different available types of micropumps, valveless micropumps are better suited for biological applications as they do not have flow-rectifying valves and are less prone to clogging and wear. However, their main drawback is low thermodynamic efficiency. This can be improved if we have a better understanding of the effects of geometry on the performance. This forms one of the objectives of this work. This chapter describes the activity on the design and development of valveless micropumps. A numerical parametric study of the performance of valveless micropumps has been carried out and is presented to bring out the effects of different geometrical parameters. Based on these design approaches, silicon-based micropumps are fabricated and characterized. The performance of one of these micropumps is compared with designed value in this work.

Keywords Piezoelectric actuator · Valveless micropump · Drug delivery · Micromachining · Numerical · Experimental characterization

1 Introduction

Microfluidics is the science and engineering of systems that manipulate small (10^{-9} *to* 10^{-18} l) *amounts of fluids using channels with dimensions of tens to hundreds of micrometers* [1]. This approach of transport and manipulation of fluids at micron

P. B. Eladi · A. DasGupta (✉)
Department of Electrical Engineering, Indian Institute of Technology Madras, Chennai, India
e-mail: adg@ee.iitm.ac.in

D. Chatterjee
Department of Mechanical Engineering, Indian Institute of Technology Madras, Chennai, India

K. J. Vinoy et al. (eds.), *Micro and Smart Devices and Systems*, Springer Tracts in Mechanical Engineering, DOI: 10.1007/978-81-322-1913-2_8,

scale has found many applications in the areas of chemical synthesis, environmental testing, electronic device cooling [2–4], and pertinent to present discussion, in the life sciences [5]. The usefulness of a microfluidic system lies in its ability to integrate different components like actuators, mixers, valves, pumps, etc., for efficient fluid transport and control, which eventually led to the development of micro total analysis systems (μTAS), commonly known as Lab-On-a-Chip (LOC) [5] and drug delivery systems [6]. At the heart of a μTAS or a microfluidic drug delivery system is a micropump, which pumps fluid with flowrates of microliters per minute (μl/min) to nanoliters per minute (nl/min). Brand [7] referred to such micropumps as the "beating heart" of microfluidics.

The first patent on a miniaturized conventional pump for drug delivery was by Thomas and Bessman [8] in 1975. However, it was not until 1984 that the first patent on a micropump based on silicon microfabrication technology was filed by Smits [9] and the results were published in 1990 [2]. The micropump was a peristaltic pump consisting of three active valves actuated by piezoelectric disks. The device was intended for drug delivery. In the meantime, the first diaphragm micropump with passive check-valves was presented by van Lintel et al. in 1988 [10]. The diameter of the glass membrane was 12.5 mm and the diameter and the thickness of the actuating piezoelectric disk was 10 mm and 0.2 mm, respectively. A single silicon wafer of 2″ diameter was used for fabricating the device that had two check-valves and a membrane. This publication demonstrated the feasibility of silicon-based micropumps and marked the beginning of extensive research in the use of a micropump in various MEMS devices across various fields of application.

Micropumps can be classified as mechanical or nonmechanical based on actuation principles. Valveless micropump, a type of mechanical micropump, does not have any check valve to regulate the flow. In comparison with micropump having moving valves, a valveless micropump has the advantages of better reliability, less risk of clogging, and a long life. Instead of valves, these pumps have flow-rectifying conduits which cause direction-dependent differential flow losses inside them, thereby ensuring a net flow out of the pump. Quite often, diffusers/nozzles are used as flow-rectifying conduits [10]. Depending on the position of the rectifying elements relative to the chamber, reciprocating pumps can be divided into two types, planar and nonplanar. A particular nonplanar pump described in this work is referred to as a pyramidal pump, because of the pyramidal shape of the nozzle/diffuser elements.

The structure and functioning of a valveless micropump can best be understood with the help of Fig. 1. A valveless micropump has an oscillating membrane driven by a piezoelectric or electrostatic actuator. Pump chamber connects to one end of the nozzle/diffuser element while the other end of nozzle/diffuser is connected with tubing. A valveless micropump is a positive-displacement pump with suction and delivery strokes. In the suction stroke, shown in Fig. 1a, the upward movement of the membrane from undeflected position increases the chamber volume and fluid enters the pump chamber due to fall in pressure inside the chamber. Because of the difference in flow losses in the inlet and outlet sections,

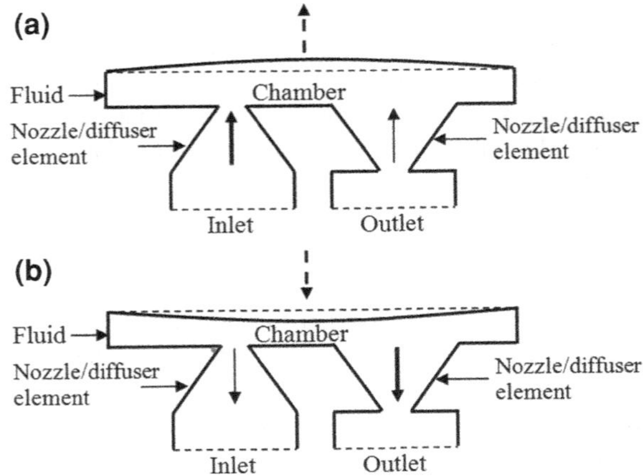

Fig. 1 The operation of valveless micropump during **a** suction and **b** delivery modes marked by the upward or downward movement of membrane from undeflected positions respectively

more fluid enters the chamber through the inlet than through the outlet. In the delivery stroke shown in Fig. 1b, the downward movement of the membrane reduces the effective chamber volume and more fluid comes out of the outlet than through the inlet. Therefore, in a complete cycle there is a net flow of fluid from the inlet to the outlet. The successful operation of this micropump, therefore, depends on the efficiency of the flow rectification process.

The first piezoelectrically actuated valveless micropump by Stemme and Stemme [11] was fabricated in brass, and its diameter was 19 mm. The pump had a nozzle and diffuser as flow-rectifying elements with a small opening angle of 20°. Gerlach and Wurmus [12] fabricated a valveless micropump by bulk micromachining of silicon. A single-crystal silicon of (100) orientation was used that gave an angle of 70.52° opening for the rectifying elements with anisotropic etching. They worked on glass membranes with different sides of 11, 7, and 5 mm. A micropump, fabricated by self-aligning the membrane unit to the valve unit was fabricated by Schabmueller et al. [13]. The pump was bubble tolerant and the PZT was fabricated using screen-printing technology. Recently valveless micropumps have been fabricated using a variety of materials and fabrication processes for various applications. The structures are silicon-glass [14], silicon-silicon [13], brass-PCB (printed circuit board)-copper [15], PMMA (polymethylmethacrylate)-PDMS magnetic membrane [16], glass-PDMS (polydimethylsiloxane) [17], etc.

The design and development of an efficient micropump requires a proper understanding of the interaction of flow with different components which, in turn, is influenced by the geometrical parameters involved. In the past, it has been observed that different researchers have undertaken experimental, analytical or numerical simulations to arrive at a working geometry of valveless micropumps. If the objective of a work is to arrive at micropumps with improved performance, then we need to vary different geometrical parameters while keeping the remaining parameters constant. This is difficult to achieve experimentally and hence, most of the work on micropump fabrication and characterization discussed above was restricted to one geometry of a micropump. Thus, this goal can only be achieved through analytical and/or numerical simulations. Though the working of a valveless micropump appears simple, yet the mathematical description of fluid flow in it is complex owing to the fact that the moving membrane influences the flow and the fluid pressure inside the pump chamber affects the displacement of the membrane. This is termed a two-way coupled fluid-structure interaction problem and needs to be addressed in analytical/numerical approaches.

Of the two methods, the analytical method makes simplifying assumptions and can quickly produce results quantifying the effect of different geometrical parameters on the performance of micropump. Such approaches have been done in the past by several researchers [18–24] in the form of an electrical network or mechanical connections. The analytical approach is simple and easy to implement but it suffers from the fact that the effects of geometrical parameters on flow losses at different operating frequencies is difficult to ascertain. This can be obtained numerically.

Hence, many researchers have attempted a complete numerical solution of the fluid-structure interaction (FSI) problem [25–31]. A careful survey of these works reveal that this approach, though accurate, poses a challenge, not only in terms of huge computational resources (memory, computational time and hard disk space) but also suffers from numerical instability, particularly at higher operating frequencies. This has prompted many researchers to assume or determine, experimentally, membrane deflection and utilize this information to decouple two-way coupled fluid-structure interaction problems into one where fluid flow is only solved with boundary conditions posed by time-dependent membrane deflection information [32–36].

The aim of this work is to bring out the effect of different geometrical parameters on the performance of the micropump and come out with a recommendation of the design of a valveless micropump. The next step is to fabricate a micropump using silicon bulk micromachining technology, which is reliable and also has fast response time. Therefore, a valveless micropump with piezoelectric actuation is chosen.

Numerical prediction and design recommendation of a piezoelectrically excited valveless micropump is briefly discussed in Sect. 2 and a detailed description of the same is available in the work of Paul et al. [37]. The fabrication process is discussed in detail in Sect. 3. In Sect. 4, the characterization results are presented. Finally, the conclusions are discussed in Sect. 5.

2 Numerical Prediction and Design of Valveless Micropump

The different significant parameters of micropump geometry investigated in this work are shown in Fig. 2. In all the numerical results reported here, the pump structure has a square membrane of silicon whose area is 5 × 5 mm with a

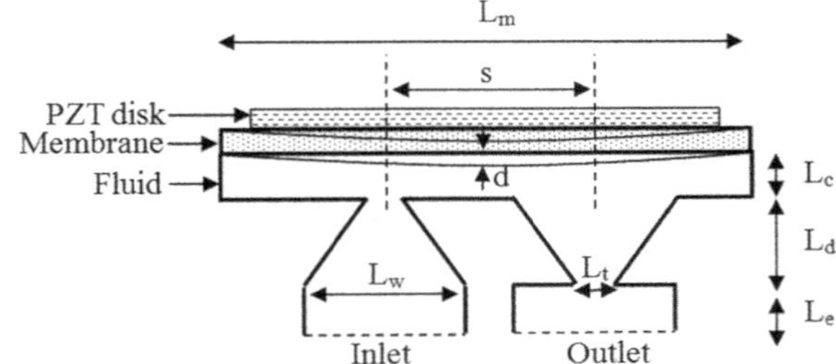

Fig. 2 Micropump with structural region formed by PZT disk and membrane and fluid region with nozzle/diffuser elements

thickness of 50 μm. The pump chamber has a base area of 5 × 5 mm. The half-divergence angle of the nozzle/diffuser elements is 35.3°, which is typical in nozzle elements fabricated using crystallographic etchants in MEMS technology. The length of extensions (L_e) is 200 μm with cross-sectional area being the same as the nozzle wider area. Water is the working fluid. The performance of the pump has been studied for geometrical variations of chamber height (L_c), pyramidal nozzle/diffuser element height (L_d), nozzle/diffuser throat width (L_t), and separation (s) between the nozzle/diffuser elements. The results [37] obtained using two types of numerical approaches—FSI and assumed membrane displacement—are reported here.

The effect of pump chamber height is shown in Fig. 3. It is seen that the maximum flowrate is obtained for pump chamber heights between 50 and 60 μm. This is because too short a height of the pump chamber offers more flow losses while too large heights produce extensive regions with recirculation. This figure also shows that assumed displacement results agree well with the FSI predictions.

Figure 4 shows the effects of nozzle height. Similar simulations were carried out for different throat areas, nozzle heights, and separation distances between nozzle and diffuser openings inside the pump chamber. Based on the simulation results and keeping practical considerations of fabrication using MEMS technology and packaging in mind, geometry of a pyramidal valveless micropump with 5 × 5 mm membrane has $L_c = 50$ μm, $L_d = 280$ μm, $L_w = 496.5$ μm, $L_e = 200$ μm, and s = 2.5 mm is recommended. This micropump was fabricated and characterized and this fabrication process is described in the next section.

3 Fabrication Process

The fabrication of a valveless micropump involves the processing of two silicon wafers and one glass wafer separately, followed by bonding the three wafers into one composite structure. The piezoelectric crystal is then fixed to the membrane on one side and the inlet/outlet tubes to the glass wafer on the other side. The details of the fabrication steps are given below.

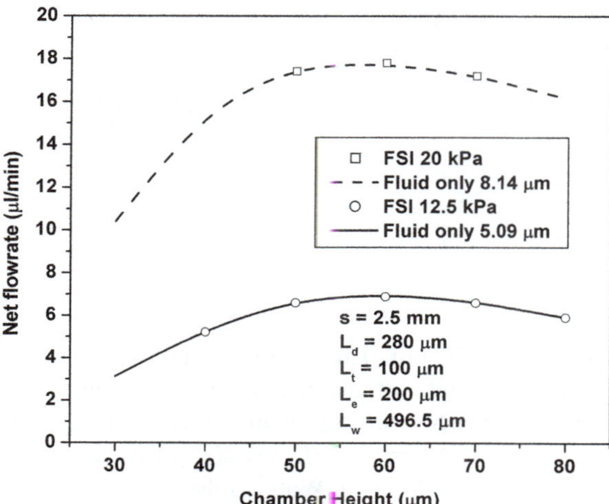

Fig. 3 Effect of chamber height on net flowrate

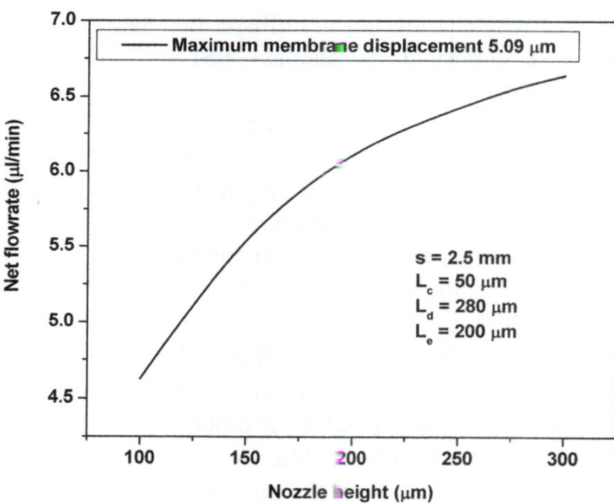

Fig. 4 Effect of nozzle height on net flowrate as predicted by assumed membrane displacement method

3.1 Fabrication of Top Silicon Wafer

The starting wafer is a 200 μm thick p-type double side polished silicon wafer of (100) orientation. After cleaning the wafers using a standard procedure, a 0.9 μm thick silicon dioxide layer is grown on all sides using wet oxidation at 1000 °C for 4 h. The cross-sectional view of the wafer after this step is shown in Fig. 5a. This is followed by photolithography using positive photoresist (PPR) and Mask #1, which has 5 × 5 mm openings. The oxide layer is etched using buffered hydrofluoric acid (BHF), followed by the stripping of PPR. The cross-sectional views of the wafer after these steps are shown in Fig. 5b, c.

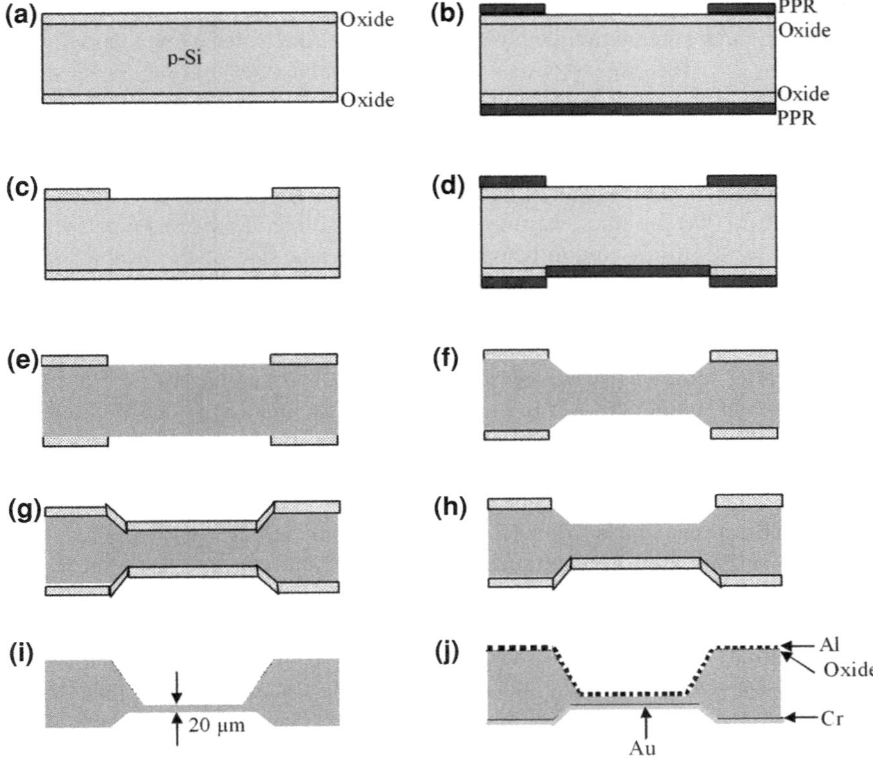

Fig. 5 Cross-sectional views of the top silicon wafer after **a** oxidation; **b** first photolithography using PPR; **c** etching of oxide; **d** flipping the wafer and lithography with back side alignment; **e** oxide etching; **f** etching 50 μm in KOH; **g** etching the oxide and growth of fresh oxide layer; **h** lithography and etching of oxide on one side; **i** etching 80 μm in KOH leaving behind a 20 μm membrane; **j** deposition of Al on one side and Cr/Au on the other side

A thick layer of PPR is now coated to protect the etched surface. The wafer is now flipped and PPR is coated on the other side. Using Mask #1 again, the PPR is patterned using a double-sided mask aligner (MA6, Suss Microtech) to align with the pattern on the bottom side of the wafer. The cross-sectional view of the wafer after this step is shown in Fig. 5d. The oxide is now etched using BHF to realize the cross-sectional view shown in Fig. 5e.

Silicon is now etched using a 25 % KOH solution at 70 °C, which is an anisotropic etchant. The etchant exposes <111> surfaces which make an angle of 54.74° with the silicon surface. The concentration of the etchant and temperature is chosen to obtain smooth etched surfaces as well as a reasonable etch rate. The etch rate is 45 μm/h. The silicon wafer is dipped in the KOH solution kept in a constant temperature bath for 67 min for etching 50 μm, which corresponds to the chamber height of the micropump. The cross-sectional view of the wafer after this step is shown in Fig. 5f.

The wafer is dipped in BHF to etch silicon dioxide, followed by a fresh growth of a 0.5 μm thick silicon dioxide layer on all sides of the wafer by wet oxidation at 1000 °C for 2 h. The cross-sectional view of the wafer after this step is shown in Fig. 5g. PPR is now coated on both sides of the wafer. Photolithography is carried out again using Mask #1 on one side of the wafer. The cross-sectional view of the wafer after etching of the oxide using BHF and stripping of PPR is shown in Fig. 5h. Silicon is now again etched using a 25 % KOH solution at 70 °C for 107 min (for a 200 μm thick starting wafer) using silicon dioxide as a mask. This will result in 80 μm of silicon being etched from one side of the wafer leaving behind a 20 μm thick silicon membrane. It may be mentioned, here, that the etching time has to be adjusted to obtain a 20 μm thick membrane, since the initial wafer thickness may vary between 190 and 210 μm. The cross-sectional view of the wafer after removing the oxide layers using BHF is shown in Fig. 5i.

A thin oxide layer of 0.3 μm is then grown by wet oxidation at 1000 °C for 1 h. This insulating oxide layer electrically isolates the fluid in the chamber of a working micropump, the membrane and the piezoelectric crystal. The chamber side of the wafer was then coated by a Cr/Au layer (100 nm Au on 50 nm Cr) by e-beam metallization. This is followed by electroplating a thick Au layer of ~1 μm by using a sulfite-based Au electroplating solution (Transene Inc., USA) at 60 °C. The electroplating is carried out using platinum mesh as an anode and the Cr/Au-coated silicon wafer as a cathode with a current density of 1 mA/cm^2 maintained using a current source. This Au layer will be used for eutectic bonding to the middle silicon wafer containing the nozzle and diffuser elements. The other side of the wafer is coated with a 150 nm thick Al layer deposited by thermal evaporation. This layer is required for electrical contact with the piezoelectric crystal. The cross-sectional view of the wafer after this step, which completes the processing of the top wafer, is shown in Fig. 5j.

3.2 Fabrication of Middle Silicon Wafer

The starting wafer in this case is a 280 μm thick p-type double side polished silicon wafer of (100) orientation. After cleaning the wafers using a standard procedure, a 0.9 μm thick silicon dioxide layer is grown on all sides using wet oxidation at 1,000 °C for 4 h. The cross-sectional view of the wafer after this step is shown in Fig. 6a. This is followed by photolithography using positive photoresist (PPR) and Mask #2, which has 496.5 × 496.5 μm openings. The oxide layer is etched using buffered hydrofluoric acid (BHF), followed by the stripping of PPR. The cross-sectional view of the wafer after these steps is shown in Fig. 6b.

A thick layer of PPR is now coated to protect the etched surface. The wafer is now flipped and PPR is coated on the other side. Using Mask #2 again, the PPR is patterned using a double-sided mask aligner (MA6, Suss Microtech) to align with the pattern on the bottom side of the wafer. It may be mentioned here that care must be taken in designing the mask plate containing the patterns for a large

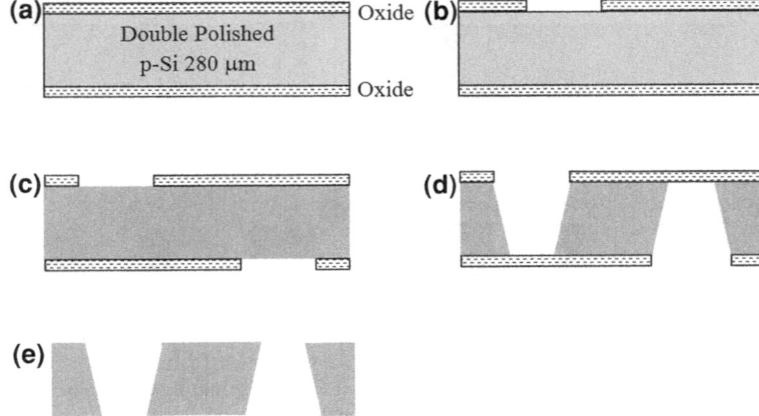

Fig. 6 Cross-sectional views of middle silicon wafer after **a** oxidation; **b** lithography and etching of oxide on one side; **c** lithography and etching of oxide on the other side; **d** silicon etching using KOH; **e** etching of oxide

number of devices to ensure that the same plate can be used for both sides. The oxide is now etched using BHF and PPR stripped to realize the cross-sectional view shown in Fig. 6c.

Silicon is now etched using a 25 % KOH solution at 70 °C for 370 min to etch 280 μm. Etching stops when the etchant encounters the silicon dioxide layer on the other side of the wafer. The cross-sectional view of the wafer after this step is shown in Fig. 6d. The wafer is dipped in BHF to etch silicon dioxide, resulting in openings of 100 × 100 μm. The separation between the centers of the two openings on each side is 3 mm. The cross-sectional view of the wafer after this step, which completes the processing of the middle wafer, is shown in Fig. 6e.

3.3 Fabrication of Bottom Glass Wafer

The starting wafer in this case is a 500 μm thick double side polished Borofloat 33 glass wafer. The etch holes in the glass provide both guidance and mechanical strength to the inlet/outlet tubes. Seed layers of Cr/Au (100 nm Au on 50 nm Cr) are deposited by e-beam metallization on both sides of the glass wafer. This is followed by electroplating thick Au layers of ~1 μm on both sides of the glass wafer by using a sulfite-based Au electroplating solution. The cross-sectional view of the wafer after this step is shown in Fig. 7a. Subsequently, thick layers of 18 μm of PPR are coated and photolithography is carried out on both sides to open circular windows of 0.8 mm diameter separated by 3 mm. The Au and Cr layers are sequentially etched with the PPR mask using Au (Potassium iodide and iodine based) and Cr (Ceric ammonium nitrate based) etchants. The cross-sectional view

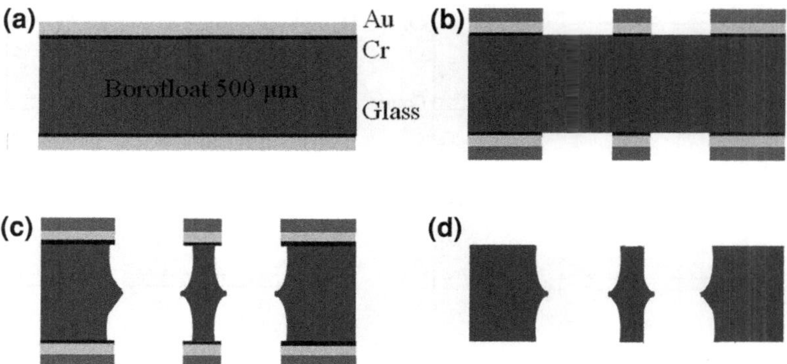

Fig. 7 Cross-sectional view of the glass wafer after **a** Cr/Au coating; **b** lithography and etching metal layers; **c** etching of glass; **d** removal of PPR and metal layers

of the wafer after this step is shown in Fig. 7b. This is followed by etching through holes in the glass using a 48 % HF solution for 40 min with Cr/Au/PPR as a mask. The glass is overetched to ensure that the opening is greater than 1.6 µm, which is the outer diameter of the inlet/outlet tubes to be inserted. The cross-sectional view of the wafer after this step is shown in Fig. 7c. Once the through holes are realized, PPR is stripped using acetone, and Au and Cr are removed by etching. The cross-sectional view of the wafer after this step, which completes the processing of the bottom glass wafer, is shown in Fig. 7d.

3.4 Realization of Micropump

After completing the fabrication of the three wafers independently, the micropump is realized by bonding them into one composite structure. At first, the top and middle silicon wafers are aligned using a bond aligner (MA6/BA6, Suss Microtech) and then eutectically bonded at 400 °C under vacuum conditions with a pressure of 600 mbar for 15 min in a substrate bonder (SB6, Suss Microtech). The cross-sectional view of the wafer after this step is shown in Fig. 8a. The silicon wafers are then anodically bonded to the processed glass wafer at a 2 kV applied bias and a tool pressure of 600 mbar. For the purpose of actuation, a 4 × 4 mm piezoelectric transducer (PZT) plate (PSI-5H4E) of 127 µm thickness is glued manually to the Al layer on the top silicon wafer using H20E silver conductive epoxy. Electrical wires are soldered to the top surface of the PZT and the Al layer to provide the actuation voltage. Silicone rubber tubes with an inner diameter of 0.9 mm and an outer diameter of 1.6 mm are used to make inlet and outlet connections. These tubes are inserted into the etched holes of the glass structure and sealed with epoxy (Araldite) as shown in Fig. 8b. This completes the fabrication and assembly of the micropump.

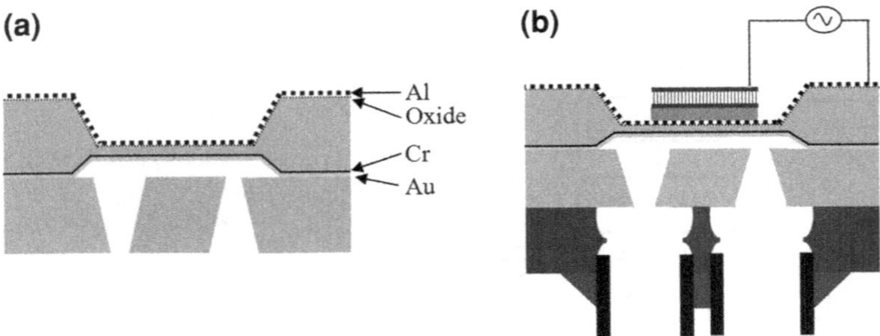

Fig. 8 Cross-sectional views of **a** the bonded top and middle silicon wafers; **b** the assembled micropump

4 Characterization of Micropump

The test setup consists of a function/arbitrary waveform generator (GW INSTEK SFG-830), a digital storage oscilloscope (GW INSTEK GDS-840C), a power amplifier (Bruel & Kjaer 2713), and a petit balance (Adair Dutt AD-60B) whose resolution is 1 mg. The amplified output of the signal generator supplies a variable voltage at variable frequencies to the micropump. The output voltage of the amplifier is measured using a digital storage oscilloscope. The weight of the collected water is measured using the petit balance. The schematic of the test setup is shown in Fig. 9a and the fabricated and assembled micropump fixed in the test jig is shown in Fig. 9b.

The performance of the micropump is tested at zero back pressure for DI water and methanol, and also at varying back pressures for DI water. The maximum flowrate occurs at zero back pressure. The water levels in the reservoir and outlet tubes are maintained at the same height for doing experiments at zero back pressure. By allowing the height of the water level in the outlet tube to be raised above the water level in the reservoir, back pressure is increased. Sinusoidally varying voltages from 50 $\pm$ 0.5 V (rms) to 80 $\pm$ 0.5 V (rms) at different frequencies are applied to characterize the pump. The flowrate variation as shown in Fig. 10a indicates a maximum flowrate of 98 μl/min obtained for 80 V at zero back pressure. Two features of the micropump depicted in Fig. 10 need some explanation. It is seen that the flowrate increases with an increase in driving voltage for any given driving frequency. This is because the membrane deflection increases with an increase in driving voltage. This implies that a larger instantaneous volume of liquid is displaced and for a given micropump geometry, this means a larger net volume flowrate is delivered by the pump. The second feature of micropump behavior is the dependence of the flowrate on the driving frequency. It is seen that for the given micropump, the peak flowrate is seen to occur at a frequency of about 250 Hz. This peak could appear because of the coupling between the fluid inertance (L) under the unsteady action of the micropump and the membrane capacitance (C) as explained in Verma and Chatterjee [24].

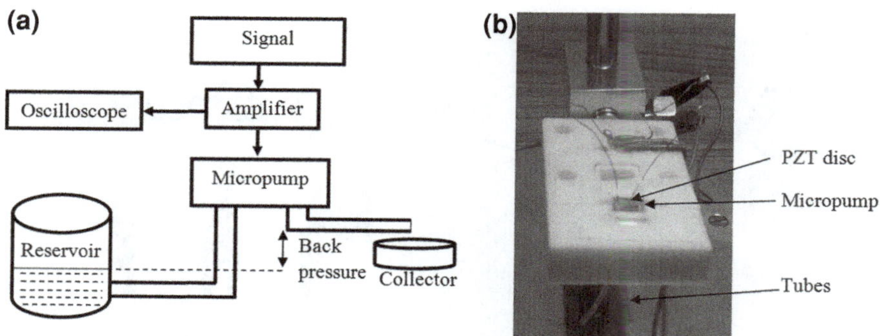

Fig. 9 **a** Schematic of the test setup; **b** test jig with the micropump

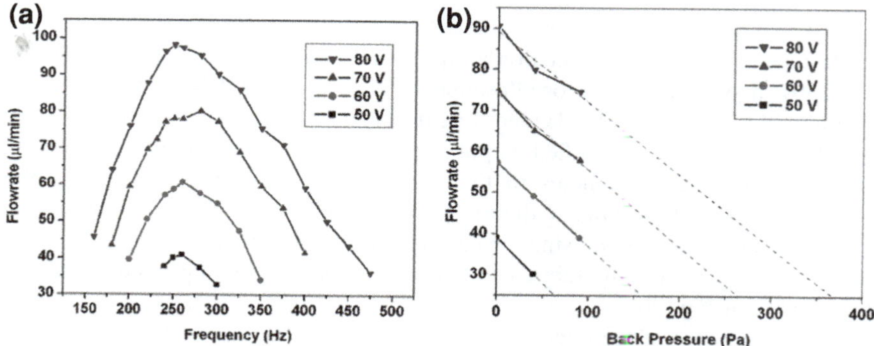

Fig. 10 **a** Flowrate at zero back pressure versus frequency for different actuation voltages; **b** Flowrate at 230 Hz versus back pressure at different actuation voltages

Experiments were carried out at varying back pressures for different voltages at 230 Hz, as shown in Fig. 10b. The linear extrapolation of the curves gives a maximum back pressure of 360 Pa at 80 V. Experiments were also done with methanol as the working fluid to compare with the flowrates for DI water at 60 V. The resonance frequency as well as the flowrate value is more for methanol as shown in Fig. 11. The density of water ($\sim$1000 kg/m^3) is more than that of methanol ($\sim$780 kg/m^3). Hence, it is expected that the fluid inertance of methanol ($L_{methanol}$) will be less than that of water under identical operating conditions. Hence, the resonant frequency ($\propto 1/\sqrt{LC}$) of a methanol-operated micropump is higher than that due to water. The kinematic viscosity of water (0.8×10^{-6} m^2/s) is slightly more than that of methanol (0.64×10^{-6} m^2/s) and, hence, there is an increase in the maximum flowrate value in the case of methanol when compared with that of water.

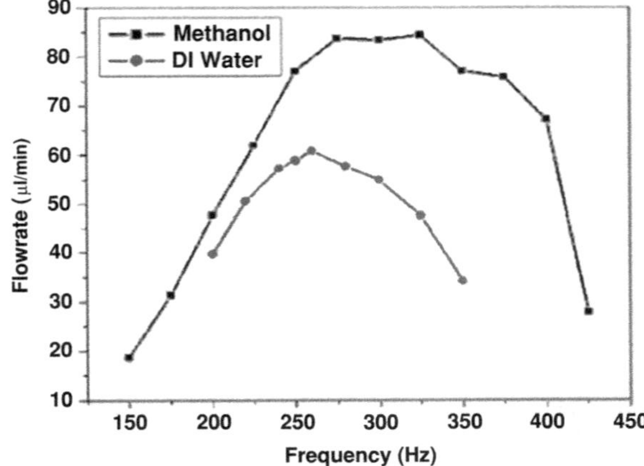

Fig. 11 Comparison of flowrates for DI water and methanol at different actuation frequencies and 60 V driving voltage. Back pressure is zero for both cases

5 Conclusions

The process for the design, analysis, and MEMS technology-based fabrication of valveless pyramidal micropumps has been presented. A valveless micropump is fabricated by bonding two micromachined silicon wafers and one glass wafer. Each fabricated micropump is assembled with a piezoelectric plate using conductive epoxy for actuation. The flowrates were measured for different actuation voltages and frequencies. A maximum flowrate of 98 μl/min is obtained for a pump with a 5 × 5 mm membrane for 80 V rms at a frequency of 250 Hz for DI water at zero back pressure. The results indicate a maximum backpressure of 360 Pa at 80 V. The pump shows a higher maximum flowrate with methanol as the fluid. Efforts are presently being made to reduce the actuation voltages by modifying the area and the thickness of the membrane.

Acknowledgments The authors would like to thank Prof. K. N. Bhat for initiating this activity and his guidance. They would also like to thank NPSM for the project on "Design and fabrication of silicon micropump for drug delivery and drug dosage control (Project No. 5:7)" and NPMASS for the project on "Upgrading facilities for MEMS design activities at National Resource Centres".

References

1. Whitesides GM (2006) The origins and the future of microfluidics. Nature 442(7101):368–373
2. Smits JG (1990) Piezoelectric micropump with three valves working peristaltically. Sens Actuators A 21(1–3):203–206
3. Garimella SV, Singhal V (2004) Single-phase flow and heat transport and pumping considerations in microchannel heat sinks. Heat Trans Eng 25(1):15–25

4. Verma P, Chatterjee D, Nagarajan T (2008) Design and development of a modular valveless micropump on a printed circuit board for integrated electronic cooling. Proc. IMechE. Part C J Mech Eng Sci 223:953–963
5. Manz A, Graber N, Widmer HM (1990) Miniaturized total chemical-analysis systems—a novel concept for chemical sensing. Sens Actuators B 1:244–248
6. Lee SJ, Lee SY (2006) Micro total analysis system (μ-TAS) in biotechnology. Appl Microbiol Biotechnol 64:289–299
7. Brand S (2006) Microdosing systems: micro-pumps the beating heart of micro-fluidics. http://www.mstonline.de/news/events/micropumps
8. Thomas LJ Jr, Bessman SP (1975) Micro pump powered by piezoelectric disk benders. US3963380, USA
9. Smits JG (1984) Piezo-electrical micropump. European patent EP0134614, Netherlands
10. van Lintel HTG, van De Pol FCM, Bouwstra S (1988) A piezoelectric micropump based on micromachining of silicon. Sens Actuators 15(2):153–167
11. Stemme E, Stemme G (1993) A valveless diffuser/nozzle-based fluid pumps. Sens Actuators A Phys 39:159–167
12. Gerlach T, Wurmus H (1995) Working principle and performance of the dynamic micropump. Sens Actuators A Phys 50:135–140
13. Schabmueller C, Koch M, Mokhtari M, Evans A, Brunnschweiler A, Sehr H (2002) Self-aligning gas/liquid micropump. J Micromech Microeng 12:420–424
14. Olsson A, Enoksson P, Stemme G, Stemme E (1997) Micromachined flatwalled valveless diffuser pumps. J Microelectromechanical Syst 6(2):161–166
15. Nguyen NT, Huang XY (2001) Miniature valveless pumps based on printed circuit technique. Sens Actuators A 88:104–111
16. Yamahata C, Lotto C, Al-Assaf E, Gijs MAM (2005) A PMMA valveless micropump using electromagnetic actuation. Microfluid Nanofluid 1:197–207
17. Kim Y-S, Kim J-H, Na K-H, Rhee K (2005) Experimental and numerical studies on the performance of a polydimethylsiloxane valveless micropump. Proc IMechE Part C J Mech Eng Sci 219:1139–1145
18. Pan LS, Ng TY, Wu XH, Lee HP (2003) Analysis of valveless micropumps with inertial effects. J Micromech and Microeng 13:390–399
19. Pan LS, Ng TY, Liu GR, Lam KY, Jiang TY (2001) Analytical solution for the dynamic analysis of a valveless micropump: a fluid-membrane coupling study. Sens Actuators A 93:173–181
20. Ullmann A (1998) The piezoelectric valve-less pump-performance enhancement analysis. Sens Actuators A 69:97–105
21. Ullmann A, Fono I (2002) The piezoelectric valve-less pump improved dynamic model. J Microelectromech Syst 11:655–664
22. Forster F, Bardell R, Afromowitz M, Sharma N (1995) Design, fabrication and testing of fixed-valve micropumps. Proc ASME Fluids Eng Div IMECE 234:39–44
23. Bardell LR, Nigel RS, Fred KF, Martin AA, Robert JP (1997) Designing high-performance micro-pumps based on no-moving-parts valves. Microelectromech Syst ASME 354:47–53
24. Verma P, Chatterjee D (2011) Parametric characterization of piezoelectric valveless micropump. Microsyst Technol 17:1727–1737
25. Fan B, Song G, Hussain F (2005) Simulation of a piezoelectrically actuated valveless micropump. Smart Mater Struct 14:400–405
26. Ha DH, Van PP, Goo NS, Han CH (2009) Three-dimensional electro-fluid-structural interaction simulation for pumping performance evaluation of a valveless micropump. Smart Mater Struct 18:104015
27. Nisar A, Nitin A, Banchong M, Adisorn T (2008) MEMS-based micropumps in drug delivery and biomedical applications. Sens Actuators B 130:917–942
28. Yao Q, Xu D, Pan LS, Melissa Teo AL, Ho WM, Peter Lee VS, Shabbir M (2007) CFD simulations of flows in valveless micropumps. Eng App Comp Fluid Mech 1(3):181–188

29. Kim Y-S, Kim J-H, Na K-H, Rhee K (2005) Experimental and numerical studies on the performance of a polydimethylsiloxane valveless micropump. Proc IMechE Part C J Mech Eng Sci 219:1139–1145
30. Dinh TX, Ogami Y (2011) A dynamic model of valveless micropumps with a fluid damping effect. J Micromech and Microeng 21:115016
31. Azarbadegan A, Eames I, Sharma S, Cass A (2011) Computational study of parallel valveless micropumps. Sens Actuators B 158:432–440
32. Nguyen NT, Huang X (2000) Numerical simulation of pulse-width-modulated micropumps with diffuser/nozzle elements. In: Proceedings of the international conference on modeling of simulator microsystems MSM2000, Santiago, CA pp 636–639
33. Tsui YY, Lu SL (2008) Evaluation of the performance of a valveless micropump by CFD and lumped system analyses. Sens Actuators A 148:138–148
34. Lu L, Wu J (2008) Flow behavior of liquid-solid coupled system of piezoelectric micropump. Frontiers Mech Eng China 3(1):50–54
35. Jeong J, Kim CN (2007) A numerical simulation on diffuser-nozzle based piezoelectric micropumps with two different numerical models. Int J Num Methods Fluids 53:561–571
36. Olsson A, Stemme G, Stemme E (1999) A numerical design study of the valveless diffuser pump using lumped-mass mode. J Micromech Microeng 9:34–44
37. Paul BE, Chatterjee D, DasGupta A (2012) An efficient numerical method for predicting the performance of valveless micropump. Smart Mater Struct 21:115012

Development and Characterization of PZT Multilayered Stacks for Vibration Control

P. K. Panda and B. Sahoo

Abstract Piezoelectric materials capable of performing both "sensing" and "actuation" belong to a class of smart materials. These materials produce electric charges on application of mechanical stress (as a sensor) or undergo dimensional change when subjected to an electric field (as an actuator). Lead zirconate titanate (PZT) is a piezoceramic material used widely due to its (i) fast response time, (ii) high frequency response, (iii) precession flow control, etc. PZT sensors and actuators are used for various applications such as vibration control of structures, for development of smart aeroplane wings/morphing structures, fuel flow control in automobile engines, etc. In this chapter, development of PZT powders, fabrication of PZT multilayered (ML) stacks, and their characterization is presented. PZT powders of high piezoelectric charge constant ($d_{33} = 590$–610 pC/N) is prepared by wet-chemical route. Simple PZT ML stacks with high block force (~ 5200 N) and amplified piezo actuators (APA) of high displacement (~ 173 μm) are fabricated by tape casting method using in-house developed PZT powder. The dynamic characterization of APA is carried out at different frequencies (100 Hz–1 kHz) and at different voltages (20–40 V). The actuator performs very well over the frequency range without attenuation of the signal, therefore, is suitable for vibration control applications.

Keywords Piezoelectric · PZT · Actuator · Tape casting · Multilayer stacks

---

P. K. Panda (✉) · B. Sahoo
Materials Science Division, CSIR-National Aerospace Laboratories, Kodihalli, Bangalore 560017, India
e-mail: pkpanda@nal.res.in

K. J. Vinoy et al. (eds.), *Micro and Smart Devices and Systems*,
Springer Tracts in Mechanical Engineering, DOI: 10.1007/978-81-322-1913-2_9,

1 Introduction

Since the discovery of piezoelectricity (pressure electricity) by Nobel laureates Pierre and Jacques Curie in 1880, there is a tremendous growth of piezo science in terms of development and fabrication of piezo materials and devices for various applications. The topic is continued to be popular and widely pursued by researchers worldwide. These materials generate electric charges on application of mechanical stress (direct piezoelectric effect), therefore, are used as "sensors". Similarly, the converse piezoelectric effect, i.e., the generation of a mechanical stress or strain on application of an electric field has been utilized for fabrication of "actuators" [1, 2]. There has been a surge on use of piezo devices after the successful fabrication of PZT in multilayered form consisting of piezo layers of thickness <100 μm which lowers driving voltage significantly to <100 V. These ML devices are fabricated by tape casting technique. Multilayer actuators have a number of advantages. These includes: (i) reproducible displacement at low driving voltage, (ii) quick response times (sub-microsecond), (iii) compact and have small dimensions, (iv) facilitates fabrication of lightweight miniaturized components, etc. Due to above advantages, piezoelectric materials are widely used for various applications such as for shape and vibration control [3, 4], structural health monitoring [5], MEMS [6], energy harvesting [7], precession flow control [8], fuel injector systems [9], etc.

At CSIR-National Aerospace Laboratories, Bangalore, R&D activities on the development of piezo materials and fabrication of ML devices have been pursued for more than a decade [10–17]. In this chapter, the development of PZT materials and fabrication of ML devices by tape casting technique and their characterization are discussed. In addition, fabrication of amplified ML actuators and their suitability for vibration control applications are also presented.

2 Experimental

2.1 Development of PZT Powders

PZT powders are generally prepared by "mixed oxide" route [18–20]. However, to improve the homogeneity, wet-chemical route is followed [21, 22]. This process consists of (i) preparation of various constituents of PZT in solution phase, (ii) mixing the stoichiometric quantity of the solutions, and (iii) precipitation/gelling of the solution. The gel/precipitate is calcined to develop PZT phase. A process flow sheet for the preparation of PZT powders is presented in Fig. 1. The calcined PZT powders are generally de-agglomerated in a ball mill/roller mill and the particle size distribution of powders which are mostly in submicron sized is presented in Fig. 2. Initially, PZT powders are prepared in lab scale ($\sim$100 g/batch), then up-scaled to (10 kg/batch). The piezo properties of the prepared powders (NAL-5H) are presented in Table 1.

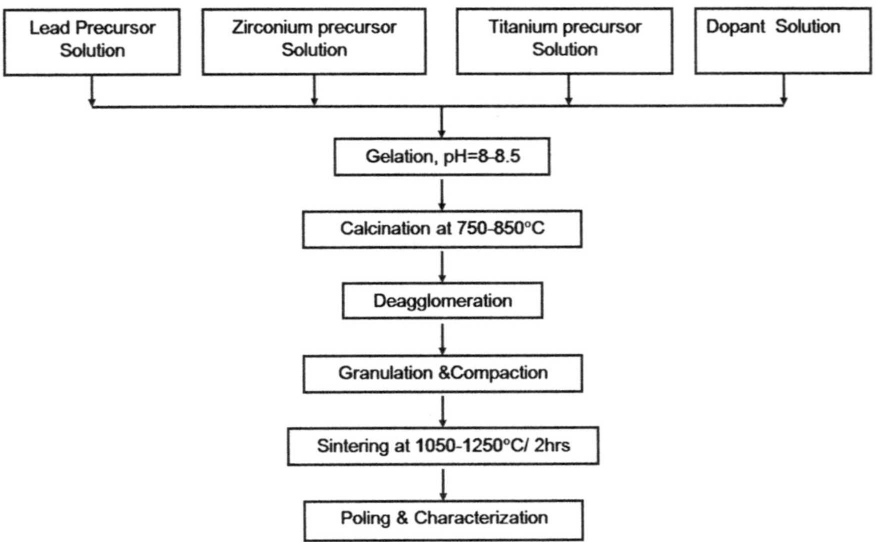

Fig. 1 Process flow sheet for the preparation of PZT powders [17]

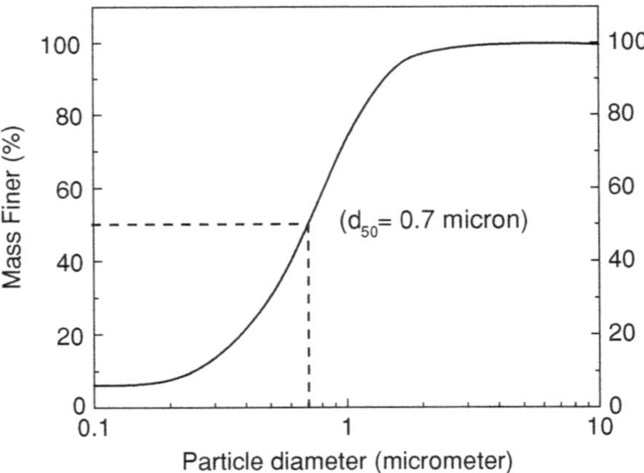

Fig. 2 Typical particle size distribution of in-house prepared PZT powder [21]

Table 1 Piezo properties of in-house prepared PZT samples

Properties	NAL-5H
Density (gm/cc)	7.6–7.7
Piezoelectric charge constant, d_{33} (pC/N)	590–610
Relative dielectric constant (K)	1700–1790
Dissipation factor (tanδ)	0.025–0.035
Particle size (median diameter, d_{50}) μm	0.6–1.2

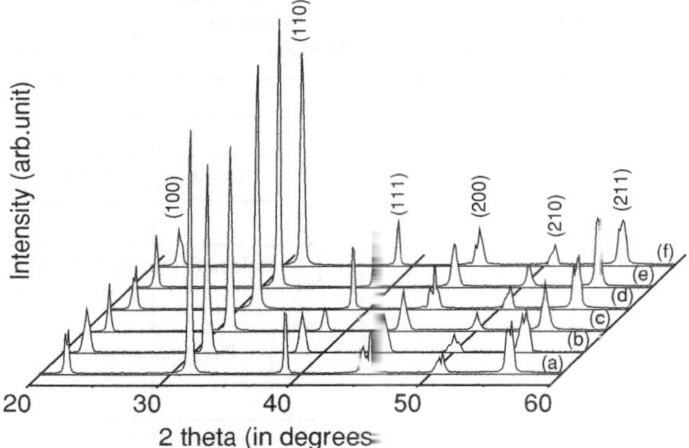

Fig. 3 XRD patterns of **a** undoped, **b** 0.03 La^{3+}, **c** 0.06 La^{3+}, **d** 0.02 Nd^{3+}, **e** 0.06 Nd^{3+}, and **f** 0.01La^{3+}+ 0.01Nd^{3+}-doped PZT samples [17]

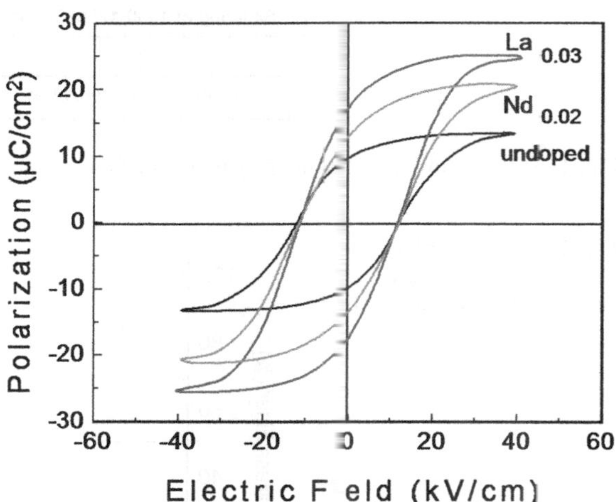

Fig. 4 Hysteresis loops of undoped, 0.02 Nd^{3+} and 0.03 La^{3+}-doped PZT samples [17]

2.2 Effect of Lanthanum and Neodymium on Properties of PZT

The effect of La^{3+} and Nd^{3+} on piezoelectric, dielectric and ferroelectric properties of PZT [$Pb(Zr_{0.53}Ti_{0.47})O_3$] was studied. The powders are prepared by "wet-chemical" route followed by calcination at 800 °C for 4 h. XRD analysis of the calcined powders confirms the tetragonal phase in the undoped PZT which gradually decreases with dopant concentration simultaneously with the appearance of rhombohedral phase (Fig. 3). The piezoelectric constant (d_{33}) and dielectric properties were maximum for 0.02 and 0.03 moles of Nd^{3+} and La^{3+} respectively. The remnant polarization of La^{3+}-doped sample was higher than Nd^{3+}-doped sample while the sample with combined dopant shows intermediate remnant polarization (P_r) (Fig. 4). SEM study of sintered pellets reveals decrease in grain

Fig. 5 SEM pictures of chemically etched **a** undoped, **b** 0.02 mole Nd^{3+}-doped, **c** 0.06 mole Nd^{3+}-doped, **d** 0.03 mole La^{3+}-doped, and **e** 0.06 mole La^{3+}-doped PZT sintered pellets [17]

size with the increase in dopant concentration beyond 0.02 and 0.03 moles of Nd^{3+} and La^{3+} respectively (Fig. 5). Based on the above study, it was concluded that La^{3+} is more effective than Nd^{3+} including from a mixture of La^{3+} and Nd^{3+} dopants.

3 Fabrication of Multilayered PZT Stacks

A multilayer stack consists of large number of PZT layers, screen printed with a suitable electrode and co-fired at high temperature for their bonding/structural integrity. A schematic diagram of a multilayered stack is presented in Fig. 6. On application of the voltage, the ML stack expands or contracts depending on the polarity of the applied voltage. This dimensional change is of about 0.1 % of the total height of the ML stack. The driving voltage required to get maximum displacement varies with the thickness of the PZT device. The driving voltage is in the order of few kV for thickness of samples in the order of few mm while it is in few volts for samples of few micron thicknesses. This can be best explained as per the equation given below:

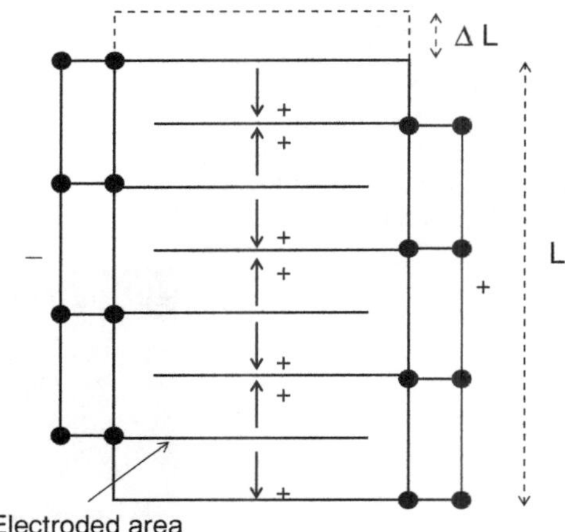

Fig. 6 Schematic diagram of a multilayered stack [21]

$$\Delta \mathbf{L} = \mathbf{n} \times \mathbf{d_{33}} \times \mathbf{V} \tag{1}$$

where

ΔL displacement of the stack
n number of PZT layers in the stack
d_{33} linear piezoelectric charge coefficient
V applied voltage

From the above equation, it is very clear that for a definite/constant displacement and for a given piezo material (given $\mathbf{d_{33}}$), the driving voltage is inversely proportional to the number of layers, i.e., as the number of layers increases, the driving voltage decreases. Therefore, multilayered stacks are preferred for high displacement actuators.

Multilayered PZT actuators are fabricated using an integrated fabrication facility consisting of tape casting unit, screen printer, laminator, etc. The process consists of preparation of homogeneous PZT slurry using required amounts of PZT powder, solvent, dispersant, binder, and plasticizers. This homogeneous slurry is poured on a tape caster and PZT tapes of required thickness are fabricated. The dried PZT tapes are screen printed with Pt electrode paste, stacked, and laminated. The green stacks are co-fired at 1,250 °C/2hrs. The co-fired stacks are then electrode, poled at 2 kV/mm. This fabrication process is shown in Fig. 7 and the fabricated ML stacks are presented in Fig. 8. These stacks are characterized for displacement and block force. The multilayered stacks of about 150 layers each of 80–100 μm thick are currently fabricated. These simple stacks/actuators with the displacement (10–12 μm) and block force up to 5,200 N meet the requirements for various applications.

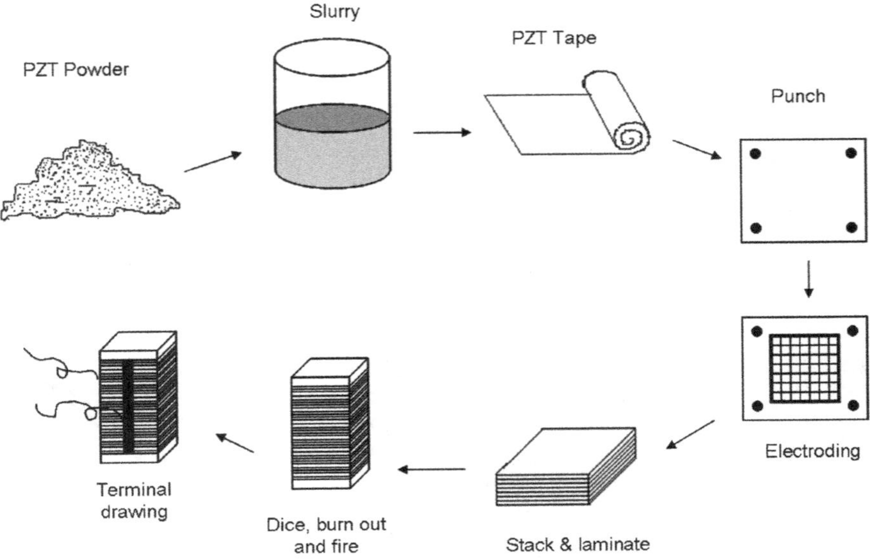

Fig. 7 The fabrication process of ML stack [16]

Fig. 8 In-house fabricated ML Stacks [16]

3.1 Fabrication of Amplified PZT actuators

The displacement of a simple ML PZT actuator is very small, typically, 1 μm per 1 mm height of the stack. For amplification of the displacement, a simple diamond-shaped metal casing is used. In this metal casing, PZT stacks are placed inside the casing horizontally in a pre-stressed condition. On application of the voltage, PZT stacks expand horizontally with a simultaneous contraction along the vertical direction. The ratio of contraction to the horizontal expansion is called amplification factor, normally lies in the range of 3–5. Some of the fabricated amplified piezo actuators (APA) are shown in Fig. 9a, b and their displacement profiles are presented in Fig. 9c, d. A simple diamond-shaped amplified piezo

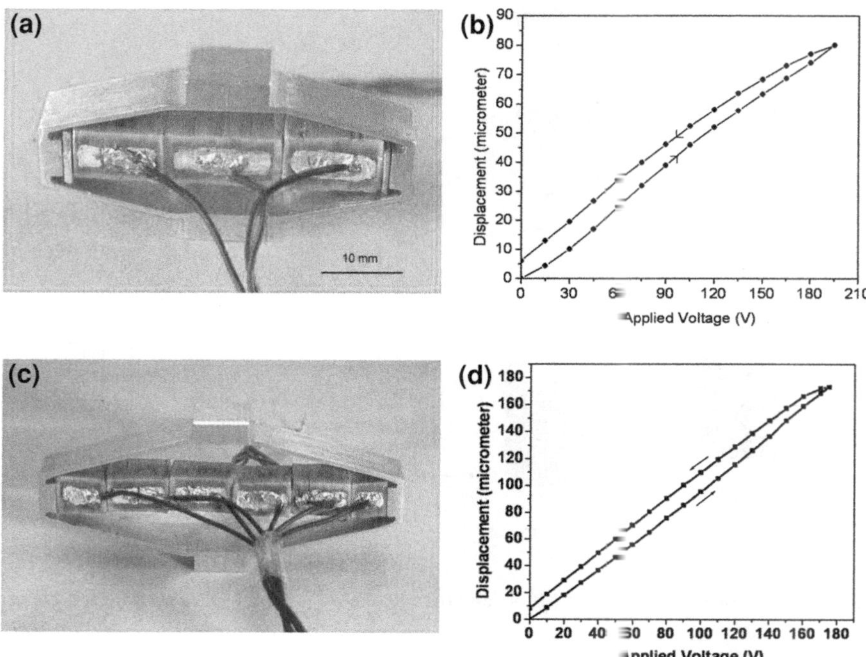

Fig. 9 **a**, **b** Amplified piezo actuator and **c, d** their corresponding displacement profiles [16]

actuator fabricated using six multilayered piezo stacks produces maximum displacement of 173 μm at 175 V with the amplification factor of 4.3.

4 Characterization of Actuators

4.1 Measurement of Displacement

The displacement of fabricated simple ML stacks and APAs was measured using a simple strain gauge. The test setup is presented in Fig. 10. The actuator is placed on a plane rigid support and the tip of the strain gauge is placed on the ML stack under pre-stressed condition. The terminals of the actuator are connected to appropriate terminals of a dc source and the voltage is gradually increased. The typical plot of the displacement versus voltage of amplified actuator is presented in Fig. 9d.

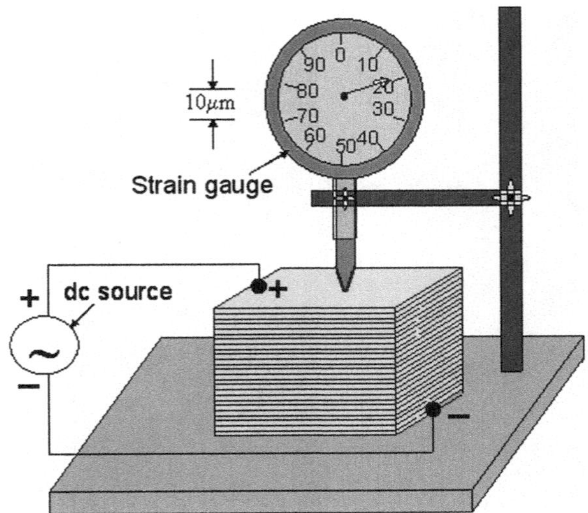

Fig. 10 Displacement measurement setup [21]

4.2 Measurement of Block Force

Block force refers to the force exerted at a given voltage level when the actuator is not allowed to move (zero displacement). The simple ML stack is characterized for block force using a block force measuring unit (TF Analyzer 2000, M/s aixACCT Systems GmbH, Germany). The actuator is placed on top of a force sensor (load cell) inside the sample holder and its positive and negative terminals are properly connected to the respective terminals of the voltage source. For measurement of block force, a constant pre-stress is applied from top of the actuator through 3–4 springs of different stiffness. The values of displacement and force generated by the actuator for all the springs are plotted by block force measurement software. It is observed that a maximum block force of 5,200 N obtained at 175 V.

4.3 Dynamic Characterization

The dynamic characterization of the actuator was carried out at different frequencies (100 Hz–1 kHz) and at different AC voltages (20–40 V). The test setup for characterization of APA-6 is presented in Fig. 11. The frequency response of the APA is presented in Fig. 12. The actuator response over this frequency range was found neat, without attenuation of the signal [10].

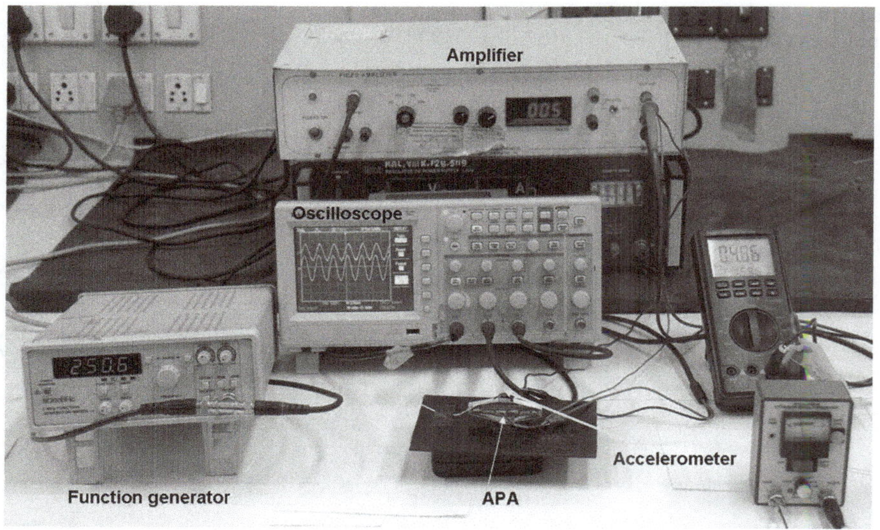

Fig. 11 Test setup for characterization of APA-6 [16]

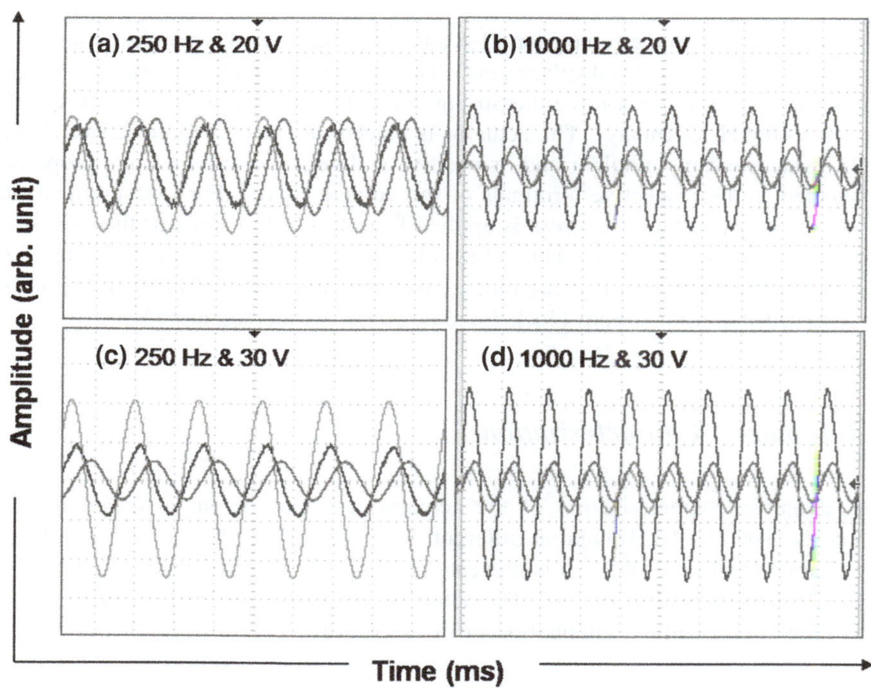

Fig. 12 Response of APA-6 at different frequencies and voltages [16]

5 Conclusions

In conclusion, this chapter discussed the technology for fabrication and characterization of PZT ML stacks starting from the powder preparation. PZT powders of high piezoelectric charge coefficient (d_{33}) in the range 590–610 pC/N are prepared by wet-chemical route. PZT multilayered stacks are fabricated by tape casting technique. Defect-free PZT tapes are prepared from well optimized slurry. Co-firing conditions are optimized to get highly dense defect free stacks. The displacement of the stacks is nearly 0.1 % of the stack height. The block force of PZT actuators was 5,200 N. The displacement of the actuator is amplified 4–5 times using a diamond-shaped metal casing. The dynamic response of fabricated actuators carried out at different frequencies (250 Hz to 1 kHz) and at different voltages (20–40 V). The actuator response is very good without attenuation of the signal; therefore, the actuator is suitable for vibration control applications. Some important issues from technological point of view are: (i) obtaining high bulk density to achieve high block force, (ii) high green tape density (>60 % of theoretical density) for obtaining high bulk density, and (iii) selection of suitable electrode material and co-firing technique addressed during the course of this study.

Acknowledgments The authors thank XRD and SEM groups of Materials Science Division, Dr. S. Raja and Mr. V. Shankar of STTD Division for dynamic characterization of amplified actuator. The authors sincerely thank NPSM, NPMASS for the financial support (PARC#1:8 and PARC#3:7). The authors are grateful to Prof. S. B. Krupanidhi and Prof. S. Gopalakrishnan for their guidance as PARC chairs. A special thank to Dr.V. K. Atre, the key person behind NPSM/ NPMASS for his interest and the encouragement in our R&D activities.

References

1. Newnham RE, Ruschau GR (1991) Smart electroceramics. J Am Ceram Soc 74:463–480
2. Uchino K (1995) Advances in ceramic actuator materials 22:1–4
3. Chandrashekhara K, Agarwal AN (1993) Active vibration control of laminated composite plates using piezoelectric devices: a finite element approach. J Intell Mater Syst Struct 4:496–508
4. Crawley EF, De Luis J (1987) Use of piezoelectric actuators as elements of intelligent structures. AIAA J 25:1373–1385
5. Giurgiutiu V, Zagrai A, Bao JJ (2002) Piezoelectric wafer embedded active sensors for aging aircraft structural health monitoring. Struct Health Monit 1:41–61
6. Trolier-McKinstry S, Muralt P (2004) Thin film piezoelectrics for MEMS. J Electroceram 12:7–17
7. Howells CA (2009) Piezoelectric energy harvesting. Energ Convers Manage 50:1847–1850
8. Cattafesta LN, Garg S, Shukla D (2001) Development of piezoelectric actuators for active flow control. AIAA J 39:1562–1568
9. Mock R, Lubitz K (2008) Piezoelectric injection systems. In: Piezoelectricity. Springer, Heidelberg, pp 299–310

10. Panda PK, Sahoo B (2005) Preparation of pyrochlore-free PMN powder by semi-wet chemical route. Mater Chem Phys 93:231–236
11. Sahoo B, Panda PK (2007) Dielectric, ferroelectric and piezo lectric properties of (1 − x) [$Pb_{0.91}La_{0.09}$ ($Zr_{0.60}Ti_{0.40}$) O_3]–x [Pb ($Mg_{1/3}Nb_{2/3}$) O_3], $0 \le x \le 1$. J Mater Sci 42:4270–4275
12. Sahoo B, Panda PK (2012) Preparation and characterization of barium titanate nanofibers by electrospinning. Ceram Int 38:5189–5193
13. Sahoo B, Panda PK (2007) Effect of CeO_2 on dielectric, ferroelectric and piezoelectric properties of PMN–PT (67/33) compositions. J Mater Sci 42:4745–4752
14. Sahoo B, Panda PK (2007) Ferroelectric, dielectric and piezoelectric properties of Pb_{1-x} Ce_x ($Zr_{0.60}Ti_{0.40}$) O_3, $0 \le x \le 0.08$. J Mater Sci 42:9684–9688
15. Sahoo B, Panda PK (2012) Fabrication of simple and ring-type piezo actuators and their characterization. Smart Mater Res 2012:821847
16. Panda PK, Sahoo B, Raja S et al (2012) Electromechanical and dynamic characterization of in-house-fabricated amplified piezo actuator. Smart Mater Res 2012:203625
17. Sahoo B, Panda PK (2013) Effect of lanthanum, neodymium on piezoelectric, dielectric and ferroelectric properties of PZT. J Adv Ceram 2:37–41
18. Jaffe B, Roth RS, Marzullo S (1954) Piezoelectric properties of lead zirconate–lead titanate solid-solution ceramics. J Appl Phys 25:809–810
19. Matsuo Y, Sasaki H (1965) Formation of lead zirconate-lead titanate solid solutions. J Am Ceram Soc 48:289–291
20. Chantreya SS, Fulath M, Pask A (1981) Reaction mechanisms in the formation of PZT solid solution. J Am Ceram Soc 64:422–425
21. Sahoo B, Jaleel VA, Panda PK (2006) Development of PZT powders by wet chemical method and fabrication of multilayered stacks/actuators. Mat Sci Engg B 126:80–85
22. Woo Lee B (2004) Synthesis and characterization of compositionally modified PZT by wet chemical preparation from aqueous solution. J Eur Ceram Soc 24:925–929

Development of Piezoelectric and Electrostatic RF MEMS Devices

Abhay Joshi, Abhijeet Kshirsagar, S. DattaGupta, K. Natarajan and S. A. Gangal

Abstract If we consider the MEMS market value breakdown, more than 4 % of the share goes to RF MEMS devices. With ever-increasing demand for optical and wireless communication, RF MEMS devices have a huge market in the coming years. Of the all RF MEMS devices, MEMS switches form the most important building block. With all the 32 types of MEMS switch configurations present, only switches with electrostatic and piezoelectric actuation are the most mature and present in the commercial market. This chapter focuses on design, simulation, optimization, and fabrication of MEMS switches. The chapter discusses two types of electrostatic switches, low process temperature, low stress electrostatic RF MEMS switch and all metal RF MEMS switches and one type of piezoelectric RF MEMS switch.

Keywords Piezoelectric switch · Electrostatic switch · Low process temperature · Surface micromachining · Design optimization · PZT · SiN

1 Introduction

In RF MEMS technology, components used for RF, microwave, and millimeter wave systems such as switches, varactors, inductors, high-Q resonators, filters, and antennas are made using micro fabrication techniques. MEMS switches are surface

A. Joshi · A. Kshirsagar · S. A. Gangal (✉)
Department of Electronic Science, University of Pune, Ganeshkhind road, Pune, India
e-mail: sagangal@gmail.com

S. DattaGupta
Department of Electrical Engineering, Indian Institute of Technology, Mumbai, India

K. Natarajan
Department of Telecommunication Engineering, MS Ramaiah Institute of Technology, Bangalore, India

K. J. Vinoy et al. (eds.), *Micro and Smart Devices and Systems*,
Springer Tracts in Mechanical Engineering, DOI: 10.1007/978-81-322-1913-2_10,

micromachined devices that use a mechanical movement to achieve an open circuit or short circuit in the RF transmission line and are designed to operate at RF to mm wave frequencies (0.1–100 GHz). The switches can be configured in series/shunt type and/or contact/capacitive type. RF MEMS switches can be implemented using different actuation, contact, and circuit configurations [1]. Selection is dependent on the application. Most mature types of RF MEMS switch actuation mechanisms are electrostatic and piezoelectric. This chapter discusses studies done on design, material, and/or material process optimization and fabrication of two different RF MEMS switches based on electrostatic actuation mechanism and one based on piezoelectric actuation mechanism.

2 Low Process-Temperature Electrostatic 2RF MEMS Switch

The type of switch configuration considered here is cantilever-based DC-contact RF MEMS electrostatic series switch. The switch is taken to be normally open. Figure 1 shows the cross-sectional view of the selected configuration. The switch is designed, the materials are selected, the process parameters are optimized, and the switch is fabricated and tested.

2.1 Design

Pull-in voltage, hold-on voltage, switching frequency, and contact force/area are the important MEMS switch performance parameters. The MEMS switch should have the lowest possible pull-in and hold-on voltages; and higher switching frequency, contact force, and area. These parameters can further be controlled using the dimensional parameters of the switch such as length, width, and thickness of cantilever, thicknesses of tip, bottom electrode and top electrode, gap between the released cantilever and substrate, width of the signal line, the overlap area of the tip onto signal line, angle of the step, distance of bottom and top electrodes from the anchor, etc. Careful analysis of all the parameters reveals that four main parameters that have maximum effect on the contact force per pull-in volt are: length of the cantilever, distance of the electrode from the anchor, thickness of tip, and thickness of the top electrode. This indicates the need for study of the dependence of these parameters on each other and finding out the optimum set of parameters for fabrication purpose. One should also have knowledge of properties of the materials to be used during fabrication.

In an attempt to study the effect of the material characteristics and dimensions of the switch on the switch parameters the simulations of dependence of spring constant, pull-in voltage, hold-down voltage, and resonant frequency on cantilever

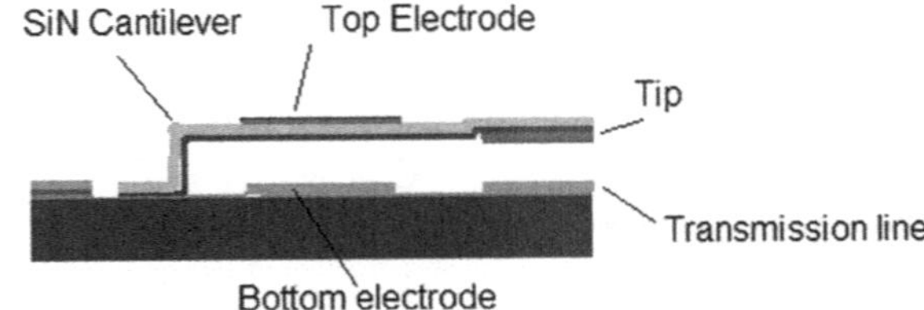

Fig. 1 Cantilever based DC-contact RF MEMS electrostatic series switch

beam length, width, and thickness with different electrode area were carried out. The MATLAB code was written for this purpose and executed. The study is reported in a research paper published in Journal of Sensors and Transducers [2] and a thesis submitted to University of Pune [3].

Following are the important observations for present case from the analysis of all the graphs:

a. Spring constant value higher than 40 N/m will increase the pull-in voltage and lower than 5 N/m will not bring the cantilever to original position after removal of actuation voltage. Therefore, the length of cantilever should be such that the spring constant lies between 5 and 40 N/m.
b. Pull-in voltage can be reduced by increasing the electrostatic force generated between the electrodes. Increasing the electrode width which in turn increases the area and therefore the electrostatic force) will reduce the pull-in voltage. Further, pull-in voltage is found to be less dependent on the cantilever beam width.
c. It is also known that the reduction in the mass of cantilever will increase the resonant frequency of the cantilever. Resonant frequency is the maximum frequency at which the structure can operate, beyond which the structure will fail. Thus, operating the switch at 50–60 % of its resonant frequency increases its switching life.
d. The placement of actuation electrode should be in between 60 and 80 % of beam length; measured from the anchor side of the beam. It is assumed that the electrode width and cantilever beam width are the same. Shifting the electrode position beyond 80 % of cantilever width will cause the beam to bend due to weight of the electrode. Shifting the electrode position below 60 % of beam length will increase the pull-in voltage.
e. Avoid placing the electrode on the 1/3 length from the fixed end, as it will have maximum stress during actuation.

On the basis of the observations mentioned above, the cantilever length between 100 and 350 μm; cantilever width between 50 and 200 μm and cantilever thickness between 0.5 and 2 μm were selected for further detailed simulations using software “CoventorWare 2009”. Detailed analysis of the bending of the cantilever for different pull-in voltages reveals some interesting facts. At low pull-in voltage the cantilever tip barely touches the two terminals, thus resulting in very less contact area, in turn less or no contact force. To increase contact area, high

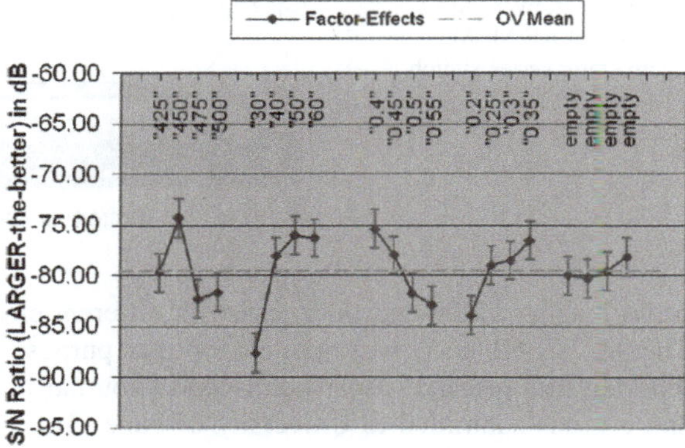

Fig. 2 Factor effect Plot: plot of (from *left* to *right*) cantilever length, distance of electrode from anchor, thickness of tip-metal and thickness top electrode. The last plot is empty, (i.e., no control factor is assigned)

pull-in voltage is applied, but it lifts the tip from the free end due to concave curving of the cantilever in the middle region of the cantilever where the electrode is located. This again results in less contact area (or less contact force). Furthermore, the high pull-in voltage produces large stress at the base of the cantilever close to the anchor. Therefore, an optimum, low pull-in voltage must exist at which the concave curving is eliminated and contact area is maximum. Taguchi method [3, 4] is well suited to solve such multiple control factor optimization problems. To prove that the optimization problem exists, deliberately longer cantilever length is chosen. In this work, four control factors selected for Taguchi analysis and their ranges are as follows: cantilever length from 425 to 500 μm, distance of electrode from anchor from 30 to 60 μm, thickness of tip from 0.4 to 0.55 μm and thickness of top electrode from 0.2 to 0.35 μm. Details of Taguchi method are given in research paper [4]. The results obtained from the Taguchi method are given in Fig. 2, called factor effect plot.

This plot is helpful in predicting trend of the control factors. The y-axis gives the signal-to-noise ratio derived from contact force computed by Taguchi method. In this case the contact force (Fig. 2) "larger-the-better" is preferred. All the values above the overall mean line are used as optimum values. The optimum values at which low pull-in voltage and maximum contact force is achieved are as follows: length of the cantilever = 450 μm; distance of electrode from anchor = 50 μm; thickness of the tip metal = 0.4 μm, and thickness of top electrode = 0.35 μm. The predicted value of contact force as per Taguchi method for the best combinations of the four control factors is obtained as 6.1×10^{-4} μN/V. The confirmation experiment is done with the optimum valves using "CoventorWare 2009". The contact force per pull-in volt obtained with the FEM tools is 6.5×10^{-4} μN/V which matches very well the predicted value by Tagichi method.

After simulation the remaining development of switch consists of selection of the materials, optimization of process parameters, and actual fabrication of the device. Development of low-temperature silicon nitride films with low surface damage and low stress is of considerable interest to get good yield and better performance of the devices. The details of selection of the materials, techniques used for their deposition, and the optimum process parameters used for deposition of low stress structural layers are available elsewhere [3]. The designed switch is fabricated and characterized.

2.2 Switch Fabrication

The process steps used for fabrication of "cantilever based DC-contact RF MEMS electrostatic series switch" are shown in Fig. 3 (**a–l**) sequentially. The device is aimed to be fabricated at processing temperatures below 100 °C. Therefore, the deposition and processing of sacrificial layer, which is an important step in device fabrication, is also to be done at temperature below 100 °C. In-house prepared PMMA is used as material for sacrificial layer. The details of its preparation and compatibility of the same as sacrificial layer with different structural layer materials, such as SiN/SiO/SiON processed at temperature ~100 °C, is described in Chap. 14 of this book [3].

The silicon wafer is first oxidized to isolate switch from the wafer. The conductor line and bottom electrode are sputter deposited (Titanium (Ti) = 20 nm and gold (Au) = 300 nm) by standard PPR lift-off process. PMMA solution prepared in-house is spin coated onto the silicon wafer with conductors. The spin speed is 2,000 rpm and the baking time is 15 min at 90 °C. The coated PMMA is not sensitive to UV or electron beam radiation so it is patterned using an aluminum (Al) hard mask. Al was deposited using Thermal Evaporation Technique. The deposited hard mask is patterned by positive photoresist and exposed Al is etched using Al etchant composed of 80 % phosphoric acid, 5 % nitric acid, 5 % acetic acid and 10 % water. The exposed PMMA is then anisotropically etched using oxygen plasma to generate the vertical wall pattern in the sacrificial layer. Al is used for just patterning PMMA and therefore, Al is etched away with etchant mentioned. Due to compatibility issues copper (Cu) material is chosen for tip fabrication of RF MEMS switch. If Au tip is chosen, then during the Au etching for patterning the tip the bottom signal lines and electrode exposed would also be etched away. Cu sputtering on the wafer was done for 500 nm thick tip. After patterning of Cu for tip, PPR protects the required region of Cu. The unprotected Cu is etched away by Cu etchant {$(NH_4)_2S_2O_8 + H_2O$; 60 mg $(NH_4)_2S_2O_8$ is added in 300 ml DI water}. The cross-compatibility of the Cu etchant was checked and found to be compatible with PMMA and Au present on the wafer. The remaining hard mask is etched away and the structural layer of low stress silicon nitride is deposited. The structural layer is deposited by inductively coupled plasma chemical vapor deposition (ICPCVD) using optimized parameters [3].

The deposited SiN film is then patterned for cantilever using positive photoresist. The same resist is used as a mask for reactive ion etching (RIE) of SiN in CF_4/O_2 plasma. The final step of the RF MEMS switch device is the releasing step,

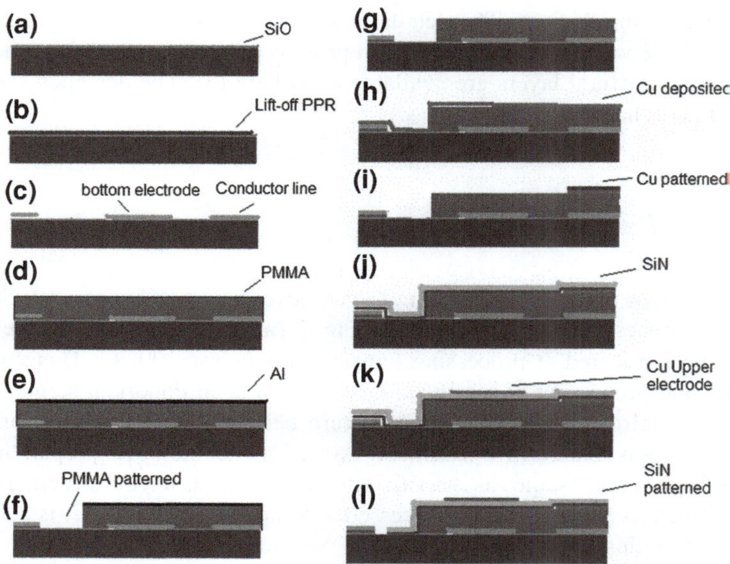

Fig. 3 Fabrication steps for DC-contact RF MEMS electrostatic series switch ((**a–i**) show variuos steps involved in fabrication.)

where the sacrificial layer is removed. Isotropic dry etching technique is used for etching PMMA (sacrificial layer) and releasing of the structures. The optimized dry parameters used are base pressure = 2×10^{-3} mbar; process pressure = 6×10^{-2} mbar; RF power = 60 W; O_2 flow = 25 sccm. Note that the fabrication process mentioned here does not exceed the thermal budget of 100 °C at any step in the process. The process also does not use any aggressive etchant which makes it compatible to CMOS technology.

2.3 Results

The fabricated wafers were observed under the SEM. The structures were found to release, however, the copper tip attached to the SiN was found missing from many devices. The typical SEM images of the fabricated devices are shown in Fig. 4a. Figure 4b shows SEM image of the released device where the copper tip was missing even before deposition of structural layer. In some of the devices where copper tip was seen is shown in Fig. 4c.

Experiments are being continued to avoid problems such as missing of Cu tip, curling of the cantilever, etc. Efforts are also being made to improve the yield of the devices.

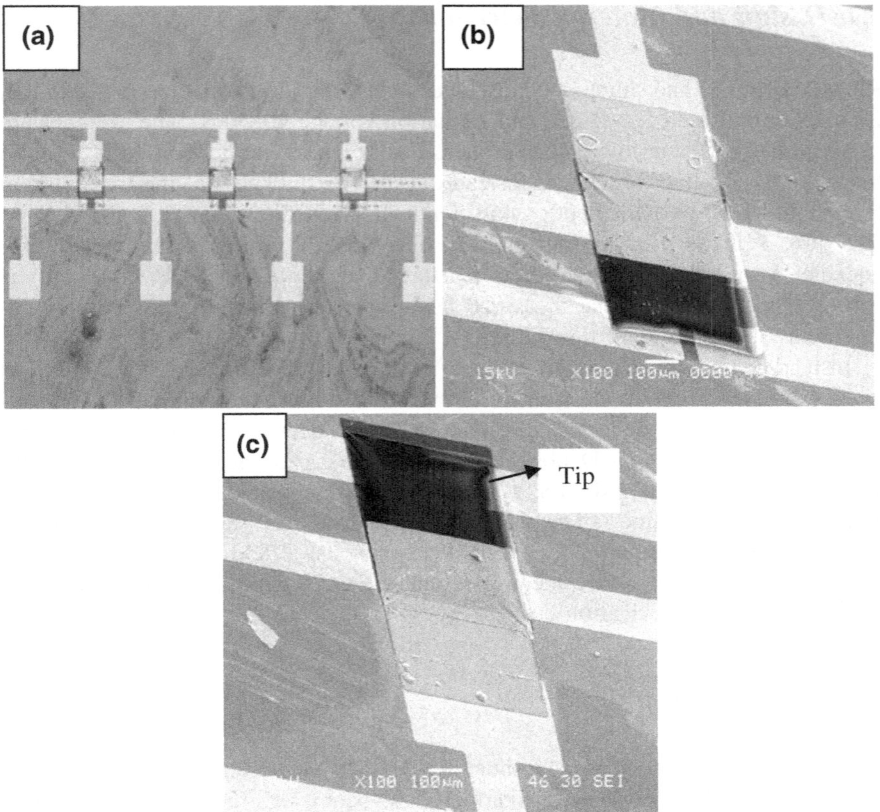

Fig. 4 **a** SEM images of fabricated devices and **b** SEM image of single device with copper tip missing **c** SEM images of single device with copper tip

3 All Metal Low Cost Electrostatic RF MEMS Switch

The metal bridge type capacitive shunt switches and cantilever-based series switches were widely tried for RF MEMS switch application [1]. Different materials have been deployed in RF MEMS switches like Al-SiO_2 bimorph-based RF MEMS switch by Siegel et al. [5], RF MEMS switch on GaAs substrate by Mullah and Karmakar [6], copper as transmission line material by Balaraman [7]. The use of Al for transmission line was demonstrated by Siegel et al. [5]. Isolation reported for Al transmission line was >15 dB.

3.1 Design and Material Selection

RF MEMS series and shunt switches are designed for High isolation, i.e., >25 dB, Low insertion loss, i.e., <1 dB, and Actuation voltage 30–50 V. The switches are to be fabricated using low cost approach. For Low cost fabrication Al is selected for transmission line fabrication instead of Au. The thickness of the transmission line should be >2x skin depth. Skin depth of Al is 2.62 μm for 1 GHz. Al of thickness 3 μm is used as transmission line material which is suitable for frequencies larger than 1 GHz. Sheet resistance of deposited Al layer (3 μm) is 0.0069 Ω/sq and Resistance measured using V/I measurement (four probe measurement) shows value of 0.00152 Ω.

Instead of widely used Si_3N_4 or metal oxides, low temperature SiO_2 is used as dielectric layer to block DC and to allow the RF signal through capacitive coupling. SiO_2 is good dielectric and can be easily deposited using Atmospheric Pressure Chemical Vapor Deposition (APCVD) and patterned using buffered HF. Typical thickness of deposited SiO_2 is 200 nm. It is deposited on DC actuation pad and transmission line. The switch fabrication is done using surface micromachining. Commercially available photoresist S2810 is used as sacrificial layer. Selection of the above-mentioned materials and deposition techniques help to bring down the fabrication cost.

3.1.1 Shunt Switch

Figure 5a shows side view of a shunt RF MEMS switch and Fig. 5b shows its equivalent circuit. It consists of fixed-fixed Al beam, i.e., bridge/membrane over co-planer waveguide (CPW) and two pull down electrodes (actuation pads) as shown in Fig. 5a. Fixed-fixed beam is anchored on ground planes of CPW line. The DC actuation voltage is applied between the membrane anchored on two ground plane and two pull down electrodes. Upon application of DC voltage, membrane gets attracted toward the signal line and touches the dielectric coating over the signal line. This configuration offers the capacitive contact, which shorts the RF signal and prevents the DC shorting between membrane and pull down electrodes.

Electrical equivalent circuit of shunt switch is shown in Fig. 5b. In this figure, R_s is resistance offered by Signal line. C is the capacitance between the membrane and Signal line. $C = C_u$ (up capacitance) in nonactuated condition and $C = C_d$ (down capacitance) in actuated condition. L is inductance offered by membrane in actuated condition only.

In upstate position of fixed-fixed beam, the shunt switch can be modeled as two short sections of transmission line and two capacitors (Cu) in series with two resistors (R_{s1}) as shown in Fig. 5b. In upstate, the insertion loss as modeled by simple R-C model is given by Eq. (1)

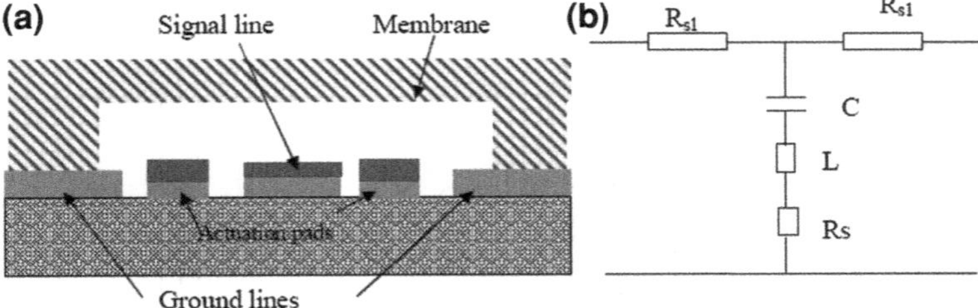

Fig. 5 **a** Side view of Shunt switch **b** Equivalent model of shunt switch

$$|S_{21}|^2 = \frac{4\left(R_{s1}^2 + 1/\left(\omega^2 C_u^2\right)\right)}{\left(2R_{s1} + Z_o\right)^2 + 4/\left(\omega^2 C_u^2\right)}. \tag{1}$$

When the bias line resistance is very low, the insertion loss is a strong function of $(1/\omega C_u)$.

In down state position, i.e., isolation state the switch is molded as C_dL model. In shunt switches, isolation is controlled by downstate capacitance and series inductance. Downstate capacitance C_d dominates the isolation up to half of resonating frequency $f_0/2$ and series inductance dominants after half of resonating frequency after $f_0/2$. When inductance is very low, the isolation is mainly dominated by down state capacitor and is given by Eq. (2)

$$|S_{21}|^2 = \frac{1}{\left(1 + (j\omega Z_o C_d)^2\right)}. \tag{2}$$

The main consideration was to get better isolation and low insertion loss with Al as transmission line material. The shunt switch was designed to give isolation of >25 dB and insertion loss <1 dB and the physical dimensions of the shunt switch are 400 (L) X 200 (W) × 1 μm (t).

3.1.2 Series Switch

RF MEMS series switch is shown in Fig. 6a. It consists of fixed-free Al beam, i.e., cantilever over Signal line of CPW structure in between ground line goes parallel to signal line and not shown in this cross section. The fixed end of cantilever is anchored on signal line connected to input port of and free end is above the signal line connected to output port. The DC actuation pad is situated below metal cantilever and covered by thin SiO_2 dielectric layer.

Upon application DC voltage, cantilever deflects down and makes signal line continuous. In series switch, cantilever-signal contact is direct contact. The dielectric layer is only applied on DC bottom electrode to prevent DC shorting.

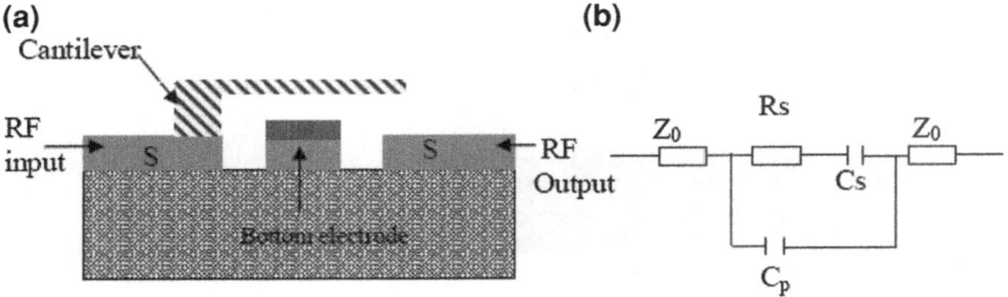

Fig. 6 **a** Schematic of series switch **b** Equivalent circuit of series switch

Electrical equivalent circuit of shunt switch is shown in Fig. 6b. In nonactuated mode, the series switch equivalent circuit consists of series resistance Rs, the capacitance C_s between cantilever and bottom signal line. The C_p represents the capacitance between signal and ground line of Co-planner waveguide (CPW). In actuated mode, Cu is zero.

In upstate position the, series switch can be modeled as two capacitance in parallel and high impedance transmission line as shown in Fig. 6b. Isolation of series switch is given by

$$S_{21} = 2j\omega C_u Z_o/(1 + 2j\omega C_u Z_o) \tag{3}$$

where $C_u = C_s + C_p$

In case of series switch, down state is modeled as

$$|S_{11}|^2 = \left(\frac{\omega L}{2Z_o}\right)^2; L \gg R_c \tag{4a}$$

$$|S_{11}|^2 = (R_s/2Z_o)^2; L \gg R_c. \tag{4b}$$

The main consideration was to get better isolation and low insertion loss with Al as transmission line material. The shunt switch was designed to give isolation of >25 dB and insertion loss <1 dB. In case of series switch cantilever parameters were 200 (L) × 120 (W) × 1 μm (t). In order, to get lower insertion loss, series resistance of switch should be low. Detail modeling can be found elsewhere [8].

3.2 Fabrication of RF MEMS Switches

RF MEMS switch fabrication was done using surface micromachining technology. The detail fabrication process flow is shown in Fig. 7.1 for series switch and Fig. 7.2 for shunt switch. The patterns for both series and shunt switches were

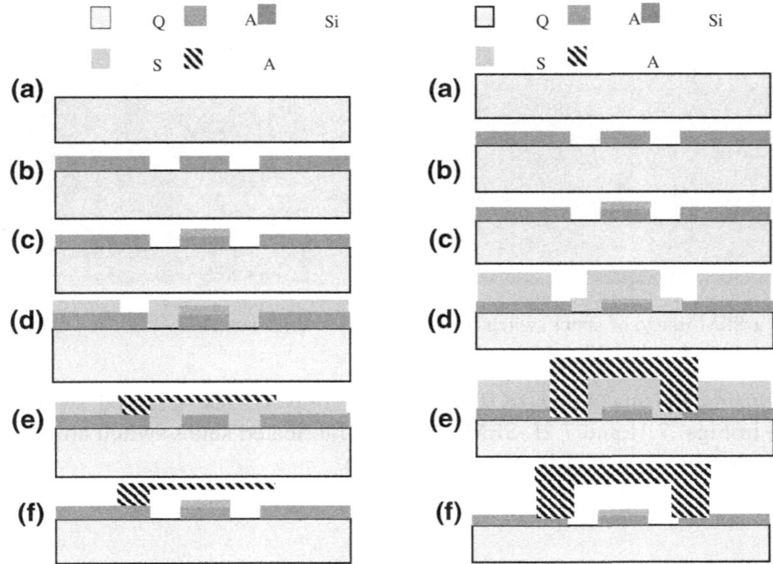

Fig. 7 **7.1** Fabrication of series switches **7.2** Fabrication of shunt switch

made on the same mask and fabrication of these switches were carried out simultaneously. Fabrication process using four masks was used for the switch fabrication.

RF MEMS switch fabrication process started with 4", 500 μm quartz substrate (Figs. 7.1a and 7.2a). 3 μm Al layer was evaporated using an e-beam and patterned in order to form Coplanar Waveguide (CPW). See Figs. 7.1b and 7.2b.

Sheet resistance of deposited Al layer was $0.69e^{-2}$ Ω/□ and resistance measured using V/I measurement (four probe measurement) shows value of $1.52e^{-3}$ Ω. To avoid DC shorting, the 200 nm LTO (Low Temperature Oxide) was deposited on DC actuation pad and transmission line using APCVD at 400 °C temperature and patterned using buffered HF (BHF) to serve as dielectric (Figs. 7.1c and 7.2c). LTO etching has to be carried out very precisely as buffered HF can attack Al layer too. Sacrificial layer was deposited using spin coating of standard positive photo-resist S2830 and patterned using photolithography (Figs. 7.1d and 7.2d). In order, to get smooth surface for structural layer the sacrificial layer deposition was carried out in two steps; first step to fill the gap in CPW line, i.e., planarization and second to serve as 3 μm sacrificial layer.

The membrane (for shunt switch) /cantilever (series switch) was formed using 1 μm sputtered Al film. The sputtering was done using 5 W/cm^2 DC power at $3.2e^{-3}$ mbar pressure with 12 sccm of argon atmosphere at room temperature (Edward Auto 500). The Al layer was patterned using wet etching (Figs. 7.1e and 7.2e). In this design, the holes were created in membrane and cantilever to ease the release and reduce air damping of the mechanical structures. Sacrificial layer was

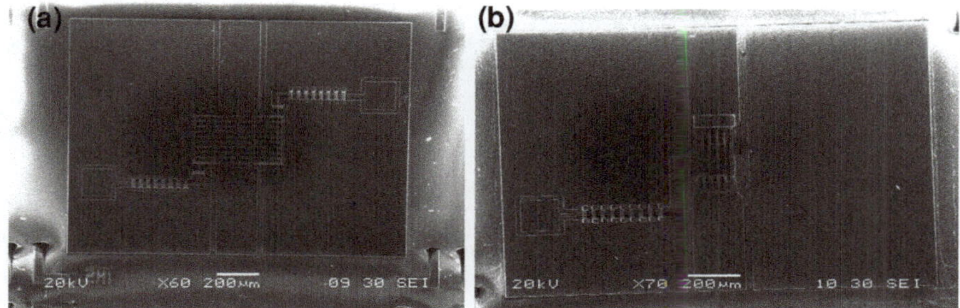

Fig. 8 **a** SEM image of shunt switch **b** SEM image of series switch

Ashed using Plasma Asher (K1050X from Emitech). The released switches are shown in Figs. 7.1f and 7.2f. SEM images of fabricated shunt switch are shown in Fig. 8a and series switch is shown in Fig. 8b.

3.3 Result and Discussion

Fabricated RF MEMS switches were tested for actuation voltage and RF performance. RF performance of RF MEMS switches was tested by measuring S parameters in ON and OFF conditions of the switch. S parameter measurement was performed on substrate using RF Prober system (From Cascade Corp.) connected to Power Network Analyzer (PNA) (from Agilent technology).

3.3.1 Shunt Switch

Typical S parameter measurement data for shunt switch is shown in Fig. 9. Shunt switch shows the actuation voltage of 30 V. Insertion loss measurement was carried out for shunt switch in ON condition. Measurement data shows insertion loss of −0.3 dB at 1 GHz and changes to −0.5 dB at 20 GHz, while isolation in OFF condition is −37 dB at 1 GHz and varied to −28 dB at 20 GHz. Return loss of shunt switch was −45 dB at 1 GHz, which reduces to −15 dB at 20 GHz.

Maximum isolation −37 db was limited by down capacitance 2.5 pF and series resistance of 0.2 Ω. The isolation was degraded to −28 dB at 20 GHz. This was due to reduction in reactance of shunt capacitance between membrane and transmission line with frequency and large via holes inductance to ground in switch structure. Isolation of this switch can be further improved by reducing the SiO_2 thickness to 100 nm from existing 200 nm or by using higher dielectric constant material.

Higher insertion loss was due to low up state capacitance. The 10 KΩ bias resistor was externally connected between the DC pad and ground line, this also

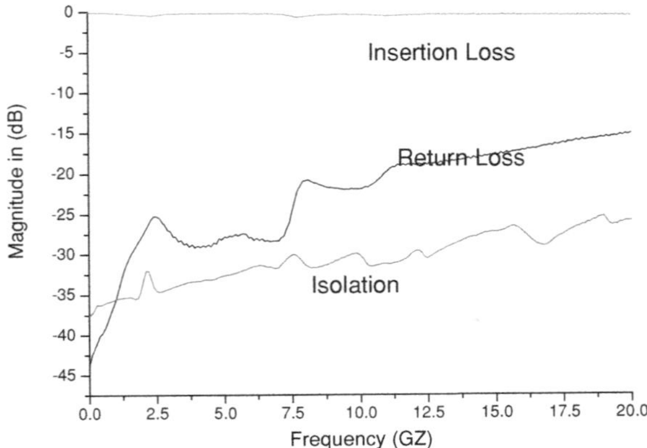

Fig. 9 RF performance of shunt switch

contributes to higher insertion loss. The return loss was due to parasitic capacitance caused by floating membrane over signal line.

Isolation value for present design is −28 dB at 20 GHz. This isolation is higher than reported value of −10 dB [9], >−15 dB [10] to >−20 dB [11, 12] and in good agreement with reported isolation data −25 to −32 dB [13]. The insertion loss of present design is bit higher as comparable with reported value of <−0.2 dB. The actuation voltage of present shunt switch is 30 V.

3.3.2 Series Switch

The S parameter measured data for series switch is shown in Fig. 10. Insertion loss of series switch was measured after switch was actuated. In case of series switch, the insertion loss was −0.5 dB at 1 GHz, and shifted to −0.7 dB at 20 GHz, while isolation was −38 dB at 1 GHz and decreased to −26 dB at 20 GHz. Series switch shows the return loss of −25 dB at 1 GHz, which shifted to −12 dB at 20 GHz.

The actuation voltage of series switch was 42 V. In case of DC in-line series switch in down bending state, the total contact resistance is given by

$$R = 2R_{s1} + R_c. \tag{5}$$

In series switch, measured total signal line resistance in down bending state was in the range of 1.8–2 Ω. Signal loss is dominated by the higher resistive loss in signal line and thus insertion loss of series switch was high. This resistive loss is independent of frequency. Surface roughness, improper contact, or presence of dust or contamination can increase the contact resistance, which increases the insertion loss of switch. This can be reduced by increasing the contact area by means of reducing the existing holes size within same cantilever dimensions.

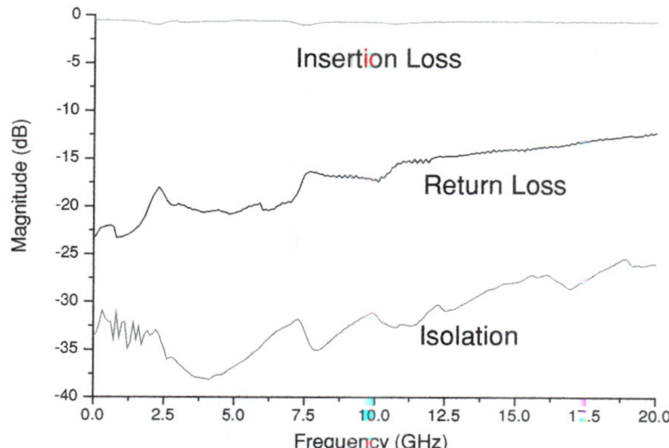

Fig. 10 RF performance of series switch

Series switch shows higher isolation as compared to shunt counterpart. Higher bias resistance of 10 KΩ is contributing to higher isolation in series switch.

Series switch performance was in good agreement with reported RF performance of RF MEMS switch. The isolation of series switch is −26 dB at 20 GHz, which is higher than reported isolation of 17 dB [14] and 25 dB [5]. Insertion loss in series switch is in good agreement with value of 0.6 dB by [15]. Isolation shown by series switch and shunt switch in present design is higher than Al-based shunt switch reported by Goldsmith [10] and Al based series switch reported by Siegal et al. [5]. This higher isolation value may be due to higher shunt capacitance in present structure.

4 Conclusions

Low cost RF MEMS switches, i.e., series and shunt switches using Al as transmission line material are successfully demonstrated. The shunt switch has actuation voltage of 30 V. The insertion loss of shunt switch was 0.3 dB and isolation was 37 dB. In case series switch, actuation voltage was 42 V. Series switch showed 31 dB isolation and 0.6 dB insertion loss. In this low cost approach, the RF performance of these switches is comparable with switches with Au transmission line.

5 PZT Thin Film-Based RF MEMS Switch

The electrostatic RF MEMS switches are at a matured stage of technology and are free from restraint of materials and good RF performances. However, electrostatically actuated RF MEMS switches operate at very high operation voltages of over 20 V. These high voltages can cause the charging of the constituent dielectric, which degrades the reliability of RF MEMS switches. Moreover, additional dc–dc converter chips are required for the mobile handset applications using low operation voltages of around 3 V. Researchers therefore are working for developing low voltage operating switches using electromagnetic actuation or by lowering spring constants of moving parts [7].

However, these methods can increase the power consumption or lower the switching speed. Piezoelectric actuation is a promising mechanism for realizing RF MEMS switches with a low operation voltage, low power consumption, and high switching speed. Piezoelectrically actuated RF MEMS switches are not reported much due to the difficulty and complexity of the fabrication.

5.1 Design and Material Selection

The proposed PZT cantilever structure was modeled and simulated using CoventorWare2009 simulation tool to find the cantilever displacement in accordance with applied voltage and its resonating frequency. The applied electric field to piezoelectric film induces strain in the material. This strain forces PZT film to contract or expand, which in turn bends the cantilever. Piezoelectric beam tip deflection (d) in terms of physical dimensions of PZT cantilever is [16];

$$d = \frac{L^2 d_{31}\left(\frac{V}{t_p}\right)(t_p + t_s)A_sE_sA_pE_p}{4\left(A_sE_s + A_pE_p\right)\left(E_pI_p + E_sI_s\right) + (t_s + t_p)^2 A_sE_sA_pE_p} \quad (6)$$

where V is the applied voltage; A_s is the area of support layer, A_p is the area of PZT layer, L is the length of the cantilever, E_p, I_p, t_p and E_s, I_s, t_s are young's modulus, moment of inertia and thickness of support layer and PZT layer, respectively, and $d31$ is the piezoelectric constant. In addition to cantilever deflection, the natural resonance frequency of a two layer cantilever structure fixed on one end is calculated using Eq. (7) [17, 18].

$$f = \frac{1.875^2}{2nL^2}\sqrt{\frac{EI}{W(\rho_s t_s + \rho_p t_p)}} \quad (7)$$

5.1.1 Physical Structure Optimization

Optimization of physical dimensions was done by considering effect of various lengths (50–400 μm), widths (50–200 μm), and the fabrication parameters. Cantilever deflection behavior was studied by simulating PZT cantilever for various length and width by applying fixed voltage of 1 V. The optimum structure dimensions are: length 200 μm and width 60 μm. The cantilever deflection is found to be 5.9 μm for applied voltage of 1 V, which is best suited for application in RF MEMS switches. Analytically calculated resonant frequency of the structure was 22.15 KHz and that obtained from simulations using CoSolve simulator of "CoventorWare 2009" was 23 KHz. The lower value of resonating frequency is appreciated for small signal application along with lower actuation voltage [19]. Detailed information on designing aspects can be found elsewhere [20].

5.1.2 Material Selection

The Lead Zirconate Titanate (PZT) is promising ferroelectric material having higher Piezoelectric coefficient d_{33}(170 pC/N) as compared to other materials. Unlike $PbTiO_3$ ferroelectric properties of PZT are independent of temperature. PZT in bulk, thin film, or thick films form is preferred as potential piezoelectric material in various applications. New materials like Lead magnesium Niobate (PMN), Lead ytterbium Niobate (PYN) have higher piezoelectric coefficient than PZT, but processing of these materials is not matured. Further, the PZT thin film deposition adds advantage of being processed along existing fabrication technology for batch device fabrication and thus preferred mostly as compared to other piezoelectric materials.

In cantilever fabrication process silicon is used as substrate material, SiO_2 is used as masking and supporting material, PZT is used as Piezo layer, Ti- Pt, and Au as bottom and top electrodes respectively. Details of selection criteria for these materials can be found elsewhere [21].

5.2 Optimization of PZT Thin Film Deposition Parameters

The PZT thin films were deposited by RF magnetron sputtering on 4-inch substrates (Pt/Ti/SiO_2/Si) using a 4-inch ceramic target of Zr: Ti ratio of 52:48 and without excess Lead. The substrates were prepared by growing 850 nm thermal oxide (SiO_2) on silicon wafer <100>. The 30 nm of titanium (Ti) layer was deposited using E-beam evaporation and platinum (Pt) layer of 150 nm was sputter deposited on Ti layer.

The PZT thin film was deposited using sputtering parameters, target size: 4". RF Power, 1–2.5 W/cm^2, sputtering gas: Ar, working pressure: 3.14–3.87 e^{-3} mbar, deposition time: 90 min on Pt/Ti/SiO2/Si substrates. Optimizations of sputtering parameters were carried out to prepare a uniform PZT film with smooth surface texture, good stoichiometry, and desired thickness of around 600 nm. To start with the PZT deposition trials were carried out by varying the RF power in the range of 80 (1 W/cm^2)–220 W (2.75 W/cm^2). Deposition pressure (3.16 e^{-3} mbar), deposition time (90 min), and Ar gas flow rate (12 sccm) were

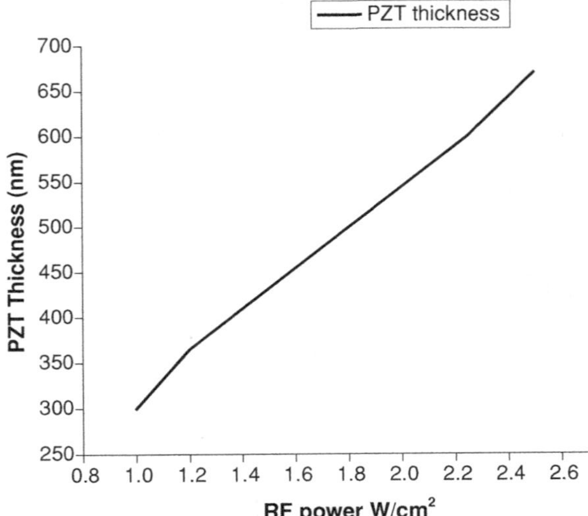

Fig. 11 Effect of RF power on PZT film thickness

kept constant. Film thickness was measured using stylus profiler. The effect of RF power on PZT thin film thickness is shown in Fig. 11.

Figure 11 shows that deposition thickness increases linearly with the applied RF power. This is attributed to high energy electrons in plasma. Increase in RF power, increases gas ionization efficiency, on the other hand, the voltage of plasma sheath decreases with increase in RF power and therefore increases deposition rate. At higher RF power of 2.5 W/cm^2 cracks were developed on target and hence the RF power applied was restricted to 2.25 W/cm^2 for all depositions.

Effect of RF power on PZT stoichiometry is studied using EDS characterization (using scanning electron microscope (SEM) (JSM 6360A) with EDS attachment.). The data obtained are shown in Table 1. During these trials deposition pressure (3.16 e^{-3} mbar), deposition time (90 min) and Ar gas flow (12 sccm) were kept constant.

At lower RF power, the sputtering yield was low, thus the element content observed for 1 W/cm^2 RF power are less. At a given pressure and power relative sputtering yield and deposition capability of different elements is a function of Mean Free Path (MFP). Further, MFP is function of atomic weight of elements. Atomic weights of these elements are Pb (207.2 g/mol), Zr (91.2 g/mol), Ti (47.867 g/mol), and O (15.9 g/mol). Pb is heaviest thus under low pressure deposition condition, the transport ability of the Pb toward substrate is larger than other elements. As a result, the Pb concentration increases in this film in comparison to the Zr and Ti with applied RF power. The Zr has bigger radius (atomic radius 206 pm). Thus increment in Zr content is observed initially. Decrement in Zr contents at higher power is attributed to scattering due to elastic collision with plasma species. While transporting from target to substrate, possibility of scattering due to collision is more.

Table 1 PZT stoichiometry as function RF power

RF power (W/cm^2)	Pb	Zr	Ti	O	Zr:Ti
1	21.5	13.5	10.73	54.26	55.7:44.3
1.5	26.51	15.77	12.97	44.74	54.87:44.12
2.25	**26.66**	**15.86**	**12.57**	**44.92**	**55.78:44.2**
2.5	26.71	9.27	12.61	47.41	42.38:57.63

Table 2 PZT stoichiometry as function of Ar gas flow rate

Ar flow (sccm)	Pb	Zr	Ti	O	Zr: Ti
6	27.53	15.5	13.86	43.11	52.79:47.21
12	**26.66**	**15.86**	**12.57**	**44.92**	**55.78:44.21**
15	29.43	16.25	13.12	41.68	55.32:44.68
18	29.22	16.76	12.42	41.6	57.43:42.56

Target to substrate transport ability for Pb is higher than Zr. With applied RF power, increase in Pb deposition rate is more than increase in Zr deposition rate. There might be reduction in Pb deposition rate as Pb is also having higher atomic radius and keen to elastic collision, but overall increase in Pb deposition rate in observed. Thus the Zr content reduces while Pb content increases in film.

From Table 1, Zr: Ti ratio of films deposited under RF power of 1.5 W/cm^2 and 2.25 W/cm^2 showed stoichiometry closer to that of target. Film deposited at 1.5 W/cm^2 gave thickness of 550 nm, which is lower than expected thickness. Film deposited at 2.2.5 W/cm^2 showed thickness of 600 nm, which is expected. Thus, RF power of 2.25 W/cm^2 was selected for further PZT film deposition.

Influence of argon flow on the PZT elements atomic ratio was studied by varying the Ar flow rate in the range of 6–18 sccm. During these experiments, the deposition time of 90 min and 2.25 W/cm^{2}applied RF power were kept constant. The element content calculated from EDS data in the film deposited under different Ar flow is shown in Table 2.

From Table 2, it is clear that the Ar gas flow in the range of 6–18 sccm has very low effect of PZT element contents. The Variations observed in the Pb, Zr, Ti and O are very less. There is slight increment observed in Pb, Zr, Ti and minor decrease in Ti contents with increase in Ar flow rate. Referring to Zr:Ti ratio in Table 2, it is clear that PZT film deposited for Ar flow of 6 sccm, showed stoichiometry more closer to that of target. But film deposited under this condition is having thickness of 550 nm, which is less than expected thickness. Film deposited under 12 sccm and 15 sccm showed same stoichiometric ratio. The film deposited at 15 sccm Ar flow rate showed 1070 nm thick film, higher than expected range of 600–700 nm.

XPS spectra were recorded (ESCA 3000 instrument with 150 W of monochromatic Mg X-ray radiation.) for Pb 4f, Zr 3d, Ti 2p and O 1s as function of Ar$^+$ ion sputtering for stoichiometry examination at surface as well as at 5 nm depth (Fig. 12). The observed XPS spectra of Pb 4f$_{7/2}$ (138.9 eV), Pb 4f$_{5/2}$ (143.8 eV), Zr 3d$_{5/2}$ (184.7 eV), Zr 3d$_{3/2}$ (182.4 eV) and Ti 2p$_{3/2}$ (458.4 eV), Ti 2p$_{1/2}$

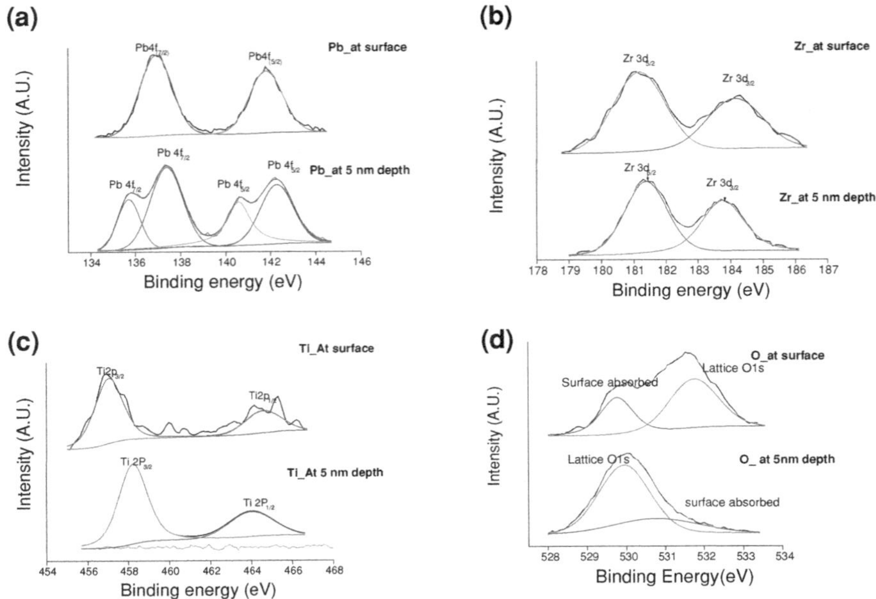

Fig. 12 XPS data of sputtered PZT thin film at surface **a** Pb **b** Zr **c** Ti **d** O elements

(464.3 eV) consist of two peaks corresponding to the angular momentum of electrons.

The peak observed for Pb spectra at 138.9 eV was attributed to metallic Pb contribution in PZT. The appearance of metallic Pb can be explained by reduction of Pb^{+2} ions to metal Pb^0 due to preferential sputtering of oxygen ion. In case of Ti, peak at 464.3 eV corresponds to Ti^{4+} bonding with 6 O atoms, while the Ti $2p_{1/2}$ peak at 458.4 eV corresponds to chemical reduced state like Ti^{3+} or Ti $^{2+}$. In oxygen, two components are observed corresponding to lattice oxygen, and surface absorbed oxygen.

Depth profiling was performed at depth of 5 nm on PZT thin film. At depth of 5 nm, two components of Pb were observed. The Pb doublet of Pb $4f_{7/2}$ and Pb $4f_{5/2}$ is clearly seen in Fig. 12a. At the depth, It is found that Pb4f peak exhibit states, Pb $4f_{7/2}$ (135.7 eV), Pb $4f_{5/2}$ (140.5 eV) and other at Pb $4f_{7/2}$ (137.3 eV), Pb $4f_{5/2}$ (142.2 eV).

The peaks observed at 135.7 and 137.38 eV are corresponding to metallic Pb and oxide Pb respectively. Lower binding energy (BE) peak was due to Pb oxide atom present in PZT thin film. The downward shift of approximately ~2 eV in BE can be assigned to presence of Pb metal. It suggests that Pb metal (PbO) coexist in PZT at depth in sputtered thin film. In case of Zr, Ti and O, peaks observed at depth are similar to peaks observed at surface, but variation atomic concentrations is observed at depth. The atomic concentration was calculated. The atomic concentration at surface and depth is given in Table 3.

Table 3 Atomic ratio calculated from XPS

Elements	Pb	Zr	Ti	O	Zr:Ti	Zr + Ti
Target atomic concentration (%)	20	10.4	9.6	60	52:48	20
Film atomic concentration (%)	20.61	11.83	8.31	59.25	58.73:41.26	20.14

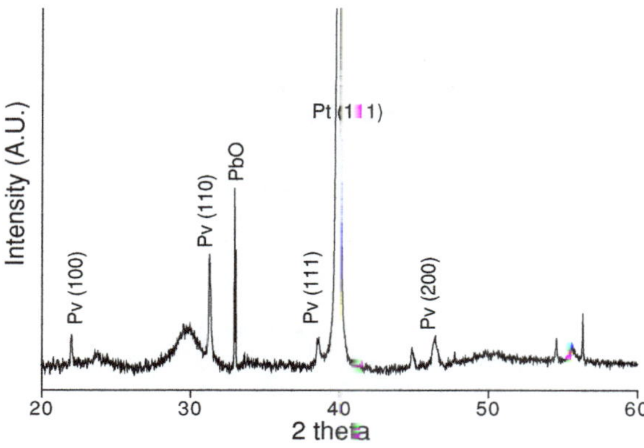

Fig. 13 X-ray diffractogram of annealed PZT thin film

The elements behavior of PZT thin film shows that the film was having higher contents of Pb at surface as compared to depth. This was attributed to diffusion of Pb from the interior of the thin film towards the surface. As Pb diffuses toward surface, the Zr, Ti content increased in bulk film. This trend may be due to Ti and Zr diffusion in bulk film. Ti is lighter element as compared to Pb and Zr, thus Ti gets diffused in bulk film easily. At depth this trend was observed as Ti concentration increases followed by Zr. The atomic percentages calculated using XPS are in good agreement with results obtained using EDS.

The PZT thin film was annealed using CFA. The effect of post annealing condition on phase formation of PZT thin film was studied using XRD. The film annealed at 650 °C for 60 min showed good crystalline behavior. The Pt peak at 40° was more dominating thus modified spectra with all peaks are plotted in Fig. 13.

Annealed film exhibited crystalline structure with ferroelectric perovskite (Pv) phase. The pyrochlore (Pyro) phase was also observed. The annealed film shows the low intensity peaks at 21.97, 31.21, 38.5, 44.7, and 46.38° corresponding to perovskite phase along with pyrochlore peak at 29.8°. The peaks observed at 40 and 32.9° corresponds to Pt and PbO.

In annealed film, perovskite <110> was more dominant; the perovskite <100> and <111> were also observed. During CFA, the temperature rise rate was fixed at 10 °C/min. This rate exposes the PZT material to lower temperature (then annealing temperature) for time sufficient to form pyrochlore phase. Thus the pyrochlore phase was observed in thin film. The same results were observed for PZT thin film annealed using CFA by researcher earlier. This XRD data revealed that the PZT is crystallized in <110> tetragonal phase with cell parameter a = 4.036 and c = 4.146 [JCPDF 33 0784].

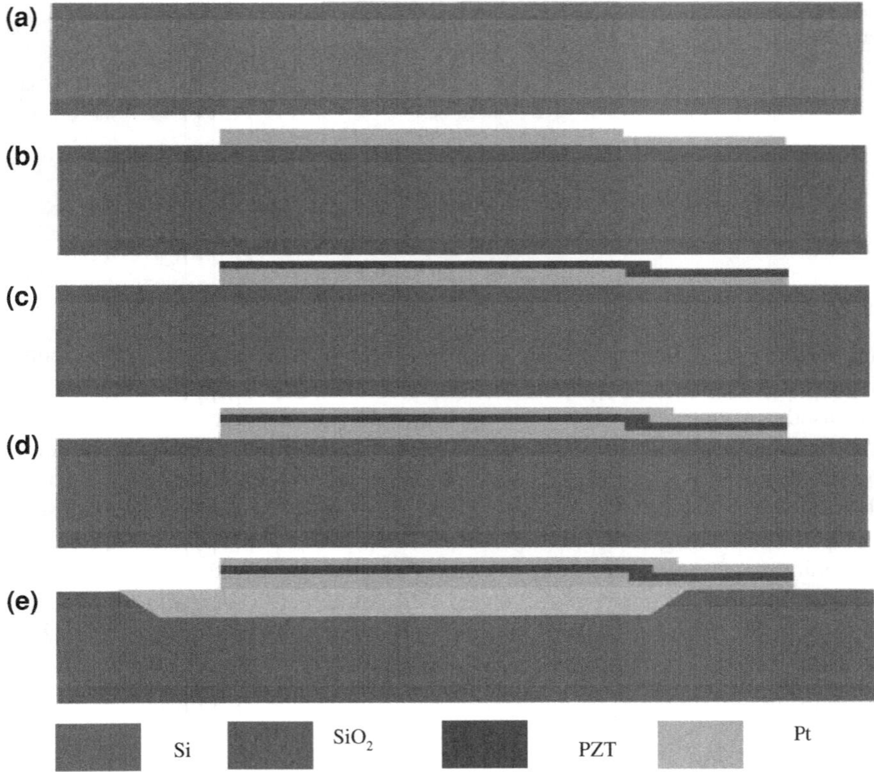

Fig. 14 Fabrication flow of PZT cantilever **a** silicon dioxide **b** patterned Pt layer for bottom electrode **c** annealed and patterned PZT layer **d** patterned Pt layer for top electrode **e** released PZT cantilever

The d_{33} measurement was done on PZT thin film annealed at 650 °C for 30 min. The PZT thin film shows the d_{33} value 450 pm/V, and is comparable to d_{33} value of bulk PZT target 407 pm/V. The PZT thin film was optimized using magnetron sputtered deposition followed by the conventional furnace annealing. The optimized film was showing good piezoelectric behavior. Detailed study of PZT film processing can be found elsewhere [20, 21].

5.3 Fabrication of Piezoelectric Switch

PZT-based cantilever fabrication was done using silicon bulk micromachining. Figure 14 shows the fabrication flow. PZT based cantilever fabrication process started with n type <100> silicon wafer. The 800 nm silicon dioxide was grown using thermal oxidation (Fig. 14a). The oxide layer was patterned in order to expose

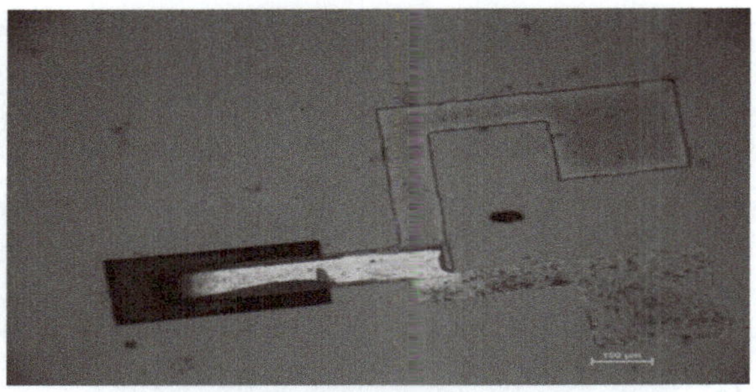

Fig. 15 Fabricated PZT cantilever

underneath silicon material to etchant. The 150 nm platinum (Pt) layer was sputter deposited and patterned to serve as a bottom electrode (Fig. 14b). The titanium layer was used as adhesive layer prior to deposition of platinum layer. The 600 nm PZT thin films was deposited on Pt surface using single target RF magnetron sputtering system. The sputtered PZT layer was patterned and annealed in perovskite phase (Fig. 14c). The 150 nm Pt layer was sputtered on PZT surface and patterned to serve as top electrode (Fig. 14d). Finally, the PZT cantilever was released using silicon bulk micromachining using KOH solution. The released PZT cantilever is shown in Fig. 14e and optical image of fabricated PZT cantilever is shown in Fig. 15. It is seen from the figure that the cantilever is released properly using silicon bulk micromachining and no much bending due to stress is observed.

6 Conclusions

The piezoelectric-based cantilever is designed. The ferroelectric material—PZT—is selected for developing piezoelectric RF MEMS switch. The PZT film is deposited using sputtering technique and the various deposition parameters are successfully optimized to get stoichiometric, smooth film of thickness 600 nm. The PZT cantilever in bulk micromachining is fabricated successfully and therefore shows the feasibility of using the same in RF MEMS switch.

Acknowledgment The authors acknowledge NPSM and NPMASS for providing funds for "Establishment of National MEMS Design Centres" (PARC#2 Project No. 2.10). The authors also acknowledge Department of Science and Technology, Govt. of India for providing funds for fabrication of devices.

References

1. Rebeiz GM (2003) RF MEMS—theory, design and technology. John Wiley and Sons, New Jersey
2. Kshirsagar AV, Duttagupta SP, Gangal SA (2010) Optimization of contact force and pull-in Voltagefor series based MEMS switch. Sens Transducers J 115(4):43–47

3. Kshirsagar AV (2011) Novel low stress silicon nitride, oxide and oxynitride surface micromachined cantilevers with PMMA sacrificial layer for low process temperature MEMS devices. PhD thesis, University of Pune
4. Kshirsagar AV, Apte P, Duttagupta SP, Gangal SA (2010) Optimization of pull-in voltage and contact force for MEMS series switch using Taguchi method. In: Proceedings of IEEE ICSE Proc Malaysia, pp 279–282
5. Siegel C, Ziegler V, Schönlinner B et al (2006) Simplified RF-MEMS switches using implanted conductors and thermal oxide. In: Proceedings of the 36th european microwave conference, Manchester UK, pp 1735–1738
6. Mullah MN, Karmakar M C (2001) RF MEMS switches: paradigms of microwave switching. In: Proceedings of APMC2001 Taipei Taiwan, pp 1024–1067
7. Balaraman D, Bhattacharya SK, Ayazi F et al (2002) Low cost low actuation voltage copper RF MEMS switches. In: Proccedings of IEEE MTT-S Digest, pp 1225–28
8. Joshi AB (2012) Deposition and optimization of PZT thin film parameters for possible use in MEMS applications. Ph.D Thesis, University of Pune, Pune, India
9. Pacheo SP, Linda P, Katehi B et al (2000) Design of low cost voltage RF MEMS switches. In: Proccedings of IEEE MTT-S Digest, pp 165–168
10. Goldsmith CL, Yao Z, Eshelman S et al (1998) Performance of low-loss RF MEMS capacitive switch. IEEE Microwave Guided Wave Lett 8:269–271
11. Newman H S (2002) RF MEMS switches and applications. In: Proceedings of 40th annual international reliability physics symposium, Dallas Texas, pp 111–115
12. Park JY, Kim GH, Chung KW et al (2001) Monolithically integrated micromachined RF MEMS capacitive switches. Sens Actuators A 89:88–94
13. Chiang S, Feng M (1999) Low actuation voltage RF MEMS switches with signal frequencies from 0.25 GHz to 40 GHz. In: Proccedings of IEDM technical digest, pp 689–692
14. Peroulis D, Sarabandi K, Linda P et al (2000) Low contact series MEMS switches. In: Proceedings of IEEE MTT-S Digest, pp 233–226
15. Yao JJ, Chien C, Mihailovich R et al (2001) Microelectromechanical system radio frequency switches in a picosatellite mission. J Smart Mater Struct 10:1196–1203
16. Hwang YS, Paek SH (1995) A study of the properties and interface of Lead Zirconate Titanate thin film with substrate temperature. J Mater Sci Lett 14:640–642
17. Liu C (2006) Foundation of MEMS, Pearson Education International Publication, USA
18. Hoffmann M, Kiippers H, Schneller T et al (2003) Theoretical calculations and performance results of a PZT thin film actuator. IEEE Trans Ultrason Ferroelectr Freq Contro 50(10):1240–1246
19. Joshi AB, Bodas D, Gangal SA (2010) Design of Piezoelectric PZT cantilever for actuator application. Sens Transducers J 123(124):15–24
20. Joshi AB (2011) Deposition and optimization of PZT thin film parameters for possible use in MEMS application. Ph.D. Thesis, University of Pune, India
21. Joshi AB, Bodas D, Rauch JY et al (2012) Optimization of RF sputtered PZT thin film for MEMS cantilever application. In: IEEE proceeding series, International conference on physics and technology of sensors, pune (ISPTS-1) Pune India, pp 173–176

Part III
Materials and Processes

Nickel–Titanium Shape Memory Alloy Wires for Thermal Actuators

S. K. Bhaumik, K. V. Ramaiah and C. N. Saikrishna

Abstract Shape memory alloys (SMAs) are a class of functional materials which exhibit unique properties of shape recovery under external stimuli such as thermal, mechanical, and magnetic fields. NiTi SMAs find wide applications as solid-state thermal actuators in a variety of aerospace, automobile, and engineering systems. Compared to conventional actuators, SMA actuators have the advantage of a high force to mass ratio, simplicity and miniaturization of design, fewer moving parts, and silent operation. This article deals with the processing and properties of NiTi SMA wires for thermal actuator applications. The effect of thermo-mechanical processing history and the employed stress–strain–temperature conditions on the functional fatigue behavior of the NiTi SMA thermal actuators has been discussed.

Keywords Shape memory alloys · Thermal actuator · Processing · Microstructure · Thermo-mechanical fatigue

1 Introduction

In recent times, research initiatives toward the development of new technologies based on shape memory alloys (SMAs) and their integration into existing systems to achieve greater functionalities have gathered momentum [1]. Thermal actuator devices using SMAs as the active elements are in demand in various automobile, aerospace and other engineering systems [2–5]. Compared to conventional actuation systems, SMA-based actuators have the advantage of high energy density, i.e., high stress and strain can be generated in a relatively low volume of material [2, 5]. As a result, the integration of SMA-based smart components and devices into systems can be accomplished in a compact arrangement. In addition, SMA-

S. K. Bhaumik (✉) · K. V. Ramaiah · C. N. Saikrishna
Materials Science Division, CSIR-National Aerospace Laboratories, Bangalore 560 017, India
e-mail: subir@nal.res.in

K. J. Vinoy et al. (eds.), *Micro and Smart Devices and Systems*,
Springer Tracts in Mechanical Engineering, DOI: 10.1007/978-81-322-1913-2_11,

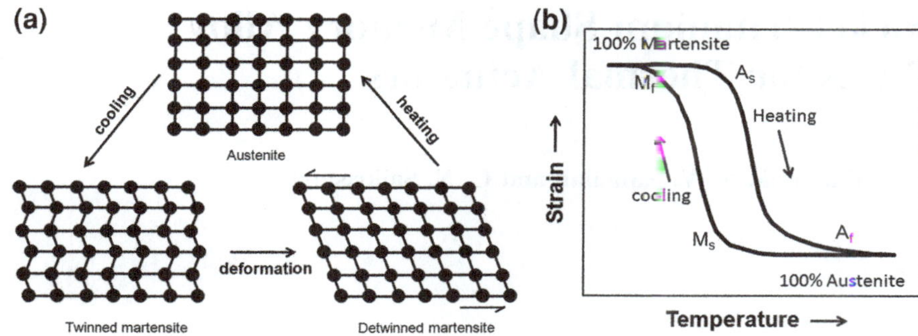

Fig. 1 Schematic of **a** shape recovery in shape memory alloys **b** strain versus temperature plot indicating characteristic transformation temperatures; M_s/M_f and A_s/A_f are martensite and austenite start/finish temperatures

based systems are simple in construction, involve fewer moving parts, have silent operation and are easy to maintain [2].

Though the science of SMAs has been known for the many decades, alloy preparation and the processing of SMAs into useful products with desired functional properties is still a challenge [6]. Further, the properties and behavior of SMAs are greatly influenced by the alloy chemistry and its subsequent thermo-mechanical processing history [4, 7]. The precise control of composition and a judicious combination of cold work and shape memory heat treatment temperature/time is essential to obtain good shape memory properties. This article provides an insight into the processing and properties of NiTi SMA wires for thermal actuator applications. The effect of thermo-mechanical processing history and the employed stress–strain–temperature conditions on the fatigue behavior of NiTi SMAs have also been discussed.

1.1 Origin of Shape Memory

The basis for shape memory behavior in SMAs is the reversible crystallographic transformation from a high temperature austenite phase (cubic B2) to a low temperature martensite phase (monoclinic B19′). Martensitic transformation is an athermal, shear-type diffusion-less phase transformation which takes place by the coordinated movement of atoms (military type), displacements being limited to a fraction of its atomic distance [8]. The shape and volume change associated with the change in crystal structure is self-accommodative in SMAs and takes place by the process of twin formation. In NiTi SMAs, as many as 24 twin-variants are formed during this accommodation process. On the application of external stress, detwinning of martensite takes place and the twin-variants which are favorably oriented in the stress direction grow at the expense of other variants, resulting in macro-strain in the material. This strain in SMAs is completely recoverable upon heating above a critical temperature, A_f (Fig. 1).

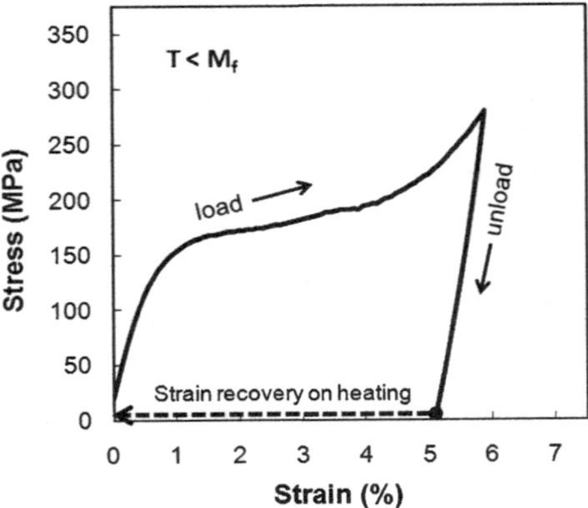

Fig. 2 Shape memory effect in a typical NiTi alloy

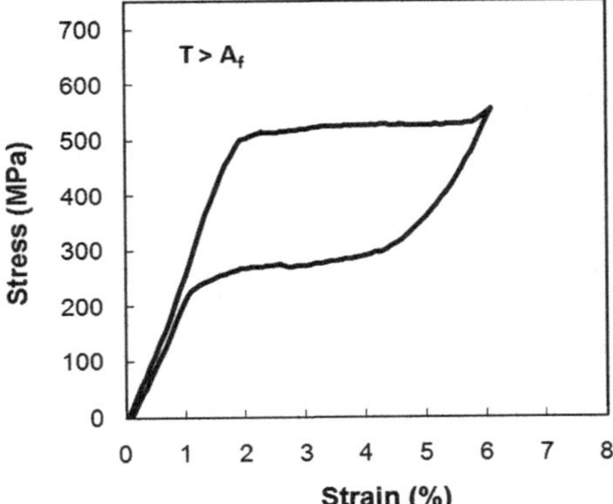

Fig. 3 Superelasticity in a typical NiTi alloy

1.2 Shape Memory Effect and Superelasticity

SMAs undergo reversible crystallographic transformation from a high symmetry austenite phase to a low symmetry martensite phase. The austenite to martensite transformation can be temperature induced or stress-induced, a decrease in temperature having a similar effect to that of an increase in stress [8]. This temperature and stress-induced phase transformations in SMAs are associated with shape memory effect (SME) and superelasticity (SE, also called pseudoelasticity), respectively. The SME refers to the ability of the alloy to be deformed in its low temperature phase (martensite), the deformation being retained in the material till it is heated above a critical temperature (A_s) (Fig. 2). The temperature of the shape recovery is characteristic of the particular alloy system and its chemistry [8]. SE refers to the ability of the alloy to undergo large deformation strains upon loading and restoration of the original shape upon the release of the applied load (Fig. 3). It is an isothermal process and occurs at temperatures in the range of A_f to M_d by a stress-induced martensitic (SIM) transformation during loading, and a subsequent reverse transformation upon unloading. The stress required for SIM formation increases linearly with temperature, and beyond M_d, it exceeds the stress required to deform the SMAs by slip mechanism, resulting in irreversible plastic deformation in the material.

1.3 SMA Thermal Actuator: Concept

The SMA thermal actuator works based on the phenomenon of the shape memory effect. The deformation of the SMA in the martensite phase and the subsequent heating to the austenite phase to recover the original shape/strain without any constraint or interference is referred to as free recovery. The applications of SMA with free recovery are limited, and mainly restricted to one-time shape recoveries. In thermal actuator applications, the SMAs recover their shape under full or partial constraints, known as constrained recovery. The recovery stress generated in the SMA due to the constraint is utilized for doing useful work (actuation). For cyclic actuation, the deformation in the martensite phase is provided through an external bias force. This force can be applied to the SMA element through a bias spring or dead load or compliance of the system. Although SMAs are capable of generating large strokes/forces, their operating frequencies are quite low (<10 Hz) [5]. This is mainly because of the limitation imposed by the necessity of cooling the SMA to the martensite phase, during temperature excursions in each cycle of actuation.

1.4 NiTi Shape Memory Alloys

Among the many alloy systems known to exhibit shape memory behavior, NiTi-base SMAs have the best combination of mechanical and functional properties [2, 6]. These properties include a shape memory strain of 6–8 %, recovery stress of 600–800 MPa, ultimate tensile strength of 1,200–1,400 MPa and elongation to failure of 12–15 % [2, 4, 6, 7]. Also, the NiTi SMAs have better stability upon thermo-mechanical cycling, a high fatigue life and superior corrosion resistance [4]. This combination of properties makes NiTi alloys a superior candidate material for the fabrication of SMA-based thermal actuators.

The composition range in which NiTi alloys exhibit good shape memory properties is quite narrow, 49.5–51.0 at.% Ni [6, 9–11]. The phase boundary for a single phase NiTi on the Ti-rich side is very steep and close to 50 at.% Ni, whereas on the Ni-rich side, the solubility of Ni decreases with a decrease in the temperature [6]. A deviation from the equi-atomic composition results in a two-phase microstructure with an equilibrium precipitate of Ti_2Ni or $TiNi_3$, depending on whether the alloy is Ti-rich or Ni-rich, respectively [6]. The presence of precipitate phases usually increases the matrix strength. But, it has a deleterious effect on alloy ductility, shape memory properties, and fatigue life. Therefore, a precise control over the composition is important to obtain the desired properties in NiTi SMAs. NiTi alloys with different A_f in the range of −20 and +100 °C can be obtained by maintaining the Ni:Ti atomic ratio in the range of 51.2:48.8 and 50:50, respectively. Generally, Ni-rich compositions are chosen for superelastic applications, and equi-atomic or slightly Ti-rich compositions for shape memory applications.

2 Fabrication of NiTi SMA Wire

2.1 Alloy Preparation

The properties of NiTi alloys, specifically the transformation temperatures (TTs), are very sensitive to chemical composition. Typically, a 0.1 at.% increase in Ni content decreases the M_s of the alloy by about 10 °C [6, 10]. Further, the pick-up of impurities such as C, O, N, H, etc., in the alloy, either from the constituent raw materials or during alloy melting, results in lower TTs due to the high affinity of Ti to form compounds with these elements. The formation of titanium oxides, oxinitrides, carbides, etc., results in an alloy matrix rich in Ni with a consequential decrease in the TTs. Also, these compounds are brittle in nature and hence, adversely affect the workability, shape memory as well as the fatigue properties of the NiTi SMAs.

The extreme sensitivity of NiTi SMAs to alloy chemistry imposes serious difficulties on process control during melting practice. Most SMA applications require control of the TTs within ± 5 °C, equivalent to a tolerance in alloy composition of ± 0.05 at.%. The most commonly used technique for the preparation of NiTi alloy is the vacuum induction melting (VIM) followed by refinement using vacuum arc re-melting (VAR). To minimize the impurities in the alloy, high purity raw materials, titanium (Ti > 99.7 wt.%) and nickel (Ni > 99.9 wt.%), are used. The melting is generally carried out in an inert atmosphere, after evacuating the furnace chamber to high vacuum levels of the order of 10^{-6} mbar. Further, to avoid carbon pick-up in the melt, the graphite crucible used during the VIM is provided with a coating of oxides of zirconium, yttrium, etc.

2.2 Secondary Processing

In spite of the extreme care exercised during melting, the cast alloy is chemically and microstructurally inhomogeneous on a micro-scale (Fig. 4a). The inhomogeneity arises because of nonequilibrium cooling during solidification and the development of microstructural features such as dendrites, columnar grains,

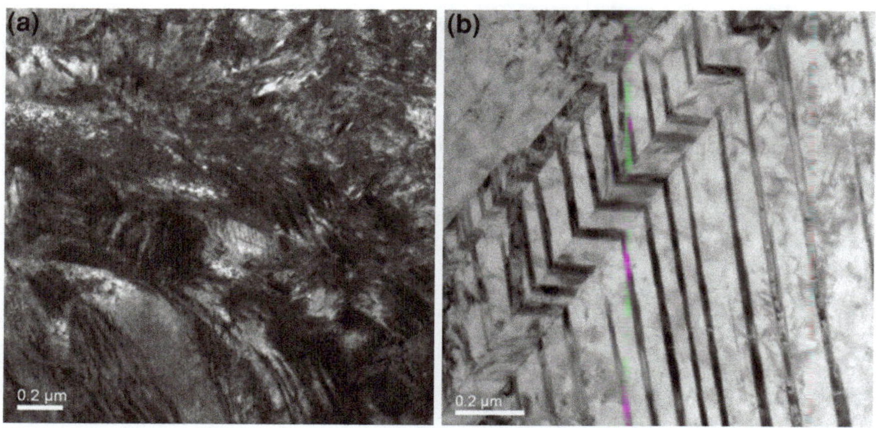

Fig. 4 TEM images of NiTi alloy: **a** cast microstructure and **b** hot worked microstructure showing well developed twins

precipitation of nonequilibrium phases, and solidification porosity. As a consequence, SMAs in a cast condition exhibit relatively low ductility and poor shape memory properties. To improve the properties, the microstructure of the cast alloy requires refinement through hot working followed by a series of cold working steps with intermediate annealing treatment.

For fabrication of wires, hot working of the cast NiTi alloy is carried out in the temperature range 850–900 °C. The selected temperature range ensures satisfactory workability of the alloy with a minimal surface oxidation. The reduction in cross-section in each step of working is limited to 20–30 % to avoid edge cracking and the generation of folds or laps in the bulk of the material. The minimum cumulative reduction in cross-section necessary during hot working to completely break the cast structure is in the range 70–80 % (Fig. 4b).

In comparison with conventional metals and alloys, the cold working of the NiTi alloy is difficult because of the high strain hardening coefficient. For example, the yield strength of the NiTi alloy in a fully annealed condition is about 600 MPa, and it increases to 1,200–1,600 MPa after about 40 % deformation [6, 7]. Therefore, a reduction in cross-section of the hot worked rod is carried out in multiple steps, interspersed with annealing after every step of cold deformation. An annealing temperature of 750–800 °C for 10–20 min is adequate to soften the material and facilitate the 20–30 % reduction in the cross-section in each step. An appropriate combination of cold work and intermediate annealing results in a material with a completely recrystallized and refined microstructure (Fig. 5).

Finally, round wires of desired dimensions are fabricated by a die drawing process. Both hot and cold drawing processes are adopted. Hot drawing is carried out at 700 °C. Because of dynamic recrystallization, the hot drawn wire has a fully recrystallized microstructure and significantly low yield strength. Alloys with such microstructure exhibit poor stability in functional properties and, hence, are not

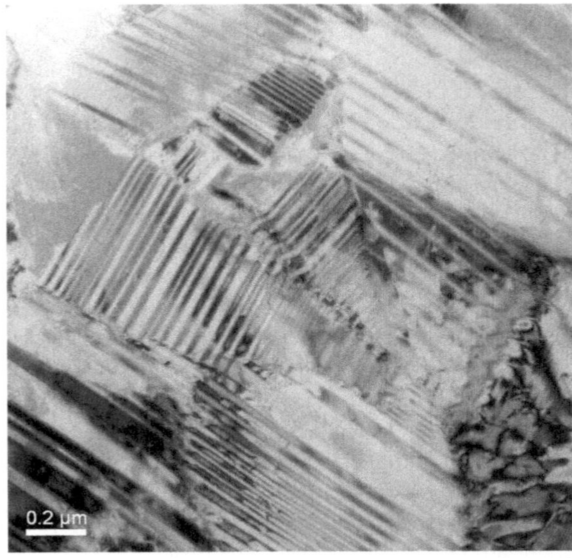

Fig. 5 TEM image of cold worked and annealed NiTi SMA wire showing uniform structure with self-accommodated twins

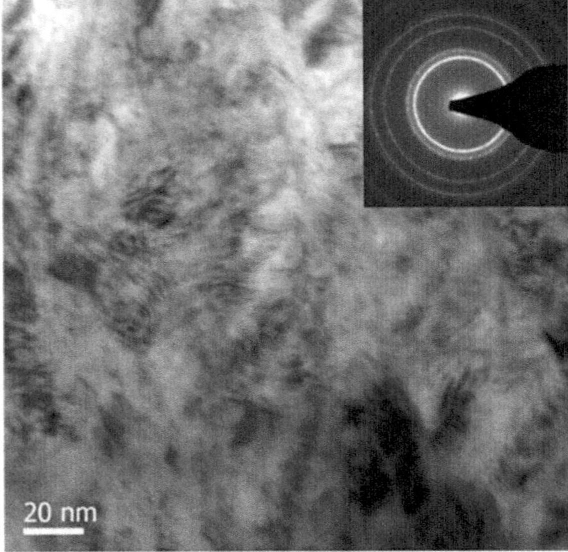

Fig. 6 TEM image of 40 % cold drawn NiTi SMA wire with heavily deformed structure and amorphisation (*inset*)

suitable for thermal actuator applications. Therefore, the hot drawing is terminated at a stage when the wire diameter is at least two times that of the final desired wire diameter. Following this, a cold drawing process is adopted in multiple steps with inter-pass annealing in the temperature range 600–650 °C. The reduction in the wire cross-section in each step is limited to 30–40 %. The cold drawn NiTi SMA wire shows a severely deformed microstructure with a high density of dislocations interspersed with regions of amorphous structure (Fig. 6).

2.3 Shape Memory Heat Treatment

The cold drawn NiTi SMA wire shows poor shape memory properties because of the microstructure shown in Fig. 6. Hence, to modify the microstructure, a moderate temperature annealing treatment of the cold drawn wire is necessary. Generally, for thermal actuator applications, the heat treatment is carried out below the recrystallization temperature, in the range 350–500 °C [6, 7, 9, 10].

To obtain a satisfactory combination of mechanical and shape memory properties in the wire, a judicious combination of percent cold work and the annealing temperature/time is required. Cold work introduces random dislocations in the material and thereby increases the yield strength of the matrix. But, a high density of random dislocations offers resistance to the movement of the twin boundaries as well as the martensite/austenite interfaces during the phase transformation. Annealing of the cold worked material helps in the rearrangement of the dislocations and facilitates the formation of cells of relatively dislocation free regions in the microstructure surrounded by dislocation networks. In such a microstructure, the martensite twin boundaries can move easily within the dislocation free regions under a relatively low stress. The dislocation networks maintain the yield strength of the matrix phase so that deformation in the material takes place preferably by the detwinning process and not by irreversible plastic deformation. Therefore, a suitable combination of retained cold work and annealing treatment is a decisive factor for obtaining wires with a balance between the twinned martensite phase and the defect concentrations in the microstructure. In general, for thermal actuator applications, this is achieved through the retention of 30–40 % cold work in the wire followed by annealing at 400–450 °C for a period of 15–30 min.

3 Properties of NiTi SMA Wires

The evaluation of NiTi SMAs for thermal actuator applications involves the study of physical, thermo-physical, mechanical and functional properties, and a dynamic response to external stimuli such as stress and temperature. SMA elements in service undergo large number of actuations and therefore, the study of degradation in the material properties upon repetitive thermo-mechanical cycling is important. The assessment of both structural and functional fatigue, in particular the latter, helps in ensuring a reliable design and performance of SMA-based actuators/systems.

3.1 Transformation Behavior

NiTi alloys undergo a temperature/stress induced phase transformation from B2 austenite to the B19′ martensite phase and vice versa (Fig. 7a). Under certain thermo-mechanical processing conditions [6, 12], a pre-martensitic phenomenon

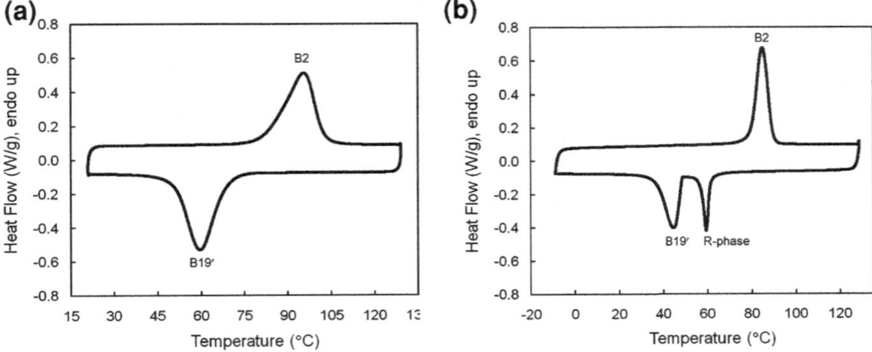

Fig. 7 DSC scans of NiTi alloy **a** fully annealed condition **b** 40 % cold work followed by annealing at 475 °C for 15 min

of the R-phase (rhombohedral crystal structure) transformation takes place in the material (Fig. 7b). The R-phase is a martensite-like first-order transformation, which displays a narrow hysteresis of 1–2 °C and a shape recovery strain of less than 1.0 %. The practical usefulness of the R-phase, however, is limited because of its low recovery strain/stress.

3.2 Stress–Strain Characteristics

The mechanical behavior of NiTi SMAs in cold worked conditions is similar to that of the conventional metallic materials and does not exhibit shape memory behavior (Fig. 8). As discussed in Sect. 2.3, a moderate temperature annealing is, therefore, necessary to restore SME in the cold worked wire. The typical stress–strain behavior of an annealed NiTi SMA wire in the martensite (25 °C) and the austenite (100 °C) phase is shown in Fig. 8. In both these phases, the SMA wire exhibits a stress plateau prior to yielding by plastic deformation. The plateau regions in the martensite and austenite phases correspond to the strains produced in the material by the detwinning of martensite and SIM, respectively. Typically, the plateau strain in the martensite phase and the plateau stress in the austenite phase are the maximum recoverable strain and recovery stress, respectively, that can be generated in the material. Therefore, an assessment of these properties is a pre-requisite for designing a SMA actuator for the envisaged application.

3.3 Dynamic Characteristics

The SMAs used in thermal actuator applications are subjected to repetitive temperature excursions (heating/cooling) under variable/constant stress, referred to as thermo-mechanical cycling (TMC). The heating and cooling of the SMA actuator

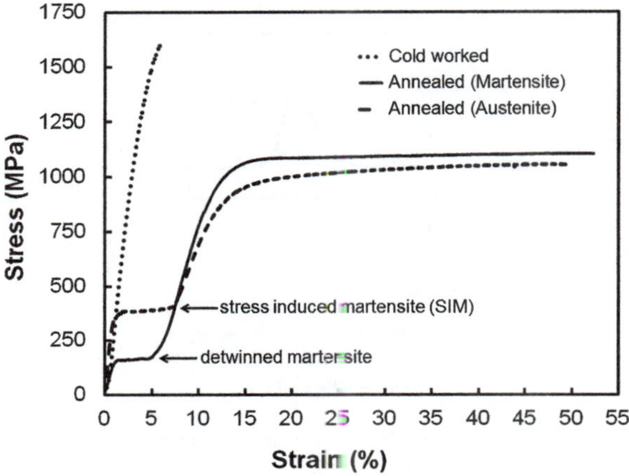

Fig. 8 Stress–strain plots of NiTi SMA wire in martensite and austenite phases; cold worked to 40 % and annealed at 475 °C for 15 min

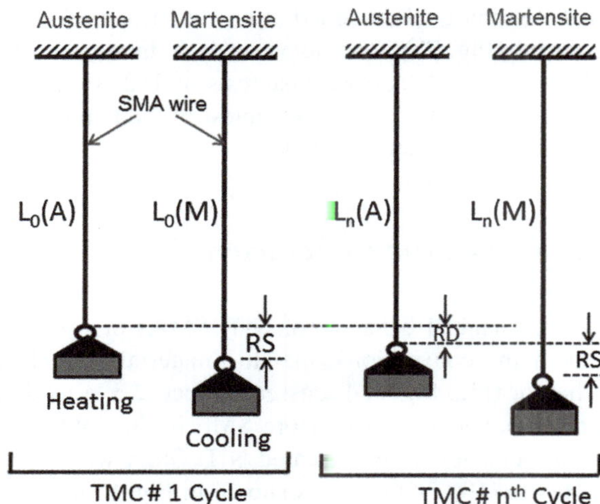

Fig. 9 Schematic showing RS and RD in SMA wire upon TMC

are generally achieved by the passage of electric current and forced/natural air cooling, respectively. The behavior of the SMA actuator during TMC is shown schematically in Fig. 9. In the figure, the recovery strain (RS) refers to the instantaneous strain recovered under load upon heating from the martensite to the austenite phase, and remnant deformation (RD) refers to the increase in the length of the wire in the austenite phase after an 'n' number of cycles.

The response of a typical NiTi SMA wire subjected to TMC under a constant stress of 200 MPa is shown in Fig. 10a. The behavior of the wire is not stable, and that the RS and RD in the wire undergo changes continuously. The changes are quite significant during the initial few hundred cycles, and the rate of change attains a steady state as the TMC progresses. Studies [13–26] have shown that the unstable functional behavior of the SMA wire is primarily because of the

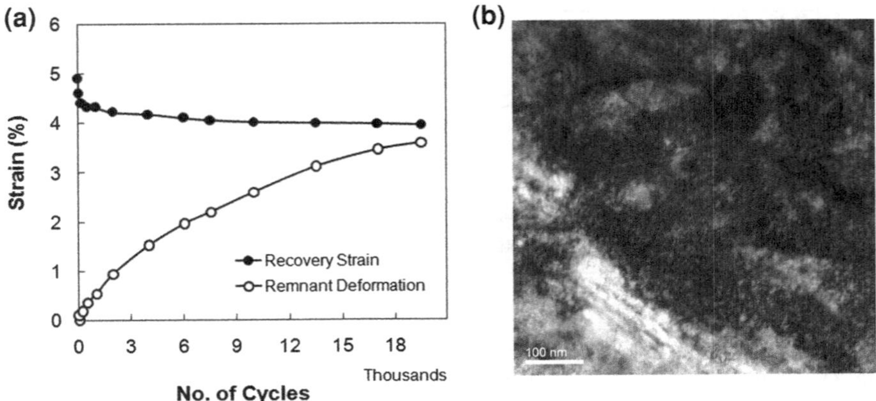

Fig. 10 **a** Strain versus number of cycles plot of NiTi SMA wire at 200 MPa **b** TEM micrograph showing generation of forest of dislocations in the microstructure of the SMA wire after 200 cycles of TMC

generation of defects in the microstructure and the stabilization of the martensite/austenite phase, which eventually does not contribute to the transformation strain during TMC (Fig. 10b).

4 Fatigue in NiTi SMA Thermal Actuators

Fatigue or cyclic loading through the transformation range is one of the generic characteristic features of SMA thermal actuators. Fatigue in SMAs is complex and does not follow the conventional fatigue life theory because of the involvement of phase transformations during stress/strain cycling. Also, since fatigue damage mechanisms vary widely in austenite and martensite phases [23–25, 27–30], the interpretation of the fatigue data in SMAs is exceedingly difficult and often not possible.

Once the composition of a NiTi alloy is chosen, the important factors that affect the fatigue behavior of the actuator wire are the thermo-mechanical processing history and the stress-strain-temperature regime of operation. These aspects have been discussed in detail in the following sections.

4.1 Thermo-Mechanical Processing History

As mentioned in Sect. 2.3, the optimization of retained cold work and annealing temperature/time is essential to achieve satisfactory shape memory properties in SMAs. It is well known [6, 7] that the increase in cold work and the decrease in the

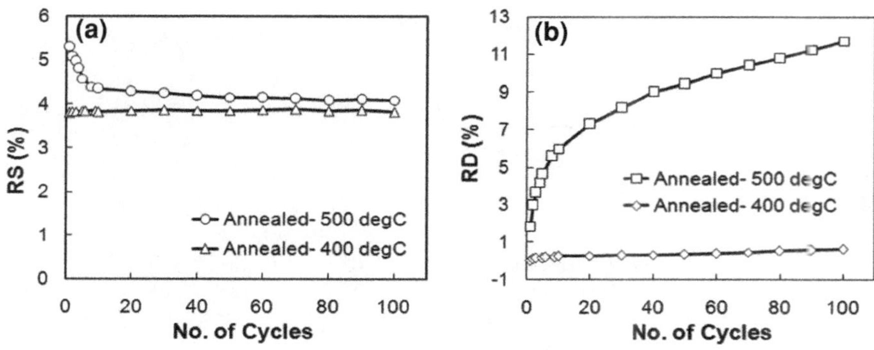

Fig. 11 Strain response of NiTi SMA wires upon TMC at 200 MPa: **a** RS versus number of cycles and **b** RD versus number of cycles; wires annealed at 400 and 500 °C for 15 min

annealing temperature have a similar effect on the mechanical and functional properties of SMAs. A higher amount of retained cold work in the material increases the yield strength of the matrix and, hence, inhibits the accumulation of plastic strain (RD) upon TMC. But, such cold worked material has poor recoverable strain. On the other hand, if the annealing temperature is high, the material exhibits high RS, but is prone to the accumulation of large RD [17–19]. The final selection of the amount of retained cold work and the annealing treatment schedule is, therefore, largely dictated by the stress-strain-temperature regime of application for the actuator.

Figure 11 shows the strain response of NiTi SMA wires during the first 100 cycles of TMC at 200 MPa. The wires have a retained cold work of 35 % and annealed at two different temperatures, 400 and 500 °C for 15 min. It can be seen that the functional response of the wire annealed at 500 °C is highly unstable. The accumulation of RD at the end of 100 cycles is as high as 11 % (Fig. 11b). Also, the decrease in the RS is substantial in the initial cycles. In comparison, the wire annealed at 400 °C shows relatively stable functional behavior with an RD less than 1.0 % and a minimal decrease in the RS.

The fatigue behavior of the two NiTi SMA wires described above can be correlated with the TTs (Fig. 12a) and the stress–strain characteristics of the individual wires (Fig. 12b). The higher the annealing temperature, the higher are the TTs and the lower is the yield strength. The low yield strength of SMAs facilitates the accumulation of defects in the material. This in conjunction with high TTs, which require higher temperature excursions in the wire during TMC, accelerates the accumulation of RD in the actuator. The reverse is true when annealing is carried out at lower temperatures. Hence, the annealing of the cold worked material at the minimum possible temperature is desirable for applications where a greater functional stability over a large number of cycles is required.

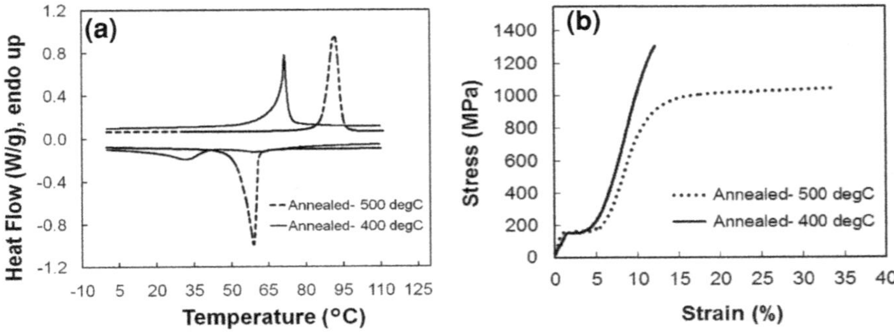

Fig. 12 NiTi SMA wires annealed at 400 and 500 °C for 15 min: **a** transformation temperatures **b** stress–strain behavior in martensite phase

4.2 Stabilization of Functional Behavior

Studies [7, 10, 16, 19] have shown that it is possible to minimize RD in the NiTi SMA wires during TMC through the appropriate selection of a thermo-mechanical processing schedule. However, it is not possible to eliminate the same completely. The accumulation of the RD in the SMA wire, irrespective of the processing schedule adopted, is quite significant in the initial cycles of TMC, and this imposes a limitation on the use of this material as an actuator. Therefore, the functional behavior of the SMA wire needs to be stabilized before it can be used for the fabrication of thermal actuators.

Instability in the behavior of the NiTi SMA wire upon TMC is primarily associated with the generation of defects in the material [19, 23–25]. As shown in Fig. 10a, the rate of accumulation of RD is significantly high in the initial few cycles and then decreases monotonically. This indicates that the generation of defects in the as-processed wire is relatively easy, and slows down drastically as the defect density in the microstructure reaches saturation. This provides an option to overcome the problem of the unstable behavior of the SMA actuator through the saturation of defects in the microstructure, prior to the application itself.

One effective method for the stabilization of NiTi SMAs is TMC of the as-annealed wires for a few cycles at stresses higher than the application stress [17, 21, 26, 31]. TMC at high stress accelerates the process of defect generation as well as the stabilization of a certain volume fraction of the martensite and austenite phases in the microstructure. Subsequently, when the TMC stress is lowered to that of the application stress, the defect generation in the material slows down considerably, and the further stabilization of phases becomes negligibly small. As a result, the SMA actuator behaves in a relatively stable manner. This is demonstrated through Fig. 13, where the dynamic strain response of a NiTi SMA wire undergoing TMC at 200 MPa stress, before and after stabilization treatment, has been shown. In this case, the wire with 35 % cold work and annealed at 400 °C for 15 min has been stabilized through TMC at 300 MPa for 25 cycles.

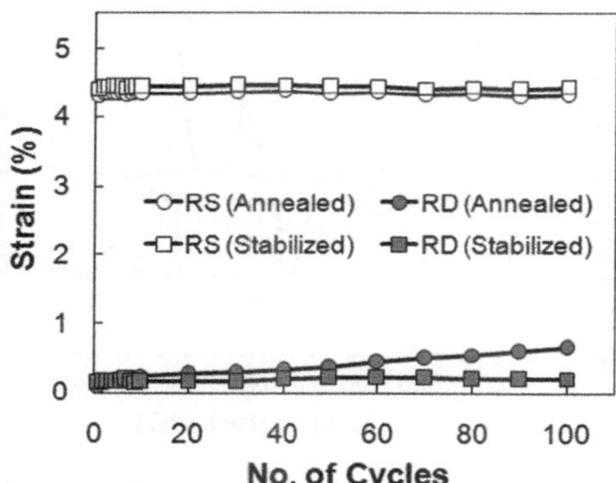

Fig. 13 Strain versus number of cycles plot upon TMC of NiTi SMA wires under a stress of 200 MPa before and after stabilization treatment, wire annealed at 400 °C for 15 min

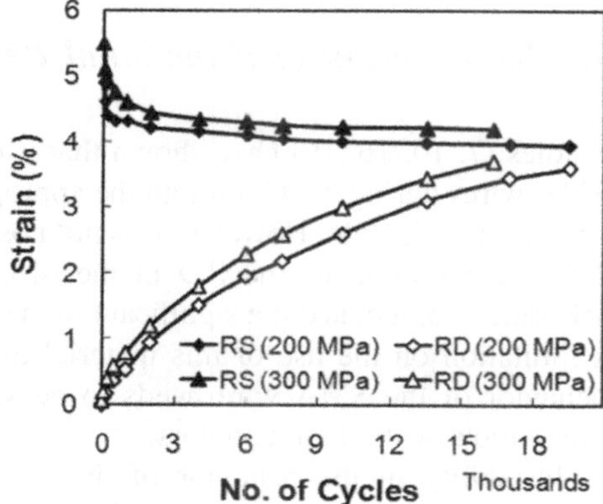

Fig. 14 Strain versus number of cycles plot of NiTi SMA wire upon TMC at 200 and 300 MPa; wire annealed at 450 °C for 15 min

4.3 Stress–Strain–Temperature

The functional as well as structural fatigue behavior of SMA actuators is strongly dependent on the stress–strain–temperature regime of application [15–19, 22]. Although the NiTi SMA wires exhibit high recovery stress and strain of the order of 600 MPa and 6 %, respectively, in practical applications requiring cyclic actuation, these are limited to 200–300 MPa and 2–4 %. In this range, the stress has very little effect on the fatigue life of the actuator. But, it has a noticeable effect on the functional fatigue behavior, the RD being greater at higher TMC stress (Fig. 14). On the other hand, the effect of recoverable strain on the fatigue behavior is drastic (Fig. 15). The wire subjected to TMC under 200 MPa stress at 4 % RS shows a fatigue life of about 17,000 cycles, significantly lower than that at 2 % RS (44,000 cycles). Also, the magnitude of RD is significantly high for the wire with higher RS.

The strain versus number of cycles plot of a NiTi SMA wire subjected to TMC under a stress of 200 MPa, and heated to different maximum temperatures (T_{max})

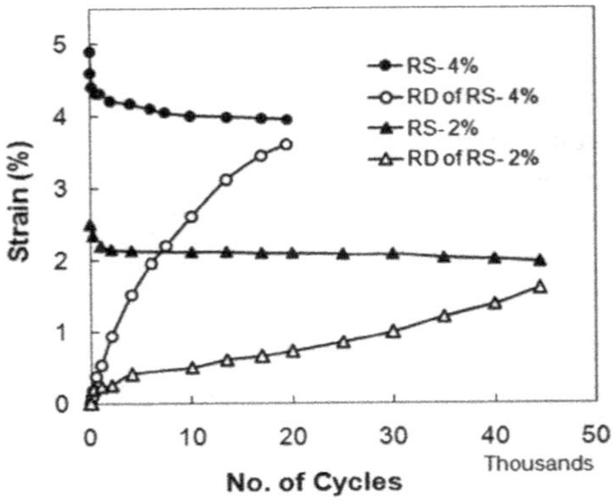

Fig. 15 Strain versus number of cycles plot of NiTi SMA wire upon TMC at 200 MPa with 2 and 4 % RS; wire annealed at 450 °C for 15 min

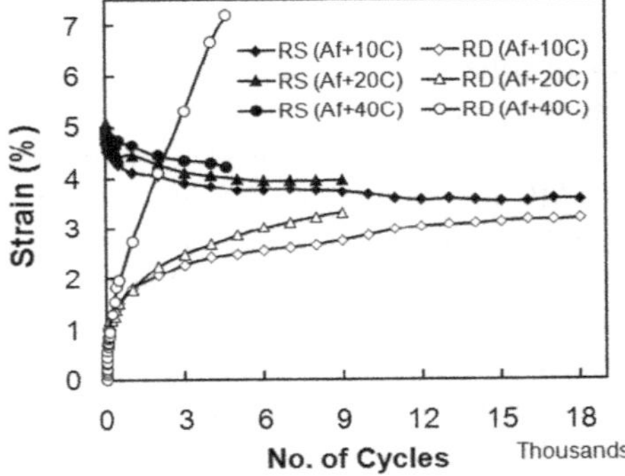

Fig. 16 Strain versus number of cycles plot showing fatigue life as a function of T_{max} in the wire; TMC was conducted at a stress of 200 MPa

is shown in Fig. 16. It can be seen that the fatigue life of the actuator wire is significantly reduced with an increase in T_{max}. The actuator wire heated to a T_{max} of A_f + 10 °C shows a fatigue life of ~18,000 cycles compared to 5,000 cycles for the wire with a T_{max} of A_f +40°C. The study shows that an increase in T_{max} by even 10 °C from A_f + 10 °C, decreases the fatigue life by about 50 %. Further, the variation in RD with an increase in T_{max} is significant, and indicative of accelerated defects generation in the material with the increase in wire temperature. It can be seen that the effect of temperature on the fatigue of the NiTi SMA actuator is far more deleterious than that of the applied stress or strain, within the limits mentioned earlier.

During TMC, depending on the application, the temperature excursion in the SMA wire can be chosen such that the material undergoes complete martensite (M) ↔ austenite (A) phase transformation or partial phase transformation such as M ↔ (M + A) or (M + A) ↔ A. Studies [18, 19, 21, 22, 32] have shown that the fatigue life of the NiTi SMA wire is strongly dependent on the thermal cycling regime. The actuator wire with TMC in the M ↔ (M + A) phase fields has a significantly higher fatigue life compared to that of the wire cycled in the phase fields (M + A) ↔ A or

M $\leftrightarrow$ A. In NiTi SMA, the degree of damage accumulation and fatigue crack growth characteristics are widely different in the martensite and austenite phases of the NiTi alloy [27–30]. The austenite phase is more prone to damage accumulation than the martensite phase. Hence, the volume fraction of these phases in the material during TMC is an important factor in determining the functional as well as structural fatigue of the NiTi SMA wire. The high fatigue life during TMC in the M $\leftrightarrow$ (M + A) phase fields can be attributed to two factors: (i) a low T_{max} in the wire during a temperature excursion and (ii) a low volume fraction of the austenite phase. Therefore, in cases where a part of the total available RS is required to be used, the fatigue life of the NiTi actuator can be enhanced significantly if the thermal cycling is restricted to the M $\leftrightarrow$ (M + A) phase fields.

5 Summary

An overview on the processing and properties of NiTi SMA wires for thermal actuator applications has been presented in this article. The effect of thermo-mechanical processing history, and the employed stress-strain-temperature regime of TMC on the functional fatigue behavior and life of the thermal actuator wires have been discussed. It has been shown that by a judicious choice of processing parameters, the properties of the NiTi SMA wires can be altered significantly. Stable functional behavior during repetitive actuation of the SMA thermal actuators over a large number of cycles can be achieved by tailoring the microstructure of the material through appropriate processing followed by a stabilization process. Also highlighted is the effect of application stress–strain–temperature on the fatigue behavior and life of the NiTi SMA actuator wire.

Acknowledgments The work presented in this study was carried out with financial support from the National Programme on Smart Materials (PARC#1 Materials and related process development; Project title: Development and processing of NiTi-base shape memory alloys for actuator applications) and XI-Five Year Plan Projects of CSIR, India. The authors acknowledge the help rendered by Mr. J. Bhagyaraj and Prof. Gouthama for the TEM study, and Mr. D. Paul for technical support.

References

1. Nespoli A, Besseghini S, Pittaccio S et al (2010) The high potential of shape memory alloys in developing miniature mechanical devices: a review on shape memory alloy mini-actuators. Sens Actuators A 158:149–160
2. Humbeeck JV (2001) Shape memory alloys: a material and a technology. Adv Eng Mater 3:837–850
3. Butera F (2008) Shape memory actuators. Adv Mater Process 166:37–40
4. Otsuka K, Wayman CM (eds) (1998) Shape memory materials. Cambridge University Press, London

5. Kumar PK, Lagoudas DC (2008) Introduction to shape memory alloys. In: Shape memory alloys: modelling and engineering applications. Springer, New York
6. Otsuka K, Ren X (2005) Physical metallurgy of Ti-Ni-based shape memory alloys. Prog Mater Sci 50:511–678
7. Saikrishna CN, Ramaiah KV, Allam Prabhu S et al (2009) On stability of NiTi wire during thermo-mechanical cycling. Bull Mater Sci 32:343–352
8. Wayman CM, Duerig TW (1990) An introduction to martensite and shape memory. Engineering aspects of shape memory alloys. Butterworth-Heinemann, London, pp 3–20
9. Russel SM (2001) Nitinol melting and fabrication. Proceedings of SMST-2000. California, USA, pp 1–9
10. Melton KN (1990) Ni-Ti based shape memory alloys. Engineering aspects of shape memory alloys. Butterworth-Heinemann, London, pp 21–35
11. Saburi T (1998) Ti-Ni shape memory alloys. Shape memory materials. Cambridge University Press, London, pp 49–96
12. Otsuka K (1990) Introduction to the R-phase transition. Engineering aspects of shape memory alloys. Butterworth-Heinemann, London, pp 36–45
13. Perkins J, Edwards GR, Such CR et al (1975) Thermomechanical characteristics of alloys exhibiting martensitic thermoelasticity. Shape memory effects in alloys. Plenum Press, New York, pp 273–303
14. Humbeeck JV (1991) Cycling effects, fatigue and degradation of shape memory alloys. J Phys IV 1:C4-189–C4-197
15. Hornbogen E (2004) Thermo-mechanical fatigue of shape memory alloys. J Mater Sci 39:385–399
16. Eggeler G, Hornbogen E, Yawny A et al (2004) Structural and functional fatigue of NiTi shape memory alloys. Mater Sci Eng A 378:24–33
17. Saikrishna CN, Venkata Ramaiah K, Bhaumik SK (2006) Effects of thermo-mechanical cycling on the strain response of Ni-Ti-Cu shape memory alloy wire actuator. Mater Sci Eng A 428:217–224
18. Lagoudas DC, Miller DA, Rong L et al (2009) Thermomechanical fatigue of shape memory alloys. Smart Mater Struct 18:085021–085033
19. Bhaumik SK, Saikrishna CN, Ramaiah KV (2012) Characteristic behaviour of NiTi SMA wire undergoing thermo-mechanical cyclic loading. In: Proceedings of ISSS-2012, IISc, Bangalore
20. Jones NG, Dye D (2011) Martensite evolution in a NiTi shape memory alloy when thermal cycling under an applied load. Intermetallics 19:1348–1358
21. Bhaumik SK, Saikrishna CN, Ramaiah KV et al (2008) Understanding the fatigue behaviour of NiTiCu shape memory alloy wire thermal actuators. Key Eng Mater 378–379:301–316
22. Bertacchini OW, Lagoudas DC, Calkins FT et al (2008) Thermo-mechanical cyclic loading and fatigue life characterization of nickel rich NiTi shape memory alloy actuators. Proc SPIE 6929:92916–92926
23. Simon T, Kröger A, Somsen C et al (2010) On the multiplication of dislocations during martensitic transformations in NiTi shape memory alloys. Acta Mater 58:1850–1860
24. Bhagyaraj J, Ramaiah KV, Saikrishna CN et al (2011) TEM studies on the microstructural changes during thermo-mechanical cycling of NiTi shape memory alloy wire. Mater Sci Forum 702–703:904–907
25. Norfleet DM, Sarosi PM, Manchiraju S et al (2009) Transformation induced plasticity during pseudoelastic deformation in Ni–Ti microcrystals. Acta Mater 57:3549–3561
26. Erbstoeszer B, Armstrong B, Taya M et al (2000) Stabilization of the shape memory effect in NiTi: an experimental investigation. Script Mater 42:1145–1150
27. Miyazaki S, Mizukoshi T, Ueki T (1999) Fatigue life of Ti–50at.% Ni and Ti–40Ni–10Cu (at.%) shape memory alloy wires. Mater Sci Eng A 273–275:658–663
28. Gollerthan S, Young ML, Baruj A et al (2009) Fracture mechanics and microstructure in NiTi shape memory alloys. Acta Mater 57:1015–1025

29. McKelvey AL, Ritchie RO (2001) Fatigue-crack growth behaviour in superelastic and shape-memory alloy nitinol. Metall Trans A 32:731–743
30. Nayan N, Buravalla V, Ramamurty U (2009) Effect of mechanical cycling on the stress-strain response of a martensitic Nitinol shape memory alloy. Mater Sci Eng A 525:60–67
31. Ramaiah KV, Saikrishna CN, Dhananjaya BR et al (2008) Stabilization of functional properties of NiTiCu shape memory alloy wire actuator. Proc ISSS 2008:P97
32. Saikrishna CN, Ramaiah KV, Bhagyaraj J et al (2013) Influence of stored elastic strain energy on fatigue behaviour of NiTi shape memory alloy thermal actuator wire. Mater Sci Eng A 587:65–71

Processing and Characterization of Shape Memory Films for Microactuators

S. Mohan and Sudhir Kumar Sharma

Abstract Titanium-rich NiTi thin films were deposited by DC magnetron sputtering from a single alloy target (Ni/Ti: 50:50 at. %). X-ray diffraction investigations indicated that the deposited films were amorphous. When annealed to 500 °C, diffraction peaks corresponding to (110) and (200) planes of Austenite phase started emerging. Annealing at 600 °C resulted in the appearance of two sharp Austenite peaks (110) and (200), along with four additional peaks (110), (020), (111), and (021) of the Martensite phase of NiTi, respectively. Scanning electron microscope studies indicated that annealing at 600 °C resulted in densification of the coating. The thermal response using differential scanning calorimeter recorded for the second thermal cycle, for films deposited at 300 and 500 °C, showed single endothermic and exothermic peaks during heating and cooling cycles. For the films deposited at 400 °C and annealed at 450 °C, the elastic modulus was found to be around 110 ± 5.5 GPa and the hardness around 7.8 ± 0.5 GPa.

Keywords Shape memory alloys · SMA films · Microactuators · NiTi · NiTi SMAs · Scanning electron microscopy · X-Ray diffraction · Differential scanning calorimeter · Transmission electron microscopy · Titanium-rich nickel titanium · Nanoindentation · Young's modulus · Hardness · Thin films processing · Sputtering

S. Mohan (✉) · S. K. Sharma
Indian Institute of Science, Bengaluru, India
e-mail: smohan46@yahoo.co.in

K. J. Vinoy et al. (eds.), *Micro and Smart Devices and Systems*,
Springer Tracts in Mechanical Engineering, DOI: 10.1007/978-81-322-1913-2_12,

1 Shape Memory Alloys

Shape memory alloys (SMAs) belong to the category of smart materials, which can regain their original shape even after several deformation cycles. This is due to the thermo-mechanical effect. The change in structural shape is due to phase changes from martensite to austenite due to applied stress or heat energy [1]. SMAs can be used both as sensors and actuators. They are capable of producing large recoverable strains. The main limitations of these materials are poor response time, loss of thermal energy, and limited temperature range. SMAs are one of the most promising materials for actuator/sensor applications and many researchers are trying to improve their capabilities to enhance their usefulness in technological applications. Nickel Titanium (NiTi) SMAs have been developed for a broad range of applications, e.g., in the fields of aerospace, industrial, and biomedical applications. This has been realized because of low manufacturing cost, good corrosion resistance, biocompatibility, and tailor-made material properties [2].

2 Shape Memory Alloy Films

Generally, bulk SMAs exhibit large strokes and high actuation forces and at the same time they suffer from poor response. However, thin film SMAs provide a larger energy density, higher frequency response and longer life time at microscopic level [3]. NiTi thin films are the most appropriate ones for micro-actuation mechanisms because of their extensive energy density, displacement, work output per unit volume, and improved frequency response. At this level, smaller mass and larger surface to volume ratio enable significant increase in the heat transfer and low power requirements. Hence, large stresses and strains can be realized. These advantages make thin film NiTi SMA a very promising actuator material for micro-device applications such as micro-valves, micro-grippers, micro-pumps, micro-cages, micro-robos, micro-sensors, microswitches, micro-positioners, and such other applications [3, 4]. In spite of several advantages as mentioned above, NiTi possesses various limitations in the extensive use because of limitation in technological processing in device fabrication. Some of those limitations are:

1. Strong sensitivity to composition.
2. Large hysteresis ($\sim 30^{\circ}$).
3. High fabrication costs.
4. High reactivity with oxygen, nitrogen, and carbon.

Hence, in the present work, most of these practical problems, in the processing of thin film NiTi alloy for SMA applications have been addressed.

3 Thin Film Processing

The DC magnetron sputtering approach has been exclusively pursued to deposit Ti-rich NiTi thin films. These films have been studied for their structural, microstructural, mechanical, electrical, and phase transformation characteristics to correlate them with the processing conditions.

Alloy target sputtering is one of the most preferred modes, used in the deposition of films from multicomponent targets such as NiTi. In this process, a proper selection of the target composition and process parameter optimization are the most important factors to deposit films with desired composition and mechanical properties. Hence, the present work concentrates on the deposition of NiTi films by alloy target DC sputtering method (Fig. 1).

The sputtering chamber houses the sputtering cathodes, substrate holder/heater, and shutter assembly. It is powered by a DC power supply (MDX, Advance Energy, 1.0 kW). The substrate holder is a 75 mm diameter heater assembly and the temperature can be raised up to 850 °C starting from room temperature. NiTi alloy targets, with composition in the ratio of 50:50 atomic percentage and 99.999 % purity, have been used to deposit the coatings. Silicon wafers (p-type 100), of 10 mm in square shape, have been used as substrates. The film deposition has been carried out keeping the substrates at different temperatures in the range ambient to 400 C. All the samples, irrespective of the deposition conditions, have been subjected to annealing at 500, 600, and 650 °C in high vacuum in the pressure range of 10^{-6} mbar. Coatings have been analyzed for their structure, composition, electrical resistivity, phase transformation, hardness, and young's modulus by appropriate techniques and presented below.

4 Properties of Ti-rich NiTi Films

4.1 XRD Studies

The XRD spectra of Titanium-rich Nickel Titanium (T-NT) film deposited at 400 °C *and annealed at* 500 *and* 600 C (D400A500) and (D400A600) have been shown in Fig. 2. The as-deposited film has been found to be amorphous. When annealed to 500 °C, diffraction peaks corresponding to (110) and (200) planes of Austenite phase started emerging. Annealing at 600 °C has resulted in the appearance of two sharp Austenite peaks (110) and (200), along with four additional peaks (110), (020), (111), and (021) of the Martensite phase of NiTi, respectively. Fernandes et al. [5] deposited T-NT films using an equiatomic target onto (111) silicon substrates at a substrate temperature of 400 °C and further annealed them at 550 °C. The XRD pattern showed the presence of peaks corresponding to Ni_3Ti precipitate along with the peaks corresponding to Austenite and Martensite phases. However, the data reported in the present context do not

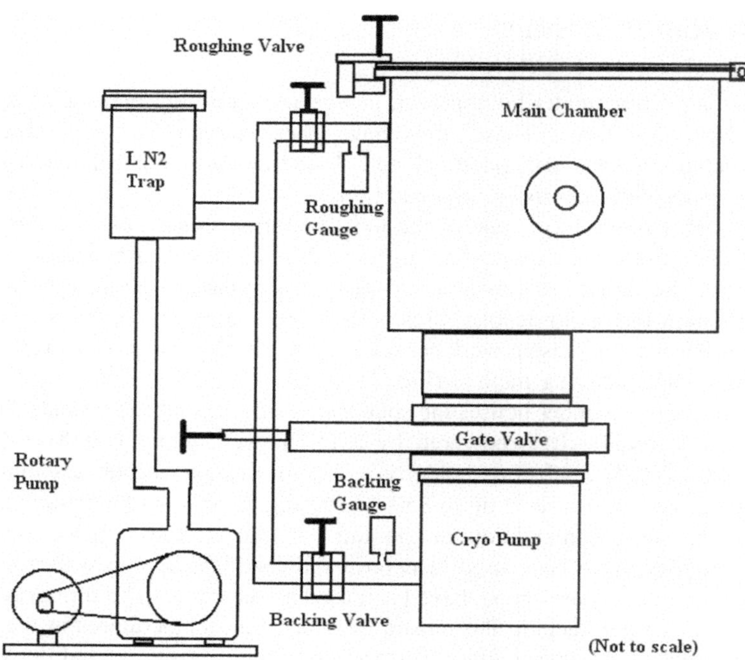

Fig. 1 Schematic of the conventional single target sputtering system. The coating system consists of a water-cooled rectangular box chamber pumped by cryo pump and rotary pump combination

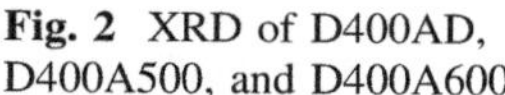

Fig. 2 XRD of D400AD, D400A500, and D400A600

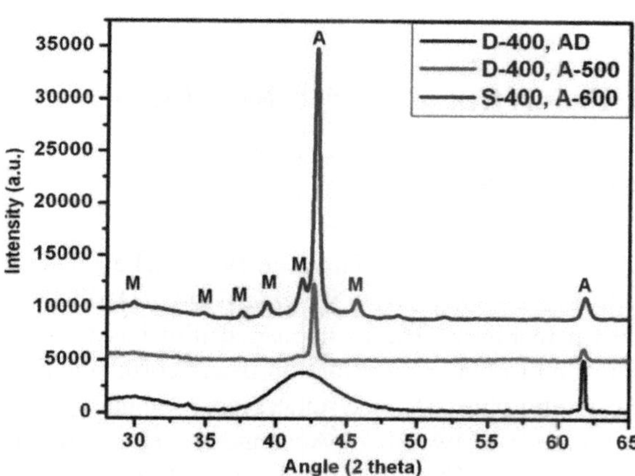

indicate the presence of any precipitates. The particle size calculations or mean crystallographic domain size have been estimated from the true broadening of the XRD lines, using Scherrer's formula [6]. Average crystallite size calculated for

D300A600 has been found to be in the range of 30 nm, corresponding to the (110) peak broadening of Austenitic phase whereas, for D400A500 and D400A600 average crystallite size of 23.76 and 36.47 nm corresponding to the same (110) diffraction peak of Austenitic phase has been observed.

4.2 SEM Studies

The surface morphologies and cross-sectional SEM images of T-NT films deposited at 400° C have been shown in Fig. 3a and b. From the micrographs it can be seen that, for the as-deposited films, the surface of the coating is very smooth and has no clear features [7, 8]. The cross-section SEM studies of the same film show a dense and featureless texture. The film thickness estimated by cross-sectional SEM has been found to be ~2.0 μm. It has also been observed that the film thickness uniformity and surface smoothness are excellent in these films.

The surface morphology and cross-sectional view of D400A600 has been shown in Fig. 3c and d. It is clear from the image that the surface appears to be rough and grainy, compared to the as-deposited film. These grains are spherical in nature and appear to have nearly the same size. The average grain size estimated by normal SEM studies has been found to be around 40 nm. The cross-sectional view of the film appears to have fine columnar microstructure [9, 10]. Such columnar structure can be the result of restricted mobility of deposited films during processing. It can be concluded that the post-annealing at 600 °C has resulted in densification of the coating.

4.2.1 Effect of Annealing Time

The effect of post-annealing time at constant post-annealing temperature have been studied for the films deposited at 400 °C and post-annealed at 500 °C for 1, 4, 8, and 12 h and are shown in Fig. 4a, b, c, and d. One hour post-annealing has resulted in the generation of small spherical coarse grains. The uniform distribution of grains has been observed throughout the coating. An increase in annealing duration for 8 h has resulted in an increase in the average grain size from 25 to 50 nm. Prolonged post-annealing for 12 h, showed a drastic change in the film morphology. It has been found that the microstructure of the film has new precipitates along with void formation. This significant change in film microstructure restricts longer post-annealing time.

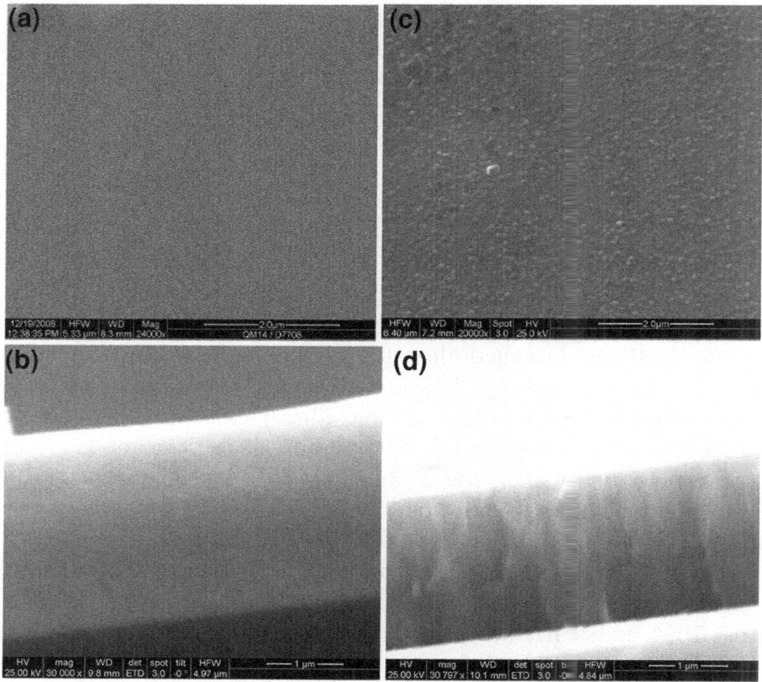

Fig. 3 SEM images of D400AD **a** Surface and **b** Cross-section and D400A600 **c** Surface, and **d** Cross-section of T-NT films

4.3 TEM Studies

The TEM micrographs along with the selected area diffraction patterns (SADP) for the films deposited at D400A600 are shown in Fig. 5a and b. It can be seen, from the TEM micrograph (Fig. 5a), that for D300A600, predominantly nanocrystalline grains with grain size in the range of from 50 to 100 nm are present. The observed diffraction spots in the SADP are due to the nanocrystals formed because of crystallization. Slight diffused rings observed, suggest the existence of amorphous phase even after annealing. The analysis of the spots in the rings has confirmed the presence of *B19'* Martensite phase. The ambiguity in identification of the crystalline phase present in the film, from XRD spectra, has been resolved from the TEM SADP. SADP analysis has also indicated the presence of some Ti_2Ni precipitates in the same film [11].

The TEM micrograph in Fig. 5b for D400A600 shows that the grain size is coarser than that of D300A600. SADP in all the TEM micrographs have also indicated the presence of fine Ti_2Ni precipitates.

This is consistent with the observations reported in the literature [12–14] regarding the presence of Ti_2Ni precipitates in T-NT films. The detailed structure

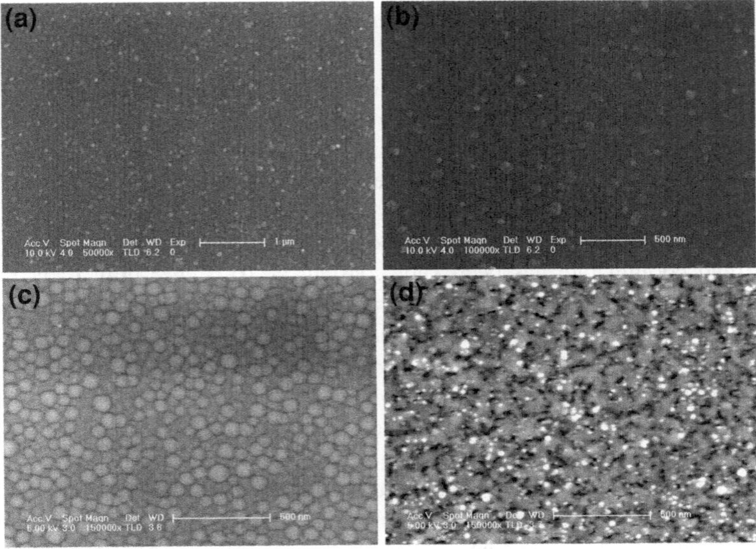

Fig. 4 SEM images of T-NT films deposited at 400 °C and annealed at 500 °C for different durations. **a** 1 h **b** 2 h **c** 8 h, and **d** 12 h

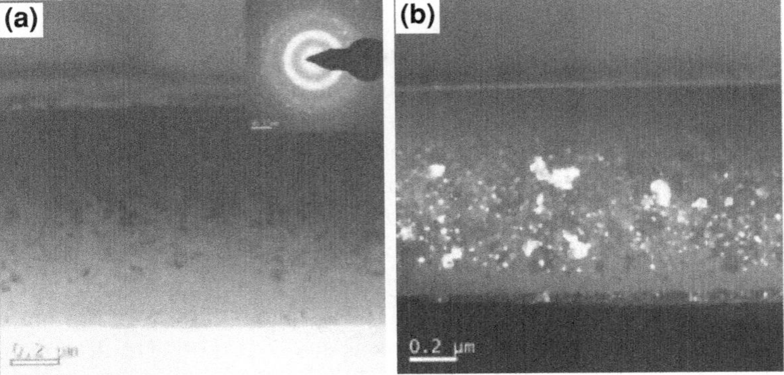

Fig. 5 Cross-section TEM images of D400A600 T-NT film **a** Bright Field (BF) image (Inset is the Selected Area Diffraction Pattern) **b** Dark Field (DF) image

of the precipitates is still controversial, but two types of structures have been reported so far in the literature: Guinier–Preston zone type structure, in which excess Ti atoms are clustered on {100} plane [15] and $C11_b$ -type ordered structure [16]. The type of precipitates formed depends on both annealing temperature and Ti content.

4.4 DSC Studies

Differential scanning calorimeter (DSC) studies have been performed to estimate the film crystallization and phase transformation temperatures of T-NT films. Film crystallization temperature has been identified by the existence of an exothermic peak during first heating thermal cycle and phase transformations during cooling cycle. All the films have been made free standing by chemical etching of the Cu coating on Si substrate. Each sample of 3–5 mg has been taken for DSC studies. Initially, the heating and cooling rate has been kept constant at 10 °C per min. Later on, the influence of heating rate on film crystallization temperature has also been investigated. Film crystallization temperature has been investigated for films deposited at *room temperature,* 100, 200, 300, 400, *and* 500 °C substrate temperatures.

All the films, in as-deposited condition deposited at *room temperature,* 100, 200, 300, *and* 400 °C have been observed to be amorphous in nature, as determined by XRD studies. DSC response of these films has also been found to be the same. Hence, the DSC response recorded for the film deposited at 300 °C alone has been shown as an example. All the samples have been thermally cycled twice. In the first thermal cycle, the free-standing films have been heated at a heating rate of 5 °C/min from room temperature to 550 °C. After reaching this temperature, the sample has been left in this condition for 30 min before starting the cooling cycle to reach the temperature up to −50 °C. In the second thermal cycle, both the heating and cooling rate have been decreased to 1 °C/min in a temperature range from −50 °C *to* +150 °C.

Figure 6a shows the existence of an exothermic peak at 445 °C, which has been identified as the crystallization temperature of the film. Further increase in the temperature, up to 550 °C, has resulted in the relaxation of the exothermic peak. In the cooling cycle, DSC response has been found to be linear up to −50 °C. Sanjabi et al. prepared equiatomic $Ni_{49.5}Ti_{50.5}$ films and annealed them, at a heating rate of 10 °C/min. They have observed crystallization at 472 °C [17, 18]. The difference in the crystallization temperature, in the present case, may be attributed to the change in film composition and heating duration. Moreover, Yang et al. have reported the DSC analysis of Ni-rich NiTi amorphous films at a heating rate of 5 °C/min. They observed the exothermic peak at 497 °C [19]. The disagreement in crystallization temperature can be attributed to the film composition, heating rate, and process parameters variations.

Films deposited at 500 °C and above, have also been heated at a heating rate of 10 °C/min. They have not resulted in any exothermic peak in the above temperature range, as it can be seen in Fig. 6b. This may be because the films have already got crystallized. It is evident from Fig. 6 that the sample deposited below the crystallization temperature has only resulted in the existence of exothermic peak, while the film deposited above crystallization temperature did not show any significant exothermic peak up to 550 °C in the first heating cycle. All the films thermally cycled by the above method showed the phase transformation temperature during heating as well as cooling cycle.

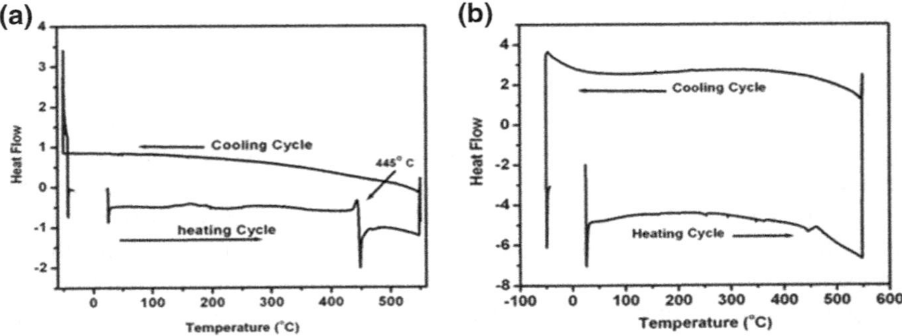

Fig. 6 DSC response of T-NT films deposited at **a** 300 °C and **b** 500 °C in first thermal cycle

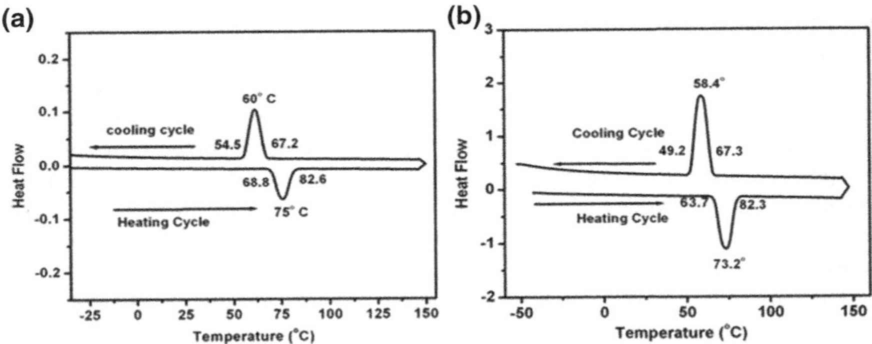

Fig. 7 DSC response of T-NT films deposited at **a** 300 °C and **b** 500 °C in second thermal cycle

In addition to this, the same samples have been thermally cycled at a slower heating and cooling rates of 3 °C/min, in the temperature range from −50 to 150 °C to determine the phase transformation temperature. The DSC thermal response, recorded for the second thermal cycle, for films deposited at 300 and 500 °C is shown in Fig. 7a and b. Both the films have shown single endothermic and exothermic peaks during heating and cooling cycles. The exothermic peaks have been observed at 75 and 73.2 °C, while the endothermic peaks have been identified at 60 and 58.4 °C. However, no R-phase transformation has been observed in this study. In this second thermal cycle, the exothermic and endothermic peaks have been identified as the phase transformation temperature [10, 17, 20–22]. Fu et al. prepared NiTi films having Ti content of 51.3 %. They studied the **M–A** and **A–M** single-phase transformation in heating and cooling cycle and found the transitions temperatures at 350 K (77 °C) and 328 K (55 °C). They have confirmed these phase transformations by high temperature XRD and stress temperature response studies [23]. This is in agreement with our observations. Similar observations have been made by other investigators [22, 24–28].

4.5 Nano-Indentation Studies

Load displacement curves have been analyzed to evaluate the elastic modulus and hardness values for T-NT films. The substrate effects have been minimized by maintaining the indentation depth to less than 10 % of the film thickness. The surface roughness values ($\sim$10 nm) have been determined by using Atomic force microscope (AFM). The selection of maximum load of 3 mN has resulted in a penetration depth of $\sim$120 nm, which is much larger than the surface roughness. The penetration depth has been found to be much smaller than the film thickness (1.8 μm), which is expected to be independent of surface roughness and substrate effects. This justifies the load for nanoindentation studies.

A representative load-displacement plot of T-NT film (D400A450) is shown in the Fig. 8a. The Oliver–Pharr approach [29] has been used to calculate the elastic modulus and hardness, at different locations (shown in Fig. 8b) over the 75 mm diameter silicon wafer. The values obtained have been listed in Table 1. It can be seen that the elastic modulus is around 110 ± 5.5 GPa and the hardness is around 7.8 ± 0.5 GPa though out the given area. This is a very important observation as the elastic modulus and hardness, especially in NiTi alloys, are very sensitive to composition and microstructure. This implies that these films have uniform composition and microstructure within the given area. As the variation in elastic modulus and hardness values have been found to be within ± 5.5 GPa and ± 0.5 GPa, respectively, it can be considered that the values are same over the entire wafer.

The variation of elastic modulus and hardness with annealing temperatures for D300AD, D300A300, D300A400, D300A500, and D300A600 has been found to be same as that of D400AD, D400A400, D400A500, and D400A600. Hence, the elastic modulus and hardness values calculated for D300AD, D300A300, D300A400, D300A500, and D300A600 have alone been shown in Fig. 9a and b. From the figure, it is clear that the elastic modulus and hardness values are independent of substrate temperatures during deposition and annealing temperatures. This indicates that in this temperature range, substrate temperature has no effect on the mechanical properties. It could be because, as there is no change in the microstructure of the films with change in substrate temperature (as discussed in TEM studies). The observed hardness and modulus for NiTi films are relatively higher [30] as compared to their bulk counterparts. This could be due to the presence of residual stresses in films [31, 32]. Both elastic modulus and hardness values have been found to increase for D300A500 and D300A600 films, irrespective of the deposition temperature. This can be attributed to the fact that the films crystallize at $\sim$483 °C [33], (which is also supported by XRD and TEM micrographs). The higher value of elastic modulus and hardness in films annealed at 600 °C as compared to annealing at 500 °C could be primarily due to the extent of crystallization with increasing annealing temperature.

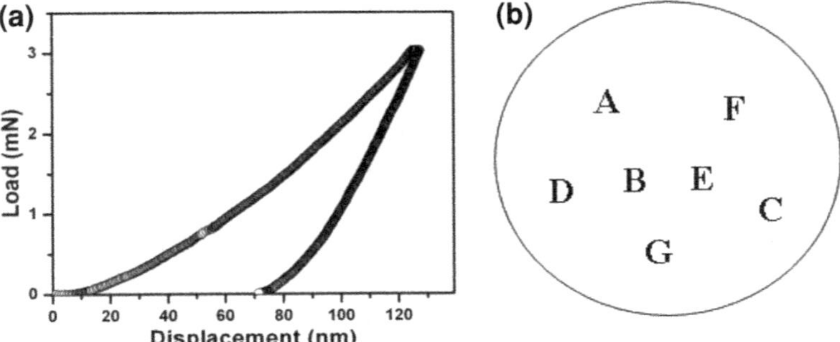

Fig. 8 **a** Load versus displacement plot for T-NT films D400A450 and **b** The different locations on 3"diameter silicon wafer chosen for nanoindentation

Table 1 Elastic modulus and hardness values obtained by nanoindentation at various locations (see Fig. 8) for Tirich NiTi film deposited on 3" diameter silicon wafer

Location on wafer (See Fig. 6.72)	Elastic modulus (GPa)	Hardness (GPa)
A	109	7.48
B	105	7.9
C	108	8.21
D	112	7.37
E	104	7.13
F	120	8.35
G	113	8.32

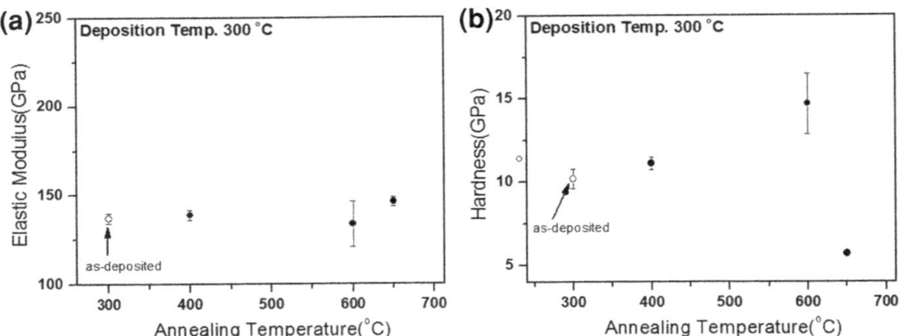

Fig. 9 Plots of **a** Elastic modulus and **b** Hardness variation of T-NT films deposited at 300 °C and post-annealed to higher temperatures

4.5.1 Effect of Applied Load and Annealing Temperature

The elastic modulus and hardness values calculated for D300A600, D350A600, D400A600, D450A600, and D500A600 at different loads of 0.5, 1.0, and 3 mN have been shown in Fig. 10a and b. With increase in load from 0.5 to 3 mN, it has

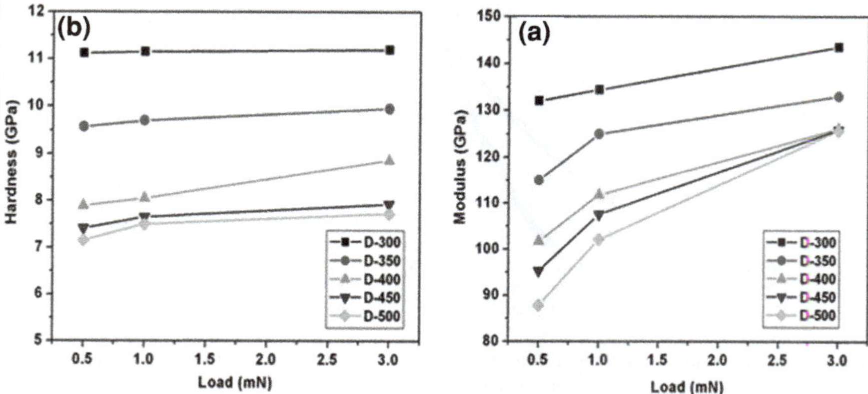

Fig. 10 **a** The elastic modulus and **b** Hardness values of the films deposited in a range of 300–500 °C substrate temperature and annealed at 600 °C at different loads

resulted in overall increase in elastic modulus from 85 to 145 GPa. However, with increase in deposition temperature, the modulus values have been found to decrease. This decrease could be explained on the basis of the extent of Austenite and Martensite phases present in the films and their corresponding contribution to modulus values. At higher loads of 3 mN, the increase in modulus values could be due to the substrate coming into picture during indentation. At a load of 3 mN, no change in the modulus has been observed above deposition temperature of 400 °C.

In case of hardness, for the same films, no change in hardness values has been observed for different loads. With increase in deposition temperature, the hardness values have been found to decrease for the given load. However, no significant change in the hardness has been noticed for D450A600 and D500A600 films. Liu and Duh [30] observed similar variation in young's modulus and hardness with load in the range from 6 to 11 GPa and they have limited the load below 10 mN to minimize the substrate effects. These results are also in agreement with the data reported by Kumar et al. [34].

5 Conclusions

Titanium-rich NiTi thin films have been deposited from alloy target by DC magnetron sputtering and the influence of temperature during film deposition and annealing at elevated temperature on the structure, microstructure, phase transformation, and mechanical properties of these films have been investigated. Annealing at temperatures above 450 °C has resulted in realization of films with required phases and mechanical properties suitable for actuation purposes.

Acknowledgments The authors acknowledge the funding agency "National Programme on Smart Materials" for supporting the project entitled "NiTi Shape Memory Alloy Thin Films for Micro Actuator Applications."

References

1. Hodgson DE, Wu MH, Biermann RJ (1996) Shape memory alloys, NiTi smart sheet. Shape Memory Applications Homepage
2. Gill JJ, Chang DT, Momoda LA, Carman GP (2001) Sens Actuators A 93:148
3. Fu YQ, Du H, Huang W, Zhang S, Hu M (2004) Sens Actuators A 112:395
4. Miyazaki S, Fu YQ, Huan WM (2009) Thin film shape memory alloys: fundamentals and device applications. Cambridge University Press, New York
5. Fernandes FMB, Martins R, Nogueeira MT, Silva RJC, Nunes P, Costa D, Ferreira I, Martins R (2002) Sens Actuators A 99:55
6. Cullity BD (1978) Elements of X-ray diffraction. Addition-Wesley, Reading, MA
7. Kumar A, Kaur D (2009) Surf Coat Technol 204(6–7):1132
8. Kumar A, Singh D, Kumar R, Kaur D (2009) J Alloy Compd 479(1–2):166
9. Chu JP, Lai YW, Lin TN, Wang SF (2000). Mater Sci Eng A 277:11
10. Kumar A, Singh D, Kaur D (2009) Surf Coat Technol 203(12):1596
11. Miyazaki S, Ishida A (1999) Mater Sci Eng A106:273–275
12. Kawamura Y, Gyobu A, Saburi T, Asai M (2000) Mater Sci Forum 327328:303
13. Otsuka K, Ren X (2005) Prog Mater Sci 50:511
14. Ishida A, Ogawa K, Sato M, Miyazaki S (1997) Metall Mater Trans A 28:1985
15. Nakata Y, Tadaki T, Sakamoto H, Tanaka A, Shimizu K (1995) J Phys IV 5:C8–671
16. Kajiwara S, Kikuchi T, Okagawa K, Matsunaga T, Miyazaki S (1996) Philos Mag Lett 74:137
17. Sanjabi S, Sadrnezhaad SK, Yates KA, Barbar ZH (2005) Thin Solid Films 491:190
18. Sanjabi S, Cao YZ, Sadrnezhaad SK, Barbar ZH (2005) J Vac Sci Technol A 23(5):1425
19. Yang YQ, Jia HS, Zhang ZF, Shen HM, Hu A, Wang YN (1995) Mater Lett 22(3–4):137
20. Busch JD, Johnson AD, Lee CH, Stevenson DA (1990). J Appl Phys 68 (12):6224
21. Ho KK, Carman GP (2000) Thin Solid Films 370:18
22. Botterill NW, Grant DM (2004) Mater Sci Eng A 378:242
23. Ishida A, Takei A, Miyazaki S (1993) Thin Solid Films 228:210
24. Fu Y, Huang W, Du H, Huang X, Tan J, Gao X (2001) Surf Coat Technol 145:107
25. Ho KK, Carman GP (2000) Thin Solid Films 370:18
26. Fu YQ, Du H (2003) Mater Sci Eng A 339:10
27. Fu YQ, Du H (2002) Surf Coat Technol 153:100
28. Fu YQ, Du H (2003) Mater Sci Eng A 342:236
29. Oliver WC, Pharr GM (1992) J Mater Res 7:1564
30. Liu KT, Duh JG (2008) Surf Coat Technol 202:2737
31. Lee YH, Kwon D (2003) Scripta Mater 49:459
32. Ko S, Lee D, Jee S, Park H, Lee K, Hwang W (2006) Thin Solid Films 515:1932
33. Otsuka K, Ren X (2003) Prog Mater Sci 50:511
34. Kumar A, Sharma SK, Bysakh S, Kamat SV, Mohan S (2010) J Mater Sci Technol 26(11):961

Piezoceramic Coatings for MEMS and Structural Health Monitoring

Soma Dutta

Abstract The advent of smart materials has given a new dimension to the field of Materials Science, resulting in many significant application-oriented developments in all fields of engineering. Piezoelectric material is an important member of the smart material family. Bulk piezo sensors and actuators have been used widely for smart structure applications. The limitation in using piezoceramic material in its bulk form as sensors and actuators in real applications is due to its brittleness, nonconformability, high temperature processing, small area coverage, and associated ill effects of using an adhesive bond layer for attaching them. In situ piezo transducers, covering a large area, provide a very good solution circumventing aforementioned problems. The indigenous development of a smart and conformal piezoceramic coating with a low process temperature makes it suitable for Silicon batch processing not only for the fabrication of piezoceramic MEMS but also for the Nondestructive Evaluation (NDE) of metals and composites in guided wave-based real-time Structural Health Monitoring (SHM). This chapter presents the development of application-specific in situ piezoeceramic coating and the technical challenges therein.

Keywords Piezoceramic coating · MEMS · Membrane · Pressure sensor · Lamb wave · SHM

1 Introduction

A significant area of ongoing research and development in the aerospace smart structures community is the implementation of an automated structural health monitoring (SHM) system using smart sensors and actuators integrated into the

S. Dutta (✉)
Materials Science Division, CSIR-National Aerospace Laboratories, Bangalore, India
e-mail: som@nal.res.in

K. J. Vinoy et al. (eds.), *Micro and Smart Devices and Systems*,
Springer Tracts in Mechanical Engineering, DOI: 10.1007/978-81-322-1913-2_13,

structure. This methodology provides a "built-in-test" diagnostic and prognostic capability for the vehicle's health-management system. A reliable SHM system enables the practice of condition-based maintenance, which can significantly reduce life cycle costs by minimizing inspection time and efforts enabling the extension of the useful life of aerospace structural components. One of the most promising techniques under this development is guided wave-based Health Monitoring, where piezoceramic wafers are generally bonded to the structure to generate and receive high frequency stress/surface waves which give information about the health condition of the structure. Brittleness, nonconformability, high temperature processing, and small area coverage are the common problems associated with piezoceramic wafers other than the issues related to the ill effect of the bonding technique. Adhesive bonding reduces the mechanical to electrical (or vice versa) energy transfer due to the loss in the transduction signal. The stiffness of piezoceramic material is intrinsically high and the presence of an intermediate adhesive layer further increases the stiffness, which is a concern for the radial mode operation of these piezoceramic wafers. The search for proper adhesive material to bond these piezoceramic sensors for high temperature applications or to use them in corrosive environments remains a great challenge. An in situ piezoceramic coating could provide a better solution circumventing aforementioned problems.

Another field of growing interest is MEMS. High performance piezoceramic films are of great demand in MEMS applications [1–3]. However, the fabrication of piezoceramic MEMS gives more challenges than ferroelectric-based micro devices. It is critical to generate a sensing and actuation response from a MEMS device. This chapter presents the development of an application-specific piezoeceramic coating emphasizing its usage (applicability) in SHM and MEMS.

2 Backgrounds

Piezoceramics are essential components of many smart sensors and actuators. Initially, these piezoceramics were limited to single crystals of quartz, which were grown under controlled conditions in a particular orientation with respect to the crystal axes. In later days, polycrystalline piezoceramic materials were processed by using cost-effective techniques common to ordinary ceramic products. Lead zirconate titanate (PZT) is one of the most important piezoceramic materials, investigated extensively in the past decades in its polycrystalline and single crystal form due to its unique functional properties [4–7]. In the late 1980s, with the advent of nonvolatile random access memory (NVRAM) devices, bulk PZT was increasingly supplemented by PZT thin/thick films due to the ease with which they could be integrated into the semiconductor chip. Thick film (10–20 μm) piezoceramics have been used in high frequency transducers, and vibration control devices for their excellent actuation properties [8]. Piezoceramic layers in thin and thick film form are advantageous for real-time applications due to their flexibility (conformability), capability of large area coverage, high strain compatibility, and adhesive free direct bonding to the substrate structure leading to a better mechanical to electrical transduction. Considering these merits of a piezoceramic coating (thin or thick), one can say it is a potential alternative of bulk piezo. But to use it as an alternative to conventional bulk piezo, suitable coating methodologies need to be developed to deposit a film of 100–500 μm thickness with a comparatively low process temperature. Probably, a coating that cures at a low temperature and gives piezoelectric properties equivalent to monolithic piezoceramic wafers would represent an advancement or contribution to this field of research. The development of such a piezoceramic (pure and doped PZT) coating, which is an integral part of the substrate, was targeted first on aerospace grade metals and alloys, carbon fiber reinforced plastics (CFRP) composites including irregular surfaces and Si wafers. Later, it was extended to various other substrates and configurations using other lead-based and lead-free piezoceramics. In the following sections, the development of application-specific piezoceramic coatings is discussed.

3 Development of Piezoceramic Coating for MEMS

3.1 Low Temperature Thick Coating (10–100 μm)

It has already been mentioned that a thick film (10–20 μm) PZT has been used in ferroelectric MEMS devices. But, to produce a PZT film of that thickness, bulk micromachining of a monolithic wafer has been preferred since fabrication methods like tape casting, and screen printing showed drawbacks in achieving a 10–20 μm PZT film due to nonuniformity, high sintering temperatures (1200 °C), and nonreproducibility in their properties [9, 10]. It has been a very challenging task to fabricate thicker films at a low temperature. Though the sol-gel-based chemical method is widely used for the fabrication of thick films, but the thickness of sol-gel-coated films are limited to a few microns (<5 μm) due to the requirement of a large number of deposition cycles. Different approaches have been proposed for the fabrication of a low sintering phase formation by the aerosol deposition method, [11, 12] laser annealing processes [13], and microwave irradiation using magnetic field [14], etc. But, these fabrication methods require very involved and expensive instrumentation.

A low temperature process of thick film deposition is suitable for the Silicon batch process which is a composite sol-gel method [15]. In this method, seeding and high energy ball milling was used to lower the phase formation temperature of PZT to 300 °C. The effect of the ball milling of the composite sol on phase formation is revealed in an X-ray diffraction plot (Fig. 1). Clear and well-established pervoskite peaks of the seeded PZT film at $2\theta = 22.5°(100)$, 32°(110), 34°(101), 38°(111), 44.2°(200), 55°(211) are observed in a 30 h milled sample

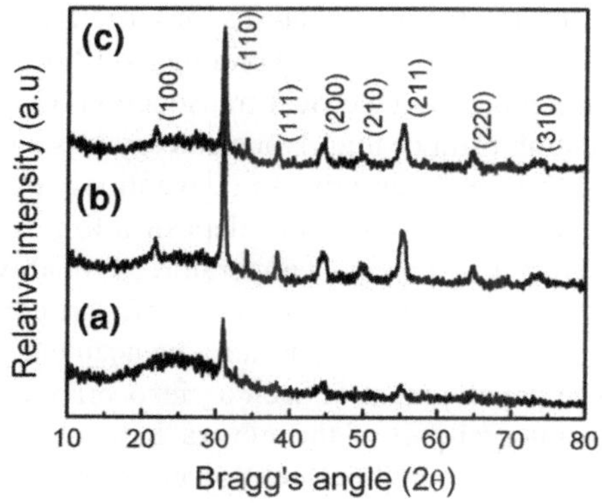

Fig. 1 XRD patterns of PZT film prepared by ball milling of composite sol for different milling times: **a** 15 h, **b** 30 h, and **c** 40 h

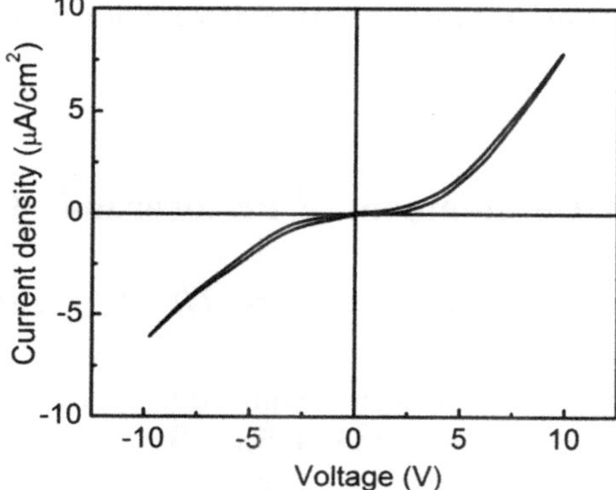

Fig. 2 Variation of leakage current density with voltage for low temperature PZT films

(Fig. 1b) [16]. With an increase in the ball milling time from 30–40 h, the peak intensity was reduced, indicating a further decrease in grain size and the disintegration of the crystal phases. The tetra splitting observed in a 30 h ball milled sample (44°-200 and 55°-211) got converted into a doublet ($2\theta = 44°$) and a singlet ($2\theta = 55°$) with 40 h milling (Fig. 1c). Only milling of the sol-gel solution without seeding, and seeding without milling was not effective in bringing down the phase formation temperature below 450 °C.

Due to the low density of the film, comparatively small dielectric polarization is observed in low temperature coatings. The leakage current is another important device parameter that ensures the reliability of the material for MEMS capacitor applications. Any dielectric material, not being a perfect insulator, allows a current to flow under an electric field which slowly discharges the capacitor. Leakage current measurement gives a qualitative estimate on imperfections and defects presents in the dielectric film (material). The characteristics plot of leakage current of the low temperature PZT film shows a current density of 7.6 μA/cm^2 at 10 V (Fig. 2) for a film of thickness 5 μm. Similar results are even reported for PZT thin films [17]. Though a higher leakage is expected in low temperature thick film structures due to porosity, the nanocrystalline grain structure sometimes prevents the leakage to a greater extent by the inter grain depletion of grain boundary limited conduction [18, 19]. This methodology of thick film preparation could easily be coupled with the conventional Si batch process.

3.2 PZT and PLZT Thin Film (200–1000 nm) Coatings

Like piezoceramic thick films, thin films are also used in surface micromachined piezo MEMS as micromotors, resonators, ultrasonic sensors, and transducers [20–23]. Thin film piezoceramics should have a low hysteresis at higher frequencies and a less coercive field for MEMS accelerometer or gyroscope applications. PZT-doped with olivalent inherits excellent piezoelectric properties. The substitution of lanthanum (La) for lead (Pb) in PZT changes its macroscopic properties and improves the fatigue resistance, enhances the spontaneous polarization, and reduces the coercive field. Epitaxial thin films have attracted much attention because of their superior directional properties and have been recognized for many technological applications. They are free from crystallographic defects like high angle grain boundaries and have a more extended life than ordinary films. In the epitaxial growth process, a specific crystallographic orientation is built up in the film as per the orientation of a substrate or by some other means. By this directional growth, anisotropy like a single crystal can be induced to the thin film with a higher mechanical strength because of polycrystalline grains. The combined advantages of the sol-gel method and epitaxial growth evolve as potential techniques in applied research with the added advantage of cost-effectiveness. For the fabrication of membrane-based piezoceramic pressure sensors, an epitaxially grown PLZT (111) thin film was used. Before going into the details of device fabrication and its performance, a quick review of the electrical and piezoelectric properties of the film would be of interest.

The hysteresis (P-E) loop shown in Fig. 3 for PLZT (111) film is obtained under switching conditions at 150 Hz with complete saturation. It shows saturation polarization P_s of 71 $\mu C/cm^2$, remnant polarization $P_r = 15$ $\mu C/cm^2$, and a coercive field $E_c = 45$ kV/cm. The coercive field increases with the frequency and at 1 MHz, it becomes 68 kV/cm.

In addition to the hysteresis loop, polarization switching by an electric field leads to the strain-electric field hysteresis in PLZT (111), as shown in Fig. 4. The strain-electric field hysteresis loop which resembles a butterfly and is called the butterfly loop, is due to the converse piezoelectric effect of the lattice. It could be due to the switching and movement of domain walls, also [24].

The reliability of piezoelectric measurement for a thin film sample is a prime concern. The effective piezoelectric behavior of thin films essentially needs to nullify the substrate effect. Surface profiles and electrode dimensions are also very important for accurate measurement. A very smooth surface with a surface profile of 5–6 nm film reproduces reliable data. An exceptionally smoothed surface

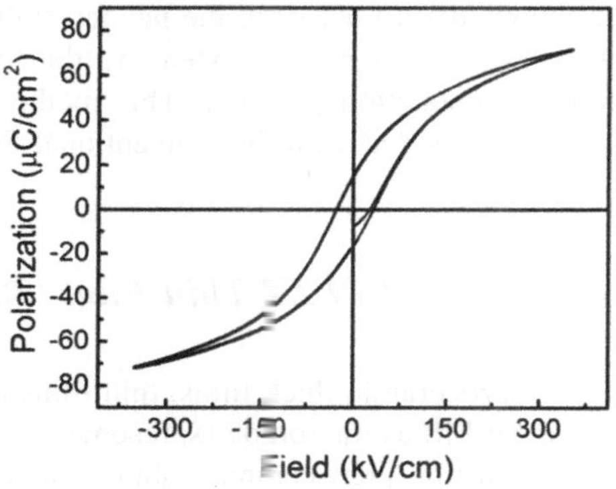

Fig. 3 The P-E hysteresis loop, measured on a (111)-oriented 1 μm thick PLZT film coated on $Si/SiO_2/Pt$

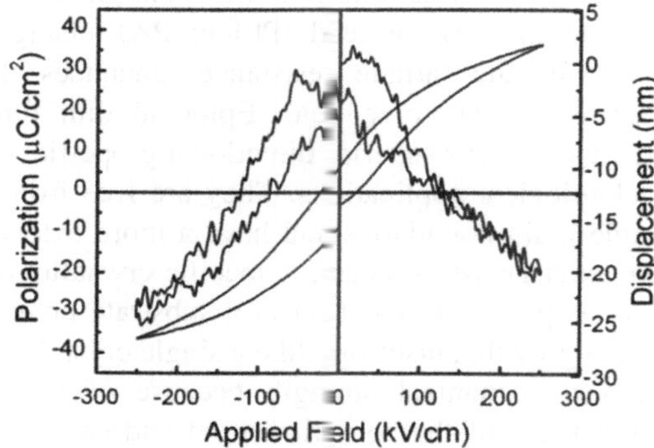

Fig. 4 Polarization and strain loops measured on a (100)-oriented, 800 nm thick PLZT (8/52/48) film at 500 Hz

profile (~2 nm) is possible to achieve by manipulation of process parameters while thin film preparation (Fig. 5).

The effective longitudinal piezoelectric coefficient $d_{33,f}$ plot shown in Fig. 6 was taken on a PLZT (111) film by applying electrical excitation voltage on the sample and measuring the generated displacement. It shows an extraordinarily high $d_{33,f}$ (380 pm/V) due to the preferential orientation of the film (111). When an electric field is applied, domains which are not exactly parallel to the field readily get aligned, resulting in a higher piezoelectric $d_{33,f}$ value

For piezoelectric transverse coefficient measurement a distributed uniform strain on the thin film sample gives reliable results. Generally, a four-point bending test fixture helps in generating distributed strain over thin films when an alternating mechanical excitation signal is applied to the sample. Under this condition, the simultaneous measurement of displacement and charge gives a linear plot. By this method, in-plane strain is precisely applied to the film and $e_{31,f}$ is determined from the slope of the displacement versus charge curve multiplied by a geometric factor. A plot of displacement versus charge shown in Fig. 7 was taken on PLZT (111) which gives e_{31} about 0.08 C/m^2. These piezoelectric properties are very useful to study before the fabrication of any device either on the macro- or microscale.

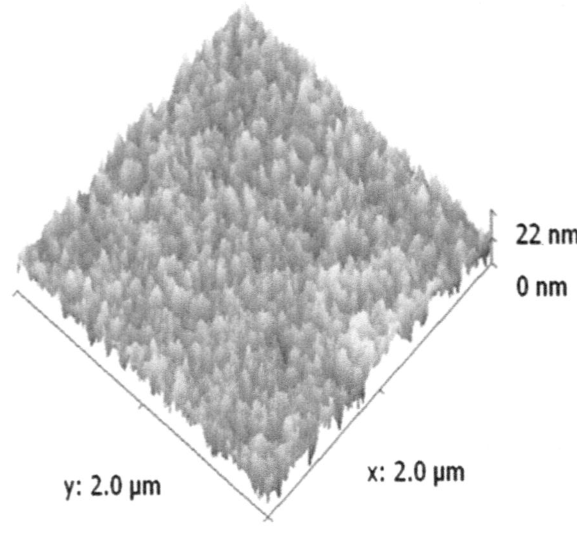

Fig. 5 Atomic force micrograph taken on 1 µm PLZT (111) having a surface profile of 2 nm

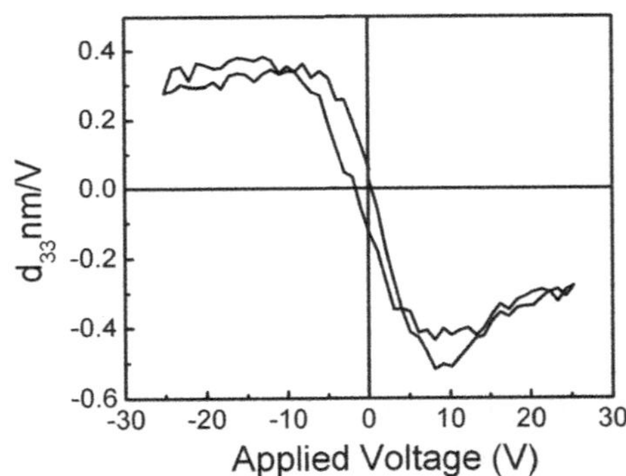

Fig. 6 Piezoelectric $d_{33,f}$ coefficient of PLZT (111) film at different electrical excitation voltages

3.3 Fabrication of a Piezoceramic (Modified PZT) Membrane-Based Pressure Sensor

A typical model of a piezoceramic MEMS sensor is shown in Fig. 8. The sensor output is closely associated with the electromechanical properties of the piezoceramic coating and final packaging.

Modeling helps in optimizing the design geometry. It is known that a reduction in the thickness of the membrane generally increases the deflection and leads to increased sensor output. But, it is necessary to keep the thickness of the membrane optimum to increase the life cycle of the device. The operational pressure range for a piezoceramic pressure sensor can be calculated by limiting the stress value below the burst pressure of the membrane using the conventional formula [25] given in Eq. (3.3.1);

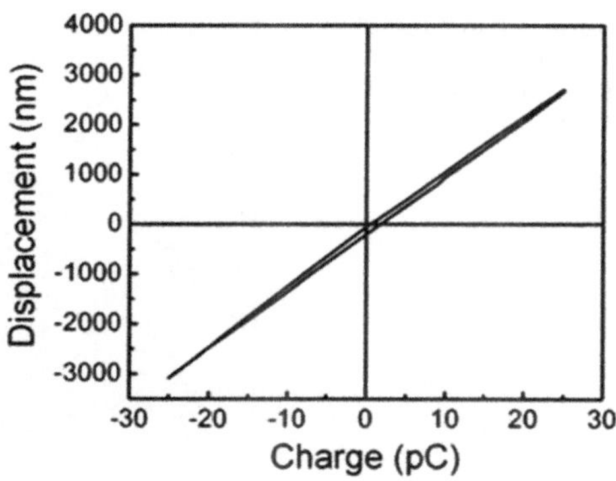

Fig. 7 Plot of displacement versus charge of PLZT film for 10 V excitation

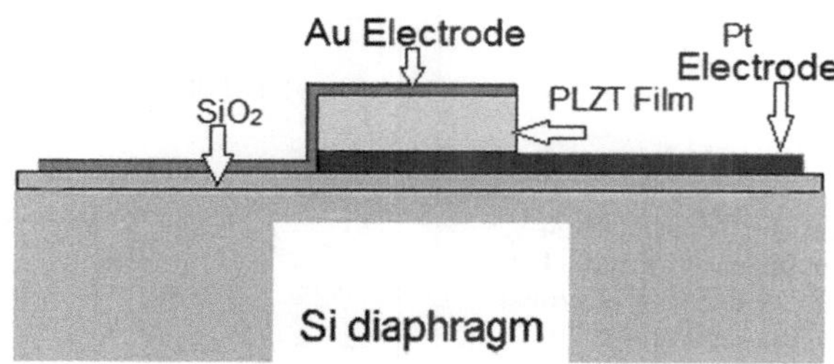

Fig. 8 Basic model of the sensor

$$\sigma_{\max} = \beta p \left(\frac{a}{h}\right)^2 \tag{3.3.1}$$

where

β Constant value which is 0.31 for a square membrane
p Applied pressure
a Width of the PZT layer
h Membrane thickness

The burst pressure defines the maximum pressure at which the membrane ruptures or undergoes mechanical failure. The rupture strength (compressive strength) for polycrystalline PZT is reported to be around 600 MPa. Though the elastic layer provides extra strength to the piezoceramic layer, it is important to operate the sensor in the pressure range where the maximum stress on the membrane is limited to 80 % of the rupture strength (Eq. 3.3.1). The rear etching of the wafer through reactive ion etching gives a perfect square released membrane. Optical micrographs of the PLZT-released membrane are shown in Fig. 9a, b.

Getting a perfect dielectric layer (PLZT) between the top (Au) and bottom electrode (Pt) is a little critical, since there is a chance of puncture or crack in the piezoceramic layer during the process. The electrical resistance measurement of

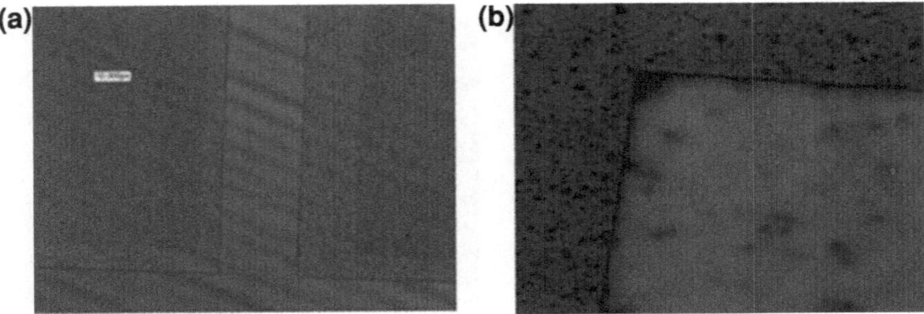

Fig. 9 **a** Top view of a PLZT membrane, **b** Rear of the released membrane

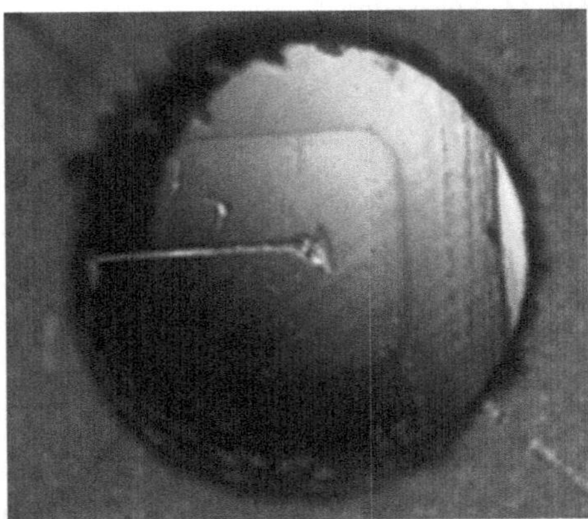

Fig. 10 Wire-bonded sensor element

the piezoelectric layer would be the right experiment to perform to detect flawless sensor elements. After this stage of fabrication, a wire bonding could create damage to this tiny membrane. Figure 10 shows a wire-bonded PLZT membrane element. The final and most critical stage of fabrication is packaging which finally gives an effective sensor output. For the measurement of above atmospheric pressure, hermetic sealing at the rear of the piezoceramic membrane would be adequate, whereas vacuum sealing is required for measuring pressures below atmospheric level.

Figure 11 shows a prototype piezoceramic MEMS pressure sensor, fabricated out of a PLZT (111) 1 μm film. A typical plot of the device performance is shown in Fig. 12. Though it was mentioned in an earlier discussion that one can go high on the operational pressure range depending on the burst pressure, the ideal cut-off value for the working pressure of the device should be decided on the linear range of the input pressure versus output voltage data.

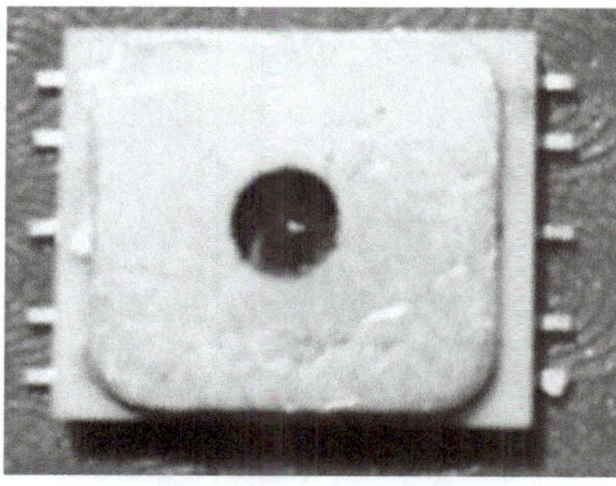

Fig. 11 Piezoceramic MEMS pressure sensor

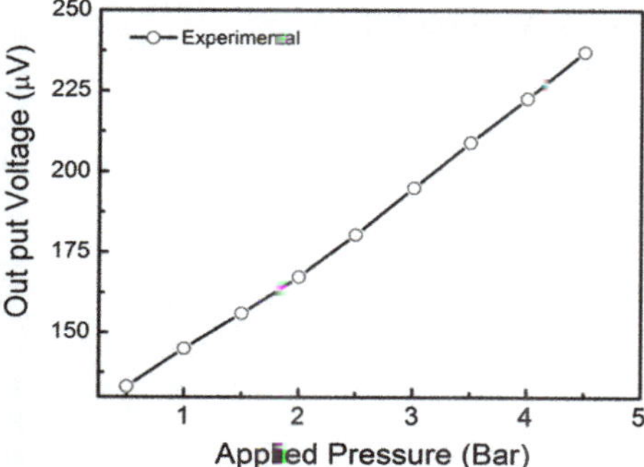

Fig. 12 Voltage response of the developed MEMS pressure sensor

4 Development of a Piezoceramic Coating for SHM

In the acousto-ultrasonic technique, a broadband ultrasonic wave is injected onto the surface at one location of the test object with the help of a piezo transducer, and a receiving transducer is coupled to the same surface at another location. When the receiver is close enough to the sender, the acousto-ultrasonic approach turns into the traditional pulse-echo ultrasonic testing method. When the tested plates are thick (compared with the wavelength of the ultrasonic wave), and the spacing between sending and receiving transducers is greater than the plate thickness, we can explain the main components of the received waves using the resonance of ultrasonic waves [26]. When the tested plates are thin, the guided waves, namely Lamb waves, may be generated in the plates. Thus, understanding the propagation characteristics of the ultrasonic wave is essential for the successful application of the technique. A piezoceramic coating directly integrated with the structure is capable of passively or actively monitoring the changes within a structure that presage a component failure, and can detect all impacts on the structure and evaluate any damage. The optimum integration strategy of piezoceramic coating has yielded a cost-effective process suitable for field applications and is discussed here.

4.1 Composition Selection and Method of Preparation

The mechanical and electrical behavior of piezoceramic materials is described by various coefficients, which determine the performance of the device. For simplicity, only the coefficients, which are most important in the application of ultrasonic, are taken into consideration. The theory becomes rather complicated if anything more than a superficial treatment is attempted. The electromechanical stress–strain relationship of the piezoceramic materials is of utmost importance for acoustic wave generation and reception. The strain produced in a transducer by the application of a unit electric field is called the piezoelectric coefficient **d**. The **d** coefficient is called the transmitting constant of the transducer. The piezoelectric coefficient **g** is defined as the electric field produced under open circuit conditions per unit-applied stress. The **g** coefficient is called the receiving constant of the transducer, as receiving amplifiers are generally voltage sensitive, rather than charge sensitive. It is necessary to choose the proper composition of piezoelectric materials to have maximum d & g coefficients for transmitting and receiving applications.

The piezoelectric properties of ferroelectric materials are highly composition and process-dependent. Lead Zirconate Titanate (PZT) has two main ferroelectric phases; rhombohedral (for $x < 0.48$ in $PbZr_{1-x}Ti_xO_3$) and tetragonal (for $x > 0.48$ in $PbZr_{1-x}Ti_xO_3$) under standard conditions. The boundary between the tetragonal and rhombohedral phase is sharply defined, and virtually independent of temperature, and this boundary is known as the morphotropic phase boundary (MPB). In bulk ceramics, maxima in the piezoelectric coefficients are generally observed at the MPB. The same behavior is often reported in thin films. But the piezoelectric properties of thin films are always smaller than those of bulk ceramics. A thick film piezoceramic coating shows a reasonably good actuation property. A spray coating of PZT slurry is suitable for thick film (500 μm) piezoceramics for the coverage of large areas, curved surfaces, and inaccessible locations (Fig. 13). An insitu piezoceramic coating on Aerospace Grade Aluminum and Composites (CFRP and GFRP) has been developed with the dual functions of transmitting and receiving of ultrasonic waves. The use of a hand-held spray and heat guns make this method suitable for coating bigger structures and field applications.

This method of coating adds flexibility to the piezoceramic layer which helps in the shear mode of actuation, axial wave propagation, and improves the bending strength. As it is known to provide electrical contact to these devices, metallization is required on top of the PZT coating. But the structure itself can serve as the bottom electrode in the case of the conducting host structure. If the host structure is nonconducting, then a wrap on electrode or IDE (Inter digited electrode) can be suitably configured (Fig. 14) depending on the mode of operation of these in situ transducers.

For practical applications of this coating, strain compatibility between the coating and the host structure (substrate) needs to be ensured. Figure 15 represents the experimental setup where a PZT-coated Carbon Fiber Reinforced Plastic

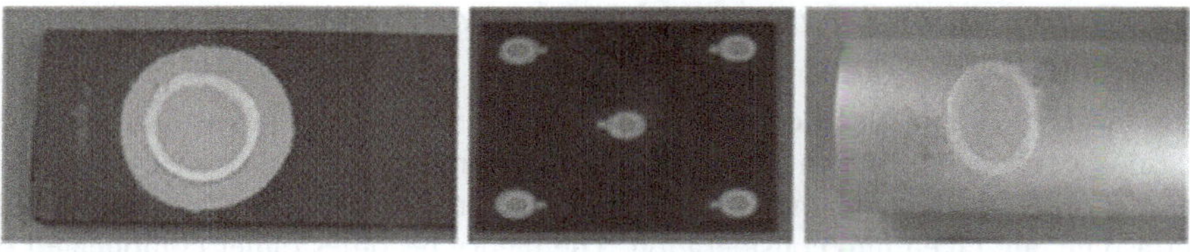

Fig. 13 In situ PZT coating on various aerospace grade materials

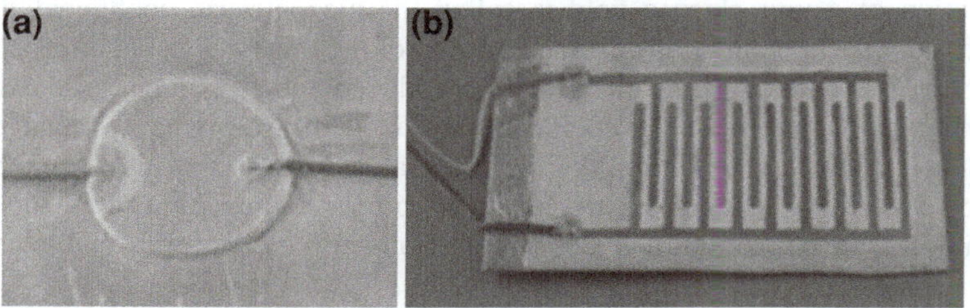

Fig. 14 PZT coating with **a** wrap on electrode and **b** IDE

(CFRP) beam is mounted on a standard UTM to perform the test of tensile loading. For compressive loading, three points or a four-point bending setup can be used in such a way that the coating area undergoes a compressive loading. The coating can withstand a strain of the same order as the substrate and improves the mechanical as well as the performance stability of the device.

4.2 Damage Detection Employing In Situ Piezoceramic Coating

As mentioned earlier, Lamb waves are a type of guided waves in doubly bounded media, and have important applications in Nondestructive Evaluation (NDE) and SHM. For plate structures of a few millimeter thickness these Lamb waves have useful properties like widely separated dispersion characteristics, less attenuation, etc., in the frequency range of 10–300 kHz. The dynamic strain sensitivity of the piezoceramic coating can be evaluated by analyzing its response to the incident Lamb waves. The Lamb waves at suitably chosen frequencies are very slightly dispersive and can propagate for very long distances without attenuation. There are two types of modes of Lamb waves: symmetric (S_0) and antisymmetric (A_0), which are also referred to as axial and flexural wave modes, respectively.

The accuracy of Lamb wave sensing using a piezoceramic coating depends on the clarity of the voltage signal acquired and the time information of the wave

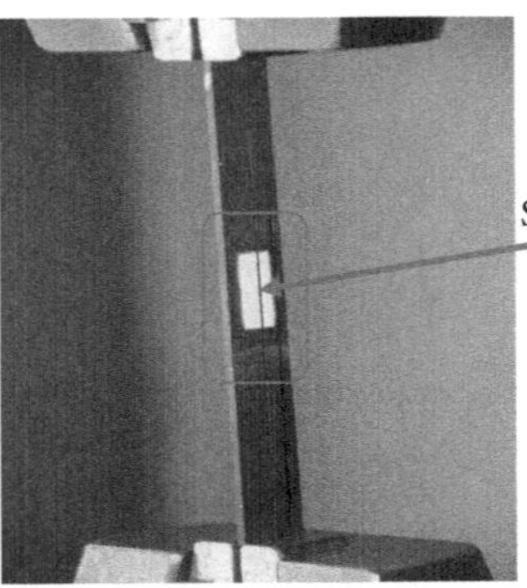

Fig. 15 CFRP specimen with PZT coating failed at a tensile load of 43KN

packets in the response. The group speeds of each mode of Lamb wave can be estimated from the respective wave packets for different frequencies using the arrival time information. The sensitivity of the piezoceramic coating as a sensor depends on its thickness, diameter, and frequency of the Lamb waves. Depending upon the diameter of the piezoceramic coating, the sensitivity can be suppressed or intensified over a particular band of frequency. Large-diameter piezoceramic coatings are very sensitive due to their large surface area but can be used at only lower frequency ranges. The sensitivity of the piezoceramic coating increases proportionately with its thickness for the entire frequency range.

Similarly, a piezoceramic coating can be used to generate the guided wave (including the Lamb wave) with two propagating wave modes (S_0 and A_0) when an excitation is sent through it. The performance of the piezoceramic coating actuators also depends on their sizes. Large-diameter piezocoating actuators are suitable for low frequency excitation, whereas small-diameter piezocoatings perform well at higher frequencies. Therefore, a careful choice of diameter of piezoceramic coating actuators is required depending upon the frequency of excitation. The stress generated in the piezoceramic coating actuator is proportional to the magnitude of the voltage applied across its electrodes.

The characterization of the sensing and actuation performance with frequency reveals the optimal frequency band at which the piezoceramic coating has to be used as a sensor, or an actuator, respectively. Damage interrogation should be taken at these optimal frequency bands using a guided wave technique. A piezoceramic coating can be used for the damage detection of various types of damages which usually occur in composite structures such as delamination, fiber cut, and fiber separation. Figure 16 shows a typical composite test object with a piezoceramic coating.

Studies on wave propagation in the presence of these damages in composite beams of identical dimensions (as shown in Fig. 16) would give information about the location and severity of such damage. Wave propagation on a healthy sample (without damage) of the same dimensions should be taken as a reference to understand the wave signatures. A generally symmetric (S_0) wave mode is used for such detection since it is free from mixed modes due to a higher group velocity and

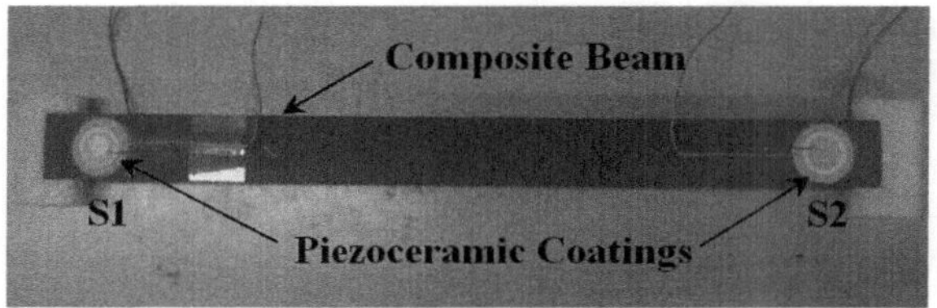

Fig. 16 Composite beam with Piezoceramic coating used in damage detection

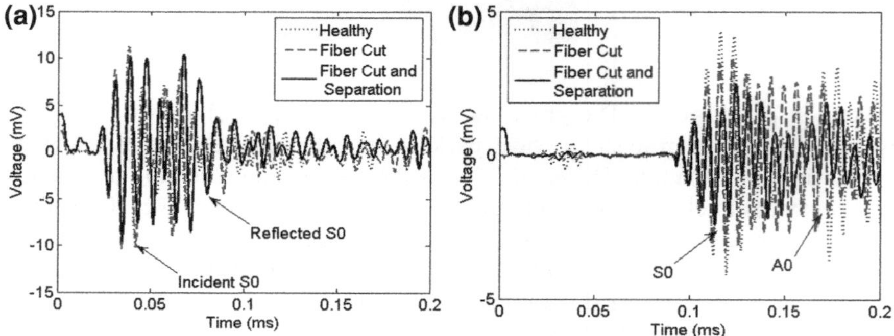

Fig. 17 Comparison of healthy and damaged signals from a piezoceramic coating **a** reflected signal, **b** transmitted signal

less dispersive when compared to an antisymmetric (A_0) wave mode. An actuation signal sent from S1 is reflected and transmitted from the damage location. The transmitted wave mode can be sensed by S2 and the reflected wave mode by S1 (Fig. 16). The effect of different damages on the reflected wave and transmitted wave signal is shown in Fig. 17a, b, respectively in comparison with healthy signals. It is seen from Fig. 17a that the magnitude of the reflected wave mode (S0) increases for fiber cut and combined damages of fiber cut plus fiber separation. The increase in the magnitude of reflected wave mode is due to the reflection of the wave energy from the damage.

Transmitted wave propagation through the damage can be monitored using the piezoceramic coating, S2. Damages such as fiber cut and fiber separation have the same effect on the transmitted signal strength. The highest impact on the signal is observed when fiber cut and fiber separation are combined as seen in Fig. 17b. The healthy and damaged signals are used to localize the damage and estimate its severity by postprocessing it using wavelet transforms. The postprocessing of the signals from the piezoceramic coatings of damaged and undamaged composite structures using wavelet coefficients gives the damage location. An accurate location estimation makes this technique very promising for the health monitoring of composite structures. The damage severity can be estimated using axial-guided wave modes.

5 Conclusion

The developed in situ piezoceramic coating for a wide range of thicknesses (200–500 μm) is a novel process combining the features of bulk piezo wafers and flexible piezo polymers. The coating, using hand-held equipment is cost-effective and resolves the issues of the adhesive bonding technology associated with the applications of stand-alone piezoceramic wafer or piezo polymer thin films as smart sensors and actuators. The technology developed is suitable for both the silicon batch process and the nondestructive evaluation of metals and composites. The in situ coating has proved to be very reliable for ultrasonic Lamb wave generation and sensing, making it a potential candidate for SHM applications and piezo MEMS which is a value addition to the Aerospace Industry and Smart Materials Research.

Acknowledgments I thank NPMASS for the funding in support of this research under the project "Acousto-ultrasonic coating for structural health monitoring" -PARC 1:2. I also thank Prof. D. Roy Mahapatra, Aerospace Engineering, IISc-Bangalore for his contribution and suggestions on SHM experiments.

References

1. Polla DL, Francis LF (1998) Processing and characterization of piezoelectric materials and integration into microelectromechanical systems. Annu Rev Mater Sci 28:563–597
2. Trolier-McKinstry S, Muralt P (2004) Thin Film Piezoelectrics for MEMS. J Electroceram 12:7–17
3. Muralt P (2000) PZT thin films for microsensors and actuators: where do we stand? IEEE T UFFC 47:903–915
4. Jang LS, Kuo KC (2007) Fabrication and characterization of PZT thick films for sensing and actuation. Sensors 7:493–507
5. Wang Z, Miao J, Zhu W (2007) Piezoelectric thick films and their application in MEMS. J Eur Ceram Soc 27:3759–3764
6. Corker DL, Zhang Q, Whatmore RW et al (2002) PZT 'composite' ferroelectric thick films. J Eur Ceram Soc 22:383–390
7. Pandey SK, James AR, Raman R et al (2005) Structural, ferroelectric and optical properties of PZT thin films. Phys B 369:135–142
8. Tyholdt F, Dorey RA, Bredesen R et al (2007) Novel patterning of composite thick film PZT. J Electroceram 19:315–319
9. Gang J, Qingxian H, Sheng L, Dongxiang Z, Qiuyun F (2012) Effect of solid content variations on PZT slip for tape casting. Process Appl Ceram 6(4):215–221
10. Dietze M, Es-Souni M (2008) Structural and functional properties of screen-printed PZT–PVDF-TrFE composites. Sensor Actuat A 143:329–334
11. Lebedev M, Akedo J, Akiyama Y (2000) Actuation properties of lead zirconate titanate thick films structured on Si membrane by the aerosol deposition method. Jpn J Appl Phys 39:5600–5606
12. Akedo J, Lebedev M (2001) Influence of carrier gas conditions on electrical and optical properties of $Pb(Zr, Ti)O_3$ thin films prepared by aerosol deposition method. Jpn J Appl Phys 40:5528

13. Chou C, Tsai S, Tu W et al (2007) Low-temperature processing of sol-gel derived Pb(Zr, Ti)O3 thick films using CO_2 laser annealing. J Sol-Gel Sci Technol 42:315–322
14. Wang ZJ, Cao ZP, Otsuka Y et al (2008) Low-temperature growth of ferroelectric lead zirconate titanate thin films using the magnetic field of low power 2.45 GHz microwave irradiation. Appl Phys Lett 92:222905–222905-3
15. Dutta S, Jeyaseelan A, Sruthi S (2013) Low temperature processing of PZT thick film by seeding and high energy ball milling and studies on electrical properties. J Electron Mater 42(12):3524–3528
16. Chen YZ, Ma J, Kong LB, Zhang RF (2002) Seeding in sol–gel process for $Pb(Zr_{0.52}Ti_{0.48})O_3$ powder fabrication. Mater Chem Phys 75:225–228
17. Zhao QL, Cao MS, Yuan J et al (2010) Thickness effect on electrical properties of $Pb(Zr_{0.52}Ti_{0.48})O_3$ thick films embedded with ZnO nanowhiskers prepared by a hybrid sol-gel route. Mater Lett 64:632–635
18. Hu SH, Hu GJ, Meng XJ et al (2004) The grain size effect of the $Pb(Zr_{0.45}Ti_{0.55})O_3$ thin films deposited on $LaNiO_3$-coated silicon by modified sol-gel process. J Cryst Growth 260:109–114
19. Boukamp BA, Pham MTN, Blan DHA et al (2004) Ionic and electronic conductivity in lead–zirconate–titanate (PZT). Solid State Ionics 170:239–254
20. Muralt P, Kohli M, Maeder T et al (1995) Fabrication and characterization of PZT thin-film vibrators for micromotors. Sensor Actuat A 48:157–165
21. Muralt P, Kholkin A, Kohli M et al (1995) Fabrication and characterization of PZT thin films for micromotors. Integr Ferroelectr 11:213–220
22. Muralt P, Ledermann N, Baborowski J et al (2005) Piezoelectric micromachined ultrasonic transducers based on PZT thin films. IEEE Trans Ultrason Ferroelect Freq Contr 52:2276–2288
23. Deshpande M, Saggere L (2007) PZT thin films for low voltage actuation: Fabrication and characterization of the transverse piezoelectric coefficient. Sensor Actuat A 135:690–699
24. Damjanovic D (2005) The Science of hysteresis. In: Mayergoyz I, Bertotti G (eds.) Hysteresis in piezoelectric and ferroelectric materials, vol 3. Elsevier, Amsterdam
25. Henning AK, Patel S, Selser M et al (2004) Factors affecting silicon membrane burst strength. Proc SPIE 5343:145–153
26. Yuanxia X, Zhenqing L (1999) Analysis of ultrasonic waves in plates by geometrical ultrasonics and its applications. Nondestruct Test 21(12):53

Cost-Effective Processing of Polymers and Application to Devices

Bhoopesh Mahale, Abhay Joshi, Abhijeet Kshirsagar, S. DattaGupta, Dhananjay Bodas and S. A. Gangal

Abstract Cost-effective and low temperature MEMS device manufacturing is driving the industry and/or laboratories for incremental innovation. A good strategy to drive Manufacturing Readiness Level (MRL) is perhaps to use more and more polymers such as PMMA, PTFE, PVdF, etc., as a masking, sacrificial, or structural layer, respectively. Use of polymers and their comparatively simple processing techniques help reducing material, processing, and manufacturing cost of the device. This chapter discusses low temperature deposition of thin films of these polymers using spin coating, plasma polymerization, and sputtering techniques. Optimization of process parameters is carried out. Applications of PMMA and PTFE as masking layer, PMMA as sacrificial layer, and PVdF as structural layer are demonstrated.

Keywords Polymers · Masking materials · Sacrificial layer · Piezoelectric sensor · PTFE · PMMA · PVdF

1 Introduction to Cost-Effective Polymers

Microelectromechanical systems (MEMS) are fabricated either by bulk or surface micromachining techniques. In bulk micromachining, the silicon substrate is used to realize the structures or devices and as a support for mechanical layers or

B. Mahale · A. Joshi · A. Kshirsagar · S. A. Gangal (✉)
Department of Electronic Science, University of Pune, Ganeshkhind Road, Pune 411007, India
e-mail: sagangal@gmail.com

A. Kshirsagar · S. DattaGupta
Department of Electrical Engineering, Indian Institute of Technology, Mumbai, India

D. Bodas
Centre of Nanobioscience, Agharkar Research Institute, GG Agarkar Road, Pune 411004, India

K. J. Vinoy et al. (eds.), *Micro and Smart Devices and Systems*, Springer Tracts in Mechanical Engineering, DOI: 10.1007/978-81-322-1913-2_14,

structures during surface micromachining. The surface micromachined MEMS devices use sacrificial layer below the mechanical/structural layer and the sacrificial layer is finally removed to achieve the required freedom of movement once the device is released. Many materials have been employed to fabricate MEMS; however, polymers are under increasing focus. The polymer-based MEMS are usually optically transparent, biocompatible, with a possibility of inexpensive prototyping, and easy to micropattern (e.g., micromolding, photopatterning, etc.). Structural layer applications based on polymers as well as its use as substrate demand low temperature processing (less than 100 °C). Polymers have attracted much attention as promising material due to their inherent properties like easy availability, temperature resistance, chemical resistance, durability, structural integrity after processing, dielectric constant, and low temperature processing.

Polymers can be deposited on the substrate by many processes like spin coating, physical vapor deposition, sputtering (Dc, RF, Magnetron), etc., which are compatible to MEMS fabrication. The processing time is very low as compared to their counterparts at low temperature. These materials are easily available and cost effective not only in terms of availability but also for their deposition techniques.

Our lab has been focusing mainly on polymers during these years and has used them in MEMS research. Polymers like poly(methyl methacrylate) (PMMA) [1–5], poly(tetrafluoro ethane) (PTFE) [6–8], poly(vinyl difluoride) PVdF [9, 10], poly(-hydroxy ethyl methacrylate) (pHEMA) [11], poly styrene [12], etc., have been used. These polymers are used as masking materials, sacrificial layers, structural materials, sensing layers, and materials for e-beam lithography. This chapter limits the usage of polymers in the area of MEMS research wherein lot of work is carried out on masking, sacrificial, and structural layers leading to many highly citied publications.

2 Masking Materials: PMMA and PTFE

The dimensional requirement of the different micromachined structures gave rise to the need of good masking materials. Masking materials are those, which protect the regions of silicon wafer from the attack of the etchant during micromachining process. The types of masking materials used in the micromachining technology are silicon dioxide, silicon nitride, copper, aluminum, gold, and polymers. The widely used masking materials are silicon dioxide and silicon nitride [13, 14]. They have very low etch rate as compared to silicon and are preferred due to the lattice matching. There are many processes developed for the deposition of silicon dioxide and silicon nitride, the latter being a better mask due to its low etch rate.

2.1 PMMA [1, 2]

PMMA was deposited using in-house fabricated parallel plate DC sputtering system [2]. The deposition power was varied from 15 to 35 W and time from 15 to

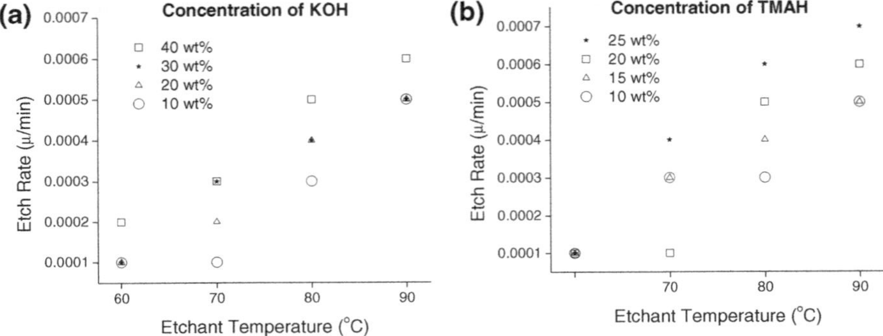

Fig. 1 Etch rate versus etchant temperature for **a** KOH and **b** TMAH etchants at different etchant concentration

35 min obtaining various thicknesses of the film. The Fourier Transform Infrared (FTIR) spectra showed all the characteristic peaks of PMMA. The peak of interest from FTIR was 842 cm^{-1} corresponding to $SiOCH_3$ which shows adhesion of the film and the substrate. X-ray Photoelectron Spectra (XPS) showed all the characteristic peaks of PMMA especially carbonyl group at binding energy of 289.0 eV. The interfacial tension was calculated from the contact angle data which showed a lower value of 0.8 dyne/cm indicating good adhesion of the substrate and the film supporting claims made by the FTIR [1, 2].

The important test of masking performance was carried out by dipping the film in aqueous KOH and TMAH. These are the known wet chemical etchants for silicon and widely used in fabrication of various microstructures. The concentration of the etchants was varied from 10 to 40 wt% for KOH and 10–25 wt% for TMAH. The temperature was varied between 60 and 90 °C to test the masking action (see Fig. 1). PMMA showed a masking time of ~300 min in which whole of the silicon substrate (500 μm) gets etched. PMMA etch rates of 0.0004 μm/min for KOH and 0.0005μ m/min for TMAH etchant were calculated. These etch rates are low as compared to the conventional masking materials viz. SiO_2 and Si_3N_4 0.004 μm/min and 0.002 μm/min, respectively. FTIR and XPS recorded after the masking action of 3 *h* reveals the presence of PMMA on the silicon substrate. Scanning Electron Micrograph (SEM) showed a pinhole free film which is a characteristic of good masking material [1, 2].

2.2 PTFE [6–8]

On the similar lines, study of PTFE deposited by RF sputtering in an in-house-developed parallel plate system, at different deposition powers of 100–200 W and at times from 15 to 90 min, was carried out for its performance as masking

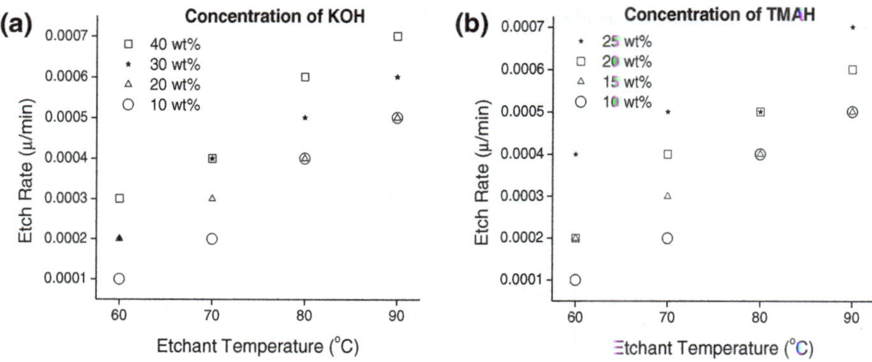

Fig. 2 Etch rate versus temperature for **a** KOH and **b** TMAH etchants at different etchant concentration

Table 1 Comparison of masking materials

Condition/Parameters	PMMA	PTFE	SiO_2	Si_3N_4
Deposition temp (°C)	RT	RT	1100	800–900
Max. deposition time (min)	35	90	240	480
Thickness of film μm	~0.18	~0.21	~1	~0.5
Etch rate in KOH (μm/min)	~0.0004	~0.0003	~0.006	~0.0004
Etch rate in TMAH (μm/min)	~0.0005	~0.0004	~0.006	~0.0004

material [6–8]. The FTIR of the RF-sputtered PTFE film showed characteristic peaks including peaks at 812 and 882 cm^{-1} corresponding to bonding between fluorine and silicon. XPS showed all the characteristic peaks of bulk PTFE. Interfacial tension calculated from contact angle data on PTFE film showed a value of 0.7 dyne/cm which is lower than the interfacial tension value of PMMA indicating better adhesion than PMMA. Further, this is also confirmed by SEM observation and supports FTIR data. Masking test at same parameters of etchants as used for PMMA showed a masking time of >300 min which is better than that for PMMA.

The etch rate of 0.0003 μm/min for KOH and 0.0004 μm/min for TMAH etchant was calculated (see Fig. 2). FTIR and XPS recorded after the masking action of 3 h reveal the presence of PTFE film on silicon substrate [6–8].

The data of masking is summarized in Table 1 giving the comparison of the masking materials studied in the present work with those conventionally available.

It can be clearly seen from the above table that polymers stand a better chance as new masking materials. They have better properties to become seminal masking material which can replace the conventional ones.

To conclude, PTFE comes out to be the best masking material due to its low etch rate of ~0.0003 μm/min in anisotropic etchants like KOH. PMMA also showed an etch rate of ~0.0004 μm/min. The issues which need to be tackled are regarding removal of masking material after pattern transfer, stress analysis of the film incorporation of hydrogen during processing in silicon substrate, and integrity of use of these masking materials with other processes in MEMS fabrication.

2.3 Transfer of Pattern

Experiments were carried out to show the feasibility of PTFE as a masking material in MEMS fabrication. RF-sputtered PTFE film was used for the transfer of pattern as it resulted in a better etch rate as compared to DC-sputtered PMMA. A simple diaphragm pattern was transferred through the PTFE layer on the silicon substrate using lift off. The silicon substrate (500 μm) with pattern on it was etched in 40 wt% KOH at 90 °C to get a diaphragm. PTFE layer was deposited on the back side of the substrate also to protect it from etching.

The diaphragm etched (5 h) in silicon using PTFE film as a masking material is shown in Fig. 3. The dimensions of the diaphragm are 1 × 1 mm and the thickness is around 80 μm.

Removal of mask layer after the completion of micromachined structure was not tried. However, references suggest that PTFE can be removed by plasma etching using a H_2/N_2 (70:30) mixture. PTFE can also be removed chemically using an etchant known as Fluoroetch (sodium napthanilide).

3 Sacrificial Layers: PMMA

In surface micromachining technology, a sacrificial layer is used to provide a support to the structural layer and is subsequently removed which leaves the structural layer suspended or free-standing. The material used for sacrificial layer should have smooth surface, uniform thickness, good adhesion, and should be chemically stable. Many types of sacrificial layers have been used, viz. metallic, nonmetallic, and polymer.

In many MEMS processes, metallic sacrificial layers such as copper, aluminum, titanum, and chromium have been used [15–19]. The metal sacrificial layer is deposited by using the techniques such as electroplating, chemical, or physical vapor deposition. Electroplating can be used for higher thickness of sacrificial layer in the range of few microns. Depositing few microns metal thin film by physical and chemical vapor deposition is time consuming and costly as the deposition rates are low. Higher thickness also causes stress in the thin film which results in pealing-off of the layer.

Nonmetallic layers such as silicon oxides, polysilicon, silicon oxynitride, phosphoric glass, borophosphosilicate glass (BPSG), phosphosilicateglass (PSG),

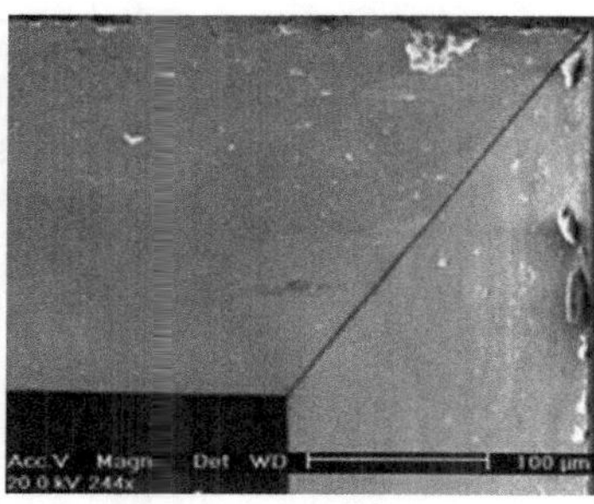

Fig. 3 Diaphragm etched in silicon using PTFE as a etch mask

porous silicon, graphite [20–27] are used as sacrificial layers. Out of these, silicon oxide, oxynitride, or polysilicon are commonly used. Deposition of nonmetallic layers is possible with the techniques such as low pressure CVD (LPCVD), RF sputtering, plasma enhanced CVD (PECVD), inductively coupled plasma CVD (ICPCVD), hot-wire CVD (HWCVD), and newly added atomic layer deposition (ALD) [28]. But all these deposition techniques have one or all of the following problems viz. high deposition temperature in the range of 300–800 °C, low deposition rate, requirement of harsh etching agents, and high cost of equipment.

Polymers are another candidate for use as sacrificial layer. Polymers proved to overcome most of the problems mentioned in metallic and nonmetallic sacrificial layers. It is because instead of wet etching, polymers can be plasma etched (dry etch) for releasing the MEMS structures which makes the process clean and IC processing compatible. Further, polymers do not react with the structural layer eliminating the cause of buckling. Polymer sacrificial layer also helps in maintaining the physical and mechanical properties of the released structures. Finally, polymers can be deposited by spin coating which is cost-effective and time-saving technique. Thus, using polymers as sacrificial layer, makes the process simple, cost efficacious, and clean.

Polymers such as photo resists, PMMA, SU-8, polystyrene, and polyimide, [29] are commonly used materials for sacrificial layer in surface micromachining. Following are the problems associated with the above polymers when used as sacrificial layer:

- These polymers are commercially available but are at very high cost and with fixed viscosity. To achieve a specific thickness of sacrificial layer the manufacturers do not provide any dilution option, it is recommended to buy a different viscosity solution.
- High thermal budget (relatively high baking temperature).
- Polymers like polyimide are highly toxic [30].
- Hard-baked SU-8 after plasma etching makes the surface hydrophilic for several months showing a moderate hydrophobic recovery [31].
- Polymers like PMMA being e-beam resist takes long time for writing larger patterns.

It is expected that the sacrificial layer could be deposited and/or cured at low temperature, should have low toxicity, can be tailored for different viscosity, is CMOS process compatible, and bares low cost. PMMA is the potential polymer which can help overcome some of the above-mentioned problems. PMMA is selected in the present work because of its comparatively low cost, less toxic, and work has been done in our group on other applications of PMMA [32, 33].

3.1 PMMA [1, 3, 5]

PMMA solution was made by dissolving PMMA granules in chlorobenzene and chloroform (CMOS grade) by ultrasonication. PMMA thin film was deposited on silicon substrate by spin coating method. When observed under SEM, the surface roughness of the film deposited using PMMA dissolved in chloroform was found to be much larger, $\approx$0.159 μm, than the film deposited using PMMA dissolved in chloro benzene which is $\approx$45 nm. The PMMA dissolved in chloro benzene was therefore not utilized for further experiments. The difference in surface roughness in case of two solvents is attributed to the difference in evaporation temperature of the solvents, 61 °C for chloroform and 131 °C for chlorobenzene. Experiments were carried out to optimize the spin speed, spinning time, concentration of liquid PMMA, and baking time, to achieve adherent film of 2 μm thickness. The Laurel spin coater WS-400-6NPP-LITE is used for spinning in two steps with speed measurement accuracy of $\pm$5 rpm. In the first step (called dispense step), the spin speed was kept at 300 rpm for 15 s. In the second step, the spin speed varied from 1,000 to 5,000 rpm in step of 1,000 rpm for 30 s. The post-spin baking time was also optimized by keeping temperature less than 100 °C. Clean room hot plate (accuracy $\pm$1 °C) was used for this purpose.

The deposited films were then optimized for isotropic and anisotropic etching in oxygen plasma. Etching parameters such as oxygen flow rate, RF power were optimized so that complete etching of PMMA is achieved. Anisotropic etching was carried out in parallel plate plasma system indigenously made which has Pfeiffer automatched 13.56 MHz RF power source with output range up to 600 W, with water-cooled parallel plates separated by 5–6 cm.

The isotropic etching was carried out in barrel type plasma system bought from EMITECH model K1050X shown in Fig. 4. It is a fully automatic system which has a dry rotary pump with embedded turbo pump, and rotameters of the accuracy of $\pm$2 sccm. It has automatched 13.56 MHz RF power source with output range up to 125 W. This was used for etching of sacrificial layer from underneath the structural layer. The ions generated follow the helical path which has more lateral etching rate. These ions generate trench underneath the structural layer. The walls created initially during anisotropic etching are etched laterally. This results in a circular wall profile with a radius equal to the etch depth.

The minimum base pressure of 4×10^{-5} mbar was achieved using a rotary-turbo pump system. Then the O_2 gas (25–30 sccm) was introduced in the chamber

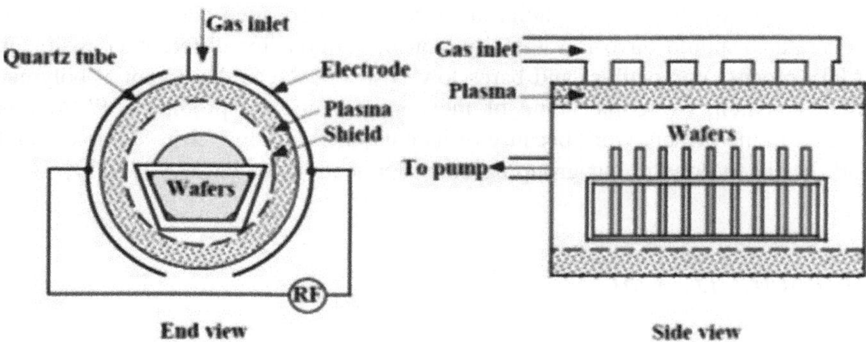

Fig. 4 Schematic diagram of barrel type plasma etching system. **a** End view, **b** Side view

so that the chamber pressure increased up to 0.2 mbar which gave stable, clean and sustained white plasma. RF power was varied from 80 to 125 W. PMMA-coated silicon substrates were kept in the reactor for initial experiments. It was observed that at lower values the of RF power most of the thickness of PMMA is etched away but a thin white layer on the wafer remained. To remove this layer we increased the RF power. It was found that around 90 W RF power gives a clean etch of PMMA with shiny silicon substrate visible.

With all process parameters known etching of PMMA was carried out through aluminum mask to form a step in PMMA for etching rate measurement by profilometer. For thickness measurement Ambios profilometer model XP2 was used with an accuracy of ± 10 nm. The etching rate was observed to be 3.12 μm/min.

PMMA has glass transition temperature of 105 °C. Therefore, for use of PMMA as sacrificial layer the structural layer deposition and processing also has to be done at temperature around 100 °C. Silicon nitride (SiN) is used as a structural layer. The silicon nitride was deposited by Oxford Instruments Inductively Coupled Plasma Chemical Vapor Deposition (ICPCVD) system with following deposition parameters (provided by the manufacturer); deposition temperature 70 °C, ICP power 1000 W, RF bias power 40 W, process pressure of 4mTorr, and 13.5, 10, 40 sccm of gas flows of SiH_4, N_2, and Ar, respectively.

Compatibility study of this in-house made PMMA as sacrificial layer in MEMS was carried out by using it for silicon nitride cantilever fabrication [3]. For deposition and processing of PMMA and silicon nitride films, above described processes were used. All the processes are low temperature Processes. The thermal budget does not exceed 100 °C at any step in the process.

For fabrication of silicon nitride cantilever, in-house prepared PMMA solution was spin coated onto the RCA cleaned silicon wafer (see Fig. 5a). The coated PMMA is not a UV-resist or electron beam resist so it is patterned using a hard mask. Aluminum (Al) was used as hard mask and was deposited by physical vapor deposition technique. The deposited hard mask is patterned by positive photoresist (PPR). The Al not protected by PPR is etched by Al etchant (see Fig. 5b).

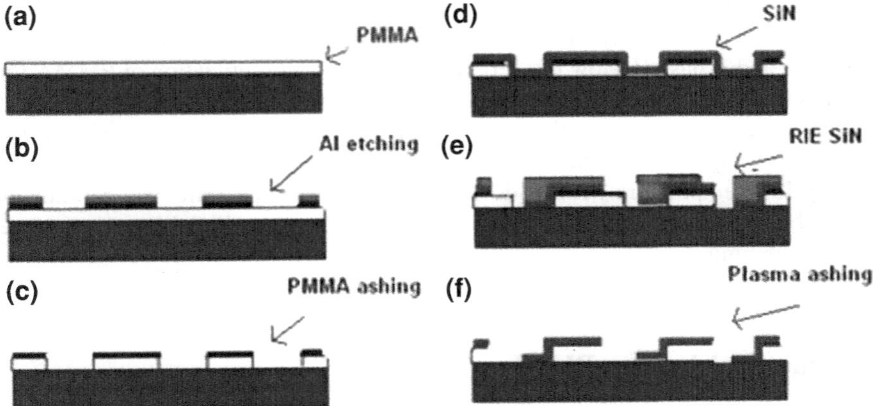

Fig. 5 Fabrication steps of SiN cantilevers

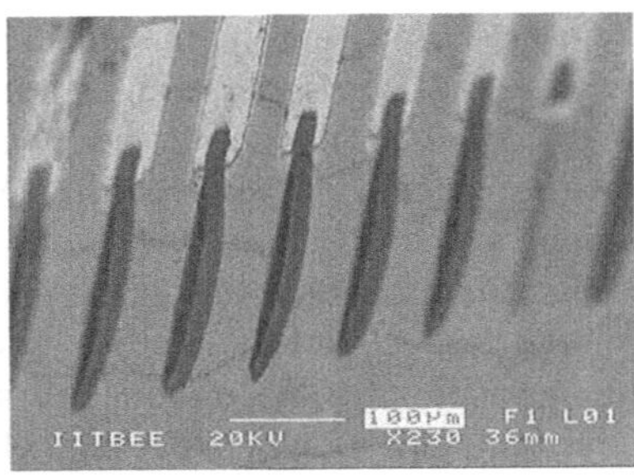

Fig. 6 SEM of low temperature SiN cantilevers

The PMMA which was not protected by Al is then etched anisotropically using parallel plate type oxygen plasma to give the vertical wall pattern to the sacrificial layer (see Fig. 5c). The remaining hard mask is etched away and then structural layer of silicon nitride is deposited. The structural layer is deposited by inductively coupled plasma chemical vapor deposition (ICPCVD) technique at temperature of 70 °C (Fig. 5d). The deposited SiN film is patterned by positive photoresist, which is used as a mask for next reactive ion etching (RIE) of SiN (Fig. 5e). CF_4/O_2 plasma is used for reactive ion etching of silicon nitride. In final step isotropic etching of PMMA is done in barrel type plasma etching system using optimum parameters to release the cantilever (Fig. 5f). The oxygen plasma not only etches the PMMA underneath but also the positive resist on top of SiN which was used as mask for RIE and releases the cantilever.

The released silicon nitride cantilevers were observed under scanning electron microscope (SEM) with a tilt angle of 65°. The released cantilevers are seen in the SEM micrographs shown in Fig. 6. It is seen that the sacrificial layer used was etched away completely. However, the cantilevers are seen to be bent down indicating that the deposited silicon nitride has stress.

In summary, the PMMA solution of different viscosity was prepared in-house using PMMA granules by dissolving in chlorobenzene. The material is low cost, less toxic and is baked at low temperature (90 °C). The optimization of PMMA for use as sacrificial layer in MEMS fabrication was done. The spinning, baking, and etching processes were optimized considering the requirements of surface micromachining devices. Making low baking temperature sacrificial layer helps in making low temperature MEMS devices. Thus, low temperature SiN cantilevers were fabricated to prove the feasibility of the PMMA as sacrificial layer in surface micromachining.

4 Structural Layer: PVdF

PVdF is a semi crystalline ferroelectric polyme. It is popular due to its properties like piezoelectricity flexibility, chemical inertness, high bandwidth of operation, low acoustic impedance, ease of fabrication, low weight, and low cost. PVdF is being used in hydraulic, pneumatic, acoustic MEMS and bioMEMS-based applications [34–39]. Because of its flexibility and versatility PVdF gained popularity over conventional ceramic piezoelectric materials like PZT and quartz.

Piezoelectric property of the PVdF is mostly dependent on β phase content among all four phases viz. α, β, γ, and δ [40–42]. β phase PVdF contains molecular chains packed in the unit cell in such way that the dipoles associated with individual molecules are parallel to one another, leading to the nonzero dipole moment of crystal. Researchers achieve β phase in the PVdF film through stretching [40], poling under high electric field [41], quenching from melt [42] etc.

PVdF-based pressure sensor is developed in our laboratory for pneumatic applications for the pressure range of 10–150 kPa [9]. The preparation of film, its packaging, and testing of the sensor is discussed in the following sections. The pressure sensor designed is aimed at measurement of pulse pressure of radial artery. Arterial pulse measurement is a basic tool for diagnosis of different diseases, in Indian Traditional Medicine (Ayurveda). The pressure sensor for such application should have accurate measurement of low frequency pulse (1–1.8 Hz) with higher sensitivity and lower resonating frequency. With available pressure sensors, pulse overlapping, flexibility and noise were main issues. Efforts are made to develop flexible, miniaturized PVdF based pressure sensor.

It is also proposed to develop array of miniaturized pressure sensors and further incorporating them in neural networks in order to get better disease diagnosis. As a first attempt, a single element pressure sensor is designed and simulated using CoventorWare 2006 [10]. The sensitivity, the maximum stress and resonating frequency of the designed devices are simulated with respect to pressure range of 10–30 kPa which a typical artery pressure range.

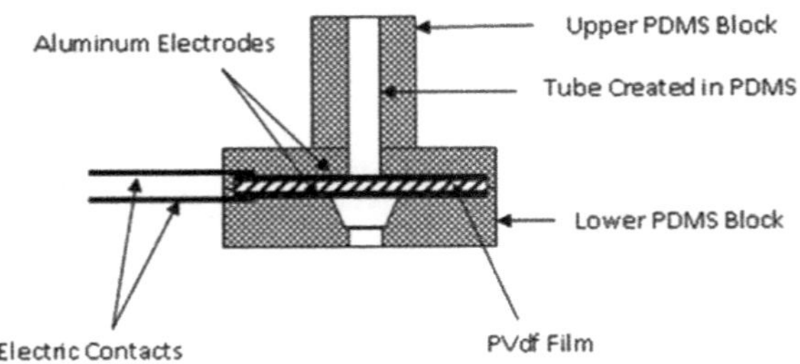

Fig. 7 Cross section of the sensor

4.1 Preparation of PVdF Film and Practicality of Sensor Fabrication

Apart from the simulation, fabrication of the PVdF based sensor is carried out using PVdF films. PVdF film is prepared by spin coating technique. 20 wt% solution is prepared by dissolving PVdF powder (Aldrich) in N-methyl pyrolidone. Solution is stirred for 30 min at 60 °C for complete dissolution of PVdF. Solution is spin coated (Laurel Technologies Corporation) on cleaned glass substrates for different spin speeds (1,000–3,000 rpm). Spin-coated films are further baked at different temperatures (60–90 °C) for 30 min. Baking of the films above 70 °C reduces the β phase content, hence baking temperature is optimized to 60 °C. The film with thickness of 20 μm spun at 2,000 rpm and baked at 60 °C is selected for further use. Baked films were peeled off from the substrates for further processing. Al electrodes (200 nm) were deposited on both sides of PVdF film using physical vapor deposition technique for contact purpose.

High electric field of ~80 MV/m is applied across the PVdF film for 2 h at 100 °C and at room temperature for 1 h. Poling ensures the permanent dipole alignment in the direction of applied electric field. Packaging was carried out using polydimethylsiloxane (PDMS). PDMS was an elastomer having good thermal stability and chemical inertness [43]. PDMS was casted in two cylindrical molds which were fabricated in acrylic. PDMS (Sylgard 184 from Dow Corning Chemicals Ltd.) was mixed in a 10:1 (base: curing agent) ratio, degassed, poured in both the molds and cured at 80 °C for 2 h. PDMS was casted in two parts (upper and lower) and used to package the sensor. Figure 7 shows the cross section of the sensor and the actual fabricated sensor is shown in Fig. 8.

To transfer air pressure to the film, in upper part of the sensor 500 μm circular channel is created using a copper wire while casting. The lower part has a groove of the size 1 mm to provide space for deflection of PVdF film. The PVdF film is mounted between these two PDMS blocks as shown in the Fig. 7. The electrical contacts to the film are taken out using conductive tape (aluminum tape 70 μm). The total size of the sensor was 6 mm in diameter.

Response of the developed pressure sensor to dynamic pressure is tested using a test setup shown in Fig. 9. Test setup includes solenoid valve, air pressure pipe, pressure gage, and sensor. Pressure gauge measures dynamic pressure applied to

Fig. 8 Photograph of the sensor

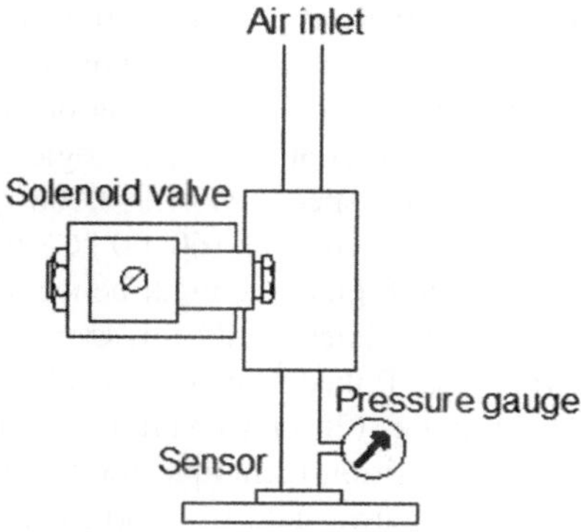

Fig. 9 Test setup

the sensor. To apply dynamic pressure on sensor, solenoid vale is switched ON and OFF, which releases the pressure on the sensor. Pressure is incremented from 0 to 150 kPa in steps of 10 kPa. Deflection of the PVdF film caused by pressure produces voltage proportional to applied pressure. Signal from sensor is further amplified by the signal conditioning unit (see Fig. 10). Output is displayed on the PC with the help of NI DAQ card 6259 and Lab VIEW program (see Fig. 11). Output of the sensor is plotted against applied dynamic pressure, which is shown in Fig. 12. Graph shows almost linear response of the sensor [9].

Response time of the sensor was measured to be 50 ms (see Fig. 13). To check the pyroelectric effect of the sensor, hot air of 40 °C is passed on sensor for 5 min. Very small change in output voltage is observed (few mV). This can be neglected because voltage produced by the sensor is much greater while testing.

To conclude easy method is adopted to fabricate pressure sensor from PVdF. Sensor is successfully tested for the dynamic pressure range from 10 to 150 kPa with negligible pyroelectric effect. Further modifications in film fabrication process may increase β phase content thereby increasing the sensitivity. Fabricated

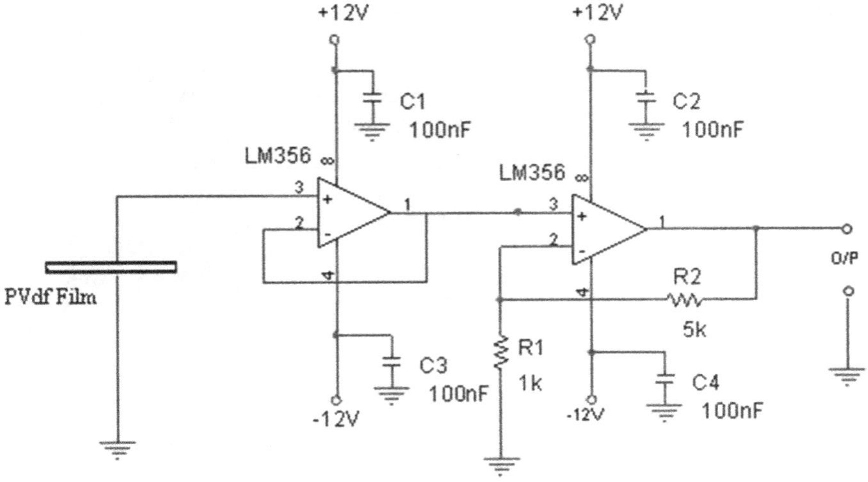

Fig. 10 Signal conditioning unit

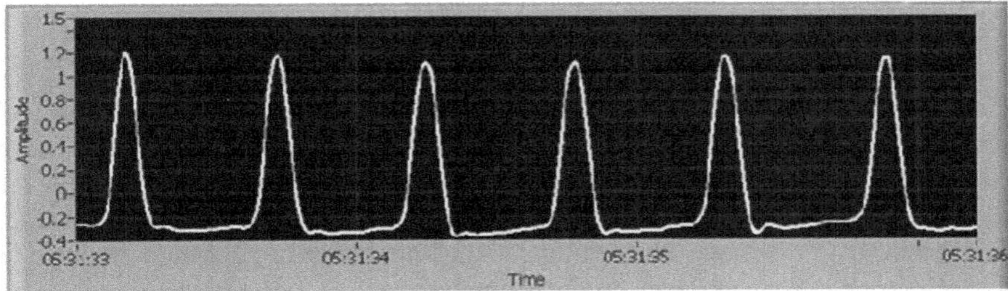

Fig. 11 Sensor o/p displayed on PC using LabVIEW

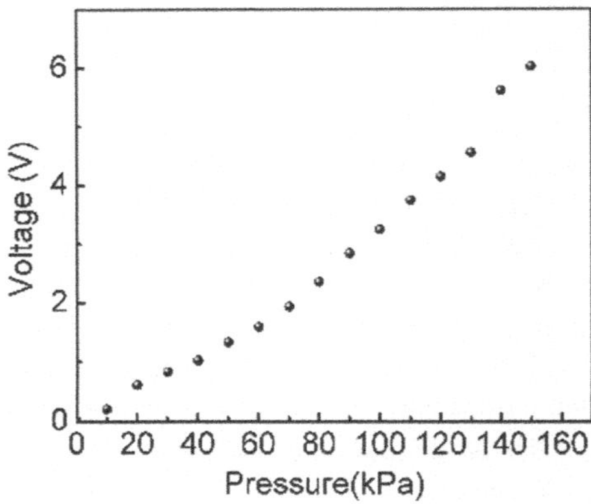

Fig. 12 Sensor response as a function of applied pressure

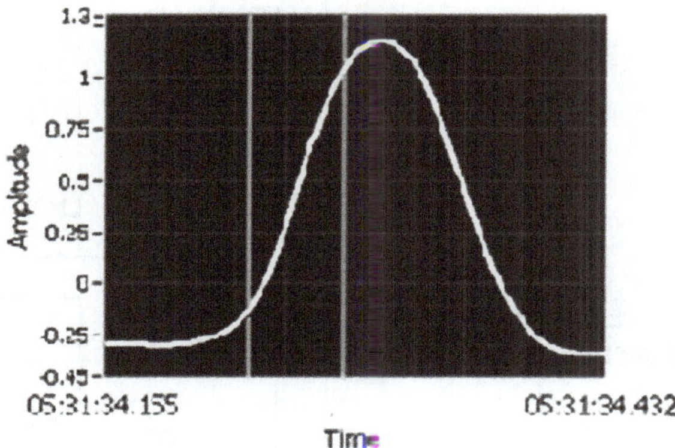

Fig. 13 Measurement of response time

sensor is small, low weight, and low cost. Efforts are being continued for achieving increase in β-phase content and in turn the sensitivity of the film. The developed films are also being tested for arterial pulse data collection.

4.2 Design and Simulations

Microcantilever is a mechanical structure used in MEMS devices for sensing and actuation applications. The fix-free cantilever-based design is chosen for pressure sensor application due to greater sensitivity obtained by the design. Cantilever structures, viz. fix-free beam offer higher deflection with respect to fix–fix beam. Applied pressure generates force on fixed-free end cantilever. Applied pressure in case of piezoelectric cantilever produces the mechanical stress in cantilever at the fixed end as well as deflects the cantilever at the free end. In case of piezoelectric cantilevers, the voltage generated on piezoelectric cantilever due to applied bending is governed by Eq. (1).

$$V = \frac{T^2 E_e z}{3 d_{31} L^2 E_p} \quad (1)$$

where V is the voltage generated on the piezoelectric layer, d_{31} the piezoelectric constant of the piezoelectric material, E_p and E_e are the Young's modulus of elasticity for the piezoelectric and elastic materials, respectively and Z is tip displacement. This equation shows that the higher displacement is advisable in order to get higher sensitivity (defined as voltage-generated/applied pressure).

In this work, the cantilever structures are designed in order to get higher deflections and stiffness. Following design objectives are set for PVdF sensor design: Miniaturized pulse sensor, i.e., <10 mm in complete size, High sensitivity ~100 mV/kPa, Low frequency operation 1–2 Hz, Working pressure range 10–30 kPa.

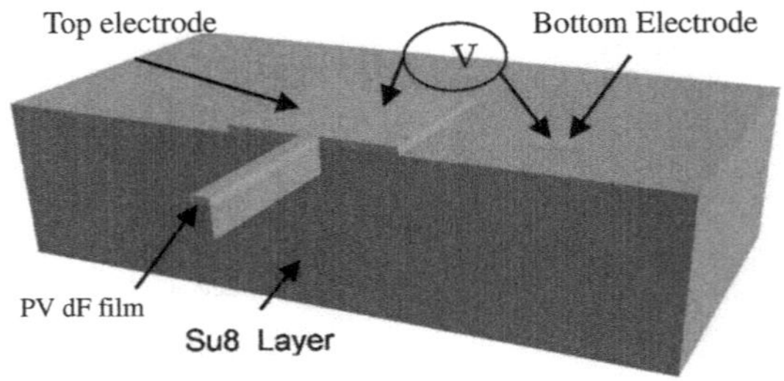

Fig. 14 Schematic of pulse pressure sensor

Pressure sensor consists of cantilever fixed at one end to the base as shown in Fig. 14. Base dimensions are fixed to 2,000 × 1,000 × 130 μm ($L \times W \times t$). Three different cantilevers are designed and simulated in order to get high sensitivity, lower resonating frequency and stress using CoventorWare 2006. The variation in width and thickness is studied to see the effect on sensitivity and stress. D1: 500 × 50 × 18.6 (μm) ($L \times W \times t$); D2: 500 × 50 × 30.6 (μm) ($L \times W \times t$); D3: 500 × 100 × 30.6 (μm) ($L \times W \times t$).

For the simulations, pressure is applied on the top surface of cantilever. The deflection as a function of applied pressure is shown in Fig. 15. The maximum deflection for D1, D2, and D3 at 30 kPa was 380, 5.6, and 75 μm, respectively, with a linear deflection pattern. D1 is more mechanically flexible and hence the device deflection is high. In case of D2, the higher cantilever thickness increases device stiffness and thus the deflection is restricted to 5.6 μm. In case of D3, the increase in cantilever width to 100 μm, increases the device's mechanical flexibility; this further increases the device deflection to 75 μm [10].

The stress calculation is more important parameter for device reliability. The maximum stress and stress location on cantilever is studied under application of pressure. The variation of stress with applied pressure is shown in Fig. 16. It can be seen that generated stress on cantilever increases linearly with applied pressure. On application of 30 kPa pressure, the maximum stress generated on D1, D2, and D3 is 840, 29, and 630 MPa, respectively. This variation shows that the D2 is stiffer and generated stress is very negligible. In case of D1 and D3, the higher stress is attributed to the higher W/t ratio. Still all the stress values are much below the fracture limit of the device 4.4 GPa.

Simulations are also carried out for the voltage variation with respect to applied pressure for three designs. D1 and D3 show higher sensitivity of 100 and 200 mV/kPa, respectively, while D2 shows sensitivity of 10 mV/kPa. In case of D2, the deflection and developed stresses are very low (3 % compared with D1, 5 % compared to D3). Low stress in piezoelectric cantilever is mainly responsible for low sensitivity of the device. Increase in PVdF layer thickness from 3 to 5 μm helps to improve the piezoelectric sensitivity, as piezoelectric coefficients are thickness dependable. In case of D3, even though the generated stresses on cantilever are less as compared to D1 but due to use of thicker PVDF layer improvement in sensitivity was observed.

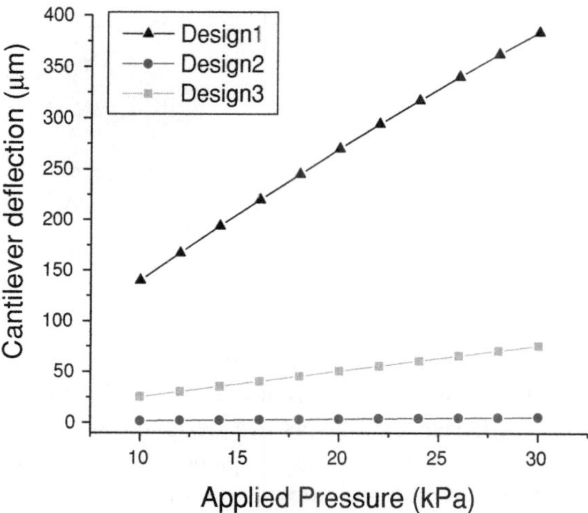

Fig. 15 Deflection verses applied pressure

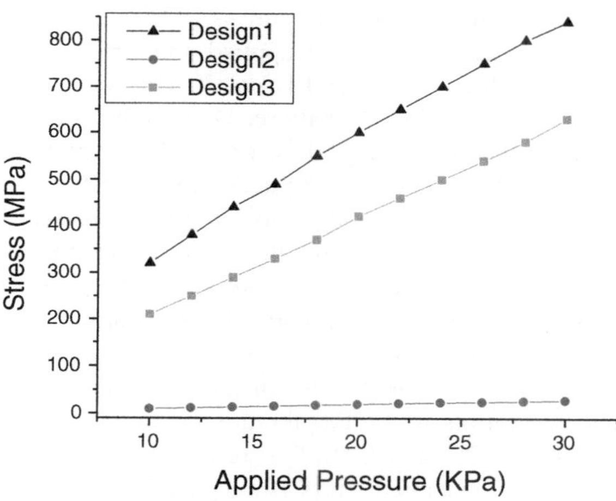

Fig. 16 Generated stress verses applied pressure

The three design simulated have the resonating frequency of 26, 58, and 58 KHz, respectively. D1 was having lower resonating frequency and high sensitivity of 100 mV/kPa. The D2 showed higher stiffness with very low stress. D3 was good for reliability and longer life as it shows highest sensitivity of 200 mV/kPa (when compared to D1 and D2); stress generated is also low as compared to D1. For all the designs simulated in this study, sensitivity values are better as compared to commercially available sensor (1 mV/kPa).

It is further planned to fabricate these cantilevers using surface micromachining in PVdF as structural layer and SU8 as sacrificial layer.

5 Conclusions

The chapter summarizes array of applications which could be targeted using polymers. Polymers exhibit a range of properties and their properties could be tailored by various means. They are very easy to handle and deposit as thin films on substrates like silicon, glass, or other polymers. Their bonding to any substrate is excellent and they are really cost effective in terms of processing cost and time. This chapter has touched on applications in the field of MEMS fabrication starting for a very crucial problem of low temperature masking material to a sacrificial layer for lift off process to a structural layer for fabrication of a pressure sensor. These polymers though today have not found a place in industry but they are seriously knocking on the door and possess potential to replace silicon in MEMS. One such example is the addition of lab-on-a-chip technologies which uses polymers in various forms. We strongly believe that polymers are the building blocks to manufacturing technologies of the future.

Acknowledgment The authors would like to acknowledge NPSM and NPMASS for providing funds for "Establishment of National MEMS Design Centres" (PARC#2 Project No. 2.10). Authors also acknowledge Department of Science and Technology, Govt. of India for providing funds for fabrication of devices.

References

1. Bodas DS, Dabhade RV, Patil SJ et al (2001) Comparative study of spin coated and sputtered PMMA as an etch mask for silicon Micromachining. In: IEEE proceedings of the international symposium on micromechatronics and human science, pp 51–56
2. Bodas DS, Mahapatra SK, Gangal SA (2005) Comparative study of spin coated and sputtered PMMA as an etch mask material for silicon micromachining. Sens Actuators A 120:582–588
3. Kshirsagar A, Duttagupta SP, Gangal SA (2011) Optimization and fabrication of low-stress, low-temperature silicon oxide cantilevers. Micro Nano Lett 6(7):476–481
4. Zambare M, Gosavi S, Gorwadkar S et al (1992) Improvement in the sensitivity of PPMMA ekectron beam resist by S and F atom doping. Jpn J Appl Phys 31:2640
5. Kshirsagar A (2011) Novel low stress silicon nitride, oxide and oxynitride surface micromachined cantilevers with PMMA sacrificial layer for low process temperature MEMS devices. Ph.D. Thesis, University of Pune, India
6. Bodas DS, Mandale AB, Gangal SA (2005) Deposition of PTFE thin films by RF plasma sputtering on (1 0 0) silicon substrates. Appl Surf Sci 245:202–207
7. Bodas DS, Gangal SA (2005) PTFE as a masking material for MEMS fabrication. J Micromech Microeng 15:802–806
8. Bodas DS, Gangal SA (2005) RF sputtered polytetrafluoroethylene—a potential masking material for MEMS fabrication process. J Micromech Microeng 15:1102–1113
9. Mahale BP, Bodas DS, Gangal SA (2011) Development of low-cost poly(vinyldifluoride) sensor for low- pressure application. Micro Nano Lett 6:540–542
10. Joshi AB, Kalange AE, Bodas DS et al (2010) Simulations of piezoelectric pressure sensor for radial artery pulse measurement. Mat Sci Eng B 168:250–253
11. Bodas DS, Desai SM, Gangal SA (2005) Deposition of plasma-polymerized hydroxyethyl methacrylate (HEMA) on silicon in presence of argon plasma. Appl Surf Sci 245:186–190

12. Gosavi S, Gangal SA, Kuruvilla B et al (1995) Plasma polymerized chlorinated α-methyl styrene (PP-C-αMS): a high performance negative electron resist. Jpn J Appl Phys 34(1)2A:630
13. Seidel H, Csepregi L, Heuberger A et al (1990) Anisotropic etching of crystalline silicon in alkaline solutions: 1. Orientation dependence and behavior of passivation layers. J Electrochem Soc 137:3612–3626
14. llie M, Marculescu B, Moldovan N, Nastase N, Olteanu M (1998) Adhesion between PMMA mask layer and silicon wafer in KOH aqueous solution. Proc SPIE 3512:422–430
15. Löchel B, Maciossek A, Knig M et al (1994) Galvanoplated 3D structures for micro systems. Microelectron Eng 23:455–459
16. Zavracky PM, Majumder S, McGruer NE (1997) Micromechanical switches fabricated using nickel surface micromachining. J Microelectromech Syst 6:3–9
17. Burbaum C, Mohr J, Bley P (1991) Fabrication of capacitive acceleration sensors by the LIGA technique. Sens Actuators A 27:559–563
18. Frazier AB, Ahn CH, Allen MG (1994) Development of micromachined devices using polymide-based processes. Sens Actuators A 45:47–55
19. Maciossek A, Löchel B, Quenzer H et al (1995) Galvanoplating and sacrificial layers for surface micromachining. Microelectron Eng 27:503–508
20. Evoy S, Carr DW, Sekaric L et al (1999) Nanofabrication and electrostatic operation of single crystal silicon paddle oscillators. J Appl Phys 86(11):6072–6077
21. Chen HR, Gau C, Dai BT et al (2003) A novel planarization process for polysilicon sacrificial layers in a micro-thermal system. Sens Actuators A 108(1–3):86–90
22. Habermehl S, Glenzinski AK, Halliburton WM et al (2000) Properties of low residual stress silicon oxynitrides used as a sacrificial layer. Mater Res Soc Symp Proc 605:49
23. Howe RT (1988) Surface micromachining for microsensors and microactuators. J Vac Sci Technol B 6:1809–1813
24. French PJ, Wolffenbuttel RF (1995) Low-temperature BPSG reflow compatible with surface micromachining. J Micromech Microeng 5:125–127
25. Bhatt V, Chandra S, Singh C (2009) Microstructures using RF sputtered PSG film as a sacrificial layer in surface micromachining. Sadhana 34(4):557–562
26. Lang W, Steinera P, Richter A et al (1994) Application of porous silicon as a sacrificial layer. Sens Actuators A 43:239–242
27. Makarova OV, Tang CM, Mancini DC et al (2002) Microfabrication of freestanding metal structures released from graphite substrates. In: IEEE international conference microelectromechanical systems (MEMS 2002), Los Vegas, NV, pp 400–402
28. Nishi Y, Doering Y (2007) Handbook of semiconductor manufacturing technology, 2nd edn. CRC Press, Boca Raton
29. Psoma SD, Jenkins DWK (2005) Comparative assessment of different sacrificial materials for releasing SU-8 structures. Rev Adv Mater Sci 10:149–155
30. Material Safety Data sheet issued by DuPont—PI2723 polyimice
31. Walther F, Davydovskaya P, Zurcher S et al (2007) Stability of the hydrophilic behavior of oxygen plasma activated SU-8. J Micromech Microeng 17:524–531
32. Bodas DS, Gangal SA (2005) Structural characterization of sputtered PMMA in argon plasma. Mater Lett 59:2903–2907
33. Dabhade RV, Bodas DS, Gangal SA (2004) Plasma treated polymer as humidity sensing material—a feasibility study. Sens Actuators B 98:37–40
34. González-Moran O, González-Allesteros R, Suaste Gómez E (2004) Cinvestav-IP polyvinylidene difluoride (PVdF) pressure sensor for biomedical applications. In: IEEE proceedings of the 1st international conference on electrical and electronics engineering, pp 473–475
35. Shirinov AV, Schomburg WK (2008) Pressure sensor from a PVDF film. Sens Actuators A 142:48–55
36. Yi J, Liang H (2008) A PVdF-based deformation sensor: modeling and experiments. IEEE Sens J 8:384–391

37. Chang WY, Chu CH, Lin YC (2008) A flexible piezoelectric sensor for microfluidic applications using polyvinylidene fluoride. IEEE Sens J 8(5):495–500
38. Gallego-Perez D, Ferrell JN, ta Castro NH, Hansford DJ (2010) Versatile methods for the fabrication of polyvinylidene-fluoride microstructures. Biomed Microdevices. doi:10.1007/s10544-010-9455-9
39. Sencadas V, Gregorio R, Lanceros-Mendez S (2009) α to β phase transformation and microstructural changes of PVdF films induced by uniaxial stretch. J Macromol Sci Part B 48:514–525
40. Sajkiewicz P, Wasiak A, Goclowski Z (1999) Phase transition during stretching of poly(vinylidene fluoride). Eur Polymer J 35:423–429
41. Davis GT, McKinney JE, Broadhurst MG et al (1978) Electric-field-induced phase changes in poly(vinylidene fluoride). J Appl Phys 49:4998
42. Yang D, Chen Y (1987) β-phase formation of poly(vinylidene fluoride) from the melt induced by quenching. J Mater Sci Lett 6(5):599–603
43. Xia Y, Whitesides GM (1998) Soft lithography. Annu Rev Mater Sci 28:153–184

Chemical Synthesis of Nanomaterials and Structures, Including Nanostructured Thin Films, for Different Applications

S. A. Shivashankar

Abstract Chemical methods for the synthesis of nanostructured materials have been found to be most versatile, not only in the range of materials that can be so formed, but also in the variety of morphologies in which they can be obtained without employing lithographic means, and in the variety of methods through which synthesis can be carried out. We present examples of how metal complexes can be used as precursors to nanostructured thin films both by the familiar chemical vapor deposition (CVD) method and by different solution routes. Composite materials—especially carbonaceous composites with interesting and useful characteristics—can also be obtained from metal complexes through different methods of processing. Equilibrium thermodynamics can be employed to analyze the metalorganic CVD (MOCVD) process to arrive predictively at nanocomposite coatings with unexpected components and morphologies.

Keywords Carbon nanotubes · MOCVD processing · Microwave processing · Metal complexes · Solution-based processing

1 Introduction

A great variety of approaches and methods have been employed to synthesize nanomaterials, including nanostructured thin films. These include physical, chemical, and physico-chemical methods, which sometimes require either elaborate and expensive protocols, such as lithography, or expensive apparatus, as in plasma-based approaches, or long durations of time, which is characteristic of solution-based methods operating at low temperatures [1]. It is desirable

S. A. Shivashankar (✉)
Centre for Nano Science and Engineering, Indian Institute of Science, Bangalore, India
e-mail: shivu@cense.iisc.ernet.in

K. J. Vinoy et al. (eds.), *Micro and Smart Devices and Systems*,
Springer Tracts in Mechanical Engineering, DOI: 10.1007/978-81-322-1913-2_15,

to develop chemical methods that are relatively rapid, simple, safe, and capable of producing nanomaterials both in powder form and as thin films and coatings, so that they may be investigated for their intrinsically interesting characteristics, as also for their practical applicability. Metalorganic complexes form a rich source of such possibilities [2], for they can serve as precursors to the chemical vapor deposition (CVD) of thin films, as is well known. In addition, because of their ready solubility in benign solvents, they are excellent starting materials solution-based synthesis of powder materials as well as thin films and coatings.

Metalorganic complexes are a class of materials in which a metal-to-heteroatom direct bond (e.g., M–O, M–S, M–N where M: metal) exists, along with the carbon chain. Figure 1 show a schematic of some metalorganic complexes with oxygen as the heteroatom that is directly bonded to the central metal atom M.

Such metalorganic complexes can be very useful as starting materials for various technologically important metal oxides, nitrides, sulfides, and also for metals themselves. The presence of direct metal-to-oxygen bonds in various metalorganic complexes such as metal isopropoxides, metal β-diketonates, and metal β-ketoesters leads to their use as starting material for the synthesis of oxides in solution-based as well as in vapor-phase techniques, specifically in the metal-organic CVD process, MOCVD [2]. Metal-to-oxygen direct bonds in such complexes sometimes allow deposition to be carried out without using any oxidizing gas. The presence of carbon in the molecular framework motivates the application of this class of materials to the growth of carbonaceous metal oxide composites. If carbon is not viewed as undesirable (as in the semiconductor industry), carbon incorporation can be employed to enhance the conductivity of the oxide materials prepared using such complexes. Moreover, the volatility of these complexes can be tuned by varying the carbon content in the molecular framework or by complexation with extra carbon-containing adducts.

In particular, metal β-diketonates are well-suited candidates as precursors in the MOCVD process and have been widely utilized for the growth of thin films of technologically important oxides (Fig. 1b). These complexes satisfy almost all the criteria needed for them to be useful as CVD precursors. Commonly used β-diketonates, such as acetyleacetonates, are stable up to about ~250 °C and are relatively inert to the ambient. They are usually crystalline solids which sublime between about 120 and 200 °C, with a rather wide window of ~40 °C separating their vaporization and decomposition temperatures [3].

2 MOCVD

One of the simplest complexes is the metal acetylacetonate, often abbreviated as "acac," whose molecular structure is represented by Fig. 1b when R = CH_3, the methyl group. Though other complexes like the dipivaloylmethanates (dpm) are more volatile, acac's are easier to synthesize and the acac complexes of most metals are crystalline, stable, and safe to handle. Such complexes have therefore

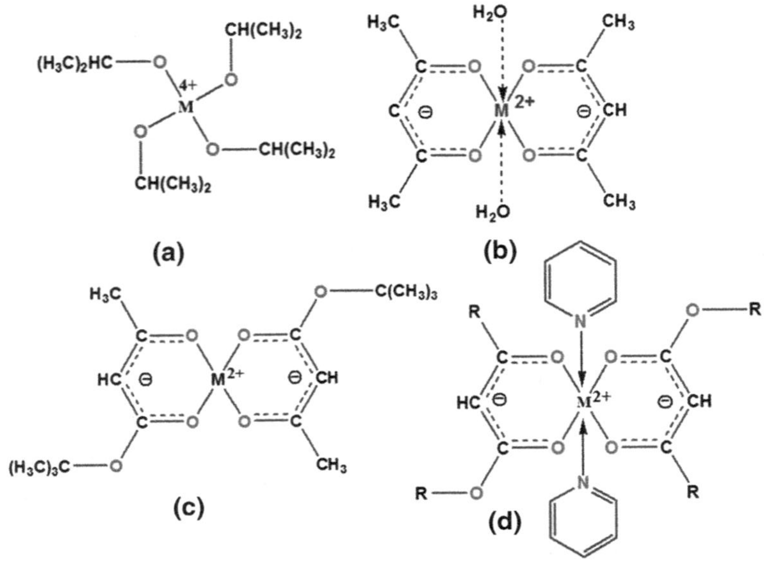

Fig. 1 Schematic structure so metal complexes, **a** alkoxide, **b** acetylacetonate, a β-diketonate, **c** β-ketoester, **d** adducted β-dikeonate

been employed as precursors for the low-pressure CVD and ALD of metal oxides and metals [4, 5], by employing an oxidizing and a reducing ambient, respectively. A question that arises is: what if the CVD process is conducted in an inert-ambient? When such a process is carried out with the acac's of aluminum and manganese, $Al(acac)_3$ and $Mn(acac)_2{\cdot}2H_2O$, very interesting results emerge [6, 7]. Depositions can be carried out in simple low-pressure, hot-wall reactors designed for subliming solid precursors, with precursor vapor carried by flowing argon. For reproducible results, it is necessary to control electronically the sublimation and substrate temperatures and the flow rate of gases, including that of oxygen, where required.

$Mn(acac)_2$ has low volatility and requires a temperature of over 200 °C to attain volatility sufficient for a CVD process. At a substrate temperature of more than 500 °C and a reactor pressure of about 5 torr, noticeable deposition begins to occur on a variety of substrates: Si(100), alumina, stainless steel (SS316). The coatings adhere strongly to the substrate when deposited at >600 °C and are black, suggestive that they are carbonaceous. This is confirmed by Raman spectroscopy (Fig. 2a), with the signature characteristic of "high temperature nanographite" [8].

Thermal analysis shows that the coatings on SS have as much as 80 at.% carbon in them. However, coatings obtained at temperatures as high as 700 °C are found to be X-ray-amorphous. TEM analysis of a piece of coating scraped from a steel substrate shows it to comprise mostly of amorphous material, with barely crystalline MnO embedded in it, as can be deduced from the diffuse rings in the SAED

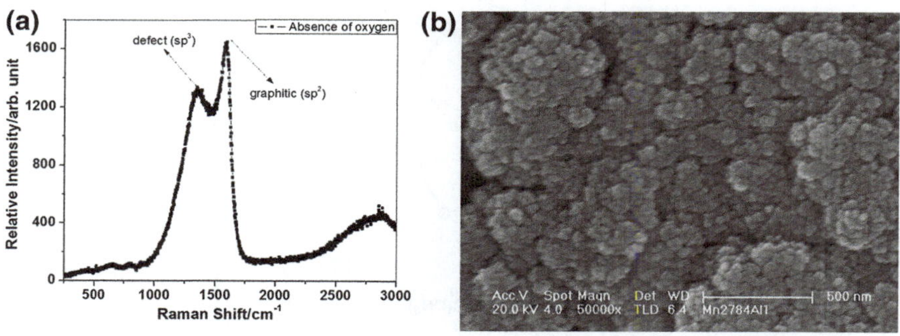

Fig. 2 **a** Raman spectrum of MnO/C composite deposited on SS316 at 680 °C. **b** SEM micrograph of the MnO/C coating on SS316, revealing the cauliflower morphology

pattern. The formation of MnO is also confirmed by XPS analysis of the coating. It is the morphology of the coating, as revealed by SEM (Fig. 2b), which is unusual and very revealing, given that it is highly carbonaceous and X-ray-amorphous. The "cauliflower morphology" with dimensions on the nanometer scale clearly indicates a very large specific surface area. Because of the difficulty in measuring the surface area of thin films by the BET method, one can only employ it to estimate the lower limit. This is found to be about 700 m^2/g.

The porous morphology and the large carbon content of the coatings, which might now be called MnO/C nanocomposites, are suggestive of the metal oxide/carbon composites developed as electrode materials for supercapacitors [9]. These are usually "bulk composites", prepared by physically mixing a metal oxide (RuO_2) with carbon. The inert-ambient MOCVD process, instead, yields a very homogeneous nanocomposite coating in a single step [7]. Although the growth temperature is high (up to 700 °C), simultaneous formation and deposition of MnO and elemental carbon occurs, as shown by thermodynamic analysis (below). Because carbon is in a much larger molar proportion, the mobility of the MnO species is so restricted as to prevent grain growth, rendering it nearly amorphous. As metal oxide/carbon composites have been found to be suitable as electrodes in supercapacitors, electrochemical measurements were made in the three-electrode configuration, with 0.5 M aqueous KOH as the electrolyte, to examine the charge (energy) storage capacity of the coatings made on SS, which serves conveniently as the current collector. As seen from the cyclic voltammetry data and charge–discharge curves (Fig. 3a, b), the nanocomposite MnO/C have very favorable characteristics as capacitor electrode materials. Measurements show that, at a current density of 1 mA/cm^2, the specific capacitance is a very appreciable 350 F/g for the coating deposited at 680 °C. More significantly, because of the intimate bonding between the composite electrode material and the current carrier (SS) resulting from deposition at an elevated temperature, charge/discharge cycles are rapid. Thus, capacitive behavior extends to a frequency exceeding 1 kHz, which is very high for such capacitors, making specialized applications possible.

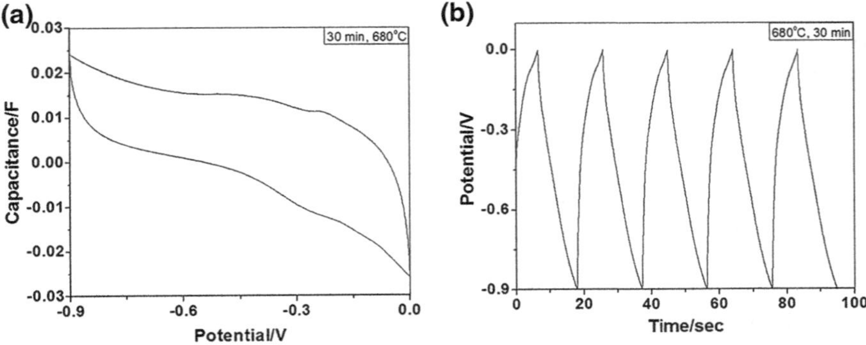

Fig. 3 **a** The nearly rectangular C–V plot of MnO/C deposited on SS316. **b** Charge–discharge cycling of the MnO/C electrode, both with 0.5 M aqueous KOH electrolyte

Given that, for example, the MOCVD process is consistent with IC fabrication processes, capacitors with MnO/C thin film electrodes (with a suitable solid electrolyte) can be "on-board" power supplies in a variety of electronic devices [10]. Such development is being pursued.

Inert-ambient MOCVD that resulted in MnO/C nanocomposite coatings was prompted by an earlier effort to employ MOCVD for the deposition of adherent coatings of alumina in the multilayer stacks on cutting tool inserts based on tungsten carbide. Such stacks, often comprising microns-thick layers of Ti(C,N), Al_2O_3, and TiN deposited by "inorganic CVD" at elevated temperatures (up to 1,100 °C), enhance tool life by nearly an order of magnitude [11]. However, as the precursors used in commercial production today are chlorides, significant corrosion occurs during CVD and expensive abatement is necessitated [11]. An effort was therefore initiated in our group (at MRC, IISc) to investigate the possibility of employing metalorganic precursors for alumina deposition, to obviate the corrosion problem and to lower the process temperature. Different aluminum complexes, including "acac" and "dpm," were employed in a low-pressure CVD process under oxygen flow. Although adherent coatings of alumina could be obtained at about 800 °C, the deposition rate was low, prompting attempts to raise it. One effective way was to conduct the process in the inert ambient of argon, which yielded very black, adherent coatings on Si(100) and WC at 800 °C. Analysis showed that the coatings on Si had about 80 at.% of carbon. The alumina embedded could not be fully characterized and appeared to be amorphous [6]. Nevertheless, the coatings were found to have hardness nearly 60 % of that of α-Al_2O_3.

Recently, more comprehensive work on inert-ambient MOCVD with $Al(acac)_3$ as the precursor has confirmed and extended these findings [11]. It has been found that strongly adherent coatings of carbonaceous alumina can be obtained on carbide substrates at temperatures as low as 650 °C and that the alumina formed is under these conditions is nanocrystalline γ-Al_2O_3, with grain size of about 10 nm, embedded in amorphous carbon as matrix (Fig. 4a). The carbon content of the coatings is about 80 at.% and is expectedly of the "nanographite" character, as in the MnO/C composite coatings. Measurements show that the coatings have a hardness of about 12 GPa, about half that of α-Al_2O_3. Nevertheless, detailed measurements of the cutting performance of coated carbide tools under actual "shop floor conditions" show that tools coated by MOCVD with carbonaceous alumina perform better than commercially produced CVD-coated tools in all respects, including longevity, depth of cut, and the smoothness of cut (Fig. 4b). Analysis of the coated tool bit before and after a cutting operation shows suggests that the superior performance of the MOCVD-coated tool is attributable to the lubricant role played by the extensive carbonaceous envelope around the alumina nanograins. Furthermore, because of the strong adhesion of the coating to the carbide substrate, the heat generated by the cutting operation is dissipated well, thanks also to the thermal characteristics of the carbon-rich coating. Thus, counter-intuitively, the carbon-rich coating does not "burn off" in air and, indeed, stays intact for the life of the tool. However, under the high-pressure conditions created during the cutting operation, γ-Al_2O_3 is found to be transformed to α-Al_2O_3 [11].

Though inert-ambient CVD with the metal-acac complexes is found to result in carbon-rich coatings, it is not axiomatic that the coatings should comprise *elemental* carbon, rather than carbides or other carbon-containing species. The composition of the solid products of a CVD process may, however, be deduced by thermodynamic analysis, assuming that equilibrium conditions prevail [12]. In the absence of experimental knowledge of the products of chemical reaction at elevated temperatures, the mass spectrum of the metal complex is used as proxy to deduce a list of likely reaction products, based on an understanding of the chemical pathways for pyrolysis. Using an abbreviated list of reactants and products so prepared, equilibrium analysis based on the minimization of the total Gibbs free energy of the system is employed, using commercially available software and the thermodynamic database embedded in it [12]. Through iteration, the composition of the products can be deduced as a function of temperature, pressure, and the molar proportions of the precursor(s). The results can then be represented as "CVD phase stability diagrams," which show the range of conditions under which solid deposits of the "allowed" compositions can be obtained predictably [13]. Phase stability diagrams deduced in this manner for CVD processes conducted with $Al(acac)_3$, $Co(acac)_2$, and $Fe(acac)_3$ are shown in Fig. 5a, b, c. It can be seen that, under oxygen-starved conditions and at low temperatures, the deposit would comprise $Al_2O_3 + C$ and $Fe_3O_4 + C$, i.e., the simultaneous formation of the oxide and carbon, leading to a composite material of composition that varies with CVD conditions.

The veracity of the "equilibrium assumption" is testified by experiment, while it also reveals an intrinsic limitation of thermodynamic analysis. Whereas the formation of the $Al_2O_3 + C$ composite is effectively independent of temperature (reduction to metal being impossible), the possibilities with $Fe(acac)_3$ are more varied. Under certain conditions, Fe_2O_3 and Fe_3O_4 are simultaneously formed, yielding an oxide composite. The SEM of such a deposit (Fig. 6) shows the

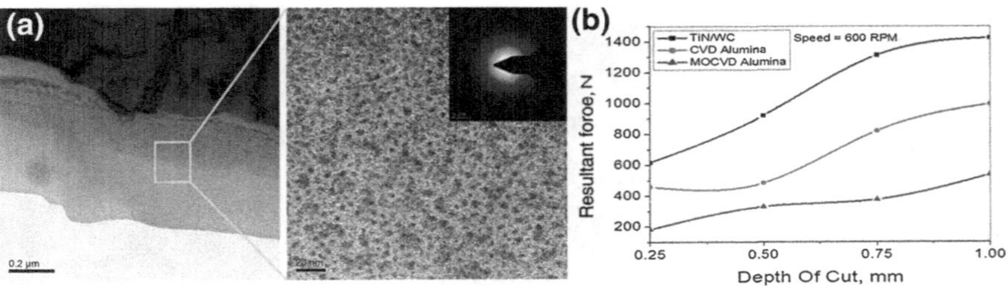

Fig. 4 **a** Cross-sectional SEM of the Al_2O_3/C coating on a tungsten carbide cutting tool substrate; high-resolution TEM of coating showing very fine nanocrystals of alumina in an amorphous carbonaceous matrix, as illustrated by SAED (*inset*). **b** Cutting force versus depth of cut for a hard steel work piece, showing that the MOCVD alumina-coated tool is the best

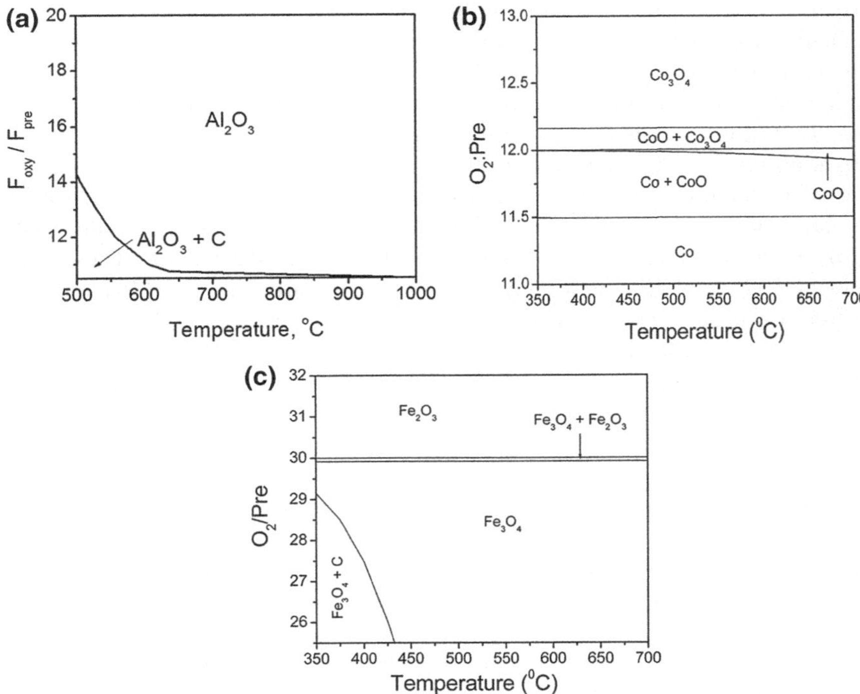

Fig. 5 CVD "phase stability diagram" plotted as a function of temperature and the molar ratio of oxygen to the metalorganic precursor, the "acac" in each case **a** $Al(acac)_3$, **b** $Co(acac)_2$, **c** $Fe(acac)_3$. Inert-ambient CVD is represented by extrapolating the *bottom left hand corner*. The simultaneous formation of (oxide + C) (Co + CoO) ($CoO + Co_3O_4$) and ($Fe_2O_3 + Fe_3O_4$) is to be noted

Fig. 6 SEM of a film with simultaneously deposited α-Fe_2O_3 and Fe_3O_4, revealing morphology characteristic of differing crystal structure

contrasting morphology of the oxides with different crystal structures. However, thermodynamics cannot distinguish between the different phases of Fe_2O_3 if the enthalpies of formation of the phases are not far apart. (For the same reason, the formation of a specific polymorph of alumina cannot be predicted.)

Neither can thermodynamic analysis predict the *polymorph of carbon* that is formed in inert-ambient MOCVD. This is well illustrated by the unexpected formation of carbon nanotubes under certain conditions of inert-ambient CVD (Fig. 7). While calculations do predict the formation of Fe_3O_4, Fe, and C, the formation of CNTs is facilitated by the availability of elemental nanocrystals of iron simultaneously formed, which is known to be an excellent catalyst for the formation of CNTs from hydrocarbon precursors. The formation of Fe is due to the highly reducing condition that argon flow produces in the reactor. Thus, inert-ambient CVD provides a new path to CNTs and, in particular, to a CNT/oxide/Fe composite whose composition can be varied through a choice of CVD conditions. The composites are also found to be good candidate materials for capacitor electrodes and for sensors, but further effort is needed to develop them [14].

A very interesting possibility arises from the phase stability diagram for CVD, with $Co(acac)_2$ as precursor. The formation of an oxide composite of CoO + -Co_3O_4 is predicted (as with Fe) and has been experimentally verified [4]. Under more reducing conditions, and over a narrow range of oxygen partial pressure, simultaneous deposition of metallic Co and CoO is predicted (Fig. 7b). This has been found to take place under stringent and experimentally difficult conditions, resulting in a remarkable nanocomposite in which the percolation threshold for both Co and CoO are exceeded (Fig. 8). This yields a surprisingly good metallic conductor (due to Co) that is also optically transparent (due to the insulating CoO), amounting to a new "algorithm" for transparent, conducting thin films [4]. As Co is a ferromagnet and CoO is an antiferromagnet (below 290 K) the Co/CoO composite thin film may be a viable spintronic material as well.

Lithium-ion batteries (LIB) have become the preferred power sources of the ubiquitous portable electronic devices today. To fulfill the demands of long cycle life, high specific capacity, and good rate capability, numerous novel anode electrode materials have been examined for LIBs, thin-film LIBs in particular. Presently, there is a strong effort in developing nanostructured metal oxides as

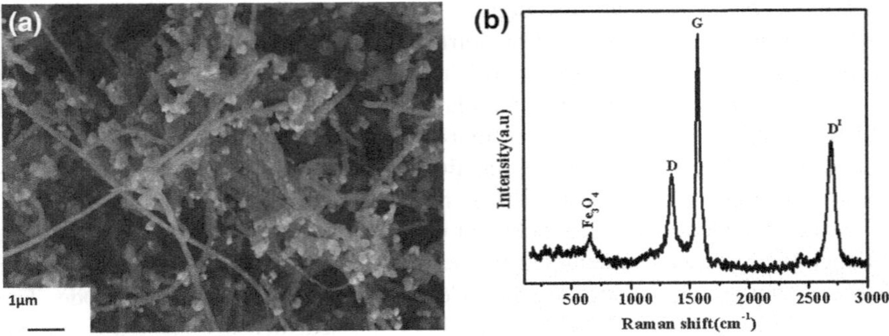

Fig. 7 **a** SEM of coating on SS316, showing simultaneous deposition at 700 °C of multi-walled CNTs and Fe_3O_4 in inert-ambient CVD with $Fe(acac)_3$. **b** Raman spectrum showing both CNTs and Fe_3O_4

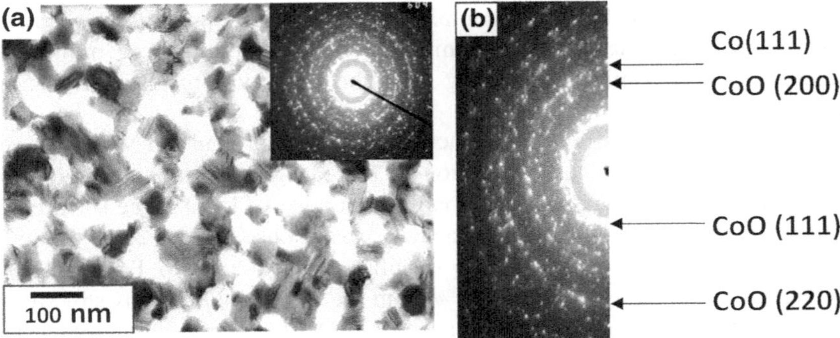

Fig. 8 **a** TEM image showing interleaved, growth of nanocrystals of two phases with the percolation threshold crossed for each phase. **b** SAED confirming the phases to be Co and CoO

anode materials for LIBs, as these have a theoretical specific capacity higher than those with conventional graphite electrodes [15]. Among the transition metal oxides, the spinel cobalt oxide Co_3O_4 in particular has been investigated for its suitability. We have therefore investigated the CVD of thin films of Co_3O_4 using $Co(acac)_2$ as the precursor, with a view to examining the films as anode material for thin-film LIBs.

It is found that well-crystallized Co_3O_4 films can be deposited on a variety of substrates with $Co(acac)_2$ as precursor and under oxygen flow, at temperatures ranging from 450 to 550 °C and a reactor pressure of 10 torr. In addition to Si(100) and fused quartz, stainless steel (SS316) substrates were employed, so that SS could act as the conducting support for the Co_3O_4 films for electrochemical measurements. Whereas the films deposited on Si and fused quartz display no specific morphology, the morphology of the films deposited at 500 °C on SS is very different and interesting, comprising uniform, closely spaced columnar structure, with the columns seemingly self-organized. Each column is a cylinder nearly perpendicular to the substrate surface (Fig. 9a). Examination at high resolution by field-emission SEM reveals that each of the "cylinders" is made of a stack of inter-grown, nanometer-sized platelets. Selective-area electron diffraction (SAED) reveals that each such cylindrical stack behaves as a single crystal, indicating clearly that oriented aggregation takes place during growth on a conducting substrate. The reason for this is not clear.

To study the electrochemical behavior of Co_3O_4 films as the anode in thin LIBs, "Swagelok" cells were assembled in the inert ambient of a glove box, with Li ribbons as the counter-electrode and a Co_3O_4 film as the working electrode. The as-grown film by MOCVD with its well-crystallized, densely packed columnar morphology makes it possible to employ it as the binder-free anode without further treatment. The electrolyte was 1 M $LiPF_6$ dissolved in ethylene carbonate, diethyl carbonate, and dimethyl carbonate [16].

The Li storage capacity and cycling performance of the thin film electrode were further studied by constant-current charge/discharge measurements. The measurement was repeated for 20 complete cycles in the potential window of 0.1–3.1 V at a specific current 100 mAg^{-1}. The charge/discharge profiles are qualitatively very similar, indicating that the electrochemical pathways are unchanged.

For the first charging, a long characteristic plateau around 1.0 V is observed, which is associated with the conversion of Li to Li_2O. The sloping region that follows might be attributed to the formation of a layer of Li_2O at the surface. For the first charging profile, the specific capacity was calculated to be 2,295 $mAhg^{-1}$. Beginning with the second cycle, the specific capacity drops to 1,690 $mAhg^{-1}$. This amounts to a decay of about 28 % from the initial value, but is still significantly higher than the theoretical capacity of 890 $mAhg^{-1}$ for Co_3O_4. This enhanced capacity can be attributed to the additional Li storage in the grain boundaries of Li_2O and the metal formed during the reduction cycle. More importantly, the thin film electrode allows 71.4 % of the stored Li^+ to be recovered from the fully charged state during its discharge. This low loss in the irreversible capacity can be attributed to the favorable, stacked columnar morphology. Charge/discharge cycles were repeated at constant current to check the reproducibility of the observed higher capacity. A maximum 4 % fluctuation has been calculated.

The stability of the thin film electrode grown at 500 °C has been evaluated from the specific capacity measured for 20 charge/discharge cycles at 100 mAg^{-1} (Fig. 9b). A maximum loss of 4 % in the discharging capacity due to cycling is found. The coulombic efficiency of the electrode material was also calculated from the ratio of discharging capacity to the charging capacity and plotted against the cycle number (Fig. 7). A maximum efficiency decay of 11 % is observed for the first few cycles, with efficiency stabilizing in later cycles at 98 %, demonstrating the reliability of the thin film electrode [16].

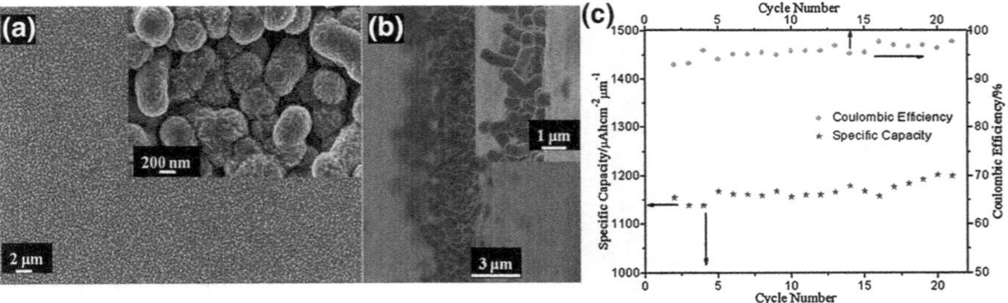

Fig. 9 **a** *Top view* of a uniform Co_3O_4 film deposited at 500 °C on SS316, with close-up showing cylindrical columnar structure, each column made of stacked crystallites. **b** *Cross-sectional view* showing columns nearly perpendicular to the substrate. **c** Specific capacitance and coulombic efficiency versus cycling, of a Li-ion battery with thin-film Co_3O_4 as the anode

3 Solution-based Processing Under Microwave Irradiation

As already noted, metal complexes such as acac's dissolve readily in benign solvents such as alcohols, making it possible to employ solution-based synthesis of nanomaterials and their coatings under microwave irradiation [17, 18]. The microwave field to which a solution is subjected, even if not spatially uniform, induces chemical reactions everywhere in the solution, leading to the formation of growth species and their simultaneous nucleation into crystallites throughout the solution. Together with the low "bulk temperature" of the irradiated solution, the conditions lead to the formation of nanocrystalline powder material and/or coating. This can be accomplished in a typical "domestic" microwave oven (2.45 GHz), in which a typically nonuniform multimode radiation is present. Much more sophisticated, single-mode, hydrothermal reactors with temperature and pressure control are today available for processing with greater reproducibility. As may be expected, the structure and morphology of the oxide materials typically obtained from acac's can be altered by adducting the complex, changing the solvent (solvent mixture), and/or adding a surfactant, thereby changing the dielectric properties of the solution. This is most apparent in ZnO nanostructures obtained by the method (Fig. 10a), ZnO crystallizing readily with morphology reflecting the hexagonal symmetry of the wuertzite phase, but with numerous interesting and potentially useful variations. Coatings of ZnO can be obtained on different substrates, such as Si(100), ITO-coated glass, and spun-on PMMA, each comprising a "forest" of single-crystalline ZnO nanorods, usually no preferred orientation. However, on a smooth, disordered substrate, such as spun-on PMMA, a strongly *c-oriented* coating is obtained, in which each ZnO nanocrystal is nearly perpendicular to the substrate surface, as evidenced by X-ray pole figure analysis (Fig. 10b). This is consistent with the "theorem" that stipulates that, when a film of a crystalline material grows on the very smooth surface of a disordered substrate, strong crystallographic orientation arises due to the minimization of interfacial energy. This occurs when the

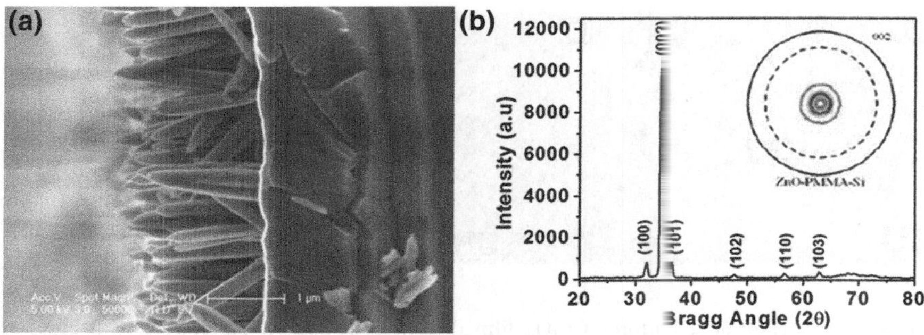

Fig. 10 **a** "Cross-sectional" SEM of ZnO "film" on spun-on PMMA/Si, revealing the hexagonal faceting of the tipped nanorods. **b** XRD pattern and X-ray pole figure, showing a sharp c-axis orientation of the ZnO film

most-densely packed plane of the film material is parallel to the substrate surface [19]; hence the c-orientation of ZnO on PMMA.

The versatility of the microwave-assisted solution process is exemplified by the facile synthesis of ternary and other complex oxides, such as the perovskites and the ferrites [17, 20], merely through the irradiation of a solution that has the different metal compounds (of similar reactivity) in the appropriate molar proportion. Thus, nanocrystalline zinc ferrite ($ZnFe_2O_4$) is readily formed from an ethanolic solution of $Zn(acac)_2$ and $Fe(acac)_3$, although the as-prepared material is barely crystalline [20]. This calls for annealing in air under appropriate conditions, leading to crystallite growth, permitting structural analysis, which reveals that significant crystallographic inversion is present in the material. It is such inversion that makes nanocrystalline $ZnFe_2O_4$ ferrimagnetic, at ordinary temperatures, bulk $ZnFe_2O_4$ being magnetically ordered but ferrimagnetic only below 10 K. Detailed magnetic measurements on annealed nanocrystalline $ZnFe_2O_4$ prepared by the microwave route can have a significant magnetic moment even at room temperature [20] and that the magnetic characteristics of such materials are rich.

As with other materials prepared by the method, $ZnFe_2O_4$ is also readily obtained in the form of an adherent coating on a variety of substrates, Si(100) being the most relevant practically. It is found that, if the solvent is either decanol or an ethanol–decanol mixture, and if the hydrothermal reaction vessel is used, the resulting higher temperature and pressure in the reaction enables the deposition of adherent zinc ferrite coatings without the need for any surfactant or other agents [21]. The coatings then formed are found to be sufficiently well crystallized, without need for annealing, and high enough magnetization because of the high degree of inversion (about 40 %).

For these reasons and because of the strong adhesion of the ferrite coating to the Si(100) wafer used as substrate, nanocrystalline $ZnFe_2O_4$ was judged to be a very good candidate as the core material for enhancing the inductance of the metal "coils" that serve as inductors in RFICs. At frequencies in the range of GHz, cores made of metal alloys no longer work very well because of losses [22]. Required is

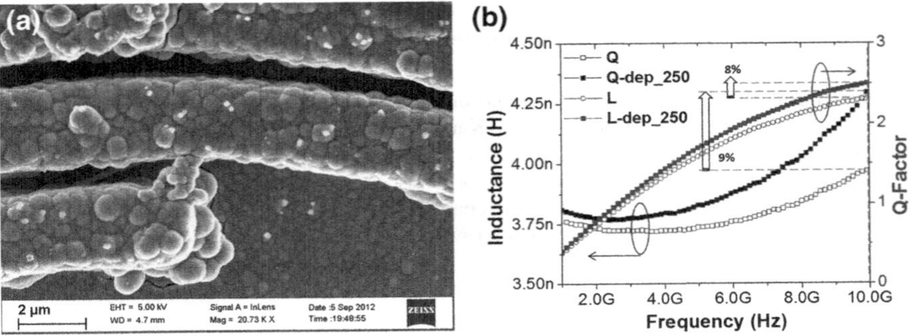

Fig. 11 **a** Conformal coverage by $ZnFe_2O_4$ film of the loops of an inductor in a sub-micron RF circuit. **b** Inductance enhancement due to $ZnFe_2O_4$ coating, as a function of frequency in GHz, with quality factor also shown

a magnetic insulator with high permeability, a condition satisfied by ferrites, including both spinels and hexaferrites. However, thin films processes for the deposition of ferrites invariably involves temperatures up to 1,200 °C, making them incompatible with the low thermal budget of VLSI devices with aggressive dimensions. Therefore, a solution-based, annealing-free process for the deposition of ferrite films is indeed very much needed. Furthermore, as the deposition process occurs in the solution medium, the coatings can be very conformal, even at sub-micron dimensions. This implies the possibility of that the coating of the ferrite "core" would envelope the coil structures fully, leading both to a higher inductance density and to reduced losses due to flux leakage.

The results of the implementation of the ferrite inductor in foundry-fabricated high-frequency circuits (Fig. 1a) show the effectiveness of the conformal coverage and the resulting enhancement of inductance even for a relatively thin ferrite core layer [23, 24]. Importantly, both inductor density and the quality factor are enhanced at frequencies up to 10 GHz, the highest reported to date (Fig. 11b). Most importantly, microwave irradiation leaves the embedded transistors in the CMOS chip completely unaffected, despite the presence of metal lines within. Furthermore, XPS analysis shows that, at the interface of Si and $ZnFe_2O_4$, a silicate is formed during irradiation, leading to the observed (and desirable) strong adhesion of film to substrate. Thus, metal complexes in solution can potentially be the basis of a technical solution for the problem of developing RF CMOS ICs to meet the communication needs of the future.

4 Conclusions

Simple and relatively safe metalorganic complexes, such as acetylacetonates, are excellent precursors to nanostructured oxides and metals, both in powder and thin film form. Examples of materials prepared by CVD and microwave-assisted reactions in the solution medium are used to show the versatility and usefulness of metal complexes as sources of nanomaterials.

Acknowledgments The manuscript is based on my work, over the years, with following colleagues: Anjana Devi, Anil Mane, Shalini Kandoor, Sukanya Dhar, Piyush Jaiswal, Gururaj Neelgund, Pallavi Arod, Ranajit Sai, Anirudha Jena, Nagendra Pratap Singh, N. Bhat, N. Munichandraiah, and S. Sampath. Part of the work was carried out under the NPMASS project, "Nanostructured metal oxide particles and coatings prepared by microwave-assisted solution processing for sensor and catalytic applications", PARC No.1, Project 1.5.

References

1. Cao G (2006) Nanostructures and nanomaterials. Imperial College Press, London
2. Singh MK, Yi Y, Christos G et al (2009) Synthesis of multifunctional multiferroic nanomaterials from metalorganics. Coord Chem Rev 253:2920–2934
3. Shalini K (2002) Ph.D. thesis, Indian Institute of Science, Bangalore
4. Mane AU (2003) Ph.D. thesis, Indian Institute of Science, Bangalore
5. Mane AU, Shivashankar SA (2005) Growth of (111)-textured copper thin films by atomic layer deposition. J Crystal Growth 275:1253
6. Devi A, Samuelason AG, Shivashankar SA (2002) MOCVD of aluminium oxide films by MOCVD using aluminium β-diketonates. J Physique IV 12:139–146
7. Dhar S, Varade A, Shivashankar SA (2011) Thermodynamic modeling to analyse the composition of carbonaceous coatings of MnO and other oxides grown by MOCVD. Bull Mat Sci 34:11–18
8. Chu PK, Li L (2006) Characterization of amorphous and crystalline carbon films. Mater Phys Chem 96:253–277
9. Wang G, Zzhang L, Zhang J (2013) A review of electrode materials for electrochemical supercapacitors. Chem Soc Rev 41:797–828
10. Varade A, Shivashankar SA, Dhar S et al (2013) Composition of electrode material in the form of a coating and process thereof, US Patent No 8343572
11. Jaiswal P (2014) Ph.D. thesis, Indian Institute of Science, Bangalore
12. Mukhopadhyay S, Shalini K, Lakshmi R et al (2002) MOCVD of Cu films from bis(t-butyl-3-oxo-butonoato)copper(II): thermodynamic investigation and experimental verification 150 205–211
13. Dhar S, Shalini K, Shivashankar SA (2008) Thermodynamic analysis of growth of iron oxide films by MOCVD using tris(t-butyl-3-oxo-butonoato)iron(III) as precursor. Bull Mat Sci 31:723–728
14. Arod P, Shivashankar SA (2011) Synthesis of CNT-metal oxide nanocomposite electrode materials for supercapacitors by low-pressure MOCVD, MRS Symp Proc MRSF10-1303—Y14-02
15. Wu HB, Chen JS, Hng HH et al (2012) Nanostructured metal oxide-based materials as advanced anodes for lithium-ion batteries. Nanoscale 4:2526–2542
16. Jena A, Munichandraiah N, Shivashankar SA (2013) Columnar metal oxide structure and a method of its coating on a substrate. Indian Patent Application No 4616/CHE/2013
17. Bilecka I, Niederberger M (2010) Microwave chemistry for inorganic nanomaterial synthesis. Nanoscale 2:1358–1374
18. Brahma S, Shivashankar SA (2010) Microwave irradiation-assisted method for the deposition of adherent oxide films on semiconducting and dielectric substrates. Thin Solid Films 518:5905–5911

19. Givargizov EI (1991) Oriented crystallization on amorphous substrates. Plenum Press, New York
20. Sai R, Kulkarni SD, Vinoy KJ et al (2012) $ZnFe_2O_4$: rapid and sub-100 °C synthesis and anneal-tuned magnetic properties. J Mater Chem 22:2149–2156
21. Gardner DS (2009) Review of on-chip inductor structures with magnetic films. IEEE Trans Magn 45:4760–4766
22. Sai R (2013) Ph.D. thesis, Indian Institute of Science, Bangalore
23. Sai R, Vinoy KJ, Bhat N et al (2013) CMOS-compatible and scalable deposition of nanocrystalline zinc ferrite thin films to improve inductance density of integrated RF inductors. IEEE Trans Magn 49:7
24. Sai R, Shivashankar SA, Bhat N et al (2013) High-frequency integrated device with enhanced inductance and a process thereof. PCT Application No PCT/IN2013/000664

A Study on Hydrophobicity of Silicon and a Few Dielectric Materials

Vijay Kumar and N. N. Sharma

Abstract Superhydrophobic surfaces of silicon and a few dielectric materials were prepared by a combination of texturing and deposition of a self-assembled monolayer. The water contact angle and surface morphology of the prepared surfaces were investigated for a number of variations during the process of texturing. The combination of surface roughness and chemical treatment renders a lotus leaves-like effect called superhydrophobicity with a water contact angle (WCA) greater than 150°. The chapter also presents the basic concepts and models of hydrophobicity and discusses the use and importance of high k-dielectric materials as superhydrophobic surfaces in microelectronics and microfluidics. The wetting behaviours of water droplets on randomly structured hydrophobic surfaces were investigated. The effects of plasma and chemical treatment on structure geometry, roughness, and relative pore fraction on the contact angles were investigated experimentally for droplets of size comparable to the size of the structures. Moreover, we have successfully prepared superhydrophobic surfaces with various texturing methods.

Keywords Hydrophobic surfaces · Hydrophilic surfaces · Superhydrophobic surfaces · Dielectric material · Silicon texturing · High-k dielectric material

1 Introduction

Lotus leaves are well known naturally occurring superhydrophobic surfaces inspiring research and consequently exciting engineering applications of artificial superhydrophobic surfaces. The superhydrophobic effect is exhibited by solid

V. Kumar · N. N. Sharma (✉)
Department of Mechanical Engineering, Nanomaterials and National MEMS Design Centre, Birla Institute of Technology and Science, Pilani, India
e-mail: nitinipun@gmail.com

K. J. Vinoy et al. (eds.), *Micro and Smart Devices and Systems*, Springer Tracts in Mechanical Engineering, DOI: 10.1007/978-81-322-1913-2_16,

surfaces and concerns with the wettability of a solid surface. The phenomenon of superhydrophobicity has been attempted from both a fundamental and practical perspectives by tailoring surface topography and surface chemical compositions. The wettability is usually determined by measuring the contact angle (CA) of the water droplet on the solid surface which is referred to in the literature as the Water contact angle (WCA). The WCA (θ) on a surface is classified into three categories, namely (a) when $\theta < 90°$, the surface is called as hydrophilic (b) for $90° < \theta < 150°$, surface is called hydrophobic (c) and for $\theta > 150°$ the surface is a superhydrophobic.

Superhydrophobic surfaces have attracted significant attention within the scientific community as well as the industrial world over the last couple of decades. These surfaces promise a wide range of uses like self cleaning [1–3], non-wetting [4–6] low adhesion [7–9], drag reducing [10, 11] and oil-water separation and anti-befouling [12, 13]. The surface energy of the materials controls the hydrophobicity of the flat surface, and, as a general rule, the hydrophobicity increases when the surface energy is lowered and decreases when the surface energy is increased. Superhydrophobicity on a surface is achieved not only by lowering surface energy, but also by micro/nano texturing of the surface. Basically, there are two possible physical modes of fluid motion on a solid surface: either the liquid follows the solid surface, or it moves on air trapped inside the micro/nano texture present on the surface (These modes of the motion of fluid over a solid surface are referred to as Wenzel or Cassie state motion after the scientists who proposed these models) [14–16]. In the Wenzel model, increased resistance to the wetting is attributed to the minimization of the surface energy by the increased surface roughness, while in the Cassie–Baxter model the hydrophobic properties of the micro/nano structured surface are attributed to the solid–gas contact area that carries or support a droplet on the surface. Therefore, the rougher the material, the more hydrophobic it will be.

Conventionally, the preparation of a superhydrophobic surface can be divided into two main categories; the first is to make a rough surface from low-surface energy materials, and the second is to chemically modify and roughen the surface of materials to produce a low surface energy surface. The chemical modification of the surface by the self assembly of organosilane is one the efficient strategies to make the surface superhydrophobic by lowering the surface energy of the surface [17–19]. In the last few decades, a large variety of self assembled monolayers (SAMs) have been deposited on substrates and studied to prepare superhydrophobic surfaces. For example, alkylthiol on gold and other metallic surfaces [20] and alykylsilane on silicon [21] are very often reported in the literature. Gold is most commonly used because it is easy to deposit as planar thin films and easy to pattern with conventional lithography tools and chemical etchants. The other class of SAMs deposited on an oxide surface includes n-alkanoic acid and phosphonic acid [22–25]. These classes of molecules have gained attention due to their ability to bind a wide range of metal oxide surfaces to form robust SAMs of similar quality to those of thiol on Au.

The other most extensively investigated SAM is alkyltrichlorosilane (ATS) which requires a hydrolyte substrate including the technologically relevant surface e.g. metal oxide and amorphous surfaces. The SAM of alkytrichlorolisilane has become important as a functional coating for micro/nano and organic electronics applications [26]. The derivative of alkyltricholrosilane e.g. Octadecytrichlorosilane (OTS, $Cl_3Si\ (CH_3)_{17}$), is commonly used for the modification of inorganic materials. The modification with organic silane on a silicon surface has been widely investigated because of its application in molecular electronic devices and microfluidics applications [27–29]. It serves as an alternative to SiO2 as a gate insulator in organic thin film transistors. Polymeric dielectric materials like SU-8, PDMS, Parylene and inorganic materials like SiO2, Si3N4 and HfO2, Al2O3 are also widely investigated for making superhydrophobic surfaces which find extensive applications in microfluidics and microelectronics applications [30–35].

Micro/Nano texturing of the silicon surface with chemical treatment is reported in the literature [36, 37]. This chapter presents a study on the synthesis of the superhydrophobicity of silicon and dielectric materials by a combination of conventional methods like texturing the materials by chemical or physical methods and by the deposition of an OTS monolayer. Semiconductor materials Si and high-k materials such as hafnium oxide (HfO_2), Aluminum oxide (Al_2O_3), Tantalum Oxide (Ta_2O_5), Titanium oxide (TiO_2) are considered for the study.

2 Classical Models of Wetting on Solid Surface

In this section, we introduce the basic concept of wettability, which describes the behavior of a liquid on a solid surface. The wetting phenomenon of surface materials is not only dependent upon their chemical compositions, but also closely related to the micro/nano texturing on their surface.

2.1 Wetting on Flat Surface

It is observed in day to day life that the smaller size droplets have a spherical shape on horizontal surface and larger size droplets flatten on a surface. The shape of liquid droplet is attributable to balance between force of gravity and surface tension at liquid–gas (γ_{lg}) interface. A microscopic drop of diameter (d), making the drop spherical, is governed solely by surface tension force acting on drop. The surface tension force scales as ($\gamma_{lg} \times d$). Whereas, the gravitational forces imposed on liquid tries to flatten the liquid. If ρ is the density of liquid and g is gravitational acceleration constant then the gravitational force scale as $\rho g d^3$. The gravitational effect can be neglect if the drop size is much smaller than that of capillary length L_C which is given as

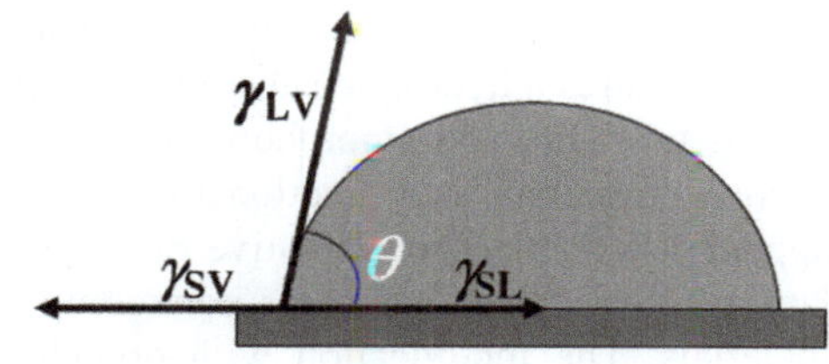

Fig. 1 Young's model for liquid droplet on a solid surface

$$L_c = \sqrt{\frac{\gamma_{lg}}{\rho g}}. \tag{1}$$

For pure water, the capillary length L_c is approximately equal to 2.5 mm at room temperature. However, if the drop dimension is smaller than L_c droplet has a shape resembling that of a sphere. Drops larger in size than the L_c are flattened by the gravitational force. The ratio between the gravitational force and surface tension is called as Bond number and is expressed as:

$$B_o = \frac{\rho g d^2}{\gamma_{lg}}. \tag{2}$$

This dimensionless number characterizes the type of drop. For $B_o < 1$, drop formation is dominated by surface tension, in which case the drop assumes nearly spherical shape, else drop formation is significantly dominated by gravitational force which tends to flatten the drops on the solid surface.

When a liquid is brought into contact with a flat homogenous solid surface, the degree of the spreading depends on the energy of the surface in contact. Young and Laplace in 1805 first established the connection between the drop shape and the interfacial energy between the solid, liquid and gas. Young and Laplace found that when there is an interface between two materials, there is specific energy turned as interfacial energy which is proportional to the surface area of the interface, and the constant of proportionality is called the surface tension. Since, in the wetting phenomenon, we typically have a liquid, solid and a surrounding gas, there are three type of surface tension: the liquid gas γ_{lg} the gas solid γ_{sg}, the liquid solid γ_{sl} surface tensions which are schematically shown in Fig. 1.

When a liquid droplet is in the air, it is spherical in shape in order to minimize the surface energy. When it is in contact with a solid, the liquid–gas interfacing surface maintains a spherical shape profile. The angle between the line tangent to the liquid drop at the extreme left and solid surface (refer Fig. 1) is called the contact angle(θ). A surface tension force balance gives:

$$\cos\theta = \frac{\gamma_{sg} - \gamma_{sl}}{\gamma_{lg}}. \tag{3}$$

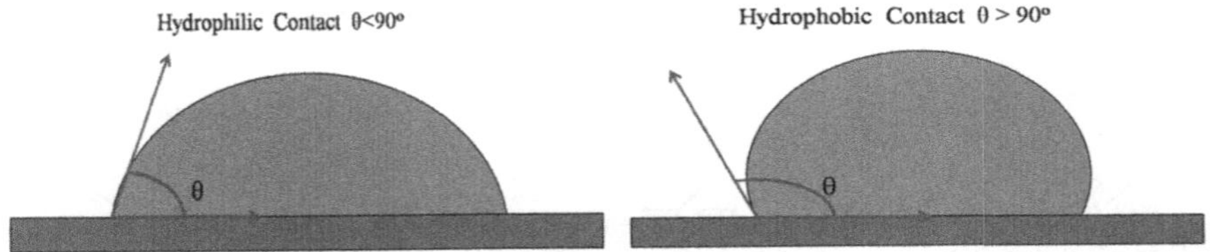

Fig. 2 Schematic of hydrophilic and hydrophobic surfaces

In Eq. (1), the measure of surface hydrophobicity i.e. WCA θ is related to the surface tension between a solid–liquid (γ_{sl}), solid-gas (γ_{sg}) and liquid–gas (γ_{lg}). When θ is smaller than 90°, the solid surface is considered intrinsically hydrophilic, and when θ is greater than 90°, the solid surface is considered intrinsically hydrophobic as illustrated schematically in Fig. 2.

2.2 Wetting on Rough surface

Young's equation is a very simplistic approximation that is used in the case of ideally flat surfaces. The effect of surface roughness on wettability was first discovered by Wenzel in 1936 and then by Cassie–Baxter in 1944. Wenzel suggested that the effective surface area increase as the surface becomes rough and hence water will tend to spread more on a rough hydrophilic substrate to develop more solid–liquid contact, while it will spread less on a rough hydrophobic substrate to decrease the contact area with the solid. A key assumption of this conclusion is that water is in complete contact with the solid surface, which is called the Wenzel state and is shown in Fig. 3a. The relationship between the apparent contact angle on a rough surface (θ_r) and Young's contact angle θ_f on a similar smooth surface is given as:

$$\cos\theta_r = r\cos\theta_f. \tag{4}$$

Equation (4) relates the parameter in the Wenzel effect state, where r is the roughness factor, defined as the ratio of the actual surface area to its plan area. Since r is always greater than 1 for a rough surface, this equations predicts that if $\theta_f > 90°$, $\theta_r > \theta_f$ and if $\theta_f < 90°$, $\theta_r < \theta_f$. Therefore, in the Wenzel state, the surface roughness will make intrinsically hydrophobic surface more hydrophobic and an intrinsically hydrophilic surface more hydrophilic.

The Cassie–Baxter state, also known as the composite or heterogeneous state considers that the water droplet sits on top of the surface protrusions and air is trapped underneath the droplet and resides in the surface porosity, as shown schematically in Fig. 3b. In this case, the liquid–surface interface is actually an interface consisting of two phases, namely a liquid–solid interface (Phase 1) and a

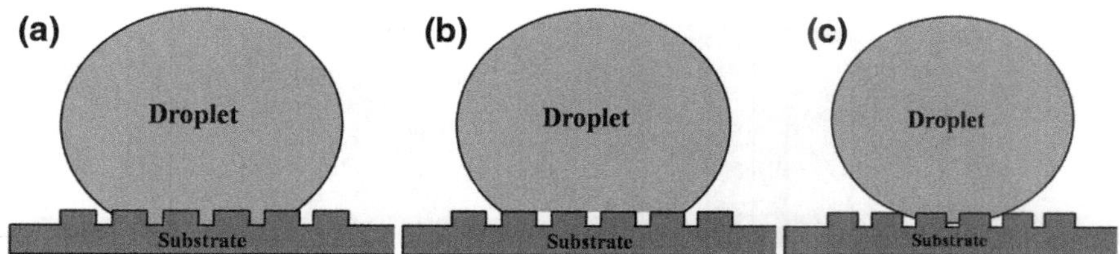

Fig. 3 **a** Wetted contact between the liquid and the rough substrate (Wenzel's model) **b** Non-wetted contact between the liquid and the rough substrate (Cassie's model) **c** Intermediate state between Cassie–Wenzel state

liquid–air interface (Phase 2). The apparent contact angle is the sum of the contribution of all the phases as defined below:

$$\cos\theta_c = f_1 \cos\theta_1 + f_2 \cos\theta_2 \tag{5}$$

where θ_c is the apparent contact angle, f_1 and f_2 are the surface fractions of phase 1 and phase 2, respectively; θ_1 and θ_2 are the contact angles in phase 1 and phase 2 respectively. This Eq. (5) is the general form, which also applies when there is no roughness. For an air–liquid interface, considering f as the fraction defined as the fraction of the solid surface that is wet by the liquid, the fraction corresponding to an air–liquid interface is $(1-f)$. The water droplet has a 180° contact angle with air, so Eq. (3) becomes:

$$\cos\theta_c = f\left(1 + \cos\theta_f\right) - 1 \tag{6}$$

The parameter f ranges from 0 to 1, where at $f = 0$ the droplet does not touch the surface at all and at $f = 1$ the surface is completely wet; the same is true of a flat surface. The movement of water across a surface is also affected by the wetting of the surface. A Wenzel type wetting mechanism means that there is complete contact between the surface and a water droplet at any point of coverage, with no air trapped underneath. The greater area of contact between the water and the surface renders water movement across the surface relatively hard.

The surface often exhibits wetting behaviour intermediate to those of the Wenzel and the Cassie–Baxter models with partial liquid penetration of the rough surface called Cassie–Wenzel transition as shown in Fig. 3c. The Cassie–Baxter surfaces allow the water to roll-off if tilted slightly, but the water droplets may fill the roughness and becomes hydrophilic. The transition from the Cassie–Baxter state to the Wenzel state depends upon the hydrophobicity of the surface and shape of the textured features. These two states are usually separated by energy barrier between them. Thus for wetting transition, the energy barrier must be overcome. Such transition from one state to another state is induced under certain external stimuli, such as by application of pressure or a force on the droplet or by applying an electric field on the surface. Recently, it is understood from the experimental observations that the transition from Cassie–Baxter to Wenzel region can be

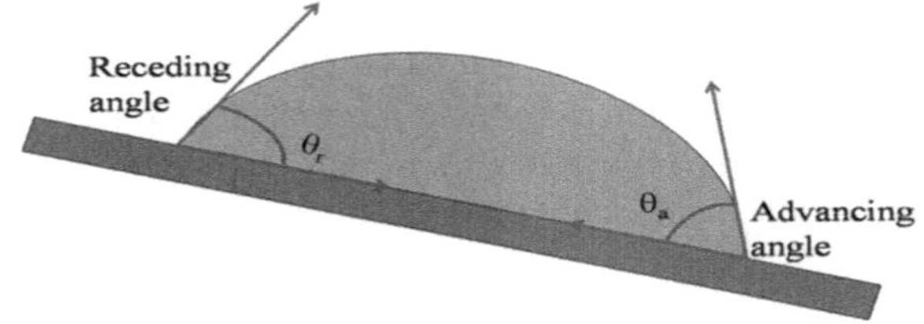

Fig. 4 Schematic of a droplet on a tilted substrate showing advancing (θ_a) and receding (θ_r) contact angles

irreversible event [38, 39]. From the literature, it has been also suggested that the transition takes place when the net surface energy of Wenzel state becomes equal to Cassie–Baxter state. The mechanism of transition from the Cassie state to the Wenzel is still widely debated in literatures.

Besides high static contact angles, the easy sliding-off behaviour of liquid droplets is another phenomenon of interest related to superhydrophobicity. The sliding behaviour of the droplet is again governed by the balance between the surface tension and gravity. On the tilted surface, the liquid drop becomes asymmetric and the contact angle of the lower side becomes larger and that of the upper side gets reduced. The difference between these two contact angles (hysteresis) reaches the maximum when the liquid drop begins to slide down the tilted surface. The contact angle of the forefront and the trailing edges of the liquid drop just prior to the movement of its contact lines are called the advancing (θ_a) and receding (θ_r) contact angles, respectively. When the gravity acting on the liquid drop becomes larger than the surface tension forces (F) caused by the contact angle which is holding the liquid droplet from sliding as in Fig. 4, the liquid droplet starts sliding. So, the critical angle (θ_a) for the water droplet to slide off the surface can be calculated by balancing the forces and is given as [1–4]:

$$\frac{mg\sin\alpha}{w} = \gamma_{lg}(\cos\theta_r - \cos\theta_a) \qquad (7)$$

$$F = \gamma_{lg}(\cos\theta_r - \cos\theta_a) \qquad (8)$$

where α denotes the sliding (tilt) angle, m is the mass of the droplet, g is the gravitational acceleration and W is the solid–liquid contact width. This equation predicts that for a given mass of water droplet, a smaller contact angle hysteresis will result in a smaller sliding angle and easier roll-off. When the water drops roll over the surface, it can easily trap and remove dust particles from the surface. Due to this reason, a superhydrophobic surface is often called "self-cleaning" since the rolling-off of water droplets keeps the surface clean.

The Wenzel state tends to give a larger contact angle hysteresis than the Cassie–Baxter state. As the contact line recedes, some liquid can be trapped in the surface texture if the liquid is conformal and completely filled in the texture initially. This can decrease the receding contact angle significantly, resulting in a large contact angle hysteresis and, thus, in a large critical sliding angle. This liquid trapping in the surface texture is not expected for the Cassie–Baxter state; so, the contact angle hysteresis and critical sliding contact angle are much smaller than in the Wenzel state. In other words, the total wetting Wenzel regime is "sticky" in that drops of water tend to adhere to them more than to a flat surface of the same type; those following the regime of Cassie and Baxter are "slippy" and allow drops of water to roll off more easily than an equivalent flat surface.

2.3 Formation Mechanism of OTS SAM

The formation of SAMs on silica and other dielectrics has been thoroughly studied for a few decades and now, a clear picture of the chemistry and structure of these SAMs is available. The structure, formation and properties of SAMs have attracted much interest in recent years due to their potential in organic electronics applications. In order to form a SAM it is necessary to find the self assembly capable molecules which can covalently bond to a specific surface. The SAM's molecule consists of three groups' an anchor group, a spacer group, and a terminal group. The anchor or head group (-$SiCl_3$) is responsible for the absorption of the molecule by forming a chemical bond with the adsorbing surface. The terminal group (-CH_2) determines the properties of SAM's surface. This non-polar methyl group gives a hydrophobic surface. The spacer group between the anchor and terminal group consists of a long alkyl chain of the methylene group (-CH_2)$_n$. The spacer group determines intermolecular interaction and promotes the ordering and orientation of SAM's molecules within the monolayer. The rate of formation of the Self-assembled monolayer is influenced by many factors, some of which can be controlled relatively easily, such as temperature, solvent, concentration of the absorbate and cleanliness of the substrate. In order to form high quality monolayer, it is very important to ensure a high level of cleanliness on the surface to keep the error by contamination as low as possible. There has been considerable research on the formation mechanism of SAMs on dielectric material surfaces.

The most common way of a SAM's formation in the scientific community is silanization (Refer to Fig. 5). Silanization includes a multi-process, where absorption is followed by hydrolysis of the chlorosilane (Si–Cl) groups to form a siloxy (Si–OH) group. The strong bonding of the head group to the substrate is normally essential for a stable monolayer. After hydrolysis, the OH groups in OTS also interact with the OH group on the substrate forming Si–O–Si bonds to the substrate. High quality SAMs of organosilane are not simple to deposit, because it is processed in number of steps and strongly depends upon the various parameter including the amount of water in solution, deposition time and temperature. The water is essential for formation of high quality monolayer but excess amount of water can make poly-condensed silane in the solution. Cleaning of the surface usually with a chemical or by an oxygen plasma increases the possibility of no formation of silanol group on the surface. However, by controlling the amount of water and deposition time also, good quality of monolayer can be formed on the surfaces.

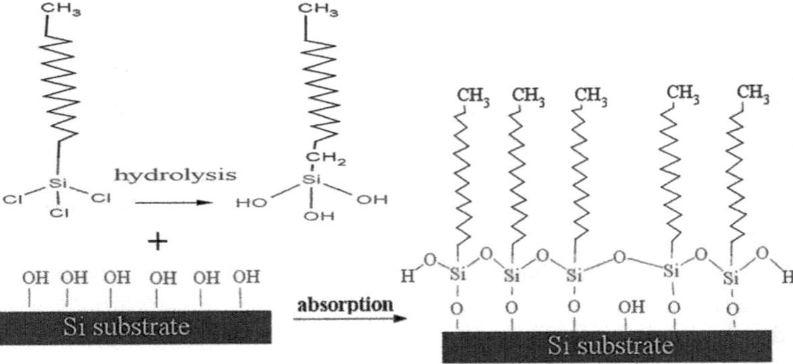

Fig. 5 Schematic of formation mechanism of OTS SAM

Making a hydrophilic and hydrophobic surface a superhydrophobic surface is an interesting challenge to the scientific community and has been widely attempted and published in the literature. From the basic principles, there are two key requirements to obtain Superhydrophobicity. One is appropriate surface chemistry and the other is appropriate surface roughness. On a flat surface, the highest water contact angle that can be obtained is about 110–115° on materials like Poly-dimethyl Siloxane (PDMS) and Poly-Tetraflouro Ethylene (PTFE). By preparing a surface with appropriate roughness and hydrophobic treatment, water contact angles higher than 150° and easy water run-off can be attained.

3 Synthesis of Superhydrophobic Surfaces

A Superhydrophobic surface can be prepared using both organic materials such as SU-8, PDMS, Polystyrene and inorganic materials such as SiO_2, HfO_2, Al_2O_3, TiO_2. Some materials which are already hydrophobic are made more rough or micro/nano textured to enter the superhydrophobic category. A wide variety of physical and chemical methods have been explored to fabricate superhydrophobic surfaces through one of the following two approaches (1) Creating a rough surface on a hydrophobic material (2) Modifying a rough surface with a hydrophobic material coating.

3.1 Making a Rough Surface on Hydrophobic Material

It is a relatively simple and one-step process to make superhydrophobic surfaces by using intrinsically hydrophobic materials and increasing the surface roughness further. In this approach, hydrophobic polymers are used to obtain superhydrophobicity. One group of materials that is of great interest is fluorinated polymers which have extremely low surface energy. For example, the water contact angle of a tetrafluoroethylene (Teflon) surface is typically 115–120°. Further roughening these polymer surfaces leads to a superhydrophobic surface. One of the most widely used methods for roughening fluorinated polymer is plasma etching. The high-energy oxygen species generated by O_2 plasma can randomly etch fluorinated polymer materials and create the surface roughness needed to increase the water contact angle to >160° [38]. The mechanical stretching of the Teflon surface can also increase the roughness of the Teflon surface. The stretched Teflon film consists of a sub micrometer diameter fibrous crystal with a large fraction of void space in the surface.

Another hydrophobic polymer with low surface energy is Poly-dimethylsiloxane (PDMS) that can easily be processed to make a rough textured hydrophobic surface. A high power laser source can make PDMS rough. The contact angle obtained is in excess of 175°. The superhydrophobicity achieved is attributed to the high porosity and chain ordering on the PDMS surface [39–41].

3.2 Modifying Rough Surface with Hydrophobic Coating

In order to make superhydrophobic surfaces on intrinsically hydrophilic materials a two-step process is usually required, i.e. making a rough surface first and then modifying it with a coating of chemicals, such as organosilane which can offer low surface energy after linking to the rough surface.

Because of the well established micro/nano fabrication technologies on silicon substrates, silicon has been widely used for making superhydrophobic surfaces through the fabrication of a wide variety of surface structures. There are several ways to obtain a texture on a silicon substrate either by chemical texturing by an alkaline solution or plasma texturing using fluorine-based plasma. Texturing can be performed on silicon with or without pattering. A study on the increase in the hydrophobicity of silicon by the use of various chemicals for texturing and by varying physical methods of increasing roughness is carried out and has been carried out and presented next.

3.2.1 Making of a Superhydrophobic Silicon Surface

The method of texturing the Silicon by a chemical alkaline solution such as Potassium hydroxide (KOH), Tetra- methyl ammonium hydroxide (TMAH) is commonly used for making surfaces for solar cell applications. A KOH etching solution is cost and time efficient but results in potassium and sodium ion contaminations. TMAH is a well-known etchant solution widely used in microelectronics due to its no alkali ions, and non-volatile, nontoxic and good anisotropic etching characteristics. The addition of Isopropyl (IPA) to alkaline etchants improves the uniformity of the random pyramid texture. The current study was

Table 1 Parameter detail of chemical and plasma textured methods

Films	Treatment methods	Time	Temp (°C)	AFM (nm)	Roughness
	Bare Silicon			Without SAM	With SAM
P	Piranha (H_2SO_4:H_2O_2)	30 min	25	0.164	9.52
PH	Piranha + HF	30 min + 10 s	25	0.092	18.6
	Chemical Methods				
K	KOH	10 min	80	758	1267
T	TMAH	10 min	80	838	880
	Plasma Methods				
S	SF_6	60 s	10	1.20	12.6
O	O_2	60 s	10	0.108	15.9

carried out for the variation of chemicals for texturing followed by the deposition of SAM. Various chemicals used for texturing by etching are piranha, Piranha + HF, KOH, TMAH and followed by two variations of a plasma environment, namely, SF_6 and O_2. The optimized average parameters used in the process and the average roughness of the surface obtained are tabulated in Table 1. The texturing process of silicon with (100) orientation were carried out in two different ways: (1) The first method uses 10 % KOH solution in de-ionized water with 5 % isopropylalcohol (IPA) and the process is done at 80°C for 10 min. (2) The second method uses a 5 % of 25 % commercial TMAH solution and 5 % IPA in deionized water and is carried out at 80 °C for 10 min. After chemical texturing of the surface, the sample is routinely cleaned using RCA1 and RCA2 to remove metallic contamination from the surface and is rinsed in 10 % dilute HF for 10 s, to remove the native oxide.

Over the past 10 years, the plasma texturing process has developed by using the RIE plasma source. Plasma texturing reduces the silicon loss due to pyramid formation via a chemical etch and requires tens of microns of materials consumption. The texturing on a silicon surface by plasma was performed using SF6 plasma. Before the plasma treatment, the bare silicon samples were piranha cleaned with H_2SO_4 and H_2O_2 in ration of 1:3 followed by the dilute HF (10 %). We employed a 25 cm^3/s gas flow rate at power 50 W, with a chamber pressure of 50 mtorr for 60 s. Table 1 gives the parameters and roughness obtained for texturing with SF_6 and O_2.

After texturing and cleaning, the self assembly of the OTS monolayer was also done to facilitate an increase in hydrophobicity of the textured surfaces. The samples were rinsed with DI water and dipped into an OTS/Toluene solution to allow the OTS to uniformly self assemble on the textured surface. Finally, the sample was rinsed with toluene and then dried in nitrogen.

The effect of surface texturing on wetting properties was studied by comparing the WCAs of the textured surfaces with the WCA of the smooth Si surface before and after OTS SAM modifications. Figure 6a and b shows that OTS modification of a smooth silicon surface increases its WCA from 80° to 132°.

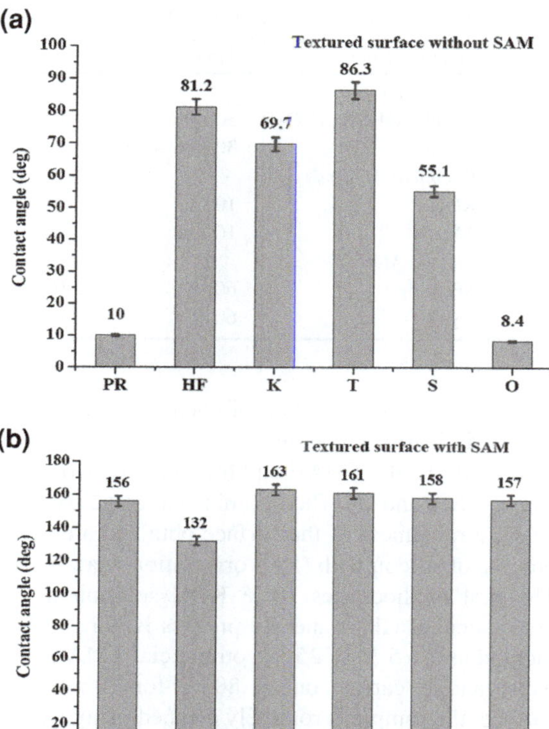

Fig. 6 Contact angle for **a** Silicon textured surface without SAM **b** Silicon textured surface with SAM

Figure 6 shows the AFM images and contact angle images of all variations in experiments. From Fig. 7a–l, it is seen that the WCA of a chemical textured silicon sample was <90°, but after coating with OTS SAM it increases to >150° which is in the superhydrophobic range. Similarly, OTS SAM for a plasma textured silicon surface also increases the WCA from 55° to 158° which is also in the regime of a superhydrophobic surface. The higher contact angle of an OTS modified textured silicon surface compared with that of the OTS SAM modified smooth surface suggests the importance of surface texturing in improving surface hydrophobicity.

The Superhydrophobicity of the OTS SAM modified textured Si surfaces is due to both chemical and topographic factors. Before the OTS self-assembly, the terminal groups of the oxidized Si surface were O atoms, which are hydrophilic by nature. When the OTS self-assembly was complete, the hydrophilic O atoms were no longer available and the surface became hydrophobic because of the hydrophobic alkyl chain of the OTS. As for topography, air was trapped between the textured features, which prevents the water droplet from wetting the surface and thus improves the hydrophobicity of the surface.

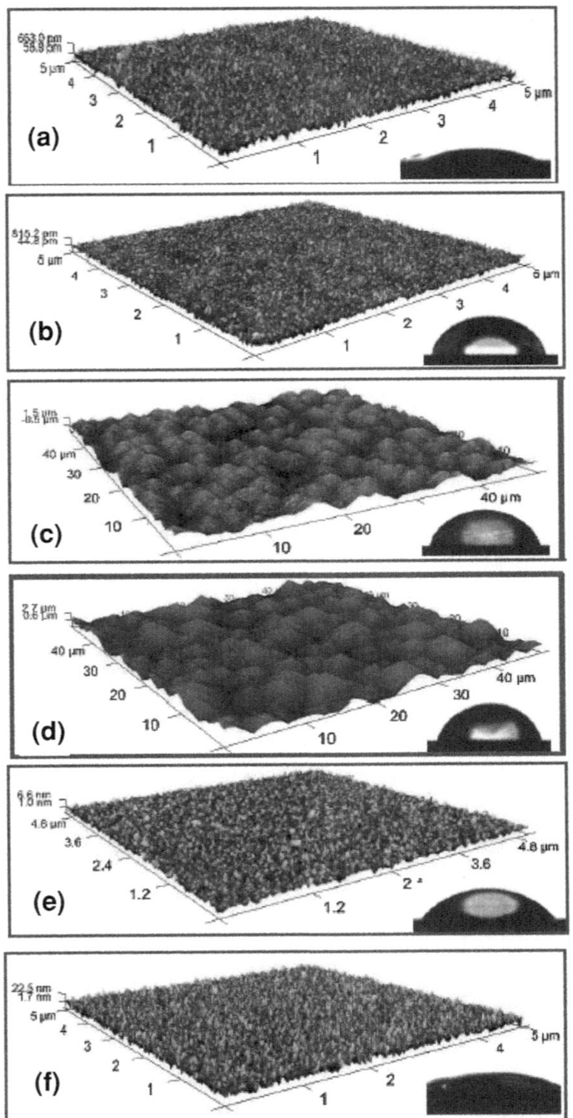

Fig. 7 Atomic force microscope Images of textured silicon **a** PR, **b** HF, **c** K, **d** T, **e** S, **f** O, **g–l** corresponding silicon surface modified with self assemble monolayer of OTS

3.2.2 Making of Superhydrophobic Dielectric Surface

Gas plasma treatment processes (glow discharges) are extensively used for increasing the surface roughness and modifying the hydrophobicity of polymeric materials. The main advantage of this versatile technique is that it is confined to the surface layer of a material without affecting its bulk properties. Moreover, it is a dry (solvent free), clean and time-efficient process with a large variety of

Fig. 7 (continued)

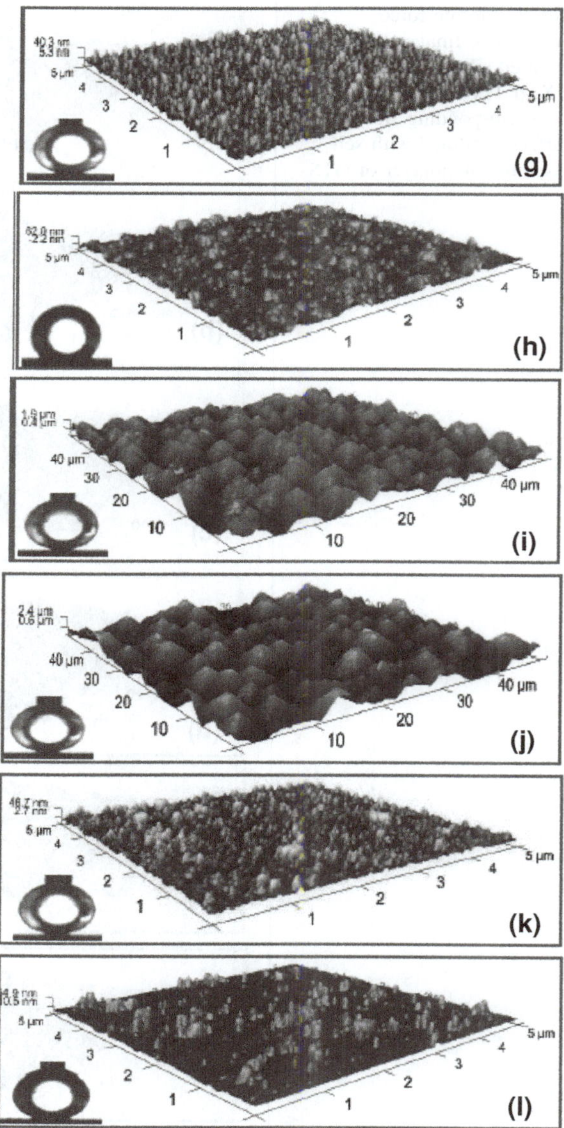

controllable process parameters (e.g. discharge gas, power input, pressure, treatment time) within the same experimental setup.

Recently, it has also been shown that the major problem with dielectric materials is the presence of a large density of electron-interface trap states near the semiconductor-dielectric interface. The dielectric materials can contain a hydroxyl group and water content, resulting in trap-sites on the surface of the dielectric. The solution for the neutralization of these sites is the deposition of SAMs on dielectric materials. A major portion of the research on SAMs since the early 1980 has continued to expand on the type of materials used for SAMs. The criteria important for selecting the type of material are dependent on the application for which the SAM is used. For example, the most popular dielectric material SiO2 is exclusively used for semiconductor device fabrications and microelectronics applications. This is mainly because thermally grown silicon dioxide with a dielectric constant of $k = 3.9$ has good compatibility and a satisfactory interface with silicon. In the past few decades, a reduction in the thickness of silicon dioxide dielectric has enabled an increase in the number of chips with enhanced circuit functionality and performance at low coast. As devices approach nanometer size scales, the effective oxide thickness of the silicon dioxide becomes less than 3 nm. Such small thickness results in high gate leakage currents due to the tunnelling effect at this scale. Also, very thin films are generally impractical for microelectronic applications because of their non-uniformity and the high density of pin holes. A thin dielectric film can be useful to improve the device performance, but the minimum thickness for which a pinhole-free film can be achieved is controlled by the deposition procedure as well as the intrinsic properties of the materials. The other important properties of dielectric materials are breakdown voltage which depends upon the thickness and dielectric strength of the dielectric materials. To continue the downward scaling, dielectric materials with a higher dielectric constant (high-k) are being suggested as an alternative solution to the aforementioned problem.

An Organosilane based monolayer on silicon dioxide and other hydroxylated surfaces is possibly the most commonly exploited self-assembly system. The chemical functionalization of silicon oxide surfaces with silane molecules is an important technique for a variety of device fabrications. High dielectric constant (k) metal oxides such as hafnium oxide, aluminum oxide (Al_2O_3), Tantalum Oxide (Ta_2O_5 and titanium oxide (TiO_2) have gained significant interest due to their applications in microelectronics. In order to study and control the surface properties of high-k dielectric materials, self-assembled monolayers (SAMs) of OTS were formed. The high-k dielectric materials are intrinsically hydrophilic material which can be modified by chemical modification by silanization. The silanization reaction is relatively easy to carry out, but the formation of a reproducible well-defined monolayer is exceedingly difficult. In order to form high quality monolayer, it is very important to ensure a high level of cleanliness to keep the error by contamination as low as possible. The silicon substrates with hafnium oxide were dipped into a 40 ml toluene with 1 % drop of OTS, and the bottles containing the submerged substrates were then sealed and held in the glove box for 2 h without disturbance. After SAM formation, the samples were rinsed sequentially with toluene and acetone and dried by argon flow. However, the monolayer quality is very sensitive to the amount of water in the system. If there is not enough water, only a partial monolayer forms, while if there is too much water the organosilane may polymerize. This can result in inconsistent surface properties.

We conducted experiments texturing dielectric surfaces by the use of plasma surface treatment in a fluorinated environment that would increase the roughness of the surface of a dielectric material. An empirical investigation of the various

Table 2 Details parameter of plasma treated dielectric materials

Films	Dielectric materials	Deposition methods	Thickness (nm)	SF_6 Plasma treatment parameter		
				Time (s)	RF power (W)	Gas flow (cm^3/s)
A	HFO_2	Sputtering	100	60	50	25
B	Al_2O_3	Sputtering	100	60	50	25
C	TiO_2	Sputtering	100	60	50	25
D	Ta_2O_5	Sputtering	100	60	50	25

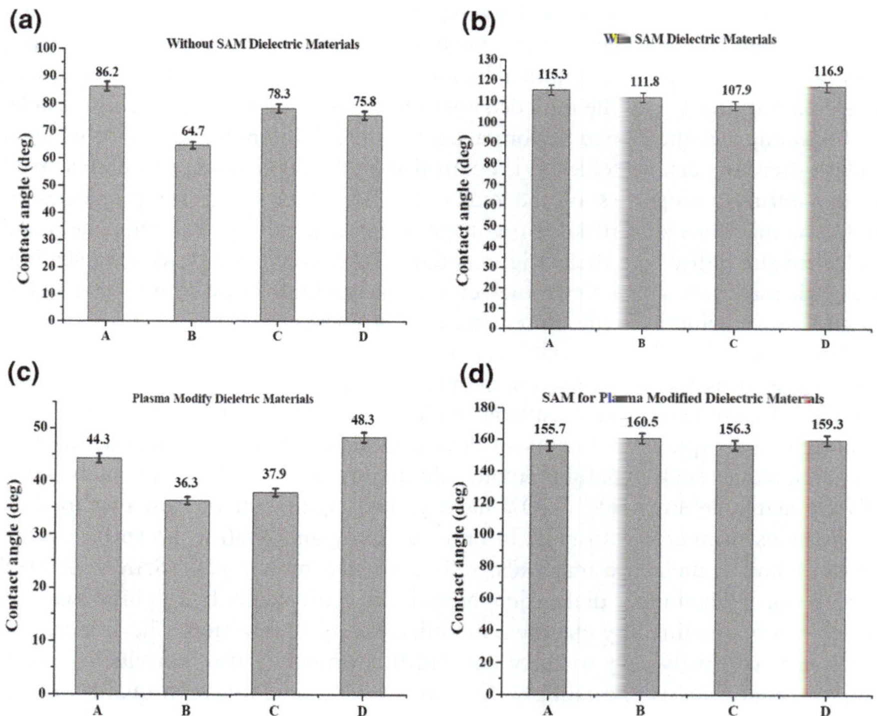

Fig. 8 Water contact angle **a** Unmodified dielectric materials. **b** OTS SAM modify dielectric materials. **c** Plasma modify dielectric materials **d** OTS SAM modify dielectric materials

parameters that affect the results of the plasma treatment such as gas composition, duration of plasma, effect of RF power etc. was carried out to optimize the surface modification process. The process gas is introduced into the plasma chambers at moderately low pressures (10–50 mtorr) at 50 W of RF power. When RF energy is supplied to the electrode pair of the plasma chamber, it accelerates stray electrons, increasing their energy to the limit where they can break chemical bonds in the process gas and generate additional ions and electrons. After the plasma treatment with SF_6 the surface become hydrophilic, later treatment of the sample by dipping it in Toluene/OTS solution for 2 h renders the surface quite superhydrophobic. The experiments were conducted on a few dielectric materials which, along with process and optimized parametric values, are tabulated in Table 2.

For all the dielectric materials enlisted in Table 2, the effect of SAM on wetting properties was studied by comparing the WCAs of the textured surfaces with the WCAs of the smooth dielectric surface before and after OTS SAM modifications. Figure 8a and b shows that OTS modification of a smooth dielectric material surface increases its WCA from a value around 86° to a value around 115°.

Figure 8c and d shows that plasma treatment makes the dielectric material hydrophilic with a WCA $< 90°$, but after coating it with OTS SAM it increases to $>150°$ which is in the superhydrophobic range [42]. Plasma treatment with fluorine atoms exists in activated states that can easily react with the oxide surface. Most of the research concentrates on using plasma surface modification to render surfaces more hydrophilic for improved adhesion properties. In some special cases, fluorination of the surfaces is also desirable to improve the chemical resistance of the surface, providing anti-stiction coatings.

4 Conclusions

This work studies the synthesis of fabricating superhydrophobic surfaces by a combination of chemically textured surfaces and an OTS monolayer. Variations in texturing were carried out using different chemicals and physical methods. The different etchants used for chemical texturing were piranha; KOH, TMAH and the plasma of SF_6 and O_2 were used for creating roughness physically. The OTS SAMs were deposited on chemical or physically textured surfaces. The surfaces used in the study were those of semiconductor (silicon) and a few dielectrics (HfO_2, Al_2O_3, TiO_2 and Ta_2O_5). It was shown that the hydrophobicity of silicon and dielectric materials significantly improved, resulting in a superhydrophobic surface with WCA $> 150°$. The combination of chemical and physical texturing with add-on SAM renders the surface of hydrophilic high k-dielectric materials like Hafnium oxide superhydrophobic. A similar observation is made on semiconductor material Silicon. The study is relevant to produce superhydrophobic surfaces which are widely used in microelectronics and microfluidics applications.

Acknowledgments This work was carried out for NPMASS National MEMS Design Centre, BITS, Pilani, India. We acknowledge the support and funding from NPMASS project sanctions in HR PARC to establish a National MEMS Design Centre at our Institute. We would also like to thank Prof. K.N. Bhat, ECE, Indian Institute of Science, Bangalore for his support and guidance.

References

1. Furstner R, Barthlott W, Neinhuis C et al (2005) Wetting and self-cleaning properties of artificial superhydrophobic surfaces. Langmuir 21:956–961
2. Bhushan B, Jung YC, Koch K (2009) Self-cleaning efficiency of artificial superhydrophobic surfaces. Langmuir 25:3240–3248
3. Sas L, Gorga RE, Joiness JA et al (2012) Literature review on superhydrophobic self-cleaning surface produced by electrospinning. J Polym Sci Part B 50:824–845
4. Gao L, McCarthy TT, Zhang X (2009) Wetting and superhydrophobicity. Langmuir 25(24):14100–14104
5. Drelich J, Milles JD, Kumar A et al (1994) Wetting characteristics of liquid drops at heterogeneous surface. J Colloids Interface 93:1–13
6. Yang Z-H, Chiu C-Y, Yang J-T et al (2009) Investigation and application of an ultra hydrophobic hybrid-structured surface with anti-sticking character. J Micromach Microeng 19:085022–085033
7. Bhushan B, Koch K, Jung YC (2008) Nanostructures for superhydrophobicity and low adhesion. Soft Matter 4:1799–1804
8. Su Y, Ji B, Huang Y et al (2010) Nature's design of hierarchical superhydrophobic surfaces of a water strider for low adhesion and low-energy dissipation. Langmuir 26(24):18926–18937
9. Koch K, Bhushan B, Jung YC et al (2009) Fabrication of artificial lotus leaves and significance of hierarchical structure for superhydrophobicity and low adhesion. Soft Matter 5:1386–1393
10. Ou J, Perot B, Rothstein JP (2004) Laminar drag reduction in microchannels using ultrahydrophobic surfaces. Phys Fluids 16(12):4635–4643
11. Samaha MA, Tafreshi HV et al (2012) Superhydrophobic surfaces: from the lotus leaf to the submarine. CR Mec 340:18–34
12. Wu L, Zhang J, Li B et al (2013) Mimic nature, beyond nature: facile synthesis of durable superhydrophobic textiles using organosilane. J Mater Chem B 1:4756–4763
13. Zhang X, Shi F, Niu J et al (2008) Superhydrophobic surfaces: from structural control to functional application. J Mater Chem 18:621–633
14. Liu J-L, Feng X-Q, Wang G et al (2007) Mechanisms of superhydrophobicity on hydrophilic substrates. J Phys Condens Matter 19:356002
15. Wang B, Zhang Y, Shi L et al (2012) Advances in the theory of superhydrophobic surfaces. J Mater Chem 22:20112–20117
16. Yan YY, Gao N, Barthlott W (2011) Mimicking natural superhydrophobic surfaces and grasping the wetting process: a review on recent progress in preparing superhydrophobic surfaces. Adv Colloid Interface Sci 169:80–105
17. Whiteside's GM, Laibinis PE (1990) Wet chemical approaches to the characterization of organic surfaces: self- assembled monolayers, wetting, and the physical-organic chemistry of the solid–liquid interface. Langmuir 6:1
18. Gooding JJ, Mearns F, Yang W et al (2003) Self-assembled monolayers into the 21st century: recent advances and applications. Electroanalysis 15(2):81–96
19. Otero R, Gallego JM, Vázquez de Parga AL et al (2011) Molecular self-assembly at solid surfaces. J Adv Mater 23:5148–5176
20. Newton L, Slater T, Clark N et al (2013) Self assembled monolayers (SAMs) on metallic surfaces (gold and graphene) for electronic applications. J Mater Chem C 1:376–393
21. Wang M, Liechti KM, Wang Q et al (2005) Self-assembled silane monolayers: fabrication with nanoscale uniformity. Langmuir 21:1848–1857
22. Jadhav SA (2011) Self-assembled monolayers (SAMs) of carboxylic acids: an overview. Cent Eur J Chem 9(3):369–378
23. Pawsey S, Yach K, Reven L (2002) Self-assembly of carboxyalkyl phosphonic acids on metal-oxide powders. Langmuir 18:5205–5212

24. McIntyre NS, Nie H-Y, Grosvenor AP et al (2005) XPS studies of octadecylphosphonic acid (OPA) monolayer interactions with some metal and mineral surfaces. Surf Interface Anal 37:749–754
25. Yildirim O, Yilmaz MD, Reinhoudt DN et al (2011) Electrochemical stability of self-assembled alkyl phosphate monolayers on conducting metal oxides. Langmuir 27:9890–9894
26. Wasserman SR, Tao Y-T, Whitesides GM (1989) Structure and reactivity of alkylsiloxane monolayers formed by reaction of alkyltrichlorosilanes on silicon substrates. Langmuir 5(4):1074–1087
27. Aswal DK, Lenfant S, Guerin D et al (2006) Self assembled monolayers on silicon for molecular electronics. Anal Chim Acta 568:84–108
28. Glass NR, Tjeung R, Chan P et al (2011) Organosilane deposition for microfluidics applications. Biomicrofluidics 5:036501–036501–036501–036507
29. Takeya J, Nishikawa T, Takenobu T et al (2004) Effects of polarized organosilane self-assembled monolayers on organic single-crystal field-effect transistors. Appl Phys Lett 85(21):5078–5080
30. Kumar V, Sharma NN (2012) SU-8 as hydrophobic and dielectric thin film in electrowetting-on-dielectric based microfluidics device. J Nanotechnol 6. doi: 10.1155/2012/312784 (Article ID 312784)
31. Dai W, Zhao Y-P (2007) The nonlinear phenomena of thin polydimethylsiloxane (PDMS) films in electrowetting. Int J Nonlinear Sci Numer Simul 8(4):519–526
32. Papadopoulou EL, Zorba V, Stratakis E et al (2012) Properties of silicon and metal oxide electrowetting systems. J Adhes Sci Technol 26:2143–2163
33. Liu H, Dharmatilleke S, Maurya DK et al (2010) Dielectric materials for electrowetting-on-dielectric actuation. Microsyst Technol 16:449–460
34. Ortiz RP, Facchetti A, Marks TJ (2010) High-k organic, inorganic, and hybrid dielectrics for low-voltage organic field-effect transistors. Chem Rev 110:205–239
35. Sun X, Di C, Liu Y (2010) Engineering of the dielectric–semiconductor interface in organic field-effect transistors. J Mater Chem 20:2599–2611
36. Martines E, Seunarine K, Morgan H et al (2005) Superhydrophobicity and superhydrophilicity of regular nanopatterns. ACS Nano Lett 5(10):2097–2103
37. Song Y, Nair RP, Zou M et al (2009) Superhydrophobic surfaces produced by applying a self- assembled monolayer to silicon micro/nano-textured surfaces. Nano Res 2:143–150
38. Bhushan B, Jung YC (2008) Wetting, adhesion and friction of superhydrophobic and hydrophilic leaves and fabricated micro/nanopatterened surfaces. J Phys Condens Matter 20:225010–225034. doi:10.1088/0953-8984/20/225010
39. Bormashenko E et al (2010) Wetting transitions on biomimetic surfaces. Philos Trans Roy Soc A 368:4695–4711. doi:10.1098/rsta.2010.0121
40. Ma M, Hill RM (2006) Superhydrophobic surfaces. Curr Opin Colloid Interface Sci 11:193–202
41. Swart M, Mallon PE (2009) Hydrophobicity recovery of corona-modified superhydrophobic surfaces produced by the electro spinning of poly (methyl methacrylate)-graft-poly(dimethylsiloxane) hybrid copolymers. Pure Appl Chem 81(3):495–511
42. Kumar V, Sharma NN (2012) Surface modification of hafnium oxide (HfO2) for microfluidics applications, Indian Patent 2455/DEL/2012 (filed)

Materials for Embedded Capacitors, Inductors, Nonreciprocal Devices, and Solid Oxide Fuel Cells in Low Temperature Co-fired Ceramic

Vivek Rane, Varsha Chaware, Shrikant Kulkarni, Siddharth Duttagupta and Girish Phatak

Abstract Low Temperature Co-fired Ceramic (LTCC) technology is evolving. New materials now offer improved embedded passive components and kindle hope of integrating devices never before imagined. This rediscovery of LTCC is making it an important vehicle for systems integration. This chapter explores the current status of embedded capacitors and inductors in LTCC, providing a glimpse of the improved materials being developed. Investigations on materials for integrated nonreciprocal devices, such as, circulators and isolators are also discussed. For the first time, we report materials development for integrated solid oxide fuel cells within LTCC structures, opening up a new area of integrated power sources, hitherto unexplored.

Keywords LTCC · Buried passives · Embedded capacitors · Embedded inductors · Integrated power source · SOFC electrolyte

---

Vivek Rane—Contributor for Materials for Integrated Ferrites.
Varsha Chaware—Contributor for Integrated Capacitors materials.
Shrikant Kulkarni—Contributor for integrated SOFC materials.

V. Rane
G. M. Vedak College of Science, Tala, Raigad 402111, Maharashtra, India

S. Duttagupta
Department of Electrical Engineering, Indian Institute of Technology Bombay (IITB), Mumbai 400076, India

V. Chaware · S. Kulkarni · G. Phatak (✉)
Centre for Materials for Electronics Technology (C-MET), Panchawati, Pune 411008, India
e-mail: gjp@cmet.gov.in

K. J. Vinoy et al. (eds.), *Micro and Smart Devices and Systems*,
Springer Tracts in Mechanical Engineering, DOI: 10.1007/978-81-322-1913-2_17,

1 Introduction

In the age of "beyond Moore's law" electronics, Low Temperature Co-fired Ceramic (LTCC) is one of the most suitable technologies, which can offer system integration solutions to sustain further miniaturization. LTCC is well-known packaging technology used for packaging of Integrated Circuits (IC), monolithic microwave ICs (MMIC), and sensors, and to fabricate multi chip modules (MCM). The low loss dielectric materials used in LTCC also allow fabrication of multilayer Microwave Integrated Circuits (MIC). The capability of making three-dimensional (3D) structures and to handle mechanical, electrical, optical, and microfluidic signals extends its usefulness to Micro Electro Mechanical (MEMS) devices and their packaging [1].

One of the major advantages of LTCC is its capability to accommodate embedded lumped passive components within the multilayer circuit. The material solutions that allow buried resistors in LTCC are relatively well developed; it is the capacitors, inductors, and other parts of circuits and systems that needs focus. Present day embedded capacitors and inductors in LTCC suffer from limited range, high tolerance, low Q factors, and limited frequency of operation. For example, the practical limit of values for embedded capacitors and inductors is a few pico Farads and nano Henry, respectively. Further, the tolerance in these values is high, ranging from 10 to 25 %. While there is surely some contribution of the fabrication process and design to these limitations, the major contribution is from the material used to realize these components. Since the present components are built around the same low permittivity (low-k) material used as interlayer dielectric, these components suffer from the limitations brought in by the materials not designed for such applications. It must be recognized that the capacitors and inductors demand specific materials properties, and these properties must be provided through different materials solutions.

Other devices which impart bulkiness to microwave circuits are the nonreciprocal devices, such as, circulators, isolators, or phase retarders used for telecommunication applications. These devices transmit microwave energy from one of its ports to an adjacent port while isolating the signal from all other ports. The present day nonreciprocal devices are quite bulky due to the need of permanent magnet in their assembly. Today, there is little work toward their integration with LTCC, may be the perceived incompatibility of the materials used to fabricate these devices with LTCC is the reason. Clearly, further improvement in embedded capacitors, inductors, and any integration of nonreciprocal devices with LTCC would be possible only through the development of specialized materials compatible with the LTCC host materials.

One important device that could bring in major space reduction and is left out of the efforts of systems integration is the power sources. Such integration can provide major advantages for portable and remotely placed systems. Small size planar Solid oxide Fuel Cells (SOFC) is one of the promising candidates in the field of in situ power generation [2]. Integration of these fuel cells with LTCC is an obvious choice for being ceramics-based technologies. LTCC, on its part brings in advantages, such as, microfluidic fabrication, sealing, and brazing process and integrated heater and temperature sensors [3]. Unwittingly, such integration can solve the long standing difficulty of sealing and interconnections faced by SOFC. Such integration would be possible only through a set of LTCC compatible SOFC materials.

This chapter presents brief treatise on LTCC compatible materials for the integration of capacitors, ferrites for embedded inductors and nonreciprocal devices, and SOFC.

2 Materials for Embedded Capacitors in LTCC

Generally, in most electronic circuits, capacitors are used for coupling/decoupling, filter, or oscillator applications; a majority of them with values up to a few nF. In LTCC, the values of the capacitors are important as they directly decide the occupied space. Generally, a low-*k* LTCC tape with a dielectric constant around 7.8 (For example, DuPont tapes) can provide a capacitance density of about 2–3 pF mm^{-2} per layer depending upon the thickness. Clearly, achieving capacitance value in the range of 10 nF would render the capacitor size impractical. In turn, using high-*k* materials with a dielectric constant of $\sim$80 and thickness of 100 μm, a capacitance density of 15 pF mm^{-2} can be expected [4]. Obviously, there is no option but to supplement the low-*k* LTCC materials with high-*k* materials with a range of dielectric constants and in various forms (tape or paste) suitable for standard LTCC fabrication process to increase the capacitance density and scale down capacitor size [5].

*2.1 Recent Status of Co-fired High-*k *and Low-*k *Materials*

There are various LTCC materials manufactures, such as, DuPont, Kyocera, NEC, Heraeus, Murata, Electro Science Laboratories (ESL), Ferro, etc, who have developed different kinds of LTCC tapes and paste. However, very few manufacturing companies have developed compatible materials for passive components. Among various manufacturers, ESL has the most products with varying permittivity in the form of tapes and pastes. ESL claims that their variety of capacitor tape and paste material is compatible with their own low-*k* tapes as well as with other manufactures, such as, DuPont 951 tape system. However, the sintering conditions are not quit clear. It is reported that the co-firing of these tape system needs constrained sintering and symmetrically placed buried high-*k* material [6]. Similarly, Heraeus also has developed the compatible low-*k* and high-*k* tapes but shrinkage compatibility may be insufficient for free sintering.

Apart from the above-mentioned reports from ESL and Heraeus, there are a few reports in the literature about development of co-sintering high-*k* and low-*k* materials, sometimes together with the commercial low-*k* tape system. Table 1 presents a compilation of these reports, which differ largely in their extent of work. Most reports present results up to co-sintering but a varying degree of results are available that extends this work to testing of compatibility with other functional materials, testing of mechanical properties, and fabrication and characterization of capacitors. None have reported re-firing tests, except Modesa et al. [7] mention possibility of refiring compatibility of their materials. In summary, it can be stated that there is a need to develop truly compatible, low loss integrated materials system with various *k* values, with following attributes:

- In tape and paste format
- Materials with closely matching sintering behavior
- Materials with matching TCE
- Materials with refiring capabilities
- Materials compatible with other functional materials, such as resistor and inductor material
- Complete study of capacitors with various designs

In our attempt to prepare compatible low-*k* and high-*k* materials for buried passive components, we have studied the low loss zinc silicate as a low-*k* ($\varepsilon_r \sim$ 6–7) material and the well-known high-*k* material, viz. $BaTi_4O_9$ (BT4). Since both the materials require high sintering temperature, we attempted adding same crystalline additive to both the materials, which act as sintering aid. This addition helped in lowering the sintering temperature to 900 °C as required by LTCC process. Due to this approach, the sintered material has shown better microwave dielectric properties. Figure 1 presents the dependence of sintering density, dielectric constant, and quality factor of Zn-silicate as a function of the weight quantity of crystalline additive. The relative permittivity for this low-*k* material is in the range of 6.8 and loss factor (tan δ) in the range of 10^{-4}. Further work on integrating BT4 with Z-Silicate is underway.

In our early attempt to embed the passive component into the LTCC, we studied the co-firing of the high-*k* ceramic BT4 by printing its thick film pastes on commercial low-*k* tape stack with pre-firing thickness of ~750 μm (DuPont 951). The sintering temperature of BT4 was reduced to 900 °C by using about 4 wt% of the above-mentioned crystalline additive. This high-*k* paste shows the good physical compatibility and no warpage. A separate experiment to measure dielectric constant and loss factor in microwave frequency range for BT4 with additives yielded dielectric constant of 37.2 and loss factor range of 10^{-4}. Detailed study about the materials interaction, microstructural properties at the interface, etc, is in progress. This material would be, subsequently, tested together with Z-Silicate composite reported above.

Table 1 Summary of reports regarding co-firing of low-*k* and high-*k* tapes

References	Code	Material system	Glass	ε_r	Q/tanδ	'C' in (pF) and TCC (ppm/°C)	Sintering temp. (°C)/method
Higuchi et al. [8]	AHT 01-004	Ba–Al–Si	Si	7.0	0.0040	–	900/free dual peak sintering
	AHT 00-023	RE–Ba–Ti		65.0	0.0014	–	
Modesa et al. [7]	K8	Al_2O_3	Ca–Al–B–Si	8.7	330	264 pF and 130	870/constrained Sinterd
	K15	Ba–(Sm, Nd)–Ti–O		15.1	940	−10	
Ko et al. [9]	M2S	Mg–Si	Li–Si–B–Ca–Al	6.6	–	–	900/free sintered
	BT4	Ba–Ti		29	–	–	
Choi et al. [10]	LG3	Al_2O_3	Si–Al–B	6.5	–	–	875/free sintered
	MG4	Zn–Nb–Ti		19.2	–	–	
Cho et al. [11]	K8	Al_2O_3	Ca–Al–B–Si	8.2	0.004	–	850/free sintered
	K48	Ba–Ti	Ba–Nd–Ti–b	48.2	0.001	–	
	K1340	Ba–Ti	LiF, Bi–Si	1337	0.024	–	
Randall [12]	DuPont 951	Al–Ca–Zr	–	7.0	300	97 pF and <±40	875/free sintered
	BZT	Bi–Zn–Ti	Zn–B	60.0	0.001	–	
	BZN	Bi–Zn–Nb		100		–	

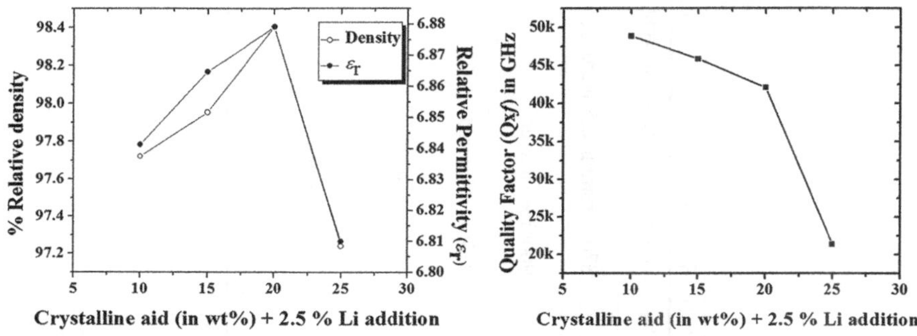

Fig. 1 Relative density, permittivity and Quality factor (Q $\times$ f) at resonant frequency after sintering at 900 °C for 3 h with different crystalline material loading with Z-silicate

2.2 Conclusions: Materials for Embedded Capacitor in LTCC

The above paragraphs underline the need of compatible low-*k* and high-*k* LTCC dielectric materials for embedded capacitors. It is seen that the commercial solutions available today are inadequate. We found only two manufactures listing compatible high-*k* and low-*k* materials. Heraeus and ESL list tape materials with a range of *k* values that are said to be capable of co-firing, although other reports on processing of these materials indicate constraints. There are only a handful of reports in open literature discussing co-fired low-*k* and high-*k* LTCC materials. The reports concentrate on only a few aspects among dielectric properties, type of material (tape, paste), sintering conditions (free or constrained), compatibility with other materials (low-*k* dielectric, conductors or other functional materials), capacitor properties (capacitance per unit area) or re-firing capabilities of the materials system. Clearly, comprehensive development for truly compatible low-*k* and high-*k* dielectric materials is still lacking. We have initiated simultaneous development of low-*k* and high-*k* dielectric material aimed at full compatibility covering all the above listed issues, so that embedded lumped capacitors with higher range of capacitance are made possible. Admittedly, our work has also not yet covered all aspects of compatibility.

3 Materials for Embedded Inductor and Nonreciprocal Devices

Over the years, most of the research in embedded passives technology is focused on resistors and capacitors. This may be because of the availability of materials solutions closer to the need as well as the impact. We note that these components account for a major part of the total count of passive components and their integration would best fulfill the primary goal of miniaturization. In contrast, share of the inductors and transformers is only 10–20 % of the total passives count, although these components acquire disproportionately larger area on the substrate board. There are two main difficulties in embedding these components in LTCC. First is the issue related to device geometry, its interaction with electromagnetic wave and consequently its placement in an embedded environment. Nonetheless, the integration of inductors, transformers, and nonreciprocal devices in LTCC has been accomplished within the host dielectric material. Second is the range of compatibility requirements in integrating the currently used ferrite materials with LTCC. Surely, creation of embedded solutions for inductors and non-reciprocal devices in LTCC need development of compatible, specialized magnetic materials, and processes of their use. The following Sections discuss both the issues, with an emphasis on the development of ferrite materials for integrated LTCC solutions.

3.1 Status and Issues of Buried Inductors and Nonreciprocal Devices in LTCC Dielectrics

The integrated inductors in LTCC are accomplished within the host dielectric material and built using the multilayer signal lines and a ground plane. These inductors primarily rely upon the mutual inductance between multilayer signal lines and ground plane. They are typically formed using one of two design topologies viz. planar and 3D [13]. The planar inductors are divided into meander and spiral (square or circular) structures, while 3D helical coils can be arranged in both series and shunt connection either horizontally or vertically on the LTCC layer stack. Both the structures are easy to design, simulate, and fabricate. The inductance values are limited to nano-Henries (usually <5 nH) at operating frequencies above 1 GHz. Both the design approaches face significant design challenges with increasing operating frequencies. These difficulties include conductor losses, increase in nonlinear resistance of the conductor, wide inductor area, distortion of inherent magnetic field in the vicinity of the device, multiple resonance, etc., and result in low inductance, low self resonating frequency, low Q-factor, and larger foot-prints of the device. Therefore, these inductors can be used for limited applications, such as, for example, microwave delay lines.

Another important requirement in LTCC is the realization of nonreciprocal components, such as, circulators, isolators, etc. An integrated front-end module in LTCC has been developed for wireless local area network (WLAN) antenna with high isolation characteristics [14]. The circulator referred here is a 15–17 GHz micro-strip line circulator of size of 7.5 × 6 mm. It occupies more space than MMIC and indicates a major drawback of such devices in terms of size of the component.

All these aspects of the embedded inductors and nonreciprocal devices in LTCC affect the performance at RF/microwave frequencies and miniaturization of the LTCC modules. Furthermore, the micro-Henry inductors are practically difficult to realize in LTCC due to large profile or thickness. Such state of affairs calls for innovative approach for embedding these devices into LTCC. The technology of choice to circumvent the above-mentioned challenges is to use the ferrimagnetic ceramics, i.e., ferrites. The ferrites offer twin advantages of high permeability and high resistivity, which are beyond the reach of dielectric materials. However, the insertion of ferrite into LTCC creates for materials related difficulties. Thus, the integration of the targeted devices in LTCC can be tackled through:

- Development of novel ferrite materials with repeatable materials properties
- Providing solution to the issues due to insertion of new materials into LTCC
    - Difference in sintering temperature
    - Mismatch of temperature coefficient of expansion
    - Mismatch of shrinkage of different materials
    - Chemical compatibility of different materials

The approach of integrating ferrite materials in LTCC is discussed below with help of available literature and the results obtained in our quest in developing LTCC ferrite materials. The focus is on LTCC ferrite materials for high performance, miniaturized components for higher frequencies of operation than those available today.

3.2 LTCC Ferrites for Inductor and Nonreciprocal Devices

There are not many commercially available LTCC ferrite tapes which are said to be compatible with LTCC dielectric material. Ferrite tapes are available from ESL and Ferro for LTCC applications. These tapes are said to be co-sintered with dielectric tapes, but the difference in shrinkage properties suggest requirement of constraint sintering. The inductors fabricated using these tapes have operating frequency in the range from kHz to below 5 MHz and are mainly used in power electronic applications. These devices show 50 % reduction in the size vis-à-vis their MLCI counterpart [15]. Thus, ferrites with extended operating frequency and better compatibility with the host LTCC materials are necessary. The materials of interest are MnZn ferrite, NiZn ferrite, and NiCuZn ferrite, NiCoZn ferrite, M-hexaferrite, and Z-hexaferrite, chosen on the basis of specific requirements for inductor and nonreciprocal devices. In order to use these materials in LTCC applications, these materials need to be sintered with the help of sintering aids, such as, low melting, reactive BBSZ glass (Bi_2O_3–B_2O_3–SiO_2–ZnO glass) [16]. The use of BBSZ glass fulfills low temperature sintering as well as shrinkage matching with that of LTCC. In the following, we discuss our efforts to obtain improved LTCC compatible materials for embedded inductor, transformer, and nonreciprocal devices.

3.2.1 LTCC Integrated Inductors for VHF and UHF Frequencies

LTCC demands embedded inductors, transformers, and chokes operating in the range essentially above 100 MHz (i.e., in VHF to UHF range) to cater the next generation applications. NiCoZn ferrite, which is a soft ferrite material, can be used to fabricate LTCC integrated inductors [17, 18]. The resonance frequency increases with the addition of cobalt ions in NiZn ferrite mainly due to increase in the anisotropy constant K_1. The NiCoZn ferrite shows maximum permeability in the range of 18 at 300 MHz [18]. Figure 2a shows the permeability spectra of NiCoZn ferrite sintered at 900 °C using 2 wt% BBSZ glass, showing increase in resonance frequency.

In order to explore the applications of LTCC inductors in ultra-high frequency range (UHF; 300 MHz–3 GHz), a material system known as hexaferrite can be used. Hexaferrites, such as CoTi substituted $BaFe_{12}O_{19}$ (M-hexaferrite) and $Ba_3Co_2Fe_{24}O_{41}$ (Co_2-Z-hexaferrite) are explored in LTCC. M-hexaferrite, if doped with Co^{2+} and Ti^{4+} induces planar anisotropy, and provides high inductance density and high resonance frequency (up to 1 GHz) vis-à-vis their MLCI counterparts [19]. Co_2-Z-hexaferrite can also be used in LTCC in similar manner [17].

In our quest to develop the LTCC integrated inductors for UHF frequency range, we developed Zn substituted Co_2-Z-hexaferrite with stoichiometry $Ba_3Co_{0.8}Zn_{1.2}Fe_{24}O_{41}$. This material shows soft magnetic properties, permeability up to 8, and gyro-magnetic resonance frequencies well above 1 GHz [18]. Figure 2b presents the permeability spectra of $Ba_3Co_{0.8}Zn_{1.2}Fe_{24}O_{41}$ hexaferrite ferrite sintered at 900 °C using BBSZ glass developed by us.

3.2.2 LTCC Integrated Nonreciprocal Devices

As mentioned in Sect. 3.1, LTCC can be used to integrate nonreciprocal devices such as circulators, isolators, etc. Nonreciprocal devices are bulky and are used in *stand-alone* form, which are usually fabricated using ferrite materials. Clearly, ferrites are the best choice of materials in realizing these applications. These devices also require permanent magnet in their assembly. The remedy to solve the constraints posed by the present generation nonreciprocal components is to use different LTCC ferrites, as discussed in the following.

Jensen et al. have demonstrated a fully functional integrated circulator in LTCC using compatible ferrite tape (ESL40012) and showed promising results in terms of miniaturization of the device [20]. The result signifies the benefit of using ferrite material over LTCC dielectrics, as it provides miniaturization along with good performance of the device at microwave frequencies. However, the system uses external magnetic bias field. Nonetheless, this a major step forward in realizing integrated nonreciprocal devices in LTCC. The way forward, mainly to remove the

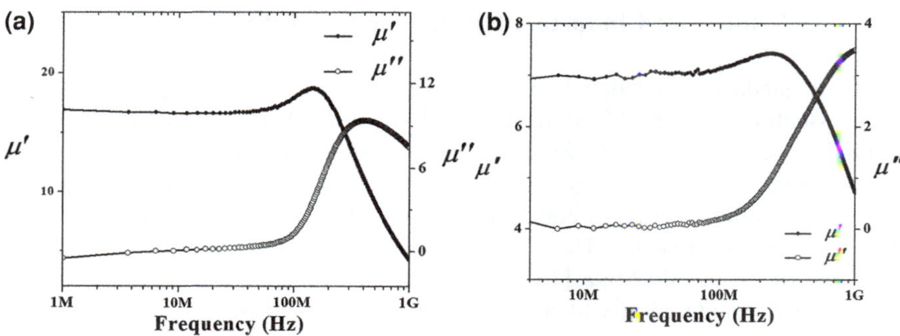

Fig. 2 Permeability spectra of **a** NiCoZn ferrite, and **b** $Ba_3Co_{0.8}Zn_{1.2}Fe_{24}O_{41}$ hexaferrite, sintered at 900 °C with help of 2 wt% BBSZ glass

external bias field, is to fabricate these devices with materials having *self-biased* properties, such as, $BaFe_{12}O_{19}$ (M-hexaferrite).

We have developed M-hexaferrite material for LTCC integrated applications with low temperature sintering (900 °C) and similar shrinkage property with that of LTCC [18, 21, 22].

Figure 3 presents the high frequency permeability spectra of $BaFe_{12}O_{19}$ hexaferrite prepared by us. This sample shows permeability in the range of 1.8 and resonance at 150 and 750 MHz. This material was prepared using combustion synthesis and sintered with help of 2 wt% BBSZ glass at 900 °C. These results indicate that M-hexaferrite can be used to fabricate integrated nonreciprocal components in LTCC.

3.3 *Conclusions: Materials for Embedded Inductors and Non-reciprocal Devices*

The report presented in this part of the chapter deals with ferrite materials developed for integrated inductors and nonreciprocal devices embedded in LTCC for high frequency applications. The driving factor behind this development is the need of miniaturized and high performance integrated devices, which can be satisfied only by integrating ferrite materials with LTCC. We have developed soft ferrites, viz. NiCoZn ferrite and Z-hexaferrite material for integrated inductor, transformer, chokes or filter applications in LTCC for the frequency range in VHF and UHF, respectively. The results presented here also indicate that the choice of a single optimum ferrite material for every application is not possible. The LTCC compatible hard ferrite viz. $BaFe_{12}O_{19}$ developed by us can be used to fabricate integrated nonreciprocal devices in LTCC and eliminate the use of external permanent magnet. While the preliminary materials compatibility with LTCC has been demonstrated, planar or 3D passive components such as inductors,

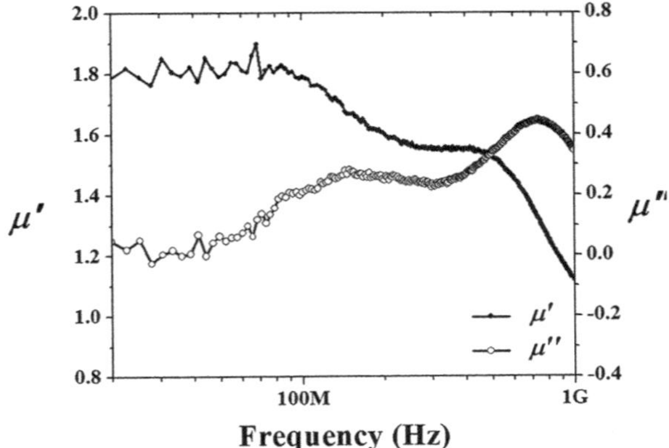

Fig. 3 Permeability spectra of $BaFe_{12}O_{19}$ (M-hexaferrite), sintered at 900 °C with help of 2 wt% BBSZ glass

transformer, circulators, etc., are yet to be demonstrated. Nevertheless, the work presented here has opened new avenues in LTCC-based integration of devices for HF and microwave applications.

4 Materials for Integrated SOFC in LTCC

The previous discussion in the Introduction (Sect. 1) of this chapter has highlighted the need and importance of integrating power source within LTCC structures. It was reasoned that power source with independent fuel supply may help hand-held and remote applications. It was also argued that the ceramics-based SOFC technology would be best suited for such integration, mainly due to similarity of the materials technology. While SOFCs are the most efficient power sources providing close to 90 % efficiency if the heat generated from exothermic reactions is also utilized, fuel flexible and are most environmentally clean, these fuel cells suffers from major difficulties caused by the high operating temperature of about 1,000 °C. One, is of sealing for fuel, oxidant, and exhaust, and second is of electronic conductors, which pose reliability issues [23]. Integration with LTCC imposes need to reduce the operating temperature, but at the same time provides various advantages intrinsic to LTCC. The high operating temperature of these fuel cells is a necessity due to the need for a minimum ionic conductivity of the oxide electrolyte. This conductivity can be improved by optimizing materials and their respective properties. Another, better way of reducing the operating temperature is through innovative electrolyte materials that conduct ions at lower temperatures.

It is proposed here that these issues faced by SOFCs can be resolved using various advantages offered by the LTCC technology, which include sealing, thermal stability, chemical inertness, reliable interconnections. In the following Sections we present a brief account of our efforts to develop LTCC compatible SOFC materials.

4.1 SOFC Integration with LTCC

Electrolyte and electrodes are among the important SOFC components and made up of ceramic (oxide) materials that require high sintering temperature beyond the range of LTCC (above 1,200 °C). Apart from this, there are several other parameters that underline the mismatch between LTCC and SOFC materials. The LTCC materials are highly insulating and chemically inert glass-ceramics sintered below 900 °C, can be operated up to 600 °C, show shrinkage of about 12–13 %. On the other hand, the SOFC materials are pure ceramics, need to be ionic and mixed conductors, are sintered at temperatures above 1,200 °C, operate at 800–1,000 °C and show shrinkage of about 14 %. Clearly, the physical, chemical and electrical properties of materials used in SOFC are quite different from that of LTCC. Nevertheless, there are many advantages in integrating SOFC and LTCC materials and processes, such as, easy fabrication of microfluidic channels, buried heaters and temperature sensors, connection of metallic tubing by brazing, well developed sealing technology, and use of the thick film technology for fabrication. Mainly due to such enormous advantages we undertook the difficult task of integrating these two dissimilar materials technologies. The following sections present the current status of our quest to develop such materials.

4.2 Selection of Materials

Electrolytes in SOFC are highly dense oxides that allow diffusion of oxygen or proton ions, while blocking electron flow through it. Electrolyte materials should be highly dense, must have higher ionic conductivity, lower electronic diffusion, and must possess appropriate vacancies for ionic conduction. Presently, Yttria Stabilized Zirconia (YSZ) is used commercially as electrolyte for SOFCs that shows oxygen ion conductivity of the order of 0.1 S cm^{-1} at 1,000 °C. According to literature, 8 mol.% Y_2O_3 doped in ZrO_2 (8YSZ) and Sr and Mg doped $LaGaO_3$ (LSGM) shows comparable ionic conductivity in intermediate range of operating temperature (600–800 °C). The CeO_2-based materials doped by Samarium (Sm) or Gadolinium (Gd) are good oxygen ion conductor in temperature range of 600–800°C [24].

Ceria in its pure form is insulating due to unavailability of oxygen ion vacancies. When trivalent lanthanides, such as, Sm, Gd, Yttrium (Y), etc., are added to Ceria, oxygen vacancies are created at the octahedral sites due to replacement of Ce^{4+} ions that enabled oxygen ion conduction by ion hopping mechanism at 600–800 °C. It is seen from the reports that ionic radius of Gd is suitable to Ceria compared to Sm, and about 10–20 mol.% doping of Gd is required for ionic conductivity [25]. Based upon the above discussed facts, GDC was found to be most suitable for further studies aiming at integration with LTCC.

4.3 Synthesis of GDC

Available literature indicates that GDC is synthesized by different physical and chemical methods. Of these, combustion synthesis method using Glycine as a fuel is one of the best methods reported in literatures [26–28]. This is a simple wet chemical, self-propagating method with *redox* reaction that involves oxidation and reduction process simultaneously [29].

We have carried out a detailed study of GDC preparation by GNP method. The synthesized GDC powders crystallized in fluorite FCC crystal structure with *Fm3m* symmetry that matches well with the JCPDS data (card no.01-075-0162) for $Ce_{0.8}Gd_{0.2}O_{1.9}$. The pellets pressed and sintered at 1,350 °C for 8 h possessed density as high as 94 % of the theoretical density. These pellets showed ionic conductivity of 0.04 S cm^{-1} at 700 °C. The activation energy in the temperature range of 500–700 °C was found to be 0.19 eV [29] which is considered low.

It is seen from the above results that very high sintering temperature of 1,350 °C is required for GDC to achieve density >94 %. Lowering such sintering temperature without any sintering aid may be difficult. Therefore, there is a need of an appropriate sintering aid that can lower the sintering temperature of SOFC ceramic electrolyte in the range of LTCC firing temperatures. The following section describes the detailed study about effect of chosen sintering aids on physical, chemical and electrical properties of GDC.

4.4 Effect of Sintering Aids on Gadolinium-Doped Ceria

Similar to inductor materials development, GDC was also added with low temperature melting glass. We used Bismuth oxide-based phosphate glasses as sintering aid in GDC electrolytes. Phosphorous Pentaoxide (P_2O_5) is low melting temperature oxides having melting point 450 °C. It is also reported that P_2O_5 glasses are good protonic conductors [30]. Bi_2O_3 and V_2O_5 also have melting points around 817 and 690 °C, respectively, and are reported to form oxygen ion conducting phases at high temperature (400–600 °C), commonly known as BIMEVOX [31]. It is also reported that Bi_2O_3–P_2O_5 glasses exhibit phonon-assisted hopping conduction at high temperatures (200–400 °C) [30]. Potassium oxide (K_2O), an alkali oxide, helps in producing homogeneous melts at low temperature [32] and also helps in improving ionic conductivity. Hence, it was decided to add K_2O in Bi_2O_3–V_2O_5–P_2O_5 glass. A glass composition used to study was $x(Bi_2O_3–K_2O)–y(V_2O_5–P_2O_5)$ (abbreviated as BKVP glass), wherein both x and y vary from 0 to 0.5.[1]

The ionic conductivity of the BKVP glass—GDC composite was found to be 7.6×10^{-3} S cm^{-1} at 600 °C. This conductivity is the highest among the glass-

[1] The glass composition is under patenting process.

ceramic composite electrolytes. The previously reported highest conductivity is 4×10^{-3} S cm^{-1} at 600 °C [33]. This increase in bulk conductivity is mainly due to ionic conducting alkali oxides in the glass [33]. Figure 4 presents effect of glass content on shrinkage and ionic conductivity of the BKVP glass—GDC composite at 600 °C when fired at 1,000 °C. It is observed that increase in glass content decreases shrinkage as higher percentage of glass in molten state cannot agglomerate GDC particles.

It can also be seen from Fig. 4 that the ionic conductivity increases continuously with the glass content and reaches 0.1 S cm^{-1} at 40 wt% addition of BKVP glass significantly increased compared to undoped GDC (3.6×10^{-3} S cm^{-1}) at 600 °C. It is seen that shrinkage and conductivity values change in opposite direction with changing glass content. Based upon the shrinkage requirements, the 25 wt% glass content in GDC having shrinkage close to 9 % and ionic conductivity 0.04 S cm^{-1} at 600 °C was chosen. Ionic transference number for all samples is ≥0.98 which confirms absence of electronic conductivity. This ionic conductivity is 10 times higher than pure GDC at 600 °C and is equivalent to measure at 700 °C operating temperature for this material. Thus, it can be concluded from above discussion that the BKVP glass is useful in increasing both the shrinkage and ionic conductivity values of GDC. The comparison of physical properties for pure GDC and the glass–GDC composite with LTCC is tabulated in Table 2. This table clearly brings out the close match of various properties of glass—GDC composites with LTCC, while at the same time improvements as SOFC electrolyte. Clearly, this composite electrolyte is ready for integration with LTCC

4.5 Conclusions: Materials for Embedded SOFC in LTCC

Integration of SOFC in LTCC is a challenge because of dissimilar materials requirements. In spite of this, the present results of LTCC compatible SOFC materials ignites hopes for such an integration and eventual fabrication of integrated, co-fired SOFC as an embedded power source in LTCC. GDC prepared by chemical route under optimized conditions and sintered at 1,350 °C provided best ionic conductivity (3.6×10^{-3} S cm^{-1} at 600 °C) reported hitherto.

Since glasses are known to improve shrinkage, a glass composition using Bi_2O_3 and other known ionic conductor was prepared for use as sintering aid. The glass—GDC composite electrolyte developed here shows far improved performance compared to the original GDC material. A density of above 90 % and shrinkage (9 %) close to that of LTCC tapes was achieved at reduced sintering temperature of 1,000 °C with improved ionic conductivity, which is 0.04 S cm^{-1} at 600 °C. Addition of glass also brought the thermal expansion coefficient of the composite closer to that of LTCC. Having developed the composite electrolyte material with properties matching with LTCC, the next step is to develop electrode materials and integration of SOFC with LTCC. The cell design and performance study would follow.

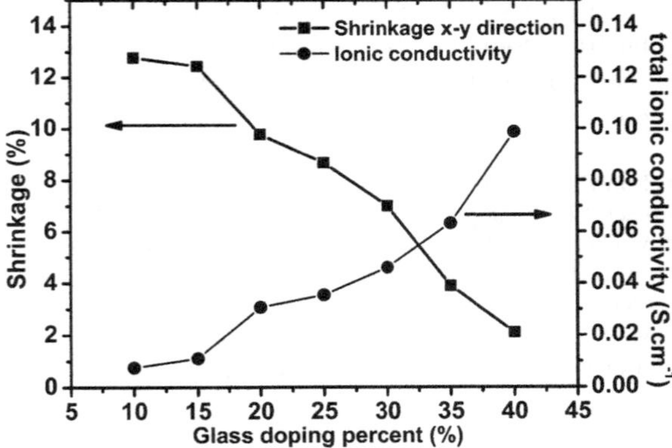

Fig. 4 Effect of glass wt% on shrinkage and ionic conductivity of glass—GDC composite

Table 2 Comparison of properties of pure GDC and BKVP glass—GDC composite

Property	Pure GDC	Glass—GDC composite	Comparison with LTCC
Firing temperature (°C)	1,350	1,000	Close to LTCC firing temperature (950 °C)
Ionic conductivity @ 600 °C (S cm^{-1})	0.014	0.035	Ionic conductivity twice higher than GDC
TCE (0–300 °C) (ppm/°C)	>12	6.8	Close to DuPont 951 LTCC tapes 5.8 ppm/°C
Density (gms/cm^3)	7.24	6.55	Almost 90 % of undoped GDC achieved at 400 °C lower temp
Shrinkage (%)	14 % @ 1,350 °C	9 % @ 950 °C	Shrinkage is less than DuPont 951 tapes but close to pellet prepared by LTCC powder

5 Conclusions

Although passive components can be integrated with LTCC today, low range of values, high tolerance and, many times limited frequency operations spoil these advantages. This calls for improvements in materials properties and, above all, better compatibility with the host LTCC material. This chapter has dealt with the materials aspects for achieving improved properties for embedded capacitors and embedded inductors/transformers. It is also identified that integration of the bulky nonreciprocal devices also needs an integrated solution and provides one through the development of ferrite with specific properties for this application. This work also makes an important breakthrough by developing LTCC compatible SOFC electrolyte that can be co-fired.

While possibility of LTCC compatible materials has been shown here, the work is far from complete. Several compatibility issues are yet to be dealt with for each of these materials and the intended devices have to be realized. This work marks a beginning.

Acknowledgments The authors gratefully acknowledge generous funding from NPSM for the project "Technology Demonstration of Multilayer LTCC Packages for MEMS Application" (PARC#3.2), NPMASS and Department of Electronics and Information Technology (DeitY), Govt. of India, for the project "Development of Advanced Processing Capabilities in LTCC" (PARC#2.14) and NPMASS for the Project "AMC and Spares for the MEMS fab equipment at C-MET, Pune" (PARC#2.6). The authors are extremely thankful to Dr. V. K. Aatre, Former advisor to Raksha Mantri, Govt. of India, for constant support and encouragement.

References

1. Golonka L (2006) Technology and applications of Low Temperature Co-fired Ceramic (LTCC) based sensors and microsystems. Bulletin Of The Polish Academy Of Sciences 54(2):221–231
2. Bo J, Thomas M, Paul M et al (2010) A new platform concept for micro-scale SOFC using low temperature co-fired ceramic technology. In: Proceedings. Power MEMS 2010, 10th international workshop on micro and nanotechnology for power generation and energy conversion applications, Leuven (BE)
3. Vasudev A, Kaushik A, Jones K et al (2013) Prospects of low temperature co-fired ceramic (LTCC) based microfluidic systems for point-of-care biosensing and environmental sensing. Microfluid Nanofluid 14(3–4):683–702
4. Bertinet J, Leleux E, Cazenave J-P et al (2007) Filtering capacitor embeded in LTCC substrate for RF and Mirowave applications substrate. Microwave 50(11):72–88
5. Mi X, Uda S (2010) Integrated passives for high-frequency applications. In: Advanced microwave circuits and systems. INtec Press, Poole, pp 249–290
6. Wahlers RL, Heinz M, Feingold AH (2014) Lead free dielectric and magnetic materials for integrated passives. www.electroscience.com/publications.html. Accessed 2014
7. Modesa C, Malkmusb S, Gora F (2002) High k low loss dielectrics co-fireable with LTCC. Active Passive Electron Compon 25:141–145
8. Higuchi Y, Sugimoto Y, Harada J et al (2007) LTCC system with new high Er and high Q material co-fired with convensional new low Er base material for wireless communication. J Eur Cerm Soc 27:2785–2788
9. Ko WJ, Choi YJ, Park JH et al (2006) Co-firing of low-and high permittivity dielectric tapes for multifunctional low-temperature co-fired ceramics. Ferroelectrics 3(1):193–202
10. Choi Y-J, Park J-H, Ko W-J et al (2006) Co-firing and shrinkage matching in low- and middle- Permittivity dielectric compositions for a low-temperature co-fired ceramics system. J Am Ceram Soc 89(2):562–567
11. Cho YS, Lim WB, Kim Byeong Kon (2009) Low temperature high-k dielectrics for embedded micro circuit system. J Korean Phys Soc 51:S181–S185
12. Randall CA (2003) Bi-pyrochlore and zirconolite dielectrics for integrated passive component applications. Am Ceram Soc Bull 82:9101–9108
13. Jackson M, Pecht M, Lee SB, Sandborn P (2003) Integral, embedded, and buried passive technologies. CALCE, University of Maryland, Maryland
14. Cha SY, Lee WS, Oh KS, Yu JW (2011) Design of a WLAN Antenna with High Isolation Characteristic. In: Proceedings of Microwwve Conference—Asia-Pacific, pp 1830–1833

15. Hahn R, Krumbholz S, Reichl H (2006) Low profile power inductors based on ferromagnetic LTCC technology. In: IEEE proceedings of electronic component and technology conference (ECTC), pp 528–533
16. Karmazin R, Dernovsek O, Ilkov N et al (2005) New LTCC-hexaferrites by using reaction bonded glass ceramics. J Eur Ceram Soc 25:2029–2032
17. Matters-Kammerer M, Mackens U, Reimann K et al (2006) Material properties and RF applications of high k and ferrite LTCC ceramics. Microelectron Reliab 46:134–143
18. Rane VA (2012) Investigations into LTCC compatible ferrite materials for integrated devices. Ph.D. thesis, University of Pune, Pune
19. Li Y, Liu Y, Yuan S, Zhang H et al (2010) The design and fabrication of LTCC Chip inductors for high frequency applications based on barium ferrites. In: IEEE proceedings of 11th international conference on electronic packaging technology. High density packaging pp 906–908
20. Jenson T, Krozer V, Kjærgaard C (2011) Realization of microstrip junction circulator using LTCC technology. Electron Lett 47(2):111–113
21. Rane VA, Meena SS, Gokhale SP et al (2013) Synthesis of low coercive $BaFe_{12}O_{19}$ hexaferrite for microwave applications in low-temperature cofired ceramic. TMS-IEEE J Electron Mater 42(4):761–768
22. Rane VA, Phatak GJ, Date SK (2013) Ultra-high-frequency behavior of $BaFe_{12}O_{19}$ hexaferrite for LTCC substrates. IEEE Trans Magn 49(9):5048–5054
23. Lai TS, Liu J, Barnettz SA (2004) Effect of cell width on segmented-in-series SOFCs. Electrochem Solid-State Lett 7(4):A78–A81
24. Hui S (Rob), Roller J, Yick S et al (2007) A brief review of the ionic conductivity enhancement for selected oxide electrolytes. J Power Sources 172:493–502
25. Zha S, Xia C, Meng G (2003) Effect of Gd (Sm) doping on properties of ceria electrolyte for solid oxide fuel cells. J Power Sources 115:44–48
26. Prasad DH, Son J-W, Kim B-K et al (2010) A significant enhancement in sintering activity of nanocrystalline $Ce_{0.9}Gd_{0.1}O_{1.95}$ powder synthesized by a glycine-nitrate-process. J Ceram Process Res 11(2):176–183
27. Jadhav LD, Chourashiya MG, Subhedar KM et al (2008) Synthesis of nanocrystalline Gd doped ceria by combustion technique. J Alloy Compd 470:383–386
28. Boskovic S, Zec S, Brankovic G et al (2010) Preparation, sintering and electrical properties of nano-grained multidoped ceria. Ceram Int 36:121–127
29. Kulkarni S (2014) Investigations on electrodes and electrolyte materials for the fabrication of integrated low temperature solid oxide fuel cells in low temperature co-fired ceramic structures. Ph.D. thesis, University of Pune, Pune
30. Chakravorty D, Ghosh A (1990) AC conduction in semiconducting $CuO–Bi_2O_3–P_2O_5$ glasses. J Phys: Condens Matter 2:5365–5372
31. Wrobel W, Krok F, Abrahams I, Kozanecka-Szmigiel A et al (2006) $Bi_8V_2O_{17}$—a stable phase in the $Bi_2O_3–V_2O_5$ system. Mater Sci-Pol 24(1):23–30
32. Balaya P, Shrikhande VK, Kothiyal GP et al (2004) Dielectric and conductivity studies on lead silicate glasses having mixed alkali and alkaline earth metal oxides. Curr Sci 86(4):553–556
33. Liu M, Zhou H, Zhu H et al (2012) Microstructure and dielectric properties of Ca–Al–B–Si–O glass/Al2O3 composites with various alkali oxides contents. J Central S Univ 19(10):2733–2739

Smart Materials for Energy Harvesting, Energy Storage, and Energy Efficient Solid-State Electronic Refrigeration

Jayanta Parui, D. Saranya and S. B. Krupanidhi

Abstract The new emerging fields of MEMS-based energy harvesting from piezoelectric materials, lead to the development of solid-state electrostatic energy storage for better power/energy distribution for renewable energy and the solid-state electrocaloric cooling for low energy and hazard free refrigeration. Among them it is being reported that on application of 8.693 $TPas^{-1}$ oscillated stress generates 10 Vs^{-1} oscillated voltage in 300 nm 0.75PMN–0.15PT thin films where 22 Jcc^{-1} s^{-1} of oscillated energy density can be harvested on application of 15 $TPas^{-1}$ oscillated pressure upon 500 nm thin film of same material. It is also described that La modified antiferroelectric $PbZrO_3$ (PZ) thin films are the potential materials that can achieve the high energy density storage density in the order of 103 J/kg. Though PZT-based antiferroelectric cooling triggered the research on the materials for electrocaloric cooling by the amount of 12 K adiabatic decrease in temperature on withdrawal of electric field, the decrease in temperature by 11.4 K in pure PZ and by 31 K in 0.63PMN–0.37PT thin film are found commendable.

Keywords Energy harvesting · Piezoelectric materials · Antiferroelectric cooling · Energy storage · Solid-state refrigeration

1 Introduction

In modern materials science, the objective of material synthesis and processing has a clear and clever objective of miniaturization of smart devices among its other possibilities and successes. In the history of material science, the aspects of material implementation in sensors, actuators, electronic displays, energy harvesting from natural resources and memory devices were realized decades after

J. Parui · D. Saranya · S. B. Krupanidhi (✉)
Materials Research Centre, Indian Institute of Science, Bangalore, India
e-mail: sbkrupanidhi@gmail.com

K. J. Vinoy et al. (eds.), *Micro and Smart Devices and Systems*,
Springer Tracts in Mechanical Engineering, DOI: 10.1007/978-81-322-1913-2_18,

decades [1–8]. In the present era, the material scientists, of course, can take pride of their contribution to the world of technologically rich civilization, where electronic gazettes are no more, the objects of yesterday's sci-fi movie rather it is a complete reality. In this consequence, as we did not abandon the use of fire knowing it is one of the major causes of air pollution, we continue the research of material science addressing the environmental issues pertained to it. There are many direct and indirect environmental issues like, *electronic displays that save papers for which a major source is tree* and *a miniaturized circuit device that consume less energy to save the natural energy*, are already the parts of present day's research programs [1, 9].

Apart from these well-addressed issues, in this chapter we like to cover materials aspect of three new emerging fields of energy management including the materials extensively studied in our group. The fields are: (i) *the development of piezoelectric MEMS devices for energy harvesting from natural mechanical movements* [10], (ii) *the developments of electrostatic capacitors for high energy-high power electronics* [11], and (iii) *the growth of solid-state electronic refrigerators for energy efficient, environment friendly device applications* [12].

2 Present Account on the Fields

The present account of the fields, energy harvesting, energy storage , and effective energy utilization are described in three different topic heads as described below. As the materials and characteristics in our research group they are restricted to the functional oxide materials, antiferroelectric (AFE) and relaxor ferroelectric (FE).

2.1 Energy Harvesting

The consumption of energy is increasing in daily life of the world's population and it demands an urgent attention to utilize the available environmental energy. Though there are different techniques are available for conventional and sustainable energy harvesting, piezoelectric materials are recently found a potential alternative route of electrical energy production converting mechanical energy like footsteps and other effortless mechanical movements. Recently, the advancement of giant energy harvesting by means of 0.375 μm/V in 34 μm length 0.67PMN–0.33PT cantilevers widened the possibilities and utilization of Micro-Electro-Mechanical-Systems (MEMS) based energy harvesting [10].

Presently, we have integrated 0.75PMN–0.15PT on Si wafer at the stage where the thickness of the SiO_2 layer can no longer be decreased. Such stage was deliberately chosen to restrict the high leakage current that is detrimental to efficiency of the device. Before directly going for the device patterning on SiO2 layer, measurement of d_{33} value of 0.75PMN–0.15PT on LSCO/(111) Pt/TiO2/SiO2/Si

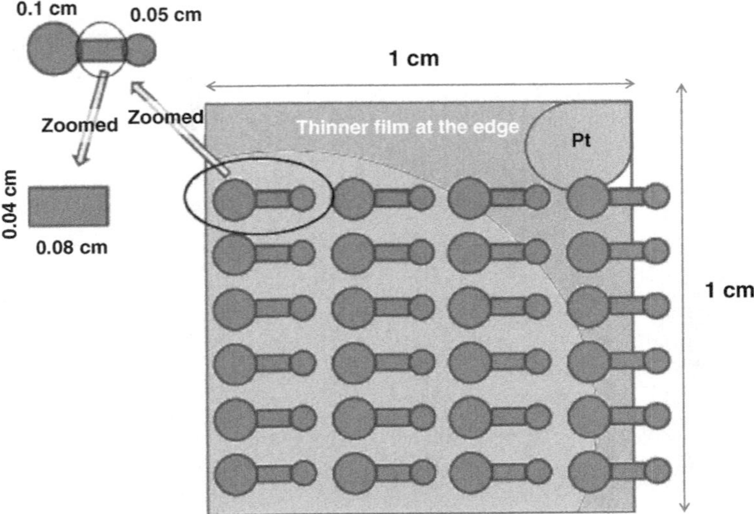

Fig. 1 Schematic diagram of electrode patterning on PLD grown 0.75PMN–0.15PT thin films for d_{33} measurement

substrate were cross-checked to confirm its reported values. The d_{33} values for different thicknesses have been measured using a Micro System Analyzer, Polytec MSA −500 by their successive analysis for energy harvesting capacity. The measurement technique is not like the conventional way of measuring, but it is measured in an indirect way using the relation

$$d_{33} = {}^{\Delta t}/_{V_{\mathrm{ac}}}. \qquad (1)$$

where Δt is the amplitude of mechanical vibration of the thin film in pm. The entire device configuration consists of the silicon substrate with Pt as the bottom electrode with LSCO buffer layer and the top gold electrode is deposited by lithography. The schematic of patterned dumb-bell shaped electrodes are shown in Fig. 1 and their digital photographs are presented in Fig. 2. Such structures are patterned for the measurement of d_{33} and they are called the Piezoelectric Micromachined Ultrasonic Transducers (pMUTs). So the photographs in Fig. 2 are also called the still photographs of pMUTs.

A schematic diagram of cantilever to be realized for the mentioned device structure in Fig. 1 is presented in Fig. 3 inset. The measured maximum Δt for 500 nm thick film is around 330 pm and respective deflections at different applied ac volt have been shown in Fig. 3. The zero-biased d_{33} values for 500 nm thick 0.75PMN–0.15PT film at different oscillation voltage with respect to frequencies are shown in Fig. 4a.

It has been seen that highest d_{33} value of 72 pm/V was observed on application 1 V of oscillating voltage at 1 kHz of frequency. Then d_{33} values at 1 kHz are

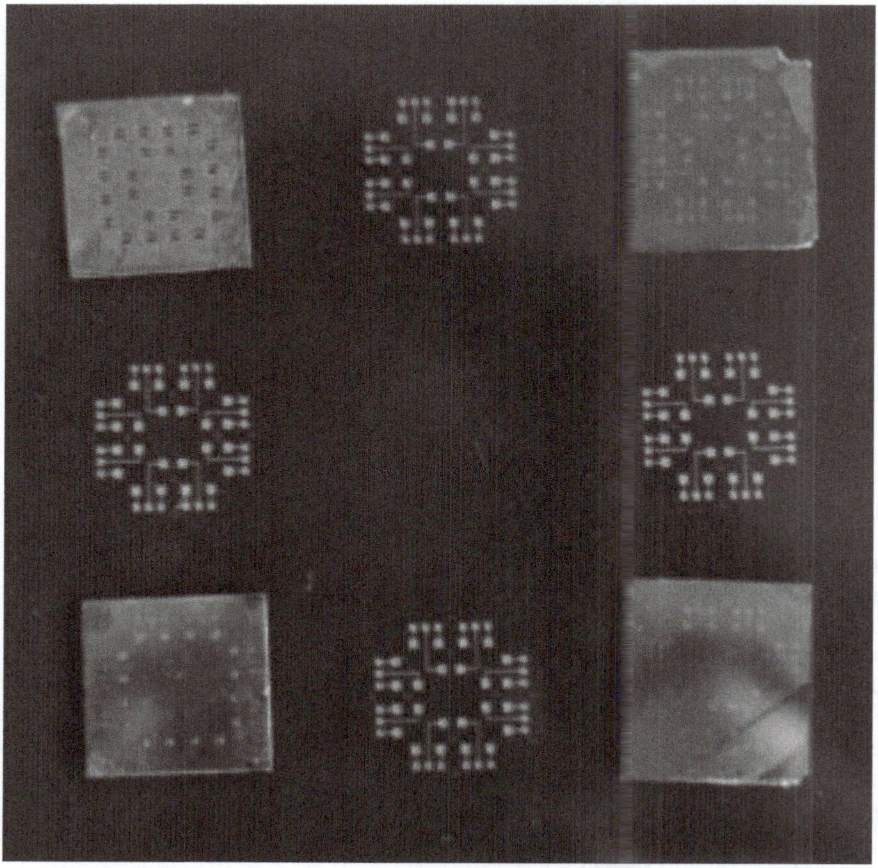

Fig. 2 0.85PMN–0.15PT thin film devices fabricated for pMUT

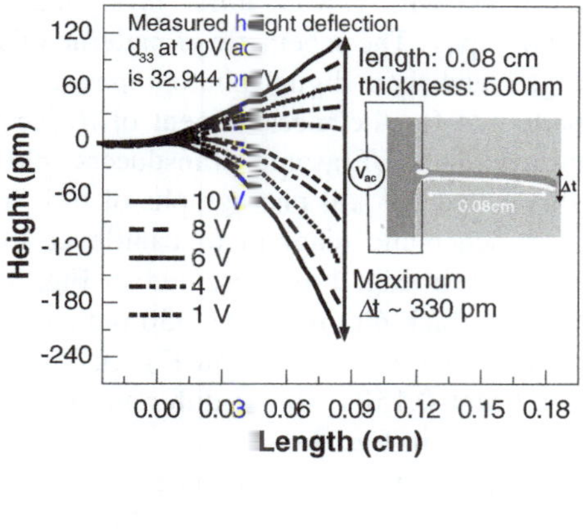

Fig. 3 Cantilever deflection on applied different ac voltage has been described with respect to length and change in height of the tip (*inset*). A schematic cantilever representation of the device

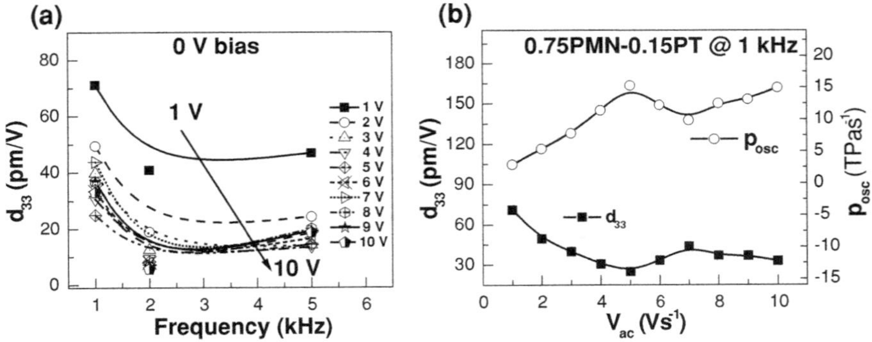

Fig. 4 **a** d_{33} measured at different oscillation voltage at frequencies 1, 2 and 5 kHz at 0 V bias and **b** d_{33} at different oscillation voltages at 1 kHz at 0 V bias along with calculated pressure or stress at the right hand y-axis

plotted against the different oscillating voltage in Fig. 4b. Here the stresses in TPas^{-1} signify that application of such oscillating stress (p_{osc}) is expected to produce corresponding voltages. Among them it is clearly being seen that on application of 15 TPas^{-1} stress maximum production of 10 Vs^{-1} is possible (Fig. 4b). Hysteresis loop (not shown) at 1 kHz, the related stresses were calculated by the generalized equation of

$$d = \frac{P}{\sigma}. \tag{2}$$

where P is the electric field driven polarization and σ is the applied stress or applied pressure (p_{osc}) on the material.

Correlating Eqs. (1) and (2), as we get a relation

$$d = \frac{P}{p_{\mathrm{osc}}} \tag{3}$$

and if we equate Eqs. (1) and (3) we get,

$$\begin{aligned} \frac{P}{p_{\mathrm{osc}}} &= \frac{\Delta t}{V_{\mathrm{ac}}} \\ \Rightarrow p_{\mathrm{osc}}\Delta t &= PV_{\mathrm{ac}}. \end{aligned} \tag{4}$$

Moreover, we know $P = \frac{C}{A}$ where C is coulomb charge and A is the area of the electrode, from Eq. (4) we get,

$$\begin{aligned} p_{\mathrm{osc}}\Delta t &= \frac{CV_{\mathrm{ac}}}{A} \\ \therefore p_{\mathrm{osc}}\Delta tA &= CV_{\mathrm{ac}} \end{aligned} \tag{5}$$

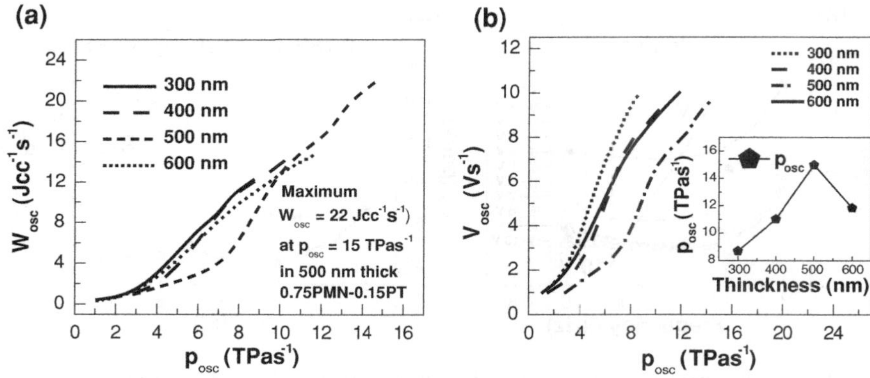

Fig. 5 The oscillated energy density W_{osc} generated from p_{osc} signifies that a mechanical movement which can create an oscillated pressure on a piezoelectric material can generate energy density pulse as oscillated energy density

Hence, as we measured d_{33} values at different frequencies and different oscillation voltage, we can easily calculate the expected energy production from mechanical movement or oscillating pressure. For the simplicity, we will present oscillated energy density W_{osc} obtained from p_{osc} and for that the Eq. (5) is rearranged as:

$$W_{osc} = \frac{CV_{ac}}{\Delta tA}. \quad (6)$$

In Fig. 5a, W_{osc} obtained from p_{osc} have been presented and the observed maximum energy density is 22 $Jcc^{-1}\ s^{-1}$ can be obtained from a 500 nm thick 0.75PMN–0.15PT film on an application of 15 $TPas^{-1}$ oscillated pressure on it at the rate of 1 kHz without any bias voltage. From the analysis of thickness-dependent oscillated energy density, it is noticed that 500 nm thick film produced maximum of it among them whereas it has been seen that to produce 10 Vs^{-1} ac voltage or oscillated voltage 300 nm thick film of the same material requires minimum p_{osc} of 8.693 $TPas^{-1}$ (Fig. 5 inset).

2.2 Energy Storage

Apart from energy by means of mechanical energy, nowadays conventional power sources like coal, gas, hydro and nuclear powers getting combined with renewable energy sources realizing that their limited source in nature. Hence, the conventional power sources getting combined with the major renewable sources like wind and solar power. Irrespective of the power sources, electrical energy storage systems (EESSs) are a necessity to integrate with the sources for temporally and

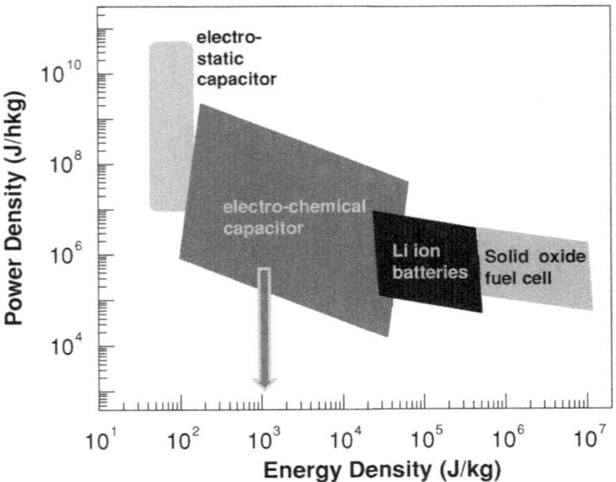

Fig. 6 A modified Ragone plot in terms of Joules after Sherrill et al. [11]. The *arrow sign* indicates the achieved energy density in the regime of electrochemical capacitor

spatially distributed load applications [11]. Though in the realm of conventional power supply fluctuation within 24 h per day can be minimized, the wind and solar power depends on natural conditions like clouds and storm. So it is obvious that the climate-dependent power supply is inefficient for the steady power supply and it has been found that so far developed materials for EESSs are not much effective when the demand of fast rate supply is high, normally during the evening hours. To visualize the type of potential materials in a modified Ragone plot has been described in the Fig. 6 in terms of Joules.

From the plot it is obvious that the synthesis of new electrostatic capacitor with high energy density (J/kg) is in high demand realizing their natural characteristics for high power density (J/hkg) electronics. Recognizing the crisis of high energy density electrostatic capacitor to be developed, we initiated to synthesize AFE $PbZrO_3$ (PZ)-based thin film materials for their application in miniaturized high energy power electronics. Though it was predicted by Chen et al. [13] that there is a possibility of $\sim$50 J/cc (6.2 kJ/kg) charging and discharging during the electric field driven phase switching of PZ, we were able to synthesize highest 14.9 J/cc (2 kJ/kg) recoverable energy density in 5 at.% La modified PZ thin film on withdrawal of 60 MV/m [14]. The variation of recoverable or dischargeable energy (W) of La modified PZ on withdrawal of 60 MV/m has been presented in Fig. 7.

From the Fig. 7, we can see there is a potential energy density storage is possible in AFE electrostatic capacitor and the value 2 kJ/kg (i.e., 14.9 J/cc) for 5 at.% La modification is recognized within the energy density regime of electrochemical capacitor (Fig. 6).

Though it is well known for the electrostatic capacitor AFEs are more capable of energy storage for a given set of polarization and electric field (Fig. 8) [15], having higher polarizability in FE materials improvement of charge density is practiced in relaxor FE for their lesser hysteresis loss. In our laboratory, we tried to improve such energy density in 0.85PMN–0.15PT thin films and we tabulated the obtained recoverable energy densities in Table 1. We found maximum 10 J/cc

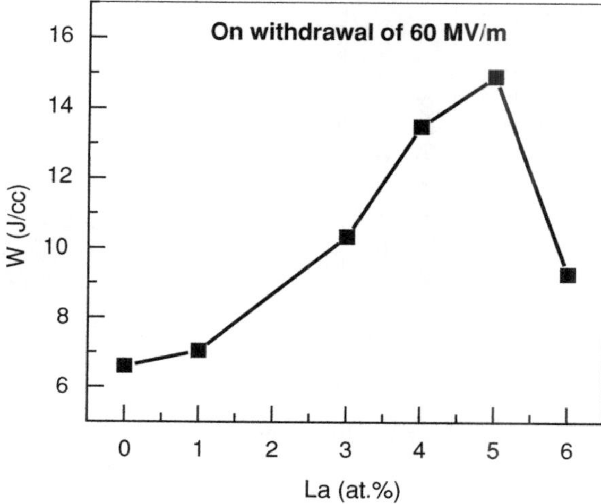

Fig. 7 Change in recoverable energy (*W*) with respect to La modification in $PbZrO_3$ on withdrawal of 60 MV/m [14]

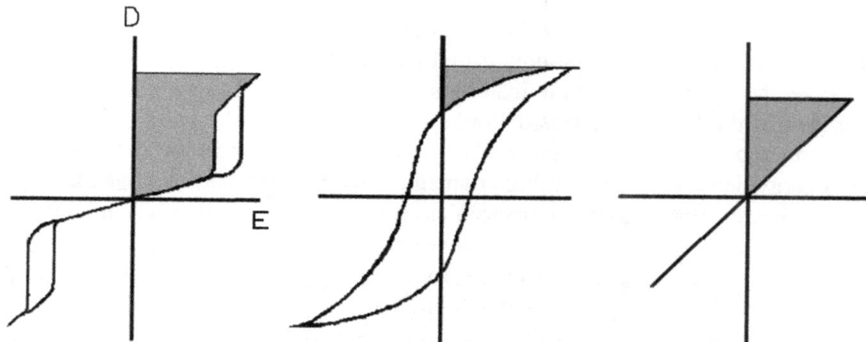

Fig. 8 The *shaded area* in antiferroelectric, ferroelectric and paraelectric electrostatic capacitor showed the recoverable energy density for a given set of polarization (*D*) and electric field (*E*)

recoverable energy density was obtained on the withdrawal of 60 MV/m. The general formula to calculate recoverable energy density from the hysteresis loops is given below,

$$W = \int_{P_0}^{P_{\max}} E \, \mathrm{d}P \qquad (7)$$

Table 1 Recoverable energy densities of 0.85PMN–0.15PT thin films at different thicknesses and applied voltages

Thickness/Voltage (nm)	5 V (J/cc)	7 V (J/cc)	10 V (J/cc)	12 V (J/cc)	18 V (J/cc)
300	0.04	2	4	5	10
400	1	2	3	4	9
500	0.8	1	2	3	7
600	0.5	0.9	1	2	5

where W = recoverable energy on withdrawal of electric field; P = polarization or charge storage density; E = applied electric field; P_{max} = maximum polarization and P_0 = polarization at zero electric field.

2.3 Solid-State Refrigeration

When refrigeration for cooling purpose is a day-to-day activity of modern era, there are subtle issues have not yet been recognized in their full concern. The energy crisis, which is already discussed in electrostatic energy storage, is nowadays a prime issue of any modern day technology and here we hardly account the energy consumption for air conditioning devices. These devices not only consume a giant amount of energy but also liberate heat to the environment and that can be a real added problem to the issue of global warming. Apart from air-conditioner, the house refrigerator is still a compressor-driven cooling device where recyclability of hazardous coolant is in general ignored and such ignorance in the past provided us a dangerous and nonrepairable ozone layer depletion [12]. Though there are other means of cooling, recently electrocaloric (EC) effect has been found a potential means of electrical cooling [16]. Though the EC performance was known since 1930 in Rochelle salt [17], recently, Mischenko et al. [16] demonstrated the possible decrease in temperature by 12 K in AFE PZT by withdrawal of 48 MV/m. Such a giant EC effect triggered a new wave of interest in EC materials. In search of such materials, the EC experiment on PZ showed a maximum EC effect of 11.4 K on withdrawal of 40 MV/m [18]. Similar attempt to 0.63PMN–0.37PT thin film we found even better EC effect of 31 K on withdrawal of 75 MV/m [19].

The principle of EC effect is similar to vapor cycle refrigeration that consists of four steps. In the first step, the vapor is compressed adiabatically, where the temperature of the vapor increases (say by ΔT_{ad}) as shown in the Fig. 4. In the second step the temperature is decreased to initial temperature (T) through heat release ($-Q$). At the third step, the pressure is released to decrease the temperature by (ΔT_{ad}). In the fourth step, the temperature is increased to the initial temperature (T) through heat absorption ($+Q$). Similarly, in electrocaloric effect electric magnetic field (E) is applied instead of pressure (P) in vapor cycle refrigeration. In AFE materials the electrocaloric effect is applicable only at the electric field greater than E_{FE-AFE} from where the dipoles can be aligned accordingly (Fig. 9).

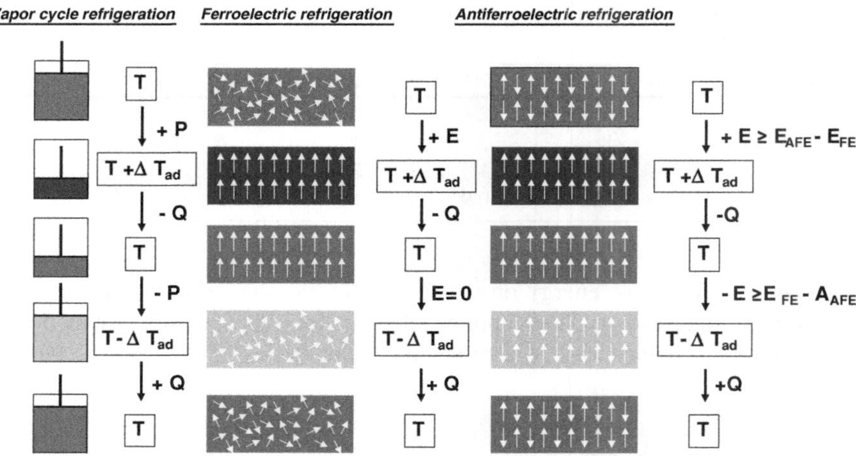

Fig. 9 Schematic presentation of vapor cycle refrigeration process, electrocaloric effect in ferroelectric materials and in antiferroelectric materials

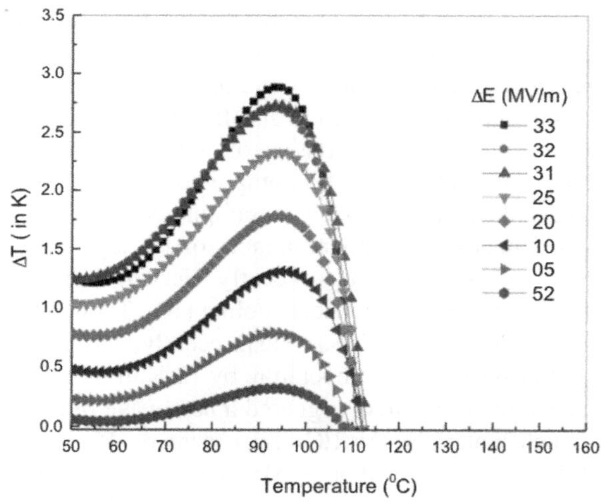

Fig. 10 Reversible adiabatic Temperature (ΔT) as a function of temperature (T) obtained at 2 kHz frequency with a maximum applied electric of 66 MV/m field for 300 nm thin film deposited on (111) Pt/TiO2/SiO2/Si substrate

In FE materials the electric field aligns the FE dipoles from their random orientation, whereas the electric field in AFEs aligns FE dipoles from antiparallel orientation.

The measured EC effect in terms of the reversible adiabatic temperature change ΔT at various $E2-E1 = \Delta E$ values have been plotted against the temperature in Fig. 10 with a noticeable maximum value of ~3.0 K at 95 °C for 400 nm film at 1 KHz on withdrawal of 50 MV/m electric field. This high value of ΔT leads to an EC effect of $\sim 0.1141 \times 10^{-6}$ Km/MV. The ΔT values were measured from the thermodynamically derived working formula of

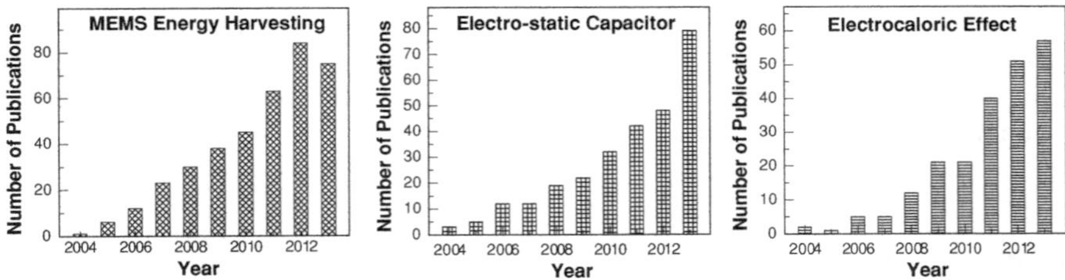

Fig. 11 The records of increasing publication from 2004 to 2013 in the field of electrocaloric effect and electrostatic capacitor

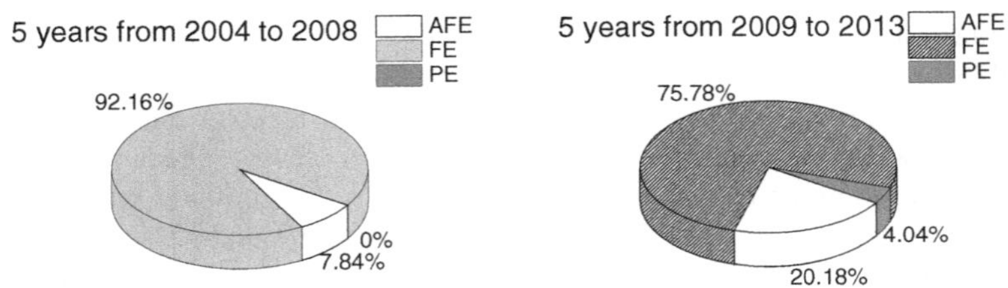

Fig. 12 3D pie chart for publication record of electrostatic capacitors distributed in three major materials, AFE, FE and PE

$$\Delta T = -\frac{1}{\rho}\int_{E_1}^{E_2} \frac{T}{c}\left(\frac{\partial P}{\partial T}\right)_E \mathrm{d}E \tag{8}$$

where ρ = the mass density of the material, c = molar heat capacity, E_2 = higher electric field, E_1 = lower electric field, P = polarization, T = absolute temperature, and E = applied electric field.

3 The Progress So Far

In the last 10 years (2004–2013) the chronicle of research in the filed of MEMS energy harvesting, electrostatic energy density and electrocaloric effect showed a sizable achievement in the scientific community. Year after year, the increasing publications (Fig. 11) in the fields are the reflection of its importance and necessity for the electronics for sustainable energy storage and solid-state electronic cooling.

Moreover, research field of electrostatic energy density is dependent on three major type materials, AFE, FE, and paraelectric (PE). Though the characteristic feature of energy storage density in AFE is better than the rest of two, the pie charts in Fig. 12 showed the research activities on AFE and PE materials are slowly taking pace within the last 10 years of publication record.

In electrostatic capacitor research, contribution of three in potential materials such as PZ, various materials has been tabulated in terms to be included the highest achievements of electrostatic energy storage and electrocaloric effects, material wise and our data in support of the achievements.

4 Summary

The device structures fabrication by lithographic techniques to form Piezoelectric Micromachined Ultrasonic Transducer (pMUT) is a technique to measure the piezoelectric coefficient of a material in its thin film form. The pMUT structure formation was found efficient to analyze the possibility of energy harvesting capability of a piezoelectric material in its form of MEMS. The energy harvesting from mechanical movements was noticed possible in relaxor FE material. The energy harvesting through MEMS device using piezoelectric effect of 0.75PMN–0.15PT has a commendable output of 22 $\text{Jcc}^{-1}\ \text{s}^{-1}$ of oscillated energy density harvested on application of 15 TPas^{-1} oscillated pressure upon 500 nm thin film, where in the same material of 300 nm thickness can provide 10 Vs^{-1} oscillated voltage on the application of 8.693 TPas^{-1}. Successively, electrostatic energy storage application in AFE, FE and PE materials are analyzed and the energy storage density of 2 kJ/kg (i.e., 14.9 J/cc) in 5 at.% La modification is recognized within the energy density regime of electrochemical capacitor. Such a high energy storage density is very important for high power—high energy device application and energy distribution. For energy efficient and environment hazard free solid-state cooling devices are found an new emerging field where the EC effect in PZ showed a maximum value of 11.4 K on withdrawal of 40 MV/m. Similar attempt in 0.63PMN–0.37PT thin film showed even better EC effect of 31 K on withdrawal of 75 MV/m which is recognized as a promising materials for solid-state refrigeration. The statistical data on the publications in each of the described field shows increasing research activities as we step toward future. This data emphasizes the importance and the demand of energy harvesting, energy storage and energy efficient, environmental friendly electrical refrigeration. The statistical data on increasing publication of AFE and PE materials for electrostatic energy storage establishes their effective energy storage density is realized.

Acknowledgments The research reported in this chapter was supported in part by National Programme on Smart Materials (NPSM).

References

1. Muralt P (2000) Ferroelectric thin films for micro-sensors and actuators: a review. J Micromech Microeng 10:136–146
2. Shahinpoory M, Bar-Cohenz Y, Simpsonx JO, Smith J (1998) Ionic polymer-metal composites (IPMCs) as biomimetic sensors, actuators and artificial muscles—a review. Smart Mater Struct 7:R15–R30
3. Chabinyc ML, Salleo A (2004) Materials requirements and fabrication of active matrix arrays of organic thin-film transistors for displays. Chem Mater 16:4509–4521
4. Fortunato E, Barquinha P, Martins R (2012) Oxide semiconductor thin-film transistors: a review of recent advances. Adv Mater 24:2945–2986
5. Atwater HA, Polman A (2010) Plasmonics for improved photovoltaic devices. Nat Mater 9:205
6. Shah A, Torres P, Tscharner R, Wyrsch N, Keppner H (1999) Photovoltaic technology: the case for thin-film solar cells. Science 285:30
7. Tehrani S, Slaughter JM, Deherrera M, Engel BN, Rizzo ND, Salter J, Durlam M, Dave RW, Janesky J, Butcher B, Smith K, Grynkewich G (2003) Magnetoresistive random access memory using magnetic tunnel junctions. Proc IEEE 91(5):703
8. Murali B, Krupanidhi SB (2014) Transport properties of CuIn1–xAlxSe2/AlZnO heterostructure for low cost thin film photovoltaics. Dalton Trans 43:1974–1983
9. Xiao L, Chen Z, Qu B, Luo J, Kong S, Gong Q, Kido J (2011) Recent progresses on materials for electrophosphorescent organic light-emitting devices. Adv Mater 23:926–952
10. Baek SH, Park J, Kim DM, Aksyuk VA, Das RR, Bu SD, Felker DA, Lettieri J, Vaithyanathan V, Bharadwaja SSN, Bassiri-Gharb N, Chen YB, Sun HP, Folkman CM, Jang HW, Kreft DJ, Streiffer SK, Ramesh R, Pan XQ, Trolier-McKinstry S, Schlom DG, Rzchowski MS, Blick RH, Eom CB (2011) Giant piezoelectricity on Si for hyperactive MEMS. Science 334:958
11. Sherrill SA, Banerjee P, Rubloff GW, Lee SB (2011) High to ultra-high power electrical energy storage. Phys Chem Chem Phys 13:20714–20723
12. Valant M (2012) Electrocaloric materials for future solid-state refrigeration technologies. Prog Mater Sci 57:980–1009
13. Chen N, Bai GR, Auciello O, Koritala RE, Lanagan MT (1999) Properties and orientation of antiferroelectric lead zirconate thin films grown by MOCVD. In: Material research society symposia proceedings, vol 541. Pittsburgh, pp 345–350
14. Parui J, Krupanidhi SB (2008) Enhancement of charge and energy storage in sol-gel derived pure and La-modified PbZrO3 thin films. Appl Phys Lett 92:192901
15. Jaffe B (1961) Antiferroelectric ceramics with field-enforced transitions: a new nonlinear circuit element. Proc IRE 49:1264
16. Mischenko AS, Zhang Q, Scott JF, Whatmore RW, Mathur ND (2006) Giant electrocaloric effect in thin-film PbZr0.95Ti0.05O3. Science 311:1270
17. Kobeco P, Kurtchatov IV (1930) Dielectric properties of Rochelle salt crystal. Z Phys 66:192–205
18. Parui J, Krupanidhi SB (2008) Electrocaloric effect in antiferroelectric PbZrO3 thin films. Phys Status Solidi (RRL) 2(5):230–232
19. Saranya D, Chaudhuri AR, Parui J, Krupanidhi SB (2009) Electrocaloric effect of PMN–PT thin films near morphotropic phase boundary. Bull Mater Sci 32(3):259–262

Part IV
Modeling and Simulation

Vibratory MEMS and Squeeze Film Effects

Rudra Pratap and Anish Roychowdhury

Abstract Vibratory MEMS devices typically consist of a plate like structure that oscillates out of plane and thus generally normal to a fixed substrate. These structures have a trapped air-film between the structure and the fixed substrate either by design or by partial vacuum packing. The trapped air behaves like a *squeeze film* offering both stiffness and damping to the structure. A good estimate of squeeze film forces is important for predicting the dynamic performance of such devices. In this chapter, we discuss the development of squeeze film modeling, going back to its origins in the lubrication theory. We explain the derivation of Reynolds' equation that governs the squeeze film flow and review some of the major analytical modeling techniques. A detailed analytical solution using the Eigen expansion technique that incorporates the effect of structural elasticity using approximate mode shapes is presented. We also discuss finite element based simulations for finding the squeeze film forces. We present a simple example analysed using an implementation of the finite element formulation discussed. We also summarize the experimental studies reported in the literature that have shaped both analytical and numerical models.

Keywords Squeezed film effects · Isothermal Reynolds' equation · Squeeze number · Quality factor

1 Introduction

Microelectromechanical systems (MEMS) devices mostly employ planar micro-mechanical structures as basic sensing elements. These structures are basically made up of beams, plates, membranes, or some combined arrangements of these

R. Pratap (✉) · A. Roychowdhury
Center for Nano Science and Engineering and Department of Mechanical Engineering, Indian Institute of Science, Bangalore, India
e-mail: pratap@mecheng.iisc.ernet.in

K. J. Vinoy et al. (eds.), *Micro and Smart Devices and Systems*,
Springer Tracts in Mechanical Engineering, DOI: 10.1007/978-81-322-1913-2_19,

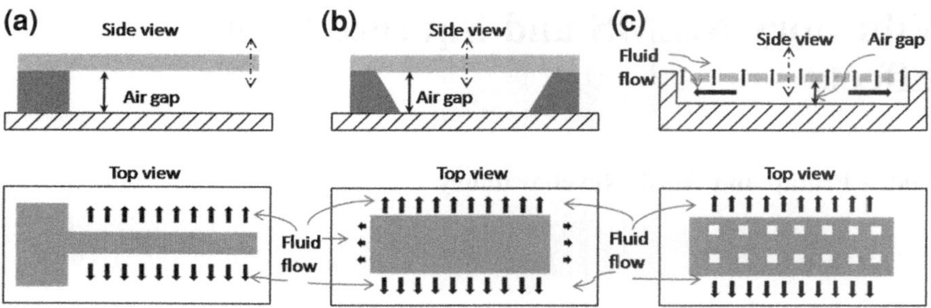

Fig. 1 Squeeze flow: **a** 1D flow of air **b** 2D flow of air **c** 3D flow of air

simple structural elements. The mechanics of these structures, therefore, is expected to be simple and intuitive. Unfortunately, at very small scales, there are many forces and phenomena that elude intuition. The subject matter of this chapter belongs to the nonintuitive type. When a very thin "film" of air—say, one micron in thickness—is pressed periodically between two sheets of solids, what kind of forces does the air film apply on the solid sheets? How does device dynamics change due to the pulsating air film? Is squeeze film induced damping the dominant energy dissipation method at all vibration frequencies? These are some of the relevant mechanics questions that must be answered if we want to ascertain the dynamic response of the vibrating solid sheets. It turns out that these questions are of central interest to all MEMS devices that use vibration of their micromechanical structures for their basic operation—sensing or actuation—and confine air or some other fluid in narrow spaces around them. As long as the surrounding fluid is in a tight confinement, it will be "squeezed" in some manner when the structure vibrates and the "squeezed film" will respond to the motion of the structure. The study of such response is the focus of this chapter.

Micromechanical structures used in MEMS devices, particularly those that are fabricated using surface micromachining, typically have a very thin cavity or air gap below the structure that is created by the sacrificial etch of some material, e.g., silicon dioxide. The thickness of these cavities range from submicron to a few microns. The overhead structure that gets suspended as a result could be as simple as a beam, a plate or a complex 2D structure with elements of 3D geometry such as etch-holes or tall side-walls as shown schematically in Fig. 1. When the suspended structure vibrates, the air in the cavity is pushed out by the structure during its downward stroke and pulled back in due to low pressure during the upward stroke. This *in* and *out* motion of the air dissipates energy and this dissipation is termed as *squeeze film damping*. Along with the damping due to the sideways motion of the air, the trapped air also acts as a spring because of its compressibility and the resulting spring force gives rise to *squeeze film stiffness*.

Both the damping and stiffness effects are dependent on the frequency of oscillation of the structure as well as the ambient pressure. MEMS devices use a fairly large range of frequencies, operating pressure conditions, and air gaps.

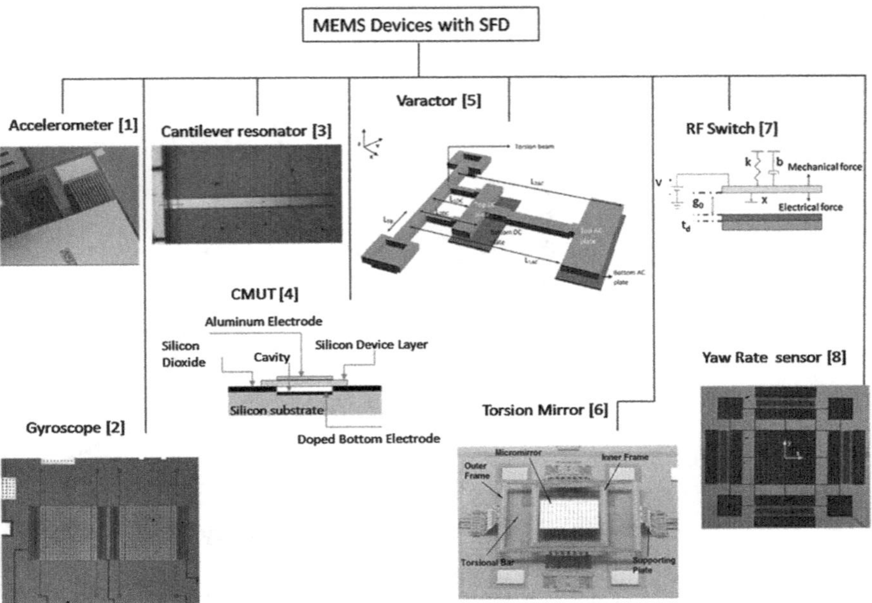

Fig. 2 MEMS devices with squeeze film damping

Therefore, the squeeze film effects vary considerably across devices and applications. Some of the MEMS devices [1–8] with a significant amount of squeeze film effects are shown in Fig. 2. The simplest way to determine the squeeze film forces is to find the steady state pressure field in the cavity from the flow conditions of the air, calculate the pressure of the air on the vibrating structure and separate the total force into a *spring force* that is in phase with the displacement, and a *damping force* that is in phase with the velocity. This is the basic calculation. The details vary according to the geometry of the structure and the cavity, the complexity of flow (1D, 2D, or 3D), the boundary conditions, the ambient pressure and the frequency of oscillation. Once the squeeze film forces are determined, the dynamic response of the structure can be determined and the effect of the squeeze film on the resonant frequency of the structure as well its quality factor, or the Q-factor, can be found.

The Q-factor (also called just Q), which is a measure of the sharpness of the resonance peak in the frequency response of a device (see Fig. 3), is directly related to the overall damping in the device. Different damping mechanisms [9] exist in MEMS devices such as thermo-elastic damping, support losses and air flow losses. For most vibratory MEMS devices fabricated from high-Q materials such as Si and operated in ambient conditions, squeeze film damping is the most dominant damping mechanism [10].

Several mathematical models exist for computing the squeeze film damping and stiffness for different geometric and flow conditions. Traditionally, in lubrication

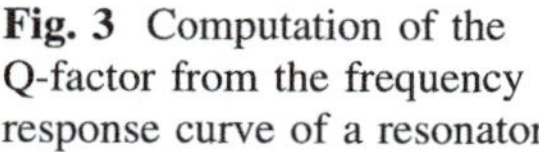

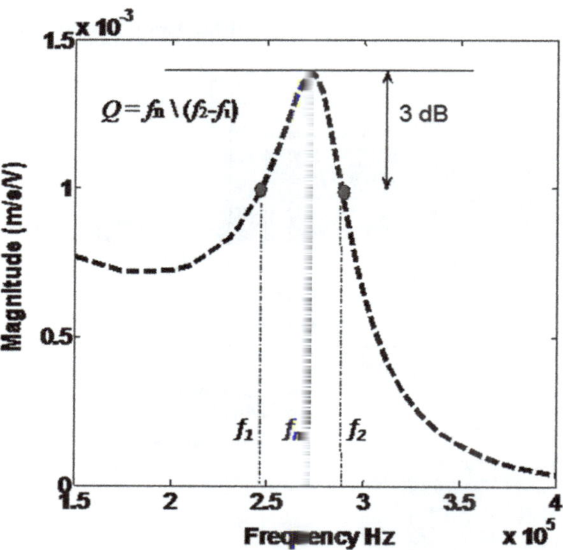

Fig. 3 Computation of the Q-factor from the frequency response curve of a resonator

theory, the squeeze film domain is modeled using Reynolds' equation [11, 12]. Blech [11] considered trivial pressure boundary conditions for a rigid plate and analytically studied the squeeze film-induced stiffness and damping using the method of series expansion. Darling [13] presented analytical solutions to the linearized Reynolds' equation for various venting conditions using the Greens function approach. For flexible structures one has to account for the variable air-gap, and the elasticity equation has to be coupled with the Reynolds' equation for accurate modeling. Closed-form analytical solutions are difficult for such problems. Various studies have approached the coupled elasticity and fluid interaction problem using semi-analytical and numerical techniques. Hung et al. [14] presented a reduced-order macromodel based on basis functions generated from finite difference simulations to model a pressure sensor. McCarthy et al. [15] studied cantilever micro-switches using a transient finite difference method. They assumed a parabolic pressure distribution along the length and invariant pressure along the width, and obtained good agreement with experimental measurements. Younis et al. [16] used perturbation methods to derive analytical expressions for pressure distribution in an electrically actuated micro-plate. Pandey et al. [17] studied the effect of flexural mode shapes on the squeeze film stiffness and damping for a cantilever resonator using a Green's function to solve the linearized compressible Reynolds' equation. They used the modal projection method available in ANSYS (www.ansys.com) to solve the coupled fluid structure problem for several flexural modes of vibration.

Here, we present a simple yet elegant analytical solution methodology using the Eigen-expansion method, and discuss a finite element scheme to numerically solve Reynolds' equation. Readers interested in the solution techniques for the fully coupled problem may refer to the work by Roychowdhury et al. [18] where a coupled solution methodology was presented to solve the elasticity equation along with Reynolds' equation in a truly coupled manner using a single step finite element formulation.

2 Mathematical Modeling

2.1 Historical Note

The origins of squeeze film analysis can be traced to the work done in the field of lubrication theory. Researchers were interested in the mechanics of thin films between rotating or squeezing shafts and journal or thrust bearings. The problem was investigated intensively in 1960s. Several researchers developed good mathematical models for squeeze film damping [11, 12]. The following two decades saw limited progress in this field. With the advent of MEMS devices in the late 1980s, the subject drew considerable attention from MEMS designers because squeeze film damping affected the Q-factor of MEMS devices. Now, several mathematical models exist for computing squeeze film forces. The governing fluid flow equations are either a linearized Reynolds' equation in continuum regime [11, 12] or a modified Reynolds' equation derived using considering slip velocities at the walls as proposed by Burgdorfer [19]. Here, we focus on modeling squeeze film flow in the continuum regime using the linearized Reynolds' equation.

2.2 Governing Equation

The governing fluid flow equations in the continuum regime are the continuity equation and the Navier Stokes momentum equations. The standard form of Navier Stokes equation in the ith direction, assuming constant viscosity, is

$$\rho \frac{\mathrm{D}u_i}{\mathrm{D}t} = F_i - \frac{\partial p}{\partial x_i} + \mu \left[\nabla^2 u_i + \frac{1}{3} \frac{\partial(\nabla.\mathbf{u})}{\partial x_i} \right] \tag{1}$$

where ρ and μ are the fluid density and viscosity, respectively, P denotes the pressure, $\boldsymbol{F_i}$ the body force per unit volume in the ith direction, $\mathrm{D}u_i/\mathrm{D}t$ the total derivative, $\partial u_i/\partial t$ the local derivative and $\mathbf{u} = u\mathbf{i} + v\mathbf{j} + w\mathbf{k}$ the fluid velocity. The dilatation term $(1/3)(\partial(\nabla.u)/\partial x_i)$ is a measure of fluid expansion and is considered to be zero when the density is assumed to be constant. The continuity equation is given as

$$\frac{\partial \rho}{\partial t} + \nabla.(\rho u) = 0 \tag{2}$$

For a flow problem depicted in Fig. 1b for a small air-gap and transverse oscillation of the plate, the flow is predominantly 2D (i.e., in the *x*, *y*-plane). Generally, we make the following assumptions:

- No external forces on the film, i.e. (F = 0),
- Small amplitude oscillations,
- Flow driven by the pressure gradient in *x* and *y* directions,
- No pressure variation across the thickness of the fluid film,
- Fully developed and laminar flow, and
- Isothermal flow, i.e., $P \alpha \rho$.

With these assumptions the Navier Stokes equation reduces to

$$\rho\left(\frac{\partial u}{\partial t}+u\frac{\partial u}{\partial x}+v\frac{\partial u}{\partial y}\right)=-\frac{\partial p}{\partial x}+\frac{\partial}{\partial z}\left(\mu\frac{\partial u}{\partial z}\right)$$
and
$$\rho\left(\frac{\partial v}{\partial t}+u\frac{\partial v}{\partial x}+v\frac{\partial v}{\partial y}\right)=-\frac{\partial p}{\partial y}+\frac{\partial}{\partial z}\left(\mu\frac{\partial v}{\partial z}\right) \tag{3}$$

Further, considering a small air-gap compared to the lateral dimensions, and the velocities u and v to be small, the convective inertial terms, $u(\partial v/\partial x), v(\partial v/\partial y), u(\partial u/\partial x)$ and $v(\partial u/\partial y)$ may be ignored. For steady flow, the unsteady inertial terms may be dropped, consequently, Eq. (3) reduces to

$$\frac{\partial p}{\partial x}=\frac{\partial}{\partial z}\left(\mu\frac{\partial u}{\partial z}\right) \quad \text{and} \quad \frac{\partial p}{\partial y}=\frac{\partial}{\partial z}\left(\mu\frac{\partial v}{\partial z}\right) \tag{4}$$

On solving Eq. (4) for the velocities u and v with no-slip boundary conditions $(u_{\pm h/2}=0, v_{\pm h/2}=0)$, we get the following expression:

$$u=\frac{1}{2\mu}\frac{\partial p}{\partial x}\left(z^2-\frac{h^2}{4}\right) \quad \text{and} \quad v=\frac{1}{2\mu}\frac{\partial p}{\partial y}\left(z^2-\frac{h^2}{4}\right) \tag{5}$$

The average radial velocities $\tilde{u}, \tilde{v}$ are then found by integrating Eq. (5) across the height of the air- gap as

$$\tilde{u}=-\frac{h^2}{12\mu}\frac{\partial p}{\partial x} \quad \text{and} \quad \tilde{v}=-\frac{h^2}{12\mu}\frac{\partial p}{\partial y} \tag{6}$$

Now, we integrate the continuity equation (Eq. 2) across the air-gap from $-h/2$ to $+h/2$ to get

$$h\frac{\partial\rho}{\partial t} + \frac{\partial}{\partial x}\left[\rho\int_{-h/2}^{h/2} u\mathrm{d}z\right] + \frac{\partial}{\partial y}\left[\rho\int_{-h/2}^{h/2} v\mathrm{d}z\right] + \rho\frac{\partial h}{\partial t} = 0 \tag{7}$$

In deriving Eq. (7), we have made a substitution for the velocity in the z direction as, $w = \partial h/\partial t$.The integrals in Eq. (7) represent the average velocities multiplied by h, and can be replaced by $\tilde{u}h$ and $\tilde{v}h$. Considering the flow to be isothermal, we finally arrive at the nonlinear compressible Reynolds' equation

$$\frac{\partial}{\partial x}\left(\frac{ph^3}{\mu}\frac{\partial p}{\partial x}\right) + \frac{\partial}{\partial y}\left(\frac{ph^3}{\mu}\frac{\partial p}{\partial y}\right) = 12\frac{\partial(ph)}{\partial t}. \tag{8}$$

For small amplitude oscillations, the Reynolds' equation given by (8) can be linearized using perturbation parameters. By substituting $p = p_a + \hat{p}$ and $h = h_a + \hat{h}$ in Eq. (8) and by neglecting higher order terms, we arrive at the following linearized compressible Reynolds' equation:

$$\left[\frac{\partial^2\hat{p}}{\partial x^2} + \frac{\partial^2\hat{p}}{\partial y^2}\right] = \frac{12\mu}{p_a h_a^3}\left[h_a\frac{\partial\hat{p}}{\partial t} + p_a\frac{\partial\hat{h}}{\partial t}\right]. \tag{9}$$

This is the governing equation for the squeeze film flow assuming small amplitude motion of the vibrating structure compared to the air-gap height.

3 Analytical Solutions

Several analytical solutions exist for the squeeze film problem for rigid plates in both torsional as well as transverse motions [17] and for different venting conditions [13]. For flexible plates, one has to account for the elastic deformation of the plate along with the fluid motion. Kumar [20] studied the effect of venting conditions on flexible plates by considering their mode shapes in accordance with venting conditions using the Greens function approach proposed by Darling [13]. We present an analytical solution technique using the Eigen expansion method wherein the venting conditions are independent of the structural boundary conditions of the plate and hence independent of the structural mode shape. As an example, we consider a plate fixed along all edges with all sides in open venting condition and use the approximate first mode shape to account for the elasticity of the plate.

3.1 Solution Using the Eigen Expansion Method

The linearized compressible Reynolds' equation given by Eq. (9) can be nondimensionalized using new variables for pressure, film thickness, time, and plate geometry as $P = \hat{p}/p_a$, $H = \hat{h}/h_a$, $T = \omega t$, $X = x/L$, $Y = y/L$ which leads to the following form of the governing equation:

$$\nabla^2 P - \alpha^2 \frac{\partial P}{\partial t} = \alpha^2 \frac{\partial H}{\partial t} \tag{10}$$

where $\left(\alpha = \sqrt{\frac{12\mu}{h_a^2 p_a}}\right)$ and L and W are the length and width of the plate, respectively.

We now solve the linearized nondimensional Reynolds' equation (Eq. 10). We carry out an Eigen function analysis on the homogeneous form of Eq. (10). Let $P = \sum\limits_{m,n} f_{mn}(x,y)T(t)$ where the sum indicates a linear combination of Eigen functions $\Psi_{mn}(x,y)$ where $f_{mn} = a_{mn}\Psi_{mn}(x,y)$, $\Psi_{mn}(x,y)$ being an Eigen function, a_{mn} a scalar coefficient, and $T(t)$ the temporal component of pressure. By substituting this expression for P in the homogeneous form of Eq. (10) and using the separation of variables, we get

$$\frac{\nabla^2 f}{f} = \alpha^2 \frac{\frac{\partial T}{\partial t}}{T} = -k_{mn}^2 \tag{11}$$

where k_{mn} is a constant corresponding to each f_{mn}. We can decouple k_{mn} for the two independent indices m and n by assuming that $k_{mn}^2 = k_m^2 + k_n^2$. Further by assuming that $\Psi_{mn}(x,y) = X_m(x)Y_n(y)$ and by substituting in Eq. (11) we find, after some simplifications, that

$$\Psi mn(x,y) = \sin\left(\frac{m\pi x}{L}\right)\sin\left(\frac{n\pi y}{W}\right). \tag{12}$$

By assuming a harmonic source term, $H = \delta\Phi e^{i\omega t}$ (where δ is the vibration amplitude), Φ is the assumed first mode shape $\left(\Phi = \sin\left(\frac{\pi x}{L}\right)^2 \sin\left(\frac{\pi y}{W}\right)^2\right)$ for a plate fixed on all sides, and ω is the oscillation frequency. By substituting the expression for Ψ_{mn} in the expression for pressure and using the orthogonality of Eigen functions, we determine the coefficients a_{mn} as

$$a_{mn} = \delta \frac{\int_0^L \int_0^W i\omega\alpha^2 \Phi \Psi_{mn} \mathrm{d}y\mathrm{d}x}{\int_0^L \int_0^W \left[-k_{mn}^2 - i\omega\alpha^2\right]\Psi_{mn}^2 \mathrm{d}y\mathrm{d}x}. \tag{13}$$

Now, the pressure is given by $P = \sum a_{mn}\Psi_{mn}(x,y)T(t) = \tilde{P}T(t)$ where $\tilde{P}$ is the spatial component of the pressure. For open boundaries on all sides, we use zero

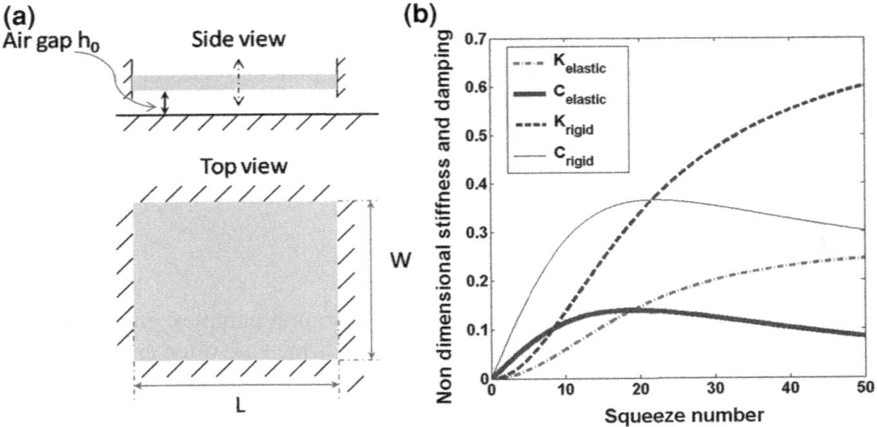

Fig. 4 Square plate in squeeze flow: **a** Schematic of the plate **b** Nondimensional stiffness and damping for rigid and elastic plates for a range of squeeze numbers

pressure boundary conditions, $P(0, y) = P(L, y) = P(x, 0) = P(x, W) = 0$, and obtain the following expressions for the nondimensional quantities, P (pressure), F_s (stiffness force), and F_d (damping force).

$$P = -64\delta \sum_{m,\, n \in odd} \frac{\left[i\pi^2\sigma(m^2\beta^2 + n^2) + \sigma^2\right] \sin\left(\frac{m\pi x}{L}\right) \sin\left(\frac{n\pi y}{W}\right)}{\left[\pi^4(m^2\beta^2 + n^2)^2 + \sigma^2\right] mn\pi^2(m^2 - 4)(n^2 - 4)} \tag{14}$$

$$F_s = \frac{-256\delta\sigma^2}{\pi^4} \sum_{m,\, n \in odd} \frac{1}{m^2n^2(m^2 - 4)(n^2 - 4)[\pi^4(m^2\beta^2 + n^2)^2 + \sigma^2]} \tag{15}$$

$$F_d = \frac{-256\delta\sigma}{\pi^2} \sum_{m,\, n \in odd} \frac{m^2\beta^2 + n^2}{m^2n^2(m^2 - 4)(n^2 - 4)[\pi^4(m^2\beta^2 + n^2)^2 + \sigma^2]} \tag{16}$$

where σ is the squeeze number $\left(\sigma = \frac{12\mu\omega W^2}{p_0 h_0^2}\right)$, and β is the aspect ratio $\beta = W/L$. The procedure may be extended for other venting conditions and corresponding mode shapes. From the preceding expressions, we can compute the nondimensional stiffness $\bar{K}_{sq} = F_s/\delta$, the nondimensional damping $\bar{C}_{sq} = F_d/\delta$. The actual stiffness and damping are given by $K_{sq} = p_a LW\bar{K}_{sq}/h$ and $C_{sq} = p_a LW\bar{C}_{sq}/h\omega$.

Example Let us consider a square plate shown in Fig. 4a with $L = 100$, $W = 100$ μm, structurally fixed along all the edges with all sides open. By assuming the nondimensional maximum displacement amplitude for the plate centre to be $\delta = 0.01$, where $\delta = w/h_0$, and w the amplitude of the place displacement in the z direction with h_0 being the initial air gap. This example considers the first mode shape of the plate as $\Phi(x, y) = \sin\left(\frac{\pi x}{L}\right)^2 \sin\left(\frac{\pi y}{W}\right)^2$. Figure 4b shows the variation of nondimensional stiffness and damping for a range of squeeze numbers. If we consider the plate to be rigid, i.e., $\Phi(x,y) = 1$, and assuming the same value of δ, the solution is shown in dotted lines in Fig. 4b. This solution was obtained by Blech [11]. As we can see, when the plate is considered to be elastic the magnitude of squeeze film stiffness and damping decrease.

3.2 Complexities

We now discuss modeling strategies for some common complex effects such as rarefaction, compressibility, and inertia of the fluid, which are often encountered in different operating conditions of MEMS devices. The linearized and simplified Reynolds' equation has to be modified appropriately to incorporate these effects.

3.2.1 Rarefaction

Rarefaction refers to the apparent thinning of air either due to decrease in pressure (partial vacuum) or due to extremely small length-scale of the characteristic flow. In either case, it has a significant effect on squeeze film flow. The Knudsen number (Kn), defined as the ratio of the mean free path of the gas molecules to the characteristic flow length, is a measure of rarefaction. Continuum theory holds for small values of the Knudsen number (Kn << 1). On the other hand, large values of the Knudsen number (Kn >> 1) necessitate the consideration of molecular motion and interactions.

For small values of Knudsen number (0.01 < Kn < 0.1), the first-order slip boundary conditions $\left(u_{\pm h/2}l = \mp\lambda\frac{\partial u}{\partial z}, v_{\pm h/2} = \mp\lambda\frac{\partial v}{\partial z}\right)$ are applied and the slip-corrected velocity distributions are obtained from Eq. (3) as

$$u = \frac{1}{2\mu}\frac{\partial p}{\partial x}\left(z^2 - \frac{h^2}{4} - Knh^2\right) \quad \text{and} \quad v = \frac{1}{2\mu}\frac{\partial p}{\partial y}\left(z^2 - \frac{h^2}{4} - Knh^2\right) \tag{17}$$

We then obtain the average radial velocity $\tilde{u}$ and $\tilde{v}$ as

$$\tilde{u} = -\frac{h^2(1+6Kn)}{12\mu}\frac{\partial p}{\partial x} \quad \text{and} \quad \tilde{v} = -\frac{h^2(1+6Kn)}{12\mu}\frac{\partial p}{\partial y} \tag{18}$$

and by substituting these in the continuity equation, we obtain the nonlinear compressible Reynolds' equation with slip correction:

$$\frac{\partial}{\partial x}\left(\frac{ph^3(1+6Kn)}{12\mu}\frac{\partial p}{\partial x}\right) + \frac{\partial}{\partial y}\left(\frac{ph^3(1+6Kn)}{12\mu}\frac{\partial p}{\partial y}\right) = \frac{\partial(ph)}{\partial t} \tag{19}$$

The factor (1 + 6Kn) in the preceding equation is referred to as the relative flow rate coefficient (Q_{pr}). The relative flow rate is included in the fluid viscosity and the combined term is called effective viscosity, defined as $\mu_{eff} = \mu/Q_{pr}$. Veijola et al. [21] presented an approximation for the effective viscosity by fitting the respective flow rate coefficients to the experimental values tabulated by Fukui and Kaneko [22]. This is given as $\mu_{eff} = \mu/(1 + 9.638Kn^{1.159})$.

In MEMS devices where the device size reaches the nanometric scale, the amplitude of surface roughness is large enough to affect the characteristics of the squeeze film damping [23]. Several researchers have developed analytical models to account for this effect [24–28]. The surface roughness effect is however coupled with rarefaction. Works by Li et al. [29] and Hwang et al. [30] have considered the coupled nature of surface roughness and rarefaction effects by modifications to the Reynolds' equation.

3.2.2 Compressibility

The linearized nondimensional compressible Reynolds' equation may be expressed in the form

$$\left[\frac{\partial^2 P}{\partial X^2} + \frac{\partial^2 P}{\partial Y^2}\right] = \sigma\left[\frac{\partial P}{\partial T} + \frac{\partial H}{\partial T}\right] \tag{20}$$

The dependence of damping and spring forces on the squeeze number, plotted as a function of frequency of oscillation, is shown in Fig. 5 for both rigid and elastic plates with all-sides-open boundary condition for the fluid flow. Initially, the damping force increases steadily with frequency, reaches a maximum value, and thereafter decreases monotonically. At low frequencies, the air can escape easily; however, with increasing frequency of oscillation the trapped air does not get enough time to escape and hence gets compressed. This compression begins to dominate over the fluid flow and the resulting spring force gradually increases and reaches a maximum value. The point of intersection of the spring force and the damping force is known as the *cut-off frequency*. For low squeeze numbers (i.e., $\sigma << 1$), the compressibility effects may be ignored. However, for high squeeze numbers the compressibility leads to a significant air-spring effect, which is generally undesirable as it can adversely affect the dynamic behaviour of a device [31].

3.2.3 Inertia

A modified Reynolds' equation [32] is derived from Eq. (3) considering the unsteady inertial terms. Thus, considering inertial effects and making the corrections for slip, the velocities of fluid flow for small amplitude harmonic oscillation $\left(\hat{h} = \delta e^{j\omega t}\right)$ are given as

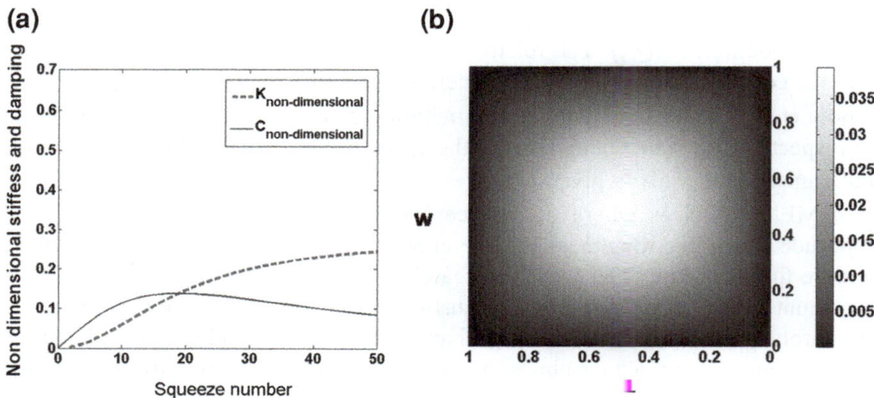

Fig. 5 FEM results: **a** Nondimensional stiffness and damping for an all sides open elastic plate **b** pressure distribution for $\sigma = 1$ for an all sides open elastic plate

$$
\begin{aligned}
u &= \frac{1}{j\omega\rho}\left(\frac{\cos(qz)}{\cos(qh/2) - \lambda q \sin(qh/2)} - 1\right)\frac{\partial p}{\partial x},\\
v &= \frac{1}{j\omega\rho}\left(\frac{\cos(qz)}{\cos(qh/2) - \lambda q \sin(qh/2)} - 1\right)\frac{\partial p}{\partial y},
\end{aligned}
\tag{21}
$$

where $q = \sqrt{j\omega\rho/\mu}$. By following the same procedure and substituting the average velocities in the continuity equation, we get a compressible Reynolds' equation with inertia and slip-correction

$$
\frac{\partial}{\partial x}\left(\frac{ph^3}{\mu}Q_{pr}\frac{\partial p}{\partial x}\right) + \frac{\partial}{\partial y}\left(\frac{ph^3}{\mu}Q_{pr}\frac{\partial p}{\partial y}\right) = \frac{\partial(ph)}{\partial t}
\tag{22}
$$

where the value of $Q_{\rm pr}$ with inertial and rarefaction effects is computed using

$$
Q_{pr} = \frac{12}{-j\mathrm{Re}}\frac{\left[(2 - \lambda j\mathrm{Re})\tan(\sqrt{-j\mathrm{Re}/2}) - \sqrt{-j\mathrm{Re}}\right]}{\left[\sqrt{-j\mathrm{Re}\left(1 - \lambda\sqrt{-j\mathrm{Re}}\tan\left(\sqrt{-j\mathrm{Re}/2}\right)\right)}\right]}
\tag{23}
$$

and $\mathrm{Re} = \frac{\rho\omega h_a^2}{\mu}$, is the Reynolds' number.

3.2.4 Modeling Perforations

Various approaches for modeling perforated geometries are available in the literature. Some employ analytical methods while some others use numerical techniques, and some others, a combination of both. The analytical approaches employ two main strategies, namely the concept of pressure cells [33, 34] and modification of the Reynolds' equation to include a flow-leakage term. For geometries where there is symmetry in the distribution of perforations, the flow problem is analyzed for each perforation cell and the result is multiplied by the number of such cells to obtain the solution for the entire structure. The second approach involves modification of the Reynolds' equation to include an additional term to account for the flow through holes. Different analytical models differ in the way the flow is modeled (2D or 3D) and the complexities included in their model (i.e., rarefaction, inertia, and compressibility).

For MEMS structures with high aspect-ratios, Bao et al. [35, 36] presented a modified Reynolds' equation having a pressure-leakage term to model flow in the z direction.

This model is valid for uniform distribution of holes where the loss through the holes is homogenized over the entire domain. Veijola et al. [37] suggested another model utilizing a specialized solver for the perforation profile for arbitrary perforations and nonuniform sizes and distribution of holes. In another work, Veijola [38] used existing analytical models in squeeze film and capillary regions and from finite element simulations, based on approximate flow resistances at the intermediate region and at the exit flow from the hole.

All the above-mentioned models cease to be valid when the hole dimensions, the air-gap and pitch are of the same order of magnitude, the assumptions used in deriving Reynolds' equation do not remain valid. Then one has to resort to computationally intensive numerical solutions for 3D Navier Stokes equation.

3.2.5 Modeling Squeeze Film in the Free Molecular Regime

For pressures much lower than the atmosphere, collisions among gas molecules are greatly reduced and the gas can no longer be considered a viscous fluid. The concept of effective viscosity becomes questionable and a free molecular model has to be considered. Christian [39] proposed a free molecular model to estimate damping in a low vacuum. This model ignores the interaction among gas molecules and the damping force on a vibrating plate is found by the momentum transfer rate from the oscillating plate to the surrounding air through collisions between the plate and the gas molecules. Christian's model showed discrepancies when compared with the experimental data of Zook et al. [40]. His model underestimated the damping force by an order of magnitude when compared with the experimental results. Bao et al. [41] proposed a new free molecular model that overcame the limitation of the Christian's model of not being able to consider the effect of a nearby substrate. Bao's model calculates the energy losses through collisions between the plate and the molecules. This model is referred to as the Energy Transfer model. It matches the experimental data much better than the Christian's model.

4 Numerical Modeling of Squeeze Film Effects

Analytical solutions exist for squeeze film flow under certain conditions as we have seen in the previous sections. However, for accurate modeling in the case of involved complex geometries and the coupled problem of structural deformation and squeeze film flow, one needs to resort to numerical or semi-numerical strategies as analytical solutions become increasingly difficult for such problems. Finite element method can be carried out using commercial tools with applicable capabilities. Some, e.g., ANSYS and COMSOL can solve Reynolds' equation over a complex fluid domain and find the steady state pressure distribution over the domain. In ANSYS, one can use either FLUID136 or FLUID142 elements for the squeeze film domain and FLUID138 for any holes in conjunction with solid elements for the vibrating structure. As there is no direct coupling, one needs to solve the structural vibration problem to get the velocity profile on the face of the structure, which is then used as the prescribed velocity at the appropriate nodes of the fluid elements. The analysis results in a pressure distribution over the entire domain that is then integrated to get the net force on the vibrating structure. The pressure computed is typically in complex form because of prescribed harmonic velocity. The net force, therefore, has real and imaginary parts. The real part is in phase with the velocity and hence represents the damping force F_{D} and the imaginary part, in phase with the displacement, represents the spring force F_{s} The damping coefficient and the spring stiffness are subsequently determined from these forces as shown earlier. One can easily account for rarefaction by simply using an appropriate value of the effective viscosity. Several meaningful results have been reported in the literature [6, 17] where the Q of the MEMS device of interest is computed using ANSYS and satisfactorily compared with experimental values.

To get a better appreciation of what is involved in FEM calculations, rather than just using a black box, we discuss here briefly the finite element formulation for the squeeze film flow problem. We use the linearized Reynolds' equation and incorporate the flexibility of the plate as an imposed mode shape to the finite element scheme.

4.1 FEM Formulation

If one does not want to use a commercial code, one can, with a little bit of effort, develop one's own finite element code to solve the governing equation, determine pressure, and compute the spring and damping forces. Here, we discuss very briefly the essential steps required for custom-developed finite element formulation.

We start with the linearized Reynolds' equation given as

$$\frac{\partial^2 P}{\partial x^2} + \frac{\partial^2 P}{\partial y^2} = \alpha^2 \left[\frac{\partial H}{\partial t} + \frac{\partial P}{\partial t}\right] \quad (24)$$

Now by assuming a solution of the form $P = \tilde{P}e^{i\omega t}$ where $\tilde{P}$stands for pressure amplitude, and using $H = \delta\Phi e^{i\omega t}$ where Φ is the mode shape of the plate, δ the displacement amplitude, and ω the vibration frequency. By substituting for different terms in Eq. (24) and nondimensionalizing using using $x = LX$ and $y = LY$, and introducing the squeeze number σ as $\sigma = \alpha^2 L^2 \omega$, we finally get the following form of the Reynolds' equation:

$$\frac{\partial^2 \tilde{P}}{\partial X^2} + \frac{\partial^2 \tilde{P}}{\partial Y^2} = \sigma\left[i\delta\Phi + i\tilde{P}\right] \quad (25)$$

For finite element formulation, we need the weak variational form of Eq. (25). By following the usual steps of variational formulation [42], e.g., assuming a variation in $\tilde{P}$ say V and using it as the weighing function to multiply with Eq. (25) and integrating over the domain we get

$$\int_{\Omega_e} \nabla.(\nabla\tilde{P})V\mathrm{d}\Omega - i\sigma \int_{\Omega_e} (\tilde{P} + \Phi\delta)V\mathrm{d}\Omega = 0 \quad (26)$$

By manipulating Eq. (26) and using the divergence theorem to change the domain integral in the first term to a boundary integral and using appropriate boundary conditions, i.e., $\tilde{P} = 0$ for an open boundary and $\nabla\tilde{P}$ or $\partial\tilde{P}/\partial n = 0$ for a closed boundary, we get

$$\int \frac{\partial \tilde{P}}{\partial x_i}\frac{\partial V}{\partial x_i}\mathrm{d}\Omega + i\sigma \int (\tilde{P} + \Phi\delta)V\mathrm{d}\Omega = 0 \quad (27)$$

Equation (27) represents the weak form of the linearized Reynolds' equation. We now use the weak form of the equation over the discretized domain with chosen finite elements. For example, let us use a four noded bilinear element to discretize the entire domain. We approximate $\tilde{P}$ over a typical finite element by the expression, $\tilde{P}(x,y) \approx \sum_{j=1}^{n} N_j^e \tilde{P}_j^e$ where $\tilde{P}_j^e$ is the pressure at the jth node of the element and N_j^e are shape functions for the element chosen [42]. Now, by substituting the FEM approximation for $\tilde{P}$ in the weak form given by Eq. (27) and following standard FE discretization, we arrive at the following relation, written compactly as:

$$\sum_{j=1}^{n} K_{i,j}\tilde{P}_j = -i\delta\sigma \sum_{j=1}^{n} f_{i,j}\Phi_j \tag{28}$$

where $K^e_{i,j} = \int_{\Omega_e}\left(\frac{\partial N_i}{\partial x}\frac{\partial N_j}{\partial x} + \frac{\partial N_i}{\partial y}\frac{\partial N_j}{\partial y}\right)\mathrm{d}x\mathrm{d}y + i\sigma\int_{\Omega_e} N_iN_j\mathrm{d}x\mathrm{d}y$ and $f^e_{i,j} = \int_{\Omega_e} N_iN_j\mathrm{d}x\mathrm{d}y$. Rearranging Eq. (28) and writing in compact form:

$$(K_1 + K_2)\tilde{P} = -\delta K_2\Phi \tag{29}$$

where $[K_1] = [S^{11}] + [S^{22}]$ and $[K_2] = i\sigma[S^{33}]$, with S^{11}, S^{22} and S^{33} given by

$$\begin{gathered} S^{11}_{i,j} = \int_{-1}^{1}\frac{\partial N_i}{\partial\sigma}\frac{\partial N_j}{\partial\sigma}\mathrm{d}\sigma\mathrm{d}\eta,\ S^{22}_{i,j} = \int_{-1}^{1}\frac{\partial N_i}{\partial\eta}\frac{\partial N_j}{\partial\eta}\mathrm{d}\sigma\mathrm{d}\eta \\ \text{and}\ \ S^{22}_{i,j} = \int_{-1}^{1}\frac{\partial N_i}{\partial\eta}\frac{\partial N_j}{\partial\eta}\mathrm{d}\sigma\mathrm{d}\eta \end{gathered} \tag{30}$$

By evaluating the integrals in Eq. (30) and using the relations for $[K_1]$ and $[K_2]$, we finally get the following element stiffness matrices:

$$[K_1] = \frac{1}{6}\begin{bmatrix} 4 & -1 & -2 & 1 \\ -1 & 4 & -1 & -2 \\ -2 & -1 & 4 & -1 \\ -1 & -2 & -1 & 4 \end{bmatrix} \quad \text{and} \quad [K_2] = \frac{i\sigma a^2}{36}\begin{bmatrix} 4 & 2 & 1 & 2 \\ 2 & 4 & 2 & 1 \\ 1 & 2 & 4 & 2 \\ 2 & 1 & 2 & 4 \end{bmatrix} \tag{31}$$

We can use the matrices in Eq. (31) to evaluate the element stiffness matrix for each element. The rest of the work is to generate the global stiffness matrix using appropriate element connectivity [42], computing the load vector q in the final matrix equation $[K]\{P\} = \{q\}$, applying appropriate boundary conditions, and finally solving the matrix equation to find the values of the pressure {P} at each node.

Example We use a code developed by following the steps outlined to solve the same plate problem used in Sect. 3.1. We find the pressure distribution for $\sigma = 1$ (a low squeeze number) and subsequently the nondimensional stiffness and damping forces. The results are shown in Fig. 5. A comparison with the analytical results is shown in Fig. 4 demonstrate that the FEM-based solution is in good agreement with the analytical solution'.

5 Experimental Studies

We now discuss some of the experimental studies repeated by several researchers in this field to validate existing analytical and numerical models and to investigate the squeeze film damping phenomenon under different ambient conditions.

Andrew et al. [43] experimentally validated Blech's [11] solution for rectangular plates using Si test structures operating in squeeze numbers ranging from zero to 1000 and found that the theoretically predicted damping was in good agreement with experimental data for low frequencies (<10 kHz). Kim et al. [44] investigated the effect of perforations on squeeze film damping. They fabricated test structures with different hole size and number of holes and compared Q-factors obtained from experiments with those from numerical FE simulations. The FE simulations considered zero pressure boundary conditions along the edges of the holes and were found to underestimate the experimental damping. Pandey et al. [6] studied the effect of pressure on the quality factor of a MEMS torsion mirror. The flow regimes from continuum to molecular flow were covered by varying the ambient pressure. A power law relation between the Q-factor and pressure was found; i.e., $Q \alpha P^{-r}$ where $r = 0$ for the continuum regime, $r = 0.5$ for slip regime and $r = 1$ for the free molecular regime. In another work, Pandey et al. [45] compared an analytical model that they proposed for computing damping in perforated structures with existing analytical models and validated the proposed model with experimental data using perforated test structures. The proposed model, which accounted for various losses associated with perforations as well as the spatial variation of pressure in 2D domains, was found to model damping more accurately as compared to the existing models. Somá and Pasquale [46] experimentally investigated the validity of reduced order finite element models for squeeze film damping using perforated test structures with varying size and number of holes. They found that the imposed velocity method, which essentially models a 2D problem, underestimated damping because the perforated regions were not accounted for. The Modal Projection method did not show considerable advantage due to the high sensitivity of the results to the meshing accuracy and the high computational requirement for large structures. Sumali [47] validated several existing analytical models with experimental studies for a large rage of Knudsen numbers. The experimental technique ensured a transverse rigid motion of the test structure. Other sources of damping were subtracted out from the measured damping. Pasquale et al. [48] studied the effect of different perforation sizes and numbers on silicon and gold structures having square and rectangular geometries. They compared experimentally obtained Q-factors with FE simulations and with a compact model. The low Q-factor value for large hole sizes in square plates was unexplained by the FE simulations and the compact model. The effect of plate elasticity was studied by Pandey and Pratap [17]. They presented analytical models for studying the effect of mode shapes on squeeze film damping. The analytical results were validated with numerical and experimental data and showed good agreement with each other. Pan et al. [49] studied the effect of squeeze film damping for MEMS torsion mirrors. They experimentally validated the dynamic response for tilting motion predicted by their analytical models by assuming harmonic dynamic response. The Q-factor and damping ratios obtained experimentally showed good agreement with analytical results.

6 Conclusions

Squeeze film effects influence affects the dynamic performance of vibratory MEMS devices that have some trapped air or gas between the vibrating body and the fixed substrate. The effects of squeeze film flow are found by solving appropriate fluid equations to compute the back pressure on the vibrating plate which, in turn, give the spring and damping components. Several analytical solutions exist in the literature, primarily based on Reynolds' equation and its modified forms. Most of these differ in the way the factors such as inertia, rarefaction, and compressibility are handled. Most of the analytical models are limited to solving simple geometries and use approximate solutions for flexibility of the plate. For accurate modeling of the squeeze film effects, one may resort to numerical simulations. The existing FEM-based commercial codes provide reasonable facilities for analysing squeeze film effects. However, the problem of coupled elastic deformation and squeeze film flow is currently solved only iteratively. The first truly coupled finite element formulation has been recently reported [18]. Squeeze film forces have also been studied experimentally for devices with varying geometric complexities, boundary conditions, and operating conditions. These experimental results have guided the development of simpler analytical models and realistic numerical models.

Acknowledgments The authors wish to acknowledge funding from NPMASS for the Computational Nano Engineering (CoNE) project that has supported the studies reported here.

References

1. Khan S, Thejas N, Bhat N (2009) Design and characterization of a micromachined accelerometer with mechanical amplifier for intrusion detection. In: 3rd national ISSS conference on MEMS, smart structures and materials, Kolkata 14–16 Oct 2009
2. Patil N (2006) Design and analysis of MEMS angular rate sensors. Master's Thesis. Indian Institute of Science, Bangalore
3. Pandey AK, Venkatesh KP, Pratap R (2009) Effect of metal coating and residual stress on the resonant frequency of MEMS resonators. Sãdhãna 34(4):651–661
4. Ahmad B, Pratap R (2011) Analytical evaluation of squeeze film forces in a CMUT with sealed air filled cavity. IEEE Sens J 11(10):2426–2431
5. Venkatesh C, Bhat N, Vinoy KJ et al (2012) Microelectromechanical torsional varactors with low parasitic capacitances and high dynamic range. J Micro/Nanolithog MEMS MOEMS 11(1):013006
6. Pandey A (2007) Analytical, numerical, and experimental studies of fluid damping in MEMS devices. Ph.D. Thesis, Indian Institute of Science, Bangalore
7. Shekhar S, Vinoy KJ, Ananthasuresh GK (2011) Switching and release time analysis of electrostatically actuated capacitive RF MEMS switches. Sens Transducers J 130(7):77–90. http://www.sensorsportal.com/HTML/DIGEST/P_826.htm
8. Venkatesh KP, Patil N, Pandey AK et al (2009) Design and characterization of in-plane MEMS yaw rate sensor. Sãdhãna 34(4):633–634

9. Pratap R, Mohite S, Pandey AK (2007) Squeeze film effects in MEMS devices. J Indian Inst Sci 87(1):75–94
10. Yang J, Ono T, Esashi M (2002) Energy dissipation in sub-micrometer thick single-crystal silicon cantilevers. IEEE J MEMS 11:775–783
11. Blech JJ (1983) On isothermal squeeze films. J Lubr Technol 105:615–620
12. Griffin WS, Richardsen HH, Yamamami (1966) A study of fluid squeeze film damping. J Basic Eng 88(2):451–456
13. Darling R, Hivick C, Xu J (1998) Compact analytical modeling of squeeze film damping with arbitrary venting conditions using a Greens function approach. Sens Actuators A 70(1):32–41
14. Hung ES, Senturia SD (1999) Generating efficient dynamical models for micro electromechanical systems from a few finite-element simulation runs. J Microelectromech Syst 8(3):280–289
15. McCarthy B, Adams GG, McGruer NE et al (2002) A dynamic model, including contact bounce, of an electrostatically actuated microswitch. J Microelectromech Syst 11(3):276–283
16. Younis MI, Nayfeh AH (2007) Simulation of squeeze film damping of microplates actuated by large electrostatic load. J Comput Nonlinear Dyn 2(3):232–240
17. Pandey AK, Pratap R (2007) Effect of flexural modes on squeeze film damping in MEMS cantilever resonators. J Micromech Microeng 17(12):2475–2484
18. Roychowdhury A, Nandy A, Jog CS et al (2013) A monolithic FEM based approach for the coupled squeeze film problem of an oscillating elastic microplate using 3D 27-node elements. J Appl Math Phys 1(6):20–25
19. Burgdorfer A (1959) The influence of the molecular mean free path on the performance of hydrodynamic gas lubrication bearings. ASME J Basic Eng 81:94–100
20. Kumar S (2010) Effect of squeeze film flow on dynamic response of MEMS structures with restrictive flow boundary conditions. Master's Thesis, Indian Institute of Science, Bangalore
21. Veijola T, Kuisma H, Lahdenpera J et al (1995) Equivalent circuit model of the squeezed gas film in a silicon accelerometer. Sens Actuators A 48:236–248
22. Fukui S, Kaneko R (1990) A database for interpolation of Poiseuille flow rates for high Knudsen number lubrication problems. J Tribol 112:78–83
23. Pandey AK, Pratap R (2004) Coupled nonlinear effects of surface roughness and rarefaction on squeeze film damping in MEMS structures. J Micromech Microeng 14:1430–1437
24. Patir N, Cheng HS (1978) An average flow model for determining effects of 3-dimensional roughness on partial hydrodynamic lubrication. J Lubr Technol 100:12–16
25. Elrod HG (1979) A general theory for laminar lubrication with Reynolds roughness. J Lubr Technol 101:8–14
26. Tripp JH (1983) Surface roughness effects in hydrodynamics lubrication: the flow factor method. J Lubr Technol 105:458–465
27. Tonder K (1986) The lubrication of unidirectional striated roughness: consequences for some general theories. ASME J Tribol 108:167–170
28. Mitsuya Y, Ohkubo T, Ota H (1989) Averaged Reynolds equation extended to gas lubrication possessing surface roughness in slip flow regime: approximate method and confirmation experiments. J Tribol 111:495–503
29. Li WL, Weng CL, Hwang CC (1995) Effect of roughness orientations on thin film lubrication of magnetic recording system. J Phys D Appl Phys 28(6):1011–1021
30. Hwang CC, Fung RF, Yang RF et al (1996) A new modified Reynolds equation for ultrathin film gas lubrication. IEEE Trans Magn 32(2):344–347
31. Starr JB (1990) Squeeze film damping in solid state accelerometers In: Proceedings of IEEE solid state sensors and actuators workshop, SC, pp 44–47
32. Veijola T (2004) Compact models for squeeze-film dampers with inertial and rarefied gas effects. J Micromech Microeng 14:1109–1118
33. Mohite SS, Kesari H, Sonti VR et al (2005) Analytical solutions for the stiffness and damping coefficients of squeeze films in MEMS devices with perforated backplates. J Micromech Microeng 15:2083–2092

34. Mohite SS, Sonti VR, Pratap R (2008) A compact squeeze-film model including inertia, compressibility, and rarefaction effects for perforated 3D MEMS structures. J Microelectro Mech Syst 17(3):709–723
35. Bao M, Yang H, Sun Y et al (2003) Modified Reynolds' equation and analytical analysis of squeeze-film air damping of perforated structures. J Micromech Microeng 13:795–800
36. Bao M, Yang H, Sun Y et al (2003) Squeeze-film air damping of thick hole plate. Sens Actuators A 108:212–217
37. Veijola T, Mattila T (2001) Compact squeezed film damping model for perforated surface. In: Proceedings of IEEE 11th international conference on solid state sensors, actuators and micro-systems, pp 1506–1509
38. Veijola T (2006) Analytical model for a MEMS perforation cell. Microfluid Nanofluid 2(3):249–260
39. Christian R (1966) The theory of oscillating-vane vacuum gauges. Vacuum 16:175–178
40. Zook J, Burns D, Guckel H et al (1992) Characteristics of polysilicon resonant microbeams. Sens Actuators A 35:51–59
41. Bao M, Yang H, Yin H et al (2002) Energy transfer model for squeeze-film air damping in low vacuum. J Micromech Microeng 12:341–346
42. Reddy JN (1993) An introduction to the finite element method, 2nd edn. McGraw-Hill, New York
43. Andrews MK, Turner GC, Harris PD et al (1993) A resonant pressure sensor based on a squeezed film of gas. Sens Actuators A 36:219–226
44. Kim ES, Cho Y, Kim M (1999) Effect of holes and edges on the squeeze film damping of perforated micromechanical structures. In: 12th IEEE international conference on micro electro mechanical systems (MEMS'99), pp 296–301
45. Pandey AK, Pratap R (2007) A comparative study of analytical squeeze film damping models in rigid rectangular perforated MEMS structures with experimental results. Microfluid Nanofluid 4(3):205–218
46. Somà A, Pasquale GD (2008) Numerical and experimental comparison of MEMS suspended plates dynamic behavior under squeeze film damping effect. Analog Integr Circ Sig Process 57(3):213–224
47. Sumali H (2007) Squeeze-film damping in the free molecular regime: model validation and measurement on a MEMS. J Micromech Microeng 17(11):2231–2240
48. Pasquale GD, Veijola T, Soma A (2009) Modelling and validation of air damping in perforated gold and silicon MEMS plates. J Micromech Microeng 20(1). doi:10.1088/0960-1317/20/1/015010
49. Pan F, Kubby J, Peeters E et al (1998) Squeeze film damping effect on the dynamic response of a MEMS torsion mirror. J Micromech Microeng 8:200–208

Streaming Potential in Microflows and Nanoflows

Jeevanjyoti Chakraborty and Suman Chakraborty

Abstract In this chapter, we discuss a wide variety of important effects due to streaming potential when fluid flow takes place through conduits of micro- and nanometric dimensions. We first introduce this as one of the four primary electrokinetic phenomena with suitable background, and describe its significance in numerous natural and engineered settings. Its practical usage and measurement being inordinately linked to predictive models, we present the theory behind streaming potential. In light of recent research findings, and recognizing their importance in micro- and nano-flows, we also highlight the influence of streaming potential when considered together with the consideration of hydrophobic, steric, and thermal effects.

Keywords Streaming potential · Electrokinetics · Hydrophobic effects · Steric effect · Soret effect · Thermoelectric effect

1 Introduction

The colloidal scientist is perennially concerned with the charge of the particles (s)he is handling as this charge is of paramount importance in determining the stability of the dispersion and its overall behavior. The geophysicist interested in a

J. Chakraborty · S. Chakraborty (✉)
Advanced Technology Development Centre, Indian Institute of Technology Kharagpur, Kharagpur 721302, West Bengal, India
e-mail: sumanchakraborty_iitkgp@yahoo.com

J. Chakraborty
Mathematical Institute, University of Oxford, Oxford OX2 6GG, UK

S. Chakraborty
Mechanical Engineering Department, Indian Institute of Technology Kharagpur, Kharagpur 721302, West Bengal, India

K. J. Vinoy et al. (eds.), *Micro and Smart Devices and Systems*,
Springer Tracts in Mechanical Engineering, DOI: 10.1007/978-81-322-1913-2_20,

broad spectrum of sub-surface investigations ranging from the mapping of pore geometry to the exploration of alternative energy sources based on harnessing geothermal energy reserves, needs to take recourse to ingenious methods that exploit underground fluid flow in newer forms. Soil and mining engineers are interested in various dewatering and decontamination processes. In a starkly different realm, biophysicists are increasingly interested in unraveling the fundamental mechanisms through which bone responds to external stress via its interaction with interstitial fluids. The membrane scientist needs to take into account the various factors contributing to the specific filtration process (s)he is trying to develop. The motivation for such developments stems from the sophistication (as far as membranes go) found in the sheath of cells, and in sub-cellular entities. The researcher trying to develop lab-on-a-chip devices at the micro- and nanoscales needs alternative fluid actuation mechanisms, or even when using traditional pressure-driven ones needs to take into consideration additional factors which might be of tremendous consequence. The common thread that binds the interests and motivation of the aforementioned scientists and engineers is the physicochemical phenomenon where the boundary layer between one charged phase and another undergoes a shearing process. This phenomenon is manifested in many different forms in various settings, and is broadly referred to as electrokinetics. In this chapter, we focus on one such particular form called the streaming potential that lies at the very core of understanding electrokinetics in different contexts.

1.1 Understanding Streaming Potential

1.1.1 The Electrical Double Layer (EDL) and Electrokinetics

The most important point to note regarding electrokinetic phenomena is that they are critically dependent on the establishment of an electrified interface between two phases, one of which is usually a solid. The first step in the formation of such an electrified interface is the generation of a surface charge. Various mechanisms through which such charge is generated include an imbalance in the number of crystal lattice cations or anions on the surface, crystal lattice defects, surface dissociation, and ion adsorption from solutions; the mechanism may also be some combination of these.

This surface charge is invariably accompanied by the presence of certain ionic species in the adjacent liquid medium. The ions in the solution of opposite charge to that of the surface are referred to as counterions while the ions of like charges are referred to as co-ions. It is intuitive to expect that the counterions would be attracted towards the surface while the co-ions would be repelled by it. If the Coulombic forces were the only factors, the counterions would stack up against the surface—thus, perfectly shielding the rest of the ions in the solution. Such a scenario is precluded, however, by the random thermal motion of the ions at any

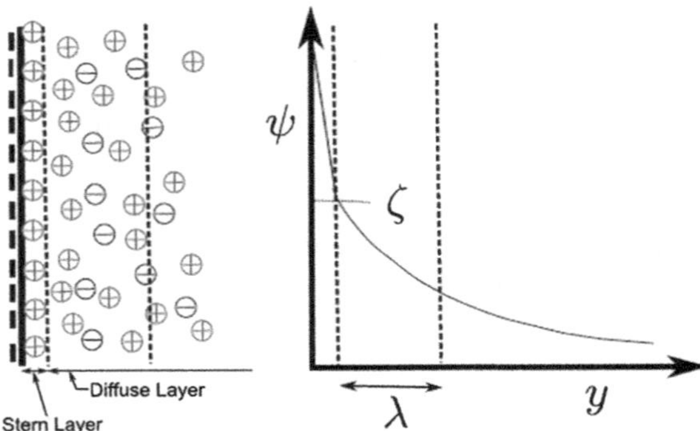

Fig. 1 Structure of the EDL and the distribution of the electric potential

finite temperature. As a result of interplay between the Coulombic forces and the thermal motion a distribution of the counterions and the co-ions is established in the vicinity of the charged surface. In this distribution, the counterions outnumber the co-ions in such a way that the charge imbalance resulting from it is perfectly neutralized by the charge on the surface. It is the surface electric charge together with the balancing charge in the solution that constitutes the electrified interface. This is referred to as the electrical double layer [1] or EDL, in short.

In the most widely accepted model for the structure of the EDL (the Guoy-Chapman-Stern model) there is a monolayer of counterions that stays attached to the charged surface (see Fig. 1). This is called the Stern layer, or the compact layer, or the Helmholtz layer. Just adjacent to the edge of the Stern layer is the shear plane. The potential at the edge of the Stern layer is called the zeta potential. The part of the EDL beyond this shear plane is called the diffuse layer because it is susceptible to motion when the fluid is sheared past the charged surface by any external actuation. The characteristic thickness of the EDL is defined to be the position from the wall where the potential drops to $1/e$ (e being the Euler number) of the zeta potential value; this characteristic thickness is referred to as the Debye length. As noted earlier, this shearing motion of the diffuse part of the electrified interface across the charged surface is what gives rise to a host of electrokinetic phenomena, the primary four among which are electroosmosis, electrophoresis, streaming potential, and sedimentation potential.

1.1.2 Origin of the Streaming Potential

When a pressure-gradient (or any other mechanical actuation) is used to set the liquid in a capillary tube (or channel) or a porous plug in motion, the diffuse part of the double layer also moves along with the flow and is sheared past the charged surface. The current that is generated as a result of the advection of the ions is called streaming current. The consequent transfer of ions downstream gives rise to an electric field in the opposite direction. The potential associated with this electric field opposes the streaming current. This happens through a back conduction of ions, and also through an electroosmotic flow to a certain extent. This current directed opposite to the streaming current is called the leak current. The term "conduction current" is often used in a sense that subsumes within it both the mechanisms of the back flow of the ions. After a very short time, a dynamic equilibrium condition is reached when the leak current balances the streaming current so that the total ionic current across the confinement through which the fluid flow is taking place is zero. The potential difference measured across the conduit in this condition is known as the streaming potential.

1.2 Many Uses of the Streaming Potential

The phenomenon of streaming potential was discovered by Georg Quincke, who first observed that he could measure an electrical potential difference between the ends of a tube when he pumped water through it [2]. Since then, streaming potential has come to be used in a wide variety of settings as outlined in the opening paragraph of Sect. 1. It is a standard practice to determine the zeta potential of various surfaces using streaming potential or streaming current measurements [3]. In the geophysical realm techniques based on downhole measurements of streaming potential are developed to monitor fluid flows in hydrocarbon reservoirs or oil fields. More generally, there is a great interest in the geophysical community on streaming potential measurements for use in pore geometry determination, for monitoring underground fluid flow, and rock/fluid interfacial chemistry [4]. Additionally, streaming potential may help in monitoring and prediction of earthquakes [5]. The study of streaming potentials at elevated temperature and pressure may also be used for geothermal exploration. These have implications in the study of volcanoes [6]. In the physiological realm, the mechanism behind how bones respond via interstitial fluid flow to external loading is believed to be based on streaming potential [7]. Streaming potential may also be able to sensitively indicate intervertebral disk degeneration [8]. In the context of filtration processes, streaming potential can be used to study the influence that fouling phenomena has on membrane surface properties [9]. A comprehensive discussion on these various aspects may be found in Ref. [10].

In this chapter, we particularly focus on studies of streaming potential mediated flows at micro- and nanofluidic scales. The motivation behind this is as follows. Streaming potential phenomena in narrow confinements is important from the perspective of a wide variety of practical applications encompassing the lab-on-a-chip (LOC) technology. A critical point to note is that LOC fluidic technology is inspired by electronic integrated circuits (interestingly, it is from "electronics" that the term "fluidic" is coined). LOC systems are created by using chip-based micromachining techniques to shrink the size of fluid handling systems aimed at improving chemical and biological analysis. Thus, this technology mimics both the fabrication technology and the overall "smaller, cheaper, faster" paradigm of the integrated circuit industry. Furthermore, just as "wires" form the basic pathways in electrical/electronic circuits, micro- and nanofluidic channels are the most fundamental structures in LOC devices [11]. These not only offer suitable settings to study the fundamental physics behind the fluidic phenomena increasingly influenced by surface effects in general, and electrokinetics, in particular, but also provide a basis for the design of practical devices [12–16]. A theoretical consideration of such channel-like geometries may serve as a good model for certain apparently complicated physical setups that are, however, characterized by symmetries or other possibilities for geometrical simplifications.

Based on the above, we present a general theoretical description of streaming potential in Sect. 2. With this background, we then present, in the rest of the chapter, an outline of the motivation of, and findings from a few recent investigations that were carried out to understand the combined influence of streaming potential and certain other physico-chemical phenomena on micro-/nanoscale fluid flows.

2 Mathematical Model of Streaming Potential

We consider the model problem of streaming potential flow of a binary symmetric electrolyte through a straight channel having either a slit or circularly symmetric cross-section. The characteristic dimension in either cross-section, generally represented by R, is considered to be much smaller than the channel length L. The fluid is primarily actuated by a pressure gradient. In the absence of surface heterogeneities, the resulting velocity field may be safely considered to be unidirectional so that $\boldsymbol{u} = u\hat{\boldsymbol{e}}_x$ where $\hat{\boldsymbol{e}}_x$ is the unit vector along the axial direction. The origin is located at the centerline of the channel. The surface is considered to be negatively charged; the counterions and the co-ions are, respectively, identified by the "+" and "−" subscripts. The valences are given by $z_+ = -z_- = z$. Then, for the case of steady flow, the species transport equation reduces to

$$-\nabla \cdot \left(D_\pm \nabla n_\pm \pm \frac{ze}{k_B T} D_\pm n_\pm \nabla \varphi \right) + \nabla \cdot (n_\pm \boldsymbol{u}) = 0, \tag{1}$$

where $D_\pm$ refers to the diffusivity of the cation, and the anion. To address the cases of the slit channel and the circular channel simultaneously in the same framework, we use a general representation. Toward that end, for the differential operators we follow the notation of Stone [17]. Thus, the gradient operator is represented by $\nabla \equiv \frac{\partial}{\partial \xi}\hat{\boldsymbol{e}}_\xi + \frac{\partial}{\partial x}\hat{\boldsymbol{e}}_x$, where ξ is the general coordinate transverse to the wall; and the divergence by $\nabla \cdot \square = \frac{1}{\xi^{d-1}}\frac{\partial}{\partial \xi}\left(\xi^{d-1}\square_\xi\right) + \frac{\partial}{\partial x}\square_x$, where $d = 1$ for the slit cross-section, and $d = 2$ for the circularly symmetric cross-section. Anticipating the use of the Laplacian operator, we also note $\nabla^2 = \frac{1}{\xi^{d-1}}\frac{\partial}{\partial \xi}\left(\xi^{d-1}\frac{\partial}{\partial \xi}\right) + \frac{\partial^2}{\partial x^2}$. We use the non-dimensional variables $\tilde{\xi} = \frac{\xi}{\mathcal{R}}$, $\tilde{x} = \frac{x}{\mathcal{L}}$, $\tilde{n}_\pm = \frac{n_\pm}{n_0}$, $\tilde{\varphi} = \frac{ez\varphi}{k_B T}$, $\tilde{\boldsymbol{u}} = \frac{u\hat{\boldsymbol{e}}_x}{\mathcal{U}}$, where n_0 is the number density of both the counterions, and the co-ions in the bulk, and U is the scale of the unidirectional velocity along the axial direction. Further, considering $D_+ = D_- = D$ to be spatially invariant, Eq. (1) reduces in non-dimensional form to

$$
\begin{aligned}
&-\frac{1}{\tilde{\xi}^{d-1}}\frac{\partial}{\partial\tilde{\xi}}\left(\tilde{\xi}^{d-1}\frac{\partial\tilde{n}_\pm}{\partial\tilde{\xi}}\right)+\left(\frac{\mathcal{R}}{\mathcal{L}}\right)^2\frac{\partial^2\tilde{n}_\pm}{\partial x^2}+Pe\left(\frac{\mathcal{R}}{\mathcal{L}}\right)\frac{\partial}{\partial x}(\tilde{n}_\pm\tilde{u}) \\
&\mp\left[\frac{1}{\tilde{\xi}^{d-1}}\frac{\partial}{\partial\tilde{\xi}}\left(\tilde{\xi}^{d-1}\tilde{n}_\pm\frac{\partial\tilde{\varphi}}{\partial\tilde{\xi}}\right)+\left(\frac{\mathcal{R}}{\mathcal{L}}\right)^2\frac{\partial}{\partial x}\left(\tilde{n}_\pm\frac{\partial\tilde{\varphi}}{\partial x}\right)\right]=0,
\end{aligned}
\tag{2}
$$

where $Pe = \frac{\mathcal{U}\mathcal{R}}{D}$ is the ionic species Péclet number. This may be viewed as the product of the Reynolds number $Re = \mathcal{U}\mathcal{R}/\upsilon$ (where $\upsilon = \eta/\rho$ is the kinematic viscosity with η being the dynamic viscosity, and ρ the density of the fluid) and the Schmidt number $Sc = \upsilon/D$. The small value of $Re(\ll 1)$, very common in micro- and nanofluidic settings, does not necessarily ensure that Pe is small because Sc which represents the strength of the momentum diffusivity relative to the species diffusivity may be significantly larger than unity. For instance, with typical ionic diffusivities, D, are of the order of 10^{-9} m^{2}s^{-1}, in aqueous solutions where $\upsilon \sim 10^{-6}$ m^{2}s^{-1}, $Sc \sim \mathcal{O}(10^3)$. It is then possible for $Pe \gtrsim \mathcal{O}(1)$. However, it is because Pe is multiplied by the factor $(\mathcal{R}/\mathcal{L})$ that the advective contribution may indeed be safely neglected as long as $\mathcal{R} \ll \mathcal{L}$. Similarly, other terms multiplied by the factor (R/L)^2 may also be neglected—thus, clearly indicating that gradients along the axial direction have negligible contribution. We further note that $\varphi = \psi - xE_S$, where E_s is the spatially invariant electric field. Thus, Eq. (2) becomes

$$
\frac{1}{\tilde{\xi}^{d-1}}\frac{\partial}{\partial\tilde{\xi}}\left(\tilde{\xi}^{d-1}\frac{\partial\tilde{n}_\pm}{\partial\tilde{\xi}}\right)\pm\frac{1}{\tilde{\xi}^{d-1}}\frac{\partial}{\partial\tilde{\xi}}\left(\tilde{\xi}^{d-1}\tilde{n}_\pm\frac{\partial\tilde{\varphi}}{\partial\tilde{\xi}}\right)=0. \tag{3}
$$

By integrating once and using the symmetry condition at the centerline, we get

$$
\partial(\ln\tilde{n}_\pm)/\partial\tilde{\xi} = \mp\partial\tilde{\psi}/\partial\tilde{\xi}. \tag{4}
$$

This is essentially the same equation that leads to the Boltzmann distribution in the purely equilibrium case. Thus, we can understand that the fluid flow does not affect the equilibrium structure of the EDL in the case of a pressure-gradient-driven flow through straight channels with large aspect ratios ($\mathcal{R} \ll \mathcal{L}$). This is complemented by the Poisson equation written for the present setting in a reduced and dimensionless form as

$$
\frac{1}{\tilde{\xi}^{d-1}}\frac{\partial}{\partial\tilde{\xi}}\left(\tilde{\xi}^{d-1}\frac{\partial\tilde{\psi}_\pm}{\partial\tilde{\xi}}\right)=-\left(\frac{\mathcal{R}^2e^2z^2n_0}{\varepsilon k_BT}\right)\tilde{\rho}_e, \tag{5}
$$

where the charge density, ρ_e, has been non-dimensionalized by ezn_0. The fluid velocity is governed by the x-direction component of the Navier-Stokes equation, which for the steady case is written as

$$\rho\left(u\frac{\partial u}{\partial x}\right) = -\frac{\partial p}{\partial x} + \eta\left[\frac{1}{\xi^{d-1}}\frac{\partial}{\partial \xi}\left(\xi^{d-1}\frac{\partial u}{\partial \xi}\right) + \frac{\partial^2 u}{\partial x^2}\right] + \rho_e E_S, \tag{6}$$

where $\rho_e E_S$ is the electrokinetic body force exerted on a unit volume of the fluid due to the streaming potential electrical field E_s. In non-dimensional terms, this is written as

$$\begin{aligned} Re\left(\frac{\mathcal{R}}{\mathcal{L}}\right)\tilde{u}\frac{\partial \tilde{u}}{\partial \tilde{x}} = & -\frac{\mathcal{R}^2 \Delta p}{\eta \mathcal{L}\mathcal{U}}\frac{\partial \tilde{p}}{\partial \tilde{x}} + \frac{1}{\tilde{\xi}^{d-1}}\frac{\partial}{\partial \tilde{\xi}}\left(\tilde{\xi}^{d-1}\frac{\partial \tilde{u}}{\partial \tilde{\xi}}\right) \\ & + \left(\frac{\mathcal{R}}{\mathcal{L}}\right)^2\frac{\partial^2 \tilde{u}}{\partial \tilde{x}^2} - \frac{\varepsilon k_B T E_0}{\eta e z \mathcal{U}}\frac{1}{\tilde{\xi}^{d-1}}\frac{\partial}{\partial \tilde{\xi}}\left(\tilde{\xi}^{d-1}\frac{\partial \tilde{\psi}}{\partial \tilde{\xi}}\right)\tilde{E}_S \end{aligned}. \tag{7}$$

The inertial term on the left-hand side in the preceding equation is clearly negligible because both Re and $(\mathcal{R}/\mathcal{L})$ are very small. The term involving the pressure-gradient is the primary actuator of the flow. Hence, the velocity scale $\boldsymbol{U}$ is set by considering this term to be $\mathcal{O}(1)$, giving $\mathcal{U} \sim (\mathcal{R}^2/\eta)(\Delta p/\mathcal{L})$. The viscous term with the prefactor $(R/L)^2$ is also neglected. The streaming potential field is non-dimensionalized by an appropriate scale E_0 (to be explicated later). Then, Eq. (7) reduces to

$$-1 + \frac{1}{\tilde{\xi}^{d-1}}\frac{\partial}{\partial \tilde{\xi}}\left(\tilde{\xi}^{d-1}\frac{\partial \tilde{u}}{\partial \tilde{\xi}}\right) - \mathcal{A}\frac{1}{\tilde{\xi}^{d-1}}\frac{\partial}{\partial \tilde{\xi}}\left(\tilde{\xi}^{d-1}\frac{\partial \tilde{\psi}}{\partial \tilde{\xi}}\right)\tilde{E}_S = 0, \tag{8}$$

where $\mathcal{A} = (\varepsilon k_B T E_0)/(\eta e z \mathcal{U})$. By integrating once, and using the symmetry condition at the channel centreline, we have

$$\frac{\tilde{\xi}}{d} + \frac{\partial \tilde{u}}{\partial \tilde{\xi}} - \mathcal{A}\frac{\partial \tilde{\psi}}{\partial \tilde{\xi}}\tilde{E}_S = 0. \tag{9}$$

By integrating again, and using the no-slip boundary condition $\tilde{u} = 0$ and $\tilde{\psi} = \tilde{\zeta}$ (the non-dimensional value of the zeta potential) at $\tilde{\xi} = 1$ (the wall), we have

$$\tilde{u} = \frac{1}{2d}\left(1 - \tilde{\xi}^2\right) + \mathcal{A}\tilde{E}_S\left(\tilde{\psi} - \tilde{\zeta}\right). \tag{10}$$

The condition used to determine the as-yet-unknown streaming potential field is to set the total ionic current across any cross-section to zero. This is given by

$$\int_{-\mathcal{R}}^{\mathcal{R}} (j_{+x}\hat{\boldsymbol{e}}_x - j_{-x}\hat{\boldsymbol{e}}_x) \cdot \hat{\boldsymbol{e}}_x (2\pi\xi)^{d-1} d\xi = 0, \tag{11}$$

where $j_{\pm x} = -D\frac{\partial n_\pm}{\partial x} \mp D\frac{ez}{k_BT} n_\pm \frac{\partial \varphi}{\partial x} + n_\pm u$ represents the flux of the counterions, and the co-ions in the x-direction. It follows directly from Eq. (4) that the ionic distributions are practically invariant along the x-direction. Then, the contribution of the $\partial n_\pm/\partial x$ term to the flux is taken to be negligible. Again, we note that $\varphi = \psi - xE_S$. But just like the ionic distribution, the contribution of $\partial\psi/\partial x$ to the flux is also negligible. With these simplifications, together with expanding the expression of u from Eq. (10), and non-dimensionalization, we get

$$\int_{-1}^{1} \begin{bmatrix} (\tilde{n}_+ - \tilde{n}_-)\left\{\frac{1}{2d}\left(1 - \tilde{\xi}^2\right) + \mathcal{A}\,\tilde{E}_S\left(\tilde{\psi} - \tilde{\zeta}\right)\right\}\mathcal{U} \\ + \frac{ezD}{k_BT}(\tilde{n}_+ + \tilde{n}_-)\tilde{E}_S E_0 \end{bmatrix} \tilde{\xi}^{d-1} d\tilde{\xi} = 0, \tag{12}$$

from which we get the nondimensional streaming potential field as

$$\tilde{E}_S = \frac{I_1}{I_2 + \mathcal{K} I_2}, \tag{13}$$

where $I_1 = \frac{1}{2d}\int_{-1}^{1} (\tilde{n}_+ - \tilde{n}_-)\left(1 - \tilde{\xi}^2\right)\tilde{\xi}^{d-1} d\tilde{\xi}$, $I_2 = \int_{-1}^{1} (\tilde{n}_+ + \tilde{n}_-)\tilde{\xi}^{d-1} d\tilde{\xi}$, $I_3 = \int_{-1}^{1} \left(\tilde{\psi} - \tilde{\zeta}\right)(\tilde{n}_+ - \tilde{n}_-)\tilde{\xi}^{d-1} d\tilde{\xi}$, and $\mathcal{K} = \frac{\varepsilon k_B^2 T^2}{\eta e^2 z^2 D}$. We note that the scale of the streaming potential field has been set as $E_0 = -\frac{k_BT}{ezD}\frac{\boldsymbol{R}^2}{\eta}\left(\frac{\Delta p}{L}\right)$. Using these in Eq. (10) gives us the velocity of the fluid flow.

3 Combined Influence of Streaming Potential and Other Effects

Notwithstanding the wide spectrum of research contributions in flows influenced by streaming potential in micro- and nanochannels, certain important aspects have remained largely unexplored until recently. In the rest of this chapter, we present a brief overview of some recent investigations that explore the concerted influence of streaming potential and certain other micro-/nanoscale physicochemical phenomena on fluid flows at such length scales.

3.1 Hydrophobic Effects

While the influence of hydrophobic effects on microscale flows in general and electrokinetic flows in particular have been studied widely, such influence is often relegated to the boundary conditions in the form of a specified slip-length. Notwithstanding the widespread success of this approach, a slip-length-based modeling framework cannot, however, resolve the actual physical mechanism through which such hydrophobic effects give rise to an overall reduction in the resistance to the fluid flow: indeed, such reduction fundamentally arises because the hydrophobic substrate induces the depletion of the fluid in the near wall region thus allowing the bulk fluid to smoothly slide over a smoothening blanket of the depleted phase with reduced viscosity.

Departing from the traditional approaches, in a recent investigation [18], we model this depletion mechanism through a phase-field model by expressing the viscosity and permittivity in terms of the phase-field variable, which results in smooth profiles of these parameters. We then utilized these in the framework described in Sect. 2 for determining the streaming potential flow with appropriate modifications to incorporate the implicit spatial dependence of the parameters. Through this framework, we are able to clearly establish that there is a sensitive interplay between the length scale of the EDL structure and that of the depleted phase. It is this interplay together with the intrinsic strength of the hydrophobic effects (captured in our framework through the specification of the contact angle) that determines the overall character of the flow. Since the overall effect of the streaming potential is to inhibit the pressure-gradient-driven flow (known as the electroviscous effect), and since the total volumetric flow rate is of tremendous importance in practical devices, we express the gross nature of the flow in terms of an effective normalized viscosity $\tilde{\eta}_{eff}$ that would result in the same (reduced) volumetric flow rate had there been only a pressure-gradient-driven flow (with no electrokinetic effects). We define: $\tilde{\eta}_{eff} = \left(\eta_{eff}/\eta_l\right) = \left(4/(3\tilde{Q})\right)$ where η_l is the viscosity of the undepleted liquid, and $\tilde{Q}\iota = \int \tilde{u}\, dy$ is the volumetric flow rate through a slit channel of height *2H*. The dependence of this effective normalized viscosity on the contact angle (representing the degree of hydrophobicity of the wall) reveals a sensitive interplay between the length scale of the EDL structure and that of the hydrophobicity-induced depletion. As shown in Fig. 2, we find that $\tilde{\eta}_{eff}$ assumes a value greater than unity (showing augmented hindrance to the flow) when the EDL is characteristically thicker than the depleted phase. For high zeta potential value, it shows an interesting transition when the length scales are equal, and it is consistently less than unity when the EDL length scale is smaller than the depleted length scale. Importantly, the thickness of the EDL is characterized by the Debye screening length, λ_l, based on the permittivity of the liquid.

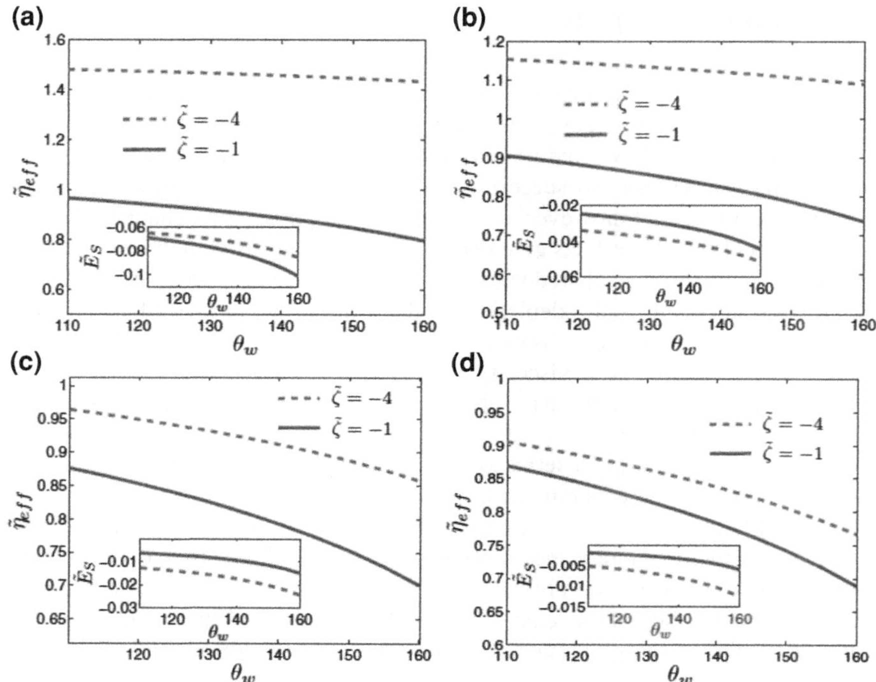

Fig. 2 Variation of the effective normalized viscosity with the contact angle for $\tilde{\zeta} = -1$, and $\tilde{\zeta} = -4$ corresponding to $H/\lambda_l = 5, 10, 25, 50$, respectively, in panels (**a**)–(**d**). The insets in each panel show the variation of the dimensionless streaming potential field with the contact angle (Reproduced from Ref. [18] with permission from the American Physical Society)

3.2 Steric Effects

It is well known that the effects of the streaming potential get progressively strong with increase in the magnitude of the zeta potential. At the same time, however, such high values of the zeta potential result in diverging values of the number density of the counterions in the near wall region as predicted by the traditional formalism described in Sect. 2 which strictly considered the ions to be point-like charges. This is taken care of by considering a modified Boltzmann approximation that incorporates the finite size of the ions—thus, precluding any potential unphysical overlap. Despite the considerably widespread use of such modified formulations, there does exist a serious theoretical inconsistency even in this modified modeling framework. This inconsistency arises from the fact that while the finite size is considered for the ionic distribution, there exists no explicit link with this finite size in the flux terms even though the diffusivity (which contributes significantly to the flux) is indeed dependent on the size of the ions. We address this fundamental theoretical issue by establishing this link [19], by using the

Stokes–Einstein relation: $D = k_BT/(6\pi\eta r)$. The need to maintain the theoretical consistency also necessitates the incorporation of the viscoelectric effect, i.e., the dependence of the dynamic viscosity on the local charge density through the relation: $1/\tilde{\eta} = 1 - \Xi$. Here, $\tilde{\eta} = \eta/\eta_0$ is the dimensionless local viscosity with η_0 being the value of the bulk viscosity, ν is the steric factor, and $\Xi = \frac{1}{2}\nu|\tilde{n}_+ - \tilde{n}_-|$. We then have $\tilde{D} = (1 - \Xi)$ where $\tilde{D} = D/D_0$ is the dimensionless diffusivity, with $D_0 = k_BT/(6\pi\eta_0 r)$ being the constant bulk value of the diffusivity. Then, by using the relation $\nu = 2n_0r^3$, we may express the bulk diffusivity directly in terms of ν. By using these and following the route outlined in Sect. 2, we obtain a modified expression of the streaming potential field corresponding to a fluid flow through a slit channel of height $2H$ as

$$\tilde{E}_s = \frac{3\pi\left(\frac{\nu}{2}\right)^{1/3}\frac{1}{\mathcal{L}_1}I_1}{I_2 - 6\pi(4\nu)^{1/3}\frac{\mathcal{L}_1^2}{K^2}I_3}, \tag{14}$$

where $\mathcal{L}_1 = Hn_0^{1/3}$ and $K = H/\lambda$. It can be seen in Eq. (14) that the steric effects influence both the ionic distributions and the factors responsible for determining the strength of transport of the ions. Since it is this transport that is ultimately responsible for inducing the streaming potential field, there is a far stronger dependence of this field on the finite size effects than could be envisaged within the formalism of prior theoretical treatments. Indeed, so much so that when the size is considered to be vanishingly small, i.e., $\nu \to 0$, the value of the streaming potential field itself becomes small. Most notably, for ν = 0 representing a situation where the finite size of the ions is not considered, and is, hence, a reflection of the unmodified Boltzmann distribution which rests upon the assumption of the ions being point-like charges, no value of $\tilde{E}_s$ other than zero is possible. This seemingly nonintuitive prediction is, however, consistent with the underlying assumption of point-like charges; point entities cannot have "friction," which is inseparably present with diffusive transport, associated with them.

This conceptual understanding is clearly seen in Fig. 3. We note, in particular, that streaming potential effects may be wrongly predicted in theoretical exercises below a particular threshold of the steric factor—as indeed the plots corresponding to K = 5 and 10 show. The correct trends, i.e., higher suppression of the volumetric flow due to streaming potential for lower value of K, are seen only for $\nu \gtrsim \mathcal{O}(10^{-1})$. Further, as an extremely important implication of this study, we show that if one were to work consistently within the framework of the traditional Poisson-Boltzmann formalism, no value of the streaming potential other than zero is possible, as indeed seen in the figure.

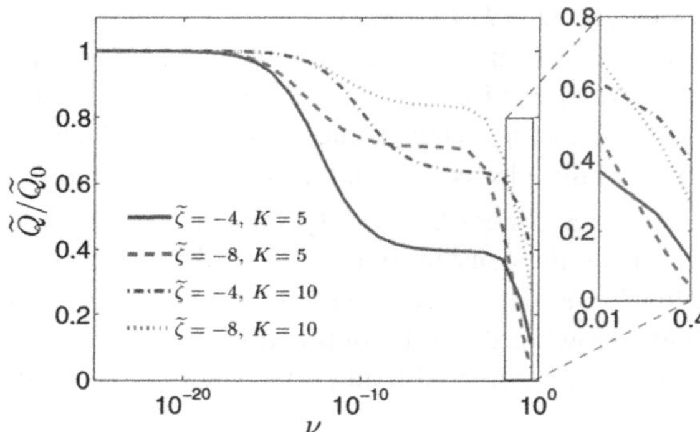

Fig. 3 Total volumetric flow rate $Q = \int_0^2 \tilde{u} d\tilde{y}$ normalized by the volumetric flow rate of the plane-Poiseuille component ($\tilde{Q}_0$) only, (i.e.,$\int_0^2 (2\tilde{y} - \tilde{y}^2) d\tilde{y}$) with the steric factor, ν, corresponding to $\tilde{\zeta} = -4, 8$; for $K = 5, 10$. As ν becomes vanishingly small, $\tilde{Q}/\tilde{Q}_0 \rightarrow 1$, showing that streaming potential effects vanish in keeping with the assumption of point-like ions in the Boltzmann formalism. The magnified region shows a reversal in the trends of $\tilde{Q}/\tilde{Q}_0$ when $\nu \gtrsim \mathcal{O}(10^{-1})$, for the two different values of$\tilde{\zeta}$ (Reproduced from Ref. [19] with permission from the American Physical Society)

3.3 Thermal Effects

Most of the modeling efforts in electrokinetics, in general, and streaming potential in particular, are carried out under the assumption of isothermal conditions. This is particularly true in the micro- and nanofluidic context. This is in spite of the fact that most natural settings are non-isothermal in nature, and also that even in controlled laboratory environments, strictly isothermal conditions are difficult to realize in practice. Recognizing this, we have developed a modeling framework where thermal effects are incorporated in a comprehensive way [20]. Notably, this framework goes beyond the traditional route of a simplistic one-way coupling between the fluid flow velocity and the temperature field used in the few cases that electrokinetic flows have been studied together with thermal effects. Importantly, in this traditional route, there is no back influence of the temperature on the velocity field. Within our model, however, a complete two-way coupling is achieved by including the influence of some additional, fundamental physical phenomena. First, we include the Soret effect, which refers to the propensity of a species to move in response to a temperature gradient, with an additional term in the flux equation: $-D_{\pm} \frac{n_{\pm} q_{\pm}}{k_B T^2} \nabla T$ that immediately augments the species transport represented in Eq. (1). Here, $q_{\pm}$ denotes the ionic heat of transport of the positive, and the negative ions, and $D_{\pm}$ denotes the now temperature-dependent diffusivity of the same. Since generally, the values of q_{+} and q_{-} are different, there is a difference in the extent to which the cations and the anions migrate in response to the temperature gradient. This generates a thermoelectric field through a

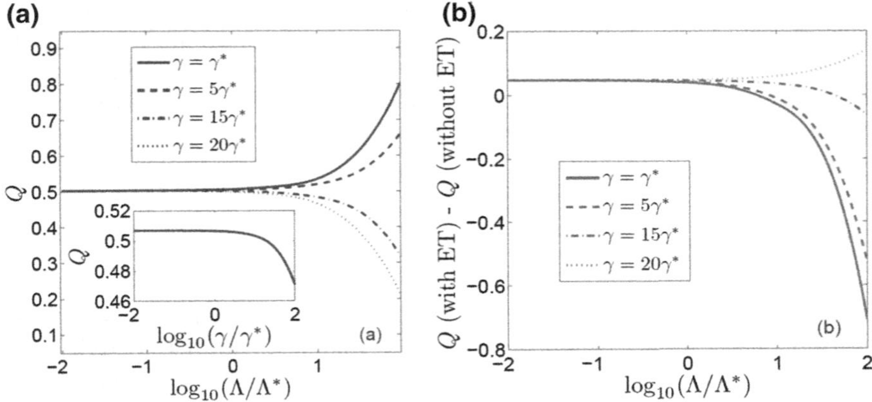

Fig. 4 Variation of the dimensionless volumetric flow rate corresponding to the variation of $\Lambda/\Lambda^* \leftarrow$ over four orders of magnitude for four different values of γ/γ^*. Inset in (**a**) shows the variation of the dimensionless volumetric flow rate with γ/γ^* varying over four orders of magnitude. In (**b**) differences in the volumetric flow rates with and without considering electrothermal effects are shown. Here, $\Lambda = q_+/(k_B T_0)$, and $\gamma = q_-/q_+$, and Λ^* and γ^* are the values of a reference alkali halide (Reproduced with slight modifications from Ref. [20] with permission from the American Physical Society)

mechanism that is analogous to the Seebeck effect. Second, we include the electrothermal effect that brings about an extra contribution to the forcing in the momentum equation due to the dependence of the permittivity on the temperature. Additionally, we incorporate the dependence of the viscosity, the thermal conductivity, and also the zeta potential on the temperature.

To study the concerted influence of these effects in a streaming potential flow, we consider again a model flow situation through a slit-channel. The thermal effects are brought about by the imposition of a linear temperature gradient on the walls.

We can observe from Fig. 4 that depending on the polarity of the generated thermoelectric field which, in turn, is determined by the relative thermo-diffusive migration strengths of the cations and the anions, the suppression of the volumetric flow rate induced by the streaming potential field may be aided or opposed. Thus, by simply imposing an external temperature gradient, we obtain extra control over the volumetric flow rate for streaming potential flows through the Soret effect and the concomitant thermoelectric field. Given a specified magnitude of the externally applied temperature gradient, we can exercise this control just by changing the nature of the electrolyte. From Fig. 4b, we can also see the role that the electrothermal effect plays in the alterations of the temperature gradient-mediated streaming potential flows. When the thermoelectric and streaming potential fields are opposing each other, this electrothermal effect basically weakens the thermoelectric field, and leads to an overall reduced flow rate compared to the case without electrothermal effects. However, when the thermoelectric and streaming potential fields are oriented along the same direction, the consideration of the electrothermal effect enhances the overall volumetric flowrate, again by weakening the thermoelectric field. So, based on the findings of our work, it can be unambiguously inferred that temperature gradients can be successfully employed for tuning flows mediated by streaming potential.

4 Conclusions

It has been shown that flows influenced by streaming potential may be sensitively tuned by changing the hydrophobicity of the confining surfaces, as well as by exerting a temperature gradient along them. It has also been shown that a theoretically consistent model of streaming potential flows necessitates the incorporation of finite size effects in the diffusivity coefficient, too. Based on these findings, the modeling of streaming potential is being currently extended on a multitude of fronts' comprehensive framework that will emerge and be able to capture the various physicochemical phenomena in a concerted and consistent manner.

References

1. Hunter RJ (2001) Foundations of colloid science. Oxford University Press, New York
2. Quincke G (1861) Ueber die fortführung materieller Theilchen durch strömende Elektricität. Ann Phys 189:513–598
3. Delgado AV, González-Caballero F, Hunter RJ et al (2007) Measurement and interpretation of electrokinetic phenomena. J Colloid Interface Sci 309:194–224
4. Reppert PM (2000) Electrokinetics in the earth: frequency dependent electrokinetics and streaming potentials at elevated temperature and pressure. Ph.D. Thesis, Massachusetts Institute of Technology, Cambridge, Massachusetts
5. Jouniaux L, Pozzi JP (1995) Streaming potentials and permeability of saturated sandstones under triaxial stress: consequences for electrotelluric anomalies prior to earthquakes. J Geophys Res 100:10197–10209
6. Zlotnicki J, Le Mouël LL (1990) Possible electrokinetic origin of large magnetic variations at La Fournaise volcano. Nature 343:633–635
7. Riddle RC, Donahue HJ (2008) From streaming potentials to shear stress: 25 years of bone cell mechanotransduction. J Orthopaedic Res 27:143–149
8. Gu WY, Mao XG, Rawlins BA et al (1999) Streaming potential of human lumbar annulus fibrosus is anisotropic and affected by disc degeneration. J Biomech 32:1177–1182
9. Nyström M, Pihlajamäki A, Ehsani N (1994) Characterization of ultrafiltration membranes by simultaneous streaming potential and flux measurements. J Memb Sci 87:245–256
10. Chakraborty J (2013) Electrokinetics in narrow fluid fluidic confinements: the role of streaming potential. Ph.D. Thesis, Indian Institute of Technology Kharagpur, India
11. Stein D, van den Heuvel M, Dekker C (2009) Transport of ions, DNA polymers, and micrtotubules in the nanofluidic regime. In: Edel JB, deMello AJ (eds) Nanofluidics. The Royal Society of Chemistry, Cambridge, Chap 1, pp 1–30

12. Schoch RB, Han J, Renaud P (2008) Transport phenomena in nanofluidics. Rev Mod Phys 80:839–883
13. Sparreboom W, van den Berg A, Eijkel JCT (2009) Principles and applications of nanofluidic transport. Nat Nanotechnol 4:713–720
14. Squires TM, Quake SR (2005) Microfluidics: fluid physics at the nanoliter scale. Rev Mod Phys 77:977–1026
15. Stone HA, Stroock AD, Ajdari A (2004) Engineering flows in small devices: microfluidics toward a lab-on-a-chip. Ann Rev Fluid Mech 36:381–411
16. Zhao C, Yang C (2012) Advances in electrokinetics and their applications in micro/nano fluidics. Microfluid Nanofluid 13:179–203
17. Stone HA (2002) Partial differential equations in thin film flows in fluid dynamics: spreading droplets and rivulets. In: Berestycki H, Pomeau Y (eds) Nonlinear PDEs in condensed matter and reactive flows. Kluwer, Dordrecth
18. Chakraborty J, Chakraborty S (2013) Influence of hydrophobic effects on streaming potential. Phys Rev E 88:043007
19. Chakraborty J, Dey R, Chakraborty S (2012) Consistent accounting of steric effects for prediction of streaming potential in narrow confinements. Phys Rev E 86:061504
20. Ghonge T, Chakraborty J, Dey R et al (2013) Electrohydrodynamics within the electrical double layer in the presence of finite temperature gradients. Phys Rev E 88:053020

A Simulation Module for Microsystems using Hybrid Finite Elements: An Overview

Kunal D. Patil, Sreenath Balakrishnan, C. S. Jog and G. K. Ananthasuresh

Abstract We present here an overview of the work done in developing a simulation module for microsystems, which entails solving coupled partial differential equations concerning multiple physical phenomena. A distinguishing feature of this work is the use of hybrid finite elements wherein displacement and stress fields are independently interpolated to mitigate the ill effects of widely known locking phenomena in finite element analysis. A beneficial consequence of hybrid elements is that a single type of 3D element is suitable for structures of any proportions. Furthermore, for the same accuracy, the number of degrees of freedom needed in the hybrid finite element model is usually much lower as compared to the displacement-based model. In this chapter, after briefly discussing the essential aspects of hybrid elements, representative results in elastic deformation under mechanical loads, coupled electrostatic-elastic simulation, and coupled piezoresistive-elastic simulation are presented. Seamless interfacing of the analysis codes with pre- and post-processing modules of any finite element software is also noted.

Keywords Hybrid finite elements · Piezoresistitvity · Multi-physics simulations · Pull-in analysis

1 Introduction

Most physical phenomena encountered in microelectromechanical systems (MEMS) are governed by partial differential equations (PDEs) in the spatial domain and ordinary differential equations in the time domain. These PDEs are invariably coupled to one another because there is interdependence of one physical

K. D. Patil · S. Balakrishnan · C. S. Jog · G. K. Ananthasuresh (✉)
Computational Nanoengineering (CoNe) group and Mechanical Engineering, Indian Institute of Science, Bangalore, India
e-mail: suresh@mecheng.iisc.ernet.in

K. J. Vinoy et al. (eds.), *Micro and Smart Devices and Systems*,
Springer Tracts in Mechanical Engineering, DOI: 10.1007/978-81-322-1913-2_21,

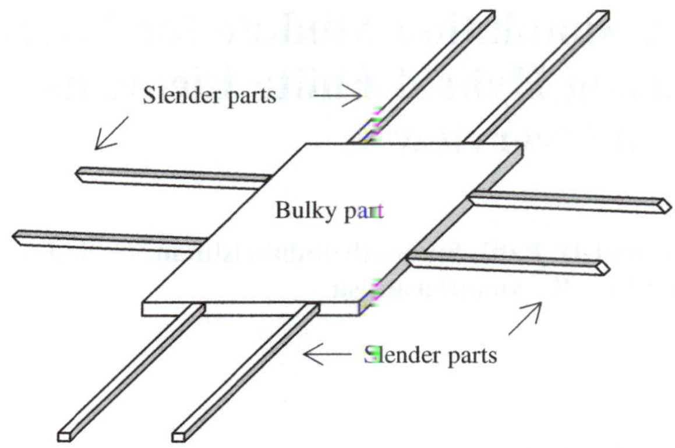

Fig. 1 A typical micromechanical structure with bulky and slender parts

energy domain to another (e.g., elastic to electrostatic and electrical to thermal) [1, 2]. Finite element analysis (FEA) is the most widely used numerical technique for solving the coupled equations in order to simulate the behavior of MEMS components and devices. However, even now in practice, there are two shortcomings to using FEA for the simulation of microsystems.

First, a large number of finite elements (i.e., a finely meshed model) is needed for good accuracy. The reason for this is twofold. First, microsystems components are often very thin in one dimension; and second, they contain relatively bulk bodies joined with slender components in the other two dimensions. Figure 1 illustrates a typical MEMS device where it can be seen that the out-of-plane thickness of the device is much smaller than the size in the in-plane direction. Therefore, a very fine mesh is needed to maintain the aspect ratio of the elements in the discretized mesh close to unity, as necessitated by the traditional finite elements. Also noticeable in Fig. 1 are very narrow portions, namely beams, connected to a wide plate. Bulky portions, which do not undergo much elastic deformation, need not be meshed as finely as the slender portions, but there should be smooth transitions in the mesh. This means that if the same type of finite element is used for slender and bulky portions, one should be careful in changing the density and size of elements while meshing the model. An alternative is to use different types of elements (e.g., beam, plate, and shell) for different portions. The mathematical implication of the type of element is the degree of the underlying interpolating function called the *shape function*. Incompatibility in the interpolating shape functions and the aspect ratios of elements leads to inaccurate results due to what are known as *shear* or *membrane locking phenomena* [3]. Deciding on the suitability of an element type for a given problem demands considerable expertise from the users. Given the multidisciplinary nature of the microsystems technology, it is not fair to expect all users to be aware of the intricacies of solving PDEs using FEA. Thus, one type of elements for any structure, however complex its geometry may be, is preferable. At the same time, too fine a mesh with one element type is an overkill because it leads to excessive computation time.

The second shortcoming of FEA arises while solving coupled PDEs; the method adopted by most simulation algorithms and commercial microsystem simulation software is a staggered approach. That is, if there are two coupled fields, the algorithm solves for one field and then takes the coupling variables to solve for the other field and iterates between the two fields in a sequential manner. As opposed to this, in a combined approach, all the PDEs related to multiple phenomena are solved simultaneously. This improves the computational efficiency.

Based on the foregoing, the aim of this chapter is to describe a simulation strategy for microsystem components and devices using a single type of hybrid finite element and an integrated strategy for solving coupled PDEs. While this approach applies to a number of problems within the MEMS field, only deformation and stress analysis under purely mechanical loads, coupled elastic-electrostatic analysis, and coupled elastic-piezoresistive analysis are explained in detail. Modal analysis, coupled electro-thermal-elastic analysis, etc., can also be solved using similar methods.

The rest of the chapter is organized as follows. Section 2 describes the theory of the hybrid finite elements. Section 3 exemplifies the efficacy of the hybrid FEA for micromechanical structures under purely mechanical loads. Sections 4 and 5 contain discussions of coupled elastic-electrostatic and elastic-piezoresistive simulations, respectively. Integration issues are briefly noted in Sect. 6. Section 7 has concluding remarks.

2 Hybrid Finite Element Procedure

Hybrid finite elements [4–8] are known to give high accuracy with only a few elements. They are not prone to become artificially stiff when the elements are thin in one direction or distorted otherwise. They are also largely free from locking phenomena. In [7], a 27-noded hybrid finite element formulation was proposed. It enables cost-effective FEA for structures that have narrow and thin parts in conjunction with bulky and wide parts. This is a common occurrence in microsystems as shown in Fig. 1. We adopt the hybrid elements for microsystem simulation as reported in [9]. A brief theory of hybrid finite element procedure is therefore pertinent here and is presented next for the linear problem by first discussing the usual displacement-based FEA. Readers interested in the nonlinear formulation of the hybrid elements may consult Ref. [10].

The PDE that governs the elastic deformation of a structural element occupying a domain Ω with the boundary $\partial\Omega$ and subjected to body forces $\mathbf{f}_b$ (e.g., gravity or inertia forces) and boundary force $\mathbf{f}_t$ (e.g., fluid pressure or electrostatic force), is as follows:

$$
\begin{aligned}
&\nabla \cdot \boldsymbol{\tau} + \mathbf{f}_b = 0 \\
&\boldsymbol{\tau} = \mathbf{D}_m : \boldsymbol{\varepsilon} \\
&\boldsymbol{\varepsilon} = \left\{\nabla \mathbf{u} + (\nabla \mathbf{u})^T\right\}/2 \\
&\mathbf{f}_t = \boldsymbol{\tau}\,\hat{\mathbf{n}} \,\text{on}\, \partial\Omega_t \\
&\mathbf{u} = \mathbf{u}^* \text{ on } \partial\Omega_u
\end{aligned}
\tag{1}
$$

where $\boldsymbol{\tau}$ is the stress, $\boldsymbol{\varepsilon}$ the strain, $\mathbf{D}_m$ the constitutive matrix relating $\boldsymbol{\tau}$ and $\boldsymbol{\varepsilon}$, and $\mathbf{u}$ the displacement vector. The force applied on the boundary $\partial\Omega_t$ is $\mathbf{f}_t$; the specified displacement on the boundary $\partial\Omega_u$ is $\mathbf{u}^*$, and $\partial\Omega = \partial\Omega_t \cup \partial\Omega_u$. The variational form of the PDE in Eq. (1) is

$$
\int_{\Omega} \boldsymbol{\varepsilon}(\delta \mathbf{u}) : \mathbf{D}_m : \varepsilon(\mathbf{u})\, d\Omega - \int_{\partial\Omega_t} \delta\mathbf{u} \cdot \mathbf{f}_t \, d(\partial\Omega) - \int_{\Omega} \delta\mathbf{u} \cdot \mathbf{f}_b \, d\Omega = 0 \;\forall \delta\mathbf{u} \tag{2}
$$

where $\delta\mathbf{u}$ is the variation of $\mathbf{u}$. In displacement-based FEA, $\mathbf{u}$ and $\delta\mathbf{u}$ are interpolated within the element using shape functions $\mathbf{N}$ and the nodal displacement degree-of-freedom vector $\mathbf{u}_e$ of an element:

$$
\mathbf{u} = \mathbf{N}\mathbf{u}_e \;\text{ and }\; \delta\mathbf{u} = \mathbf{N}\delta\mathbf{u}_e \tag{3}
$$

We differentiate the displacement as per the definition of the strain given in the third line of Eq. (1) to get the strain displacement relationship for an element:

$$
\boldsymbol{\varepsilon}_e = \mathbf{B}\mathbf{u}_e \tag{4}
$$

By substituting from Eqs. (3) and (4) into Eq. (2), performing integration over all discretized elements, and assembling the global stiffness matrix $\mathbf{K}$, we get

$$
\mathbf{K}\mathbf{U} = \mathbf{f} \tag{5}
$$

where

$$
\begin{aligned}
\mathbf{K} &= \sum_{\text{All elements}} \int_{\Omega_e} \mathbf{B}^{\mathbf{T}} \mathbf{D}_m \mathbf{B}\, \mathrm{d}\Omega_e \\
\mathbf{f} &= \sum_{\text{All elements}} \left(\int_{\partial\Omega_{et}} \mathbf{N}^T \mathbf{f}_t \, \mathrm{d}(\partial\Omega) + \int_{\Omega_e} \mathbf{N}^T \mathbf{f}_b \, \mathrm{d}\Omega \right)
\end{aligned}
\tag{6}
$$

where $\mathbf{U}$ and $\mathbf{f}$ are the global displacement and force vectors.

Thus, the usual finite element analysis described in the preceding equations, assumes only displacement as an independent variable in order to interpolate within the element using the values at the nodes. Strains and stresses are then computed using the strain-displacement and constitutive (i.e., stress-strain) relationships, respectively. On the other hand, in the hybrid finite element, both displacement and stresses are independently interpolated. Consequently, an additional condition is necessary to ensure compatibility between independently interpolated displacements and stresses. Thus, in addition to Eq. (2), we have

$$\int_{\Omega} \delta\boldsymbol{\tau} : \left\{\boldsymbol{\varepsilon}(\mathbf{u}) - \mathbf{D}_m^{-1} : \boldsymbol{\tau}\right\} d\Omega = 0 \ \forall \delta\boldsymbol{\tau} \tag{7}$$

where $\delta\boldsymbol{\tau}$ is the variation of the stress field $\boldsymbol{\tau}$. The two stress fields are interpolated as follows.

$$\boldsymbol{\tau} = \mathbf{P}\tau_e \text{ and } \delta\boldsymbol{\tau} = \mathbf{P}\delta\boldsymbol{\tau}_e \tag{8}$$

where $\boldsymbol{\tau}_e$ and $\delta\boldsymbol{\tau}_e$ denote nodal stresses of an element and their variations, respectively. Choosing the interpolating shape functions $\mathbf{P}$ in Eq. (8) is crucial here because it is to be done to avoid the so-called locking phenomena [7].

Now, Eq. (2) [after $\mathbf{D}_m$: $\boldsymbol{\varepsilon}(\mathbf{u})$ is replaced with $\boldsymbol{\tau}$)] and Eq. (7) leads to the following system of equations for the discretized model, with T denoting the global stress vector:

$$\begin{bmatrix} \mathbf{0} & \mathbf{G}^T \\ \mathbf{G} & -\mathbf{H} \end{bmatrix} \begin{Bmatrix} \mathrm{U} \\ \mathrm{T} \end{Bmatrix} = \begin{Bmatrix} \mathbf{f} \\ \mathbf{0} \end{Bmatrix} \tag{9}$$

where

$$\mathbf{G} = \int_{\Omega} \mathbf{P}^T \mathbf{B} \, d\Omega \tag{10a}$$

$$\mathbf{H} = \int_{\Omega} \mathbf{P}^T \mathbf{D}_m^{-1} \mathbf{P} \, d\Omega \tag{10b}$$

At this point, another important simplification is made to eliminate the stress degrees of freedom of all the elements (i.e., $\mathbf{T}$) by using the second equation of Eq. (9).

$$\mathbf{T} = \mathbf{H}^{-1}\mathbf{G}\mathbf{U} \tag{11}$$

Thus, we are left with

$$\mathbf{K}_h \mathbf{U} = \mathbf{f} \tag{12}$$

where

$$\mathbf{K}_h = \mathbf{G}^T\mathbf{H}^{-1}\mathbf{G} \tag{13}$$

is the stiffness matrix of the hybrid element. Now, even though the stress field was interpolated, its degrees of freedom are “condensed out” by eliminating them in terms of the usual displacement degrees of freedom. So, apart from the additional computation involved in assembling and inverting $\mathbf{K}_h$, the procedure is the same as the usual finite element method. Thus, there is no difference from the user's viewpoint because only the stiffness matrix has changed from $\mathbf{K}$ in Eq. (6) to $\mathbf{K}_h$ in Eq. (13).

Choosing appropriate number of nodes for a 3D element and suitable shape functions $\mathbf{P}$ for them [see Eq. (8)] is the key to the hybrid FEA method. The larger the number of nodes in an element is, the larger will be the element-level computation. But much fewer elements are needed to mesh a structure. That is, it should be noted that a 27-noded hexahedral hybrid brick element needs much fewer elements (i.e., a coarse mesh) for a given structure to give an accurate solution as compared to an eight-noded brick element. So, there is a trade-off between the extent of coarseness of an FE mesh and the number of nodes in an element. Based on numerical experimentation, we had reported in [9] the relative order of accuracy by implementing different elements. The elements in decreasing accuracy are: 27-noded hybrid element, 8-noded hybrid brick element, 27-noded displacement brick element, 18-noded displacement brick element, 10- and 11-noded tetrahedral displacement wedge elements, and 6-noded pentahedral hybrid wedge element.

Even though the 27-noded hybrid element gives the most accuracy, it has two limitations: (i) the meshing algorithm for 27-noded 3D elements is not common in FEA software programs, and (ii) 27-noded element entails the inversion of a 90×90 element-level matrix as can be understood from Eq. (13). So, in spite of the fact that the 27-noded element suffices to have a coarse mesh, it is not the best element in practice. In view of the computation time, accuracy, and meshing capability of the chosen software for integration (NISA from Cranes Software International Limited, here), we have implemented 8-noded hybrid hexahedral and 6-noded pentahedral hybrid wedge element in this work. These elements are shown in Fig. 2.

3 Elastic Simulation of Micromechanical Structures

In this section, we illustrate how hybrid and commonly used displacement finite elements perform relative to each other in terms of accuracy and computational efficiency. Toward this, we compare the results of our hybrid finite element code with the displacement element-based FEA in commercial software. It must be noted at the outset that this is not a comparison of our code with the commercial software because we are not comparing with the best elements of commercial

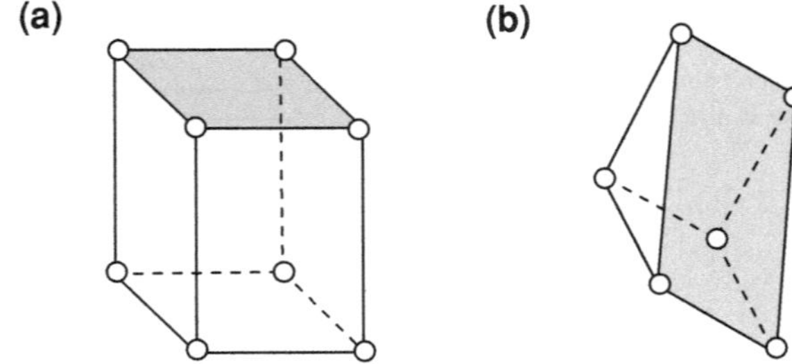

Fig. 2 **a** Eight-noded hexahedral brick element for the interior and **b** six-noded pentahedral hybrid wedge element for the boundary

software for the chosen problems. It is a known fact that commercial software adopt a variety of techniques to avoid locking phenomena. Reduced integration, hour-glass control, etc., are some of the methods [3]. But a priori knowledge of which element to use for which problem cannot be presumed. An element type that works for one problem might not work for another. This is an advantage in favor of the hybrid elements. In summary, we reiterate that in what follows we simply compare the performance of hybrid elements in our implementation with that of the displacement element-based implementations, without any ad hoc modifications, in commercial software. Here, we use ABAQUS (www.simulia.com) and NISA (www.nisasoftware.com). But similar comparisons will come true with any other software. Minor discrepancies will be there because one can never be sure of small differences in implementation.

In all examples in this chapter, we use eight-noded hexahedral elements for the interior and six-noded pentahedral wedge elements, the latter meeting the demands of corners and edges, in the hybrid finite element code in our implementation. We use the same meshed models, but with displacement-based elements in NISA and ABAQUS. NISA uses full integration, whereas ABAQUS uses selective integration in its C3D8 (brick) and C3D6 (wedge) elements.

The following points are to be noted in interpreting the results presented next. First, in tables and figures, we show the accuracy parameter of a result as a % value where the reference value is either calculated using an analytical solution (when it is available) or computed using the hybrid code with sufficiently fine mesh. Second, sometimes, the hybrid code could not be run for very fine meshes because of the limitation of the memory of the computer. But it does not matter because the hybrid code gives accurate results with just a few degrees of freedom in the meshed model. Third, computing time is also indicated for the runs on the same desktop computer for all three implementations (hybrid code, NISA, and ABAQUS).

3.1 Example 1: Folded-Beam Suspension

The geometry of the folded-beam suspension, which is a compliant substitute for a sliding joint, is shown in Fig. 3. It consists of two bulky portions connected with four slender beams in the left symmetric half. The right most edge is constrained to move

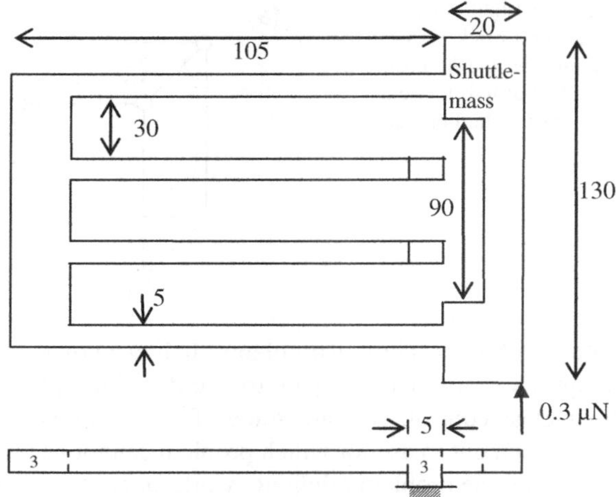

Fig. 3 Symmetric half-model of the folder-beam suspension (all dimensions are in μm)

along the edge, but not perpendicular to it in this symmetric half-model. An analytical solution is available [11] for the displacement of this model for a force applied on the bulky portion, called the *shuttle-mass*, at the bottom, as can be seen in Fig. 3.

Three different meshes were considered with 1056, 10,806, and 3,01,830 degrees of freedom (DoF). The results are shown in Table 1. The table shows the displacement of the shuttle-mass, the accuracy of this displacement relative to the known analytical solution, and the computation time. Figure 4 shows accuracy vs. DoF using results obtained by running many meshed models. In Table 1 and Fig. 4, it can be seen that the hybrid code achieves high accuracy with very few degrees of freedom. Therefore, its computation time is also much less as compared to those of the displacement elements in ABAQUS and NISA.

Similar trends were seen with other examples considered. They included a pressure-sensor diaphragm, a micromachined gripper, a gyroscope's suspension, etc. The latter two consisted of curved geometries and the last, the suspension of a ring-gyroscope, had only slender segments. The efficacy of hybrid elements was seen in all examples considered for a comparative study. The details are in [9].

The algorithm for the purely elastic analysis included geometric nonlinearity. It also has the capability to do dynamic analysis, which is not described here. We consider coupled analyses next. When we do coupled analysis, we use hybrid elements only for the elastic analysis, but not for the others. The reason for this is that the governing PDEs of the others are not known to exhibit locking phenomena.

4 Coupled Electrostatic-Elastic Simulation

Coupled electrostatic-elastic simulation is most common in microsystem simulation because electrostatic actuation is the most widely used actuation technique in microsystems. Here, we have developed algorithms for purely electrostatic

Table 1 Comparison of displacement, accuracy relative to the analytical result, and computing time for the folded-beam suspension

DoF	Hybrid	ABAQUS	NISA
1,056	0.12 μm	0.03 μm	0.02 μm
	97.5 %	20.7 %	17.8 %
	0.23 s	0.10 s	0.08 s
10,806	0.12 μm	0.09 μm	0.09 μm
	99.9 %	74.6 %	72.9 %
	1.84 s	0.80 s	0.93 s
3,01,830	Not run.	0.12 μm	0.12 μm
		99.9 %	99.9 %
		25.1 s	281.8

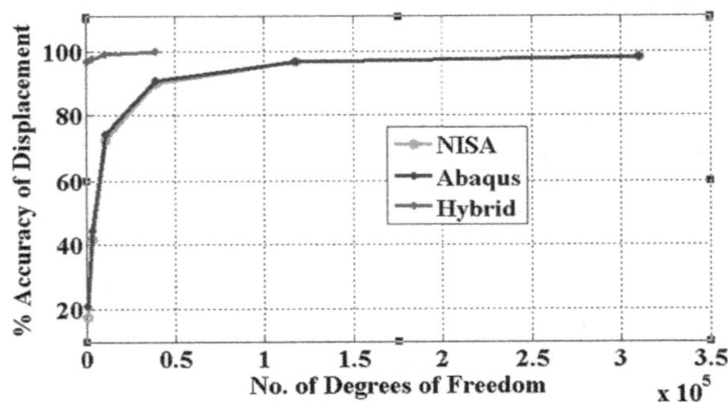

Fig. 4 Comparison of accuracy of the displacement for the example of the folder-beam suspension

simulation (i.e., capacitance calculation) and coupled electrostatic-elastic simulation. The latter is developed for geometrically nonlinear elastic behavior in static and dynamic conditions. It may be noted that only elastic analysis uses hybrid elements. Electrostatic analysis does not require hybrid elements, as we do not see the problems akin to locking here. One novel feature of the algorithm is that it is an *integrated* (some call it *monolithic* [11, 12]) procedure in that the displacement and electrical potential are solved together by combining their respective governing equations. This is in contrast to the staggered approach where the solver routines for the two are called alternately in an iterative procedure [13]. The direct coupling approach followed in the work presented in this chapter makes it computationally more efficient than the existing approaches. This approach is also useful in optimization of MEMS structures [14]. The theory of the integrated formulation is presented next.

4.1 Coupled Electrostatic-Elastic Formulation

The governing equations of the coupled electrostatic-elastodynamic problem can be written together as follows.

$$
\begin{aligned}
&\nabla \cdot (\mathbf{F}\tau) + \rho_0 \mathbf{b^0} = \rho_0 \frac{\partial^2 \mathbf{u}}{\partial t^2} \text{ on } \Omega \\
&\boldsymbol{\tau} = \boldsymbol{\tau}_{\text{mech}} + \boldsymbol{\tau}_{\text{elec}} \\
&\mathbf{t}^0 = \bar{\mathbf{t}}^0 \text{ on } \partial\Omega_t \\
&\mathbf{u} = \mathbf{0} \text{ on } \partial\Omega_u \\
&\nabla \cdot \mathbf{D}_e = 0 \\
&\mathbf{D}_e = \sigma_e J \mathbf{C}^{-1} \mathbf{E} + \mathbf{d}\varepsilon_L \\
&\mathbf{E} = -\nabla_X \phi
\end{aligned}
\tag{14}
$$

where $\mathbf{F}$ is the deformation gradient, $\boldsymbol{\tau}$ the total stress that includes elastic (i.e., mechanical stress) component $\boldsymbol{\tau}_{\text{mech}}$ and electrostatic (the so-called Maxwell stress) component $\boldsymbol{\tau}_{\text{elec}}$, ρ^0 the mass density, $\mathbf{b}^0$ the body force, $\mathbf{u}$ the displacement vector, $\boldsymbol{\Omega}$ the domain over which the problem is posed, $\mathbf{t}^0$ the surface force and $\bar{\mathbf{t}}^0$ its specified value on a portion of the boundary $\partial\Omega_t$, $\partial\Omega_u$ the boundary on which displacement is specified (i.e., anchored portion), $\mathbf{D}_e$ the electric displacement vector, σ_e the electrical conductivity, J the determinant of $\mathbf{F}$, $\mathbf{C}$ the right Cauchy-Green strain tensor, $\mathbf{d}$ the third-order piezoelectric tensor, $\boldsymbol{\varepsilon}_L$ the Lagrangian strain, ϕ the electric potential, and $\mathbf{E}$ the electric field. The expression for the Maxwell stress $\boldsymbol{\tau}_{\text{elec}}$ and the constitutive relationship for the elastic behavior are not given here. Interested readers may refer to Ref. [15]. The permittivity is also missing here, but it is taken implicitly as electrical conductivity as far as the numerical values go. There is some inherent ambiguity in this issue [12], which we take care of in the implementation.

In the usual displacement formulation, the weak form of the governing equations are:

$$
\begin{aligned}
&\int_\Omega \boldsymbol{\tau} : \delta\boldsymbol{\varepsilon}_L \, d\Omega = \int_\Omega \rho_0 \delta\mathbf{u} \cdot \mathbf{b}^0 \, d\Omega + \int_{\partial\Omega_t} \delta\mathbf{u} \cdot \bar{\mathbf{t}}^0 \, d\partial\Omega_t \\
&\int_\Omega \nabla\delta\phi \cdot \left(\sigma_e J \mathbf{C}^{-1} \nabla\phi + \mathbf{d}\varepsilon_L\right) d\Omega = -\int_{\partial\Omega_d} \delta\phi \, D_{\text{en}} \, d\partial\Omega
\end{aligned}
\tag{15}
$$

where D_{en} is the normal component of $\mathbf{D}_e$ and the symbol δ indicates the variation or the weak variable. In the hybrid formulation, one more equation is added by making stress also an independent variable.

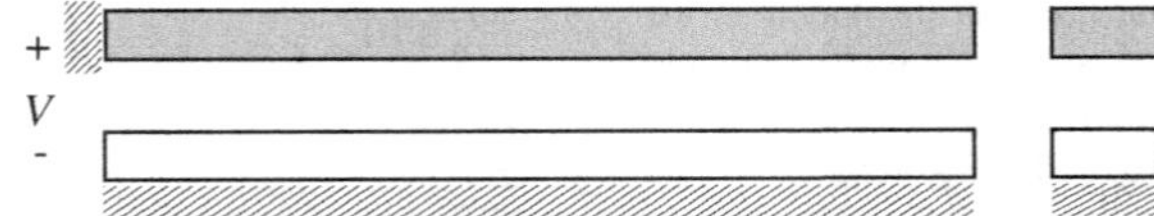

Fig. 5 A cantilever beam considered for the electrostatic pull-in analysis: lateral and side views

$$\int_{\Omega} \delta\boldsymbol{\tau}_{\text{mech}} : \left[\boldsymbol{\varepsilon}_L(\mathbf{u}) - \mathbf{D}_m^{-1} : \boldsymbol{\tau}_{\text{mech}}\right] d\Omega = 0 \; \forall \, \delta\boldsymbol{\tau}_{\text{mech}} \tag{16}$$

By using interpolation functions, which is the key to the hybrid element formulation, the discretized equations for incremental updating of displacements and electric potential are given by

$$\begin{aligned} \mathbf{K}_{uu}\Delta\tilde{\mathbf{u}} + \mathbf{K}_{u\phi}\Delta\tilde{\phi} &= \Delta\mathbf{f}_u \\ \mathbf{K}_{\phi u}\Delta\tilde{\mathbf{u}} + \mathbf{K}_{\phi\phi}\Delta\tilde{\phi} &= \Delta\mathbf{f}_\phi \end{aligned} \tag{18}$$

The cross-coupling terms involving $\mathbf{K}_{u\phi}$ and $\mathbf{K}_{\phi u}$ in Eq. (18) occur in only fully coupled integrated (i.e., "monolithic") formulation and not in the staggered formulation. It may be noticed that the stress terms are condensed out in this coupled simulation as they would be in the purely elastic formulation.

Shown in Fig. 5 are the geometric details of a sample problem pertaining to electrostatically actuated cantilever beam with an electrode underneath. The simulation result is shown in Fig. 6. It can be seen in Fig. 6 that the electrostatic pull-in result obtained using our integrated hybrid code compares well with that of the commercial code using COMSOL MultiPhysics (www.comsol.com). This confirms the accuracy of the code developed. More details of this problem and other examples can be found in [15].

5 Coupled Elastic-Piezoresistive Simulation

Piezoresistors are often used in MEMS devices. Modeling of the piezoresistive effect entails solving the electrical conduction equation because only conducting materials show piezoresistive effect, i.e., their electrical resistivity changes by measurable extent in response to mechanical stress. Thus, we begin with

$$\nabla \cdot \mathbf{J}_e = 0 \tag{19}$$

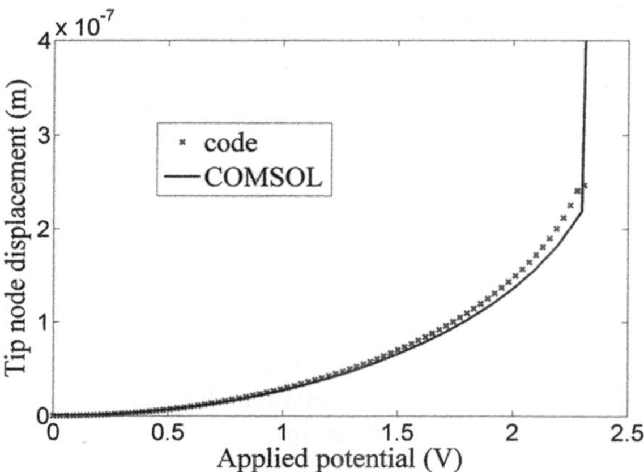

Fig. 6 The displacement of the electrostatically actuated cantilever beam against applied DC voltage until pull-in: comparison of the results of the integrated hybrid code and staggered COMSOL code

where $\mathbf{J}_e$ is the current density vector. The current density vector, at the microscopic level, is related to the electric field by Ohm's law:

$$\mathbf{J}_e = \boldsymbol{\sigma}_e \mathbf{E} \tag{20}$$

where $\boldsymbol{\sigma}_e$ is the conductivity matrix and $\mathbf{E}$ the electric field, which is given by the gradient of the electric potential, ϕ.

$$\mathbf{E} = \nabla \phi \tag{21}$$

The conductivity is the reciprocal of electrical resistivity: $\boldsymbol{\sigma}_e = \boldsymbol{\rho}^{-1}$. The resistivity of piezoresistive materials is modeled, up to first order, in terms of stress, $\boldsymbol{\tau}$.

$$\boldsymbol{\rho}_e = \boldsymbol{\rho}_e^0 \{\mathbf{I} + \pi : \tau\} \tag{22}$$

where $\boldsymbol{\rho}_e^0$ is the resistivity in the unstressed state, $\boldsymbol{\pi}$ the fourth-order piezoresistivity tensor and $\boldsymbol{\tau}$ the second-order stress tensor. This requires us to have both the piezoresistive and stress tensors expressed in the same coordinate system.

The values of the piezoresistive coefficients in the piezoresistive tensor are commonly given in a coordinate system that aligns with the <100> crystallographic directions. On the other hand, the design of a micromechanical component may be such that it is inconvenient to calculate the stress tensor in that coordinate system. For instance, if a piezoresistor is embedded in a cantilever, the piezoresistive tensor is given along the local coordinate system aligned with crystallographic directions, while it is convenient to calculate the stress tensor in the cantilever's coordinate system aligned with its longitudinal axis. The transformation of the stress to the crystallographic coordinate system is tedious because the transformation has to be carried out at every point in the domain where the change in resistivity needs to be calculated. It may be recalled that the piezoresistivity tensor is a property of the material. Thus, once transformed from the crystallographic direction to the cantilever's coordinate system, it will remain the same at every point in the domain.

Here, we observe that the transformed piezoresistive tensor can have more than three independent coefficients contrary to only three that a cubic material has along a coordinate system aligned with the <100> crystal axes. We argue here that approximation of the transformed piezoresistive tensor to one with only three independent coefficients can sometimes lead to considerable errors in the calculated piezoresistive effect. This is a noteworthy point because some of the software packages that simulate the piezoresistive effect (e.g., CoventorWare; www.coventor.com) do not have convenient interfaces for the calculation of the complete piezoresistivity tensor when the crystal lattice is arbitrarily oriented with respect to the principal directions of the micromechanical component. In what follows, we discuss how the fourth-order piezoresistive tensor can be transformed from one Cartesian coordinate system to another. The numerical details of the next two sections can be found in [16].

5.1 Coordinate Transformation of the Piezoresistive Tensor

The piezoresistivity tensor $\boldsymbol{\pi}$ in Eq. (22) has both minor and major symmetries. Consequently, it requires only 36 independent components (instead of 81 in the case of a general fourth-order tensor). It can therefore be written as a 6×6 matrix [2]. For cubic materials, there are additional symmetries when the piezoresistivity tensor is aligned with the <100> lattice directions. Then, as stated earlier, the piezoresistivity tensor contains only three independent coefficients [2]. They are as follows:

$$\boldsymbol{\pi}_{\langle 100\rangle} = \begin{bmatrix} \pi_{11} & \pi_{12} & \pi_{12} & 0 & 0 & 0 \\ \pi_{12} & \pi_{11} & \pi_{12} & 0 & 0 & 0 \\ \pi_{12} & \pi_{12} & \pi_{11} & 0 & 0 & 0 \\ 0 & 0 & 0 & \pi_{44} & 0 & 0 \\ 0 & 0 & 0 & 0 & \pi_{44} & 0 \\ 0 & 0 & 0 & 0 & 0 & \pi_{44} \end{bmatrix} \tag{23}$$

where the indices are compressed according to the usual convention, $11 \to 1$, $22 \to 2$, $33 \to 3$, $12 \to 4$, $23 \to 5$, $31 \to 6$. This means that the (1, 2, 2, 3) position of the fourth-order tensor is the coefficient at the (4, 5) position in the matrix notation. Next, consider the transformation of second-order tensors between two Cartesian coordinate systems [17]:

$$\mathbf{T}' = \mathbf{R}\mathbf{T}\mathbf{R}^{-1} \tag{24}$$

where **R** is the rotation matrix between the two coordinate systems given by

$$\mathbf{R} = \begin{bmatrix} l_1 & l_2 & l_3 \\ m_1 & m_2 & m_3 \\ n_1 & n_2 & n_3 \end{bmatrix} \tag{25}$$

and the directions of the coordinate axes of the rotated coordinate system expressed in terms of the original coordinate system is

$$\mathbf{x}' = \begin{bmatrix} l_1 \\ m_1 \\ n_1 \end{bmatrix}, \mathbf{y}' = \begin{bmatrix} l_2 \\ m_2 \\ n_2 \end{bmatrix}, \mathbf{z}' = \begin{bmatrix} l_3 \\ m_3 \\ n_3 \end{bmatrix} \tag{26}$$

Since a symmetric second-order tensor can be represented as a 6-element column vector using the convention for compression of indices, we can write

$$\mathbf{T}^T = [\mathbf{T}_1 \quad \mathbf{T}_2 \quad \mathbf{T}_3 \quad \mathbf{T}_4 \quad \mathbf{T}_5 \quad \mathbf{T}_6] \tag{27}$$

By using the vector representation of the symmetric second-order tensor shown in the preceding equation, the coordinate transformation can be carried out using a single 6 × 6 rotation matrix multiplication

$$\mathbf{T}' = \mathbf{R}_{6\times6}\mathbf{T} \tag{28}$$

where $\mathbf{R}_{6\times6}$ is given by

$$\begin{bmatrix} l_1^2 & m_1^2 & n_1^2 & 2l_1m_1 & 2m_1n_1 & 2n_1l_1 \\ l_2^2 & m_2^2 & n_2^2 & 2l_2m_2 & 2m_2n_2 & 2n_2l_2 \\ l_3^2 & m_3^2 & n_3^2 & 2l_3m_3 & 2m_3n_3 & 2n_3l_3 \\ l_1l_2 & m_1m_2 & n_1n_2 & l_1m_2 + l_2m_1 & m_1n_2 + m_2n_1 & n_1l_2 + n_2l_1 \\ l_2l_3 & m_2m_3 & n_2n_3 & l_2m_3 + l_3m_2 & m_2n_3 + m_3n_2 & n_2l_3 + n_3l_2 \\ l_3l_1 & m_3m_1 & n_3n_1 & l_3m_1 + l_1m_3 & m_3n_1 + m_1n_3 & n_1l_3 + n_1l_3 \end{bmatrix} \tag{29}$$

Similar to the case of second-order tensors, the coordinate transformation of fourth-order tensors is given by [17]

$$\pi' = \mathbf{R}_{6\times6}\pi\mathbf{R}_{6\times6}^{-1} \tag{30}$$

When the transformed coordinate system differs from the original coordinate system only through a rotation about the z-axis, the transformed piezoresistive tensor expression reduces to a 6 × 6 matrix:

$$
\boldsymbol{\pi}' = \begin{bmatrix}
\frac{3\pi_{11}+\pi_{12}+\pi_{44}+(\pi_{11}-\pi_{12}-\pi_{44})\cos(4\theta)}{4} & \frac{\pi_{11}+3\pi_{12}-\pi_{44}-(\pi_{11}-\pi_{12}-\pi_{44})\cos(4\theta)}{4} & \pi_{12} & \frac{(-\pi_{11}+\pi_{12}+\pi_{44})\sin(4\theta)}{2} & 0 & 0 \\
\frac{\pi_{11}+3\pi_{12}-\pi_{44}-(\pi_{11}-\pi_{12}-\pi_{44})\cos(4\theta)}{4} & \frac{3\pi_{11}+\pi_{12}+\pi_{44}+(\pi_{11}-\pi_{12}-\pi_{44})\cos(4\theta)}{4} & \pi_{12} & -\frac{(-\pi_{11}+\pi_{12}+\pi_{44})\sin(4\theta)}{2} & 0 & 0 \\
\pi_{12} & \pi_{12} & \pi_{11} & 0 & 0 & 0 \\
\frac{(-\pi_{11}+\pi_{12}+\pi_{44})\sin(4\theta)}{4} & -\frac{(-\pi_{11}+\pi_{12}+\pi_{44})\sin(4\theta)}{4} & 0 & (\pi_{11}-\pi_{12}-\pi_{44})\sin^2(2\theta)+\pi_{44} & 0 & 0 \\
0 & 0 & 0 & 0 & \pi_{44} & 0 \\
0 & 0 & 0 & 0 & 0 & \pi_{44}
\end{bmatrix} \tag{31}
$$

where θ is the angle of rotation about the z-axis of the transformed coordinate system with respect to the original coordinate system.

5.2 An example

We now consider an example of a representative micromechanical structure where taking only three-independent-parameter piezoresistive tensor given in Eq. (23) leads to substantial error as opposed to taking the full transformed tensor given in Eq. (31). This happens, as noted earlier, when the crystallographic directions are not aligned with the longitudinal axis of the deforming slender element. Toward this, consider a micro-mirror that twists about a single axis as shown in Fig. 7. It has a wide plate in the middle with twisting beams on either side. When there is a force on the vertical faces that are oriented along the longitudinal axis of the structure, in the opposing directions so as to cause the tilting of the plate about the

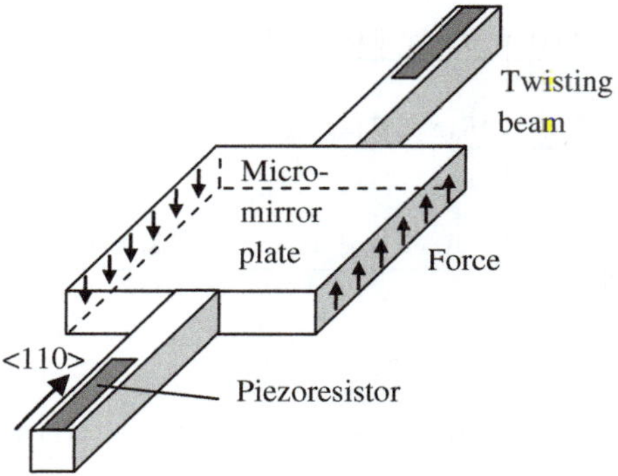

Fig. 7 A example micro-mirror structure in which the piezoreistor's <110> axis is aligned with the longitudinal axis of the twisting beams

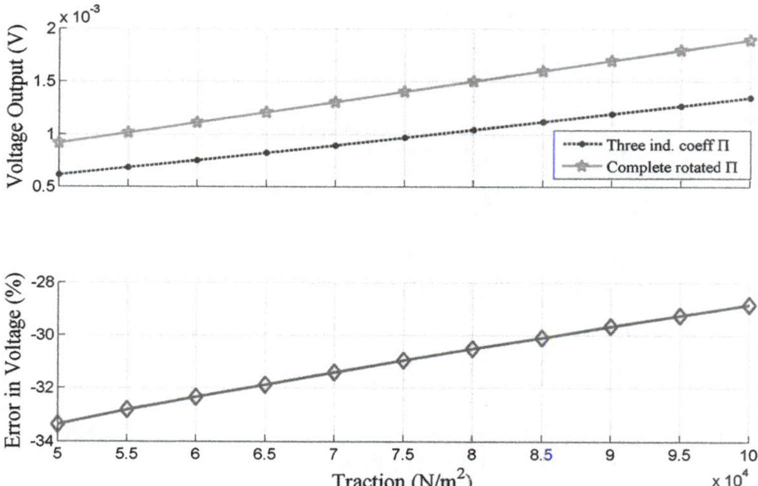

Fig. 8 The discrepancy in the computed voltage of the Wheatstone bridge circuit when the transformed piezoresistor has all parameters and only three parameters

longitudinal axis, the beams must twist. Assume that there is a pair of piezoresistors whose <110> axis are aligned with the longitudinal axis. Each one of this is one of the resistors in the respective standard Wheatstone bridge circuits. When the beams twist, there will be a change in the resistance of the piezoresistor, which results in a change in voltage of the corresponding bridge circuit.

The stress was computed using hybrid elements and it was used in calculating the resulting piezoresistive coefficients and the change in resistance. It was done using the usual three-parameter piezoresistive tensor as well as the full-parameter tensor. The difference in the computed voltage is shown in Fig. 8. It can be seen

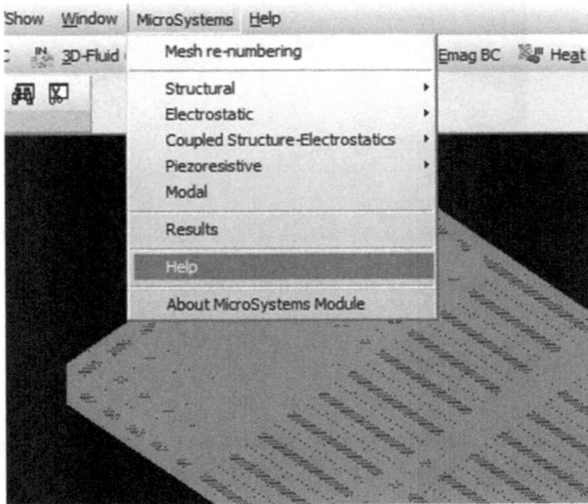

Fig. 9 Various capabilities of the microsystems simulation module

that there is as much as 33 % error between the two. Thus, in the software module we have developed, good accuracy is obtained irrespective of the alignment of the crystallographic axes of the piezoresistors with the main axes of the micromechanical structures. Hybrid elements are used in elastic analysis in this coupled code.

6 Integration of the Microsystem Module

As discussed in the preceding sections, by using the hybrid finite elements and other novel features, we have developed a simulation module for microsystems. It can be run as a stand-alone module with custom-developed graphical user interface or by interfacing it with a commercial finite element software. Here, for the purpose of illustration, we use NISA (www.cranessoftware.com). We use the DISPLAY IV module of NISA as a pre- and post-processor. That is, the model is created and material properties and boundary conditions are specified in the DISPLAY IV environment and the data is saved in the NISA environment. This file is read by our parser to write another data file that can be read and interpreted by our hybrid analysis codes. The results of the hybrid code is written to a file in the format of DISPLAY IV. The results are viewed in the GUI of NISA. This can be done, we emphasize, in any other commercial finite element software.

Figure 9 shows the current capabilities of the integrated microsystems simulation module, as can be seen in the pull-down menu item. A typical display of the result is shown in Fig. 10.

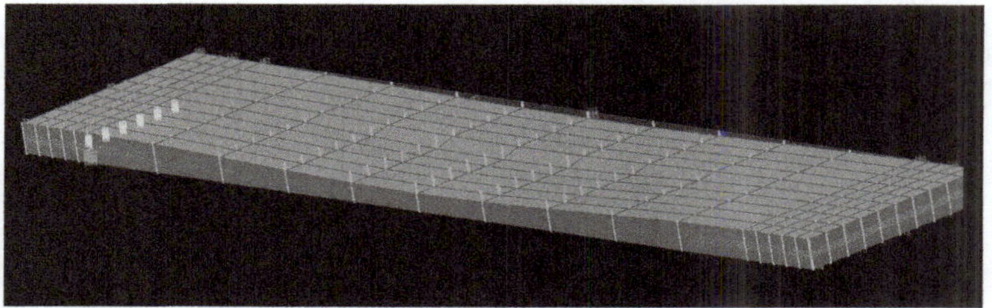

Fig. 10 Rendering of the deformed configuration of the pull-in of a fixed-fixed beam

7 Conclusions

In this chapter, we have addressed a critical need in the simulation of microsystems by allowing the users to work with a single type of finite element and obtain accurate results with a coarse mesh. It is accomplished by using hybrid finite elements where displacement and stress fields are both interpolated using suitable shape functions. It was shown that, the implementation will be seamless because the stress degrees of freedom are eliminated in terms of the nodal displacements. As a result, the new technique can be used in the same manner as the traditional displacement-based finite element simulation. The other important novel feature is the integrated solution strategy when more than one PDE is involved. Some other novel features are introduced in other capabilities of the simulation module. A representative simulation is presented for piezoresistive structures where it was shown that accuracy need not be compromised when the piezoresistor's crystallographic axes are not aligned with the coordinate system of the device. Some other capabilities of the microsystem module include modal, thermal, and electrothermal analyses.

Acknowledgments The authors gratefully acknowledge the financial support from the National Programme on Micro and Smart Systems (NPMASS), Government of India, under PARC 3.9 project entitled "Software Development and Scientific Computing for Micro and Nano Engineering". This work is a result of the synergistic efforts of numerous members of project staff and students in the Computational Nanoengineering (CoNe) group in the Indian Institute of Science, who worked on different aspects of implementing the hybrid and other codes and testing the initial versions of the simulation module. Help from Cranes Software International Limited, Bangalore, with the user-interface in NISA, is also gratefully acknowledged.

References

1. Senturia SD, Harris RM, Johnson BP et al (1992) A computer-aided design system for microelectromechanical systems (MEMCAD). J Micro Elecromech Syst 1(1):3–13
2. Senturia SD (2001) Microsystem design. Kluwer Academic Publishers, Boston
3. Bathe KJ (1996) Finite element procedures. Prentice Hall, Englewood Cliffs
4. Spilker RL, Singh SP (1982) Three-dimensional hybrid-stress isoparametric quadratic displacement elements. Int J Numer Meth Eng 18:445–465

5. Pian THH (1995) State-of-the-art development of hybrid/mixed finite element method. Finite Elem Anal Des 21:5–20
6. Pian THH, Chen DP, Kang D (1983) A new formulation of hybrid/mixed finite element. Comput Struct 16(1–4):81–87
7. Jog CS (2005) A 27-node hybrid brick and a 21-node hybrid wedge element for structural analysis. Finite Elem Anal Des 41:1209–1232
8. Jog CS (2010) Improved hybrid elements for structural analysis. J Mech Mater Struct 5(3):507–528
9. Sundaram MR, Ganesh G, Pavan K, Varun B, Jog CS, Ananthasuresh GK (2012) Static elastic simulation of microelectromechanical structures using hybrid finite elements. In: International conference on smart materials, structures and systems, 4–7 Jan, Bangalore
10. Jog CS, Kelkar PP (2006) Nonlinear analysis of structures using high-performance hybrid elements. Int J Numer Meth Eng 68:473–501
11. Rochus V, Rixen DJ, Golinval JC (2006) Monolithic modelling of electromechanical coupling in microstructures. Int J Numer Meth Eng 65:461–493
12. Yoon GH, Sigmund O (2008) A Monolithic approach for topology optimization of electrostatically actuated devices. Comput Meth Appl Mech Eng 197(45–48):4062–4075
13. Li G, Aluru NR (2002) A Lagrangian approach for electrostatic analysis of deformable conductors. J Microelecromech Syst 11:245–254
14. Alwan A, Ananthasuresh GK (2006) Coupled electrostatic-elastic analysis for topology optimization using material interpolation. J Phys Conf Ser 34:264–271
15. Patil KD, Jog, CS, and Ananthasuresh, GK (2014) Monolithic hybrid finite element strategy for coupled structure-electrostatic analysis of micromechanical structures. In: International conference on smart materials, structures and systems, 8–11 July, Bangalore
16. Balakrishnan S, Deshpande, K, and Ananthasuresh GK (2014) A note on modelling of directionality in piezoresistivity. J ISSS 3(1):1–8
17. Jog CS (2007) Continuum mechanics. Narosa Publishing House, New Delhi

Structural Health Monitoring: Nonlinear Effects in the Prognostic Analysis of Crack Growth in Structural Joints

B. Dattaguru

Abstract Structural Joints are critical locations in aerospace structures. They are potential sources of failure due to the presence of stress concentrations due to the inherent discontinuities. Health monitoring systems for aerospace structures need to concentrate on locations with joints for both diagnostic purposes and prognostic estimations. This chapter primarily deals with prognostic part of SHM. The major issue in both bonded and fastener joints, is the non-linearity present in both of them in the deformation behaviour which complicates the prognostic estimations. Typical results are presented to focus on the non-linear behaviour of damage growth characteristics of relevance to the health monitoring.

Keywords Structural health monitoring · Prognostics · Fastener and bonded joints · Non-linear analysis

1 Introduction

Structural health monitoring systems are becoming popular in safety critical engineering fields such as aerospace, nuclear and bridges. Reliable monitoring is made possible with the present day developments on the use of microsensors with embedded microprocessors in these systems. These systems are expected to participate in both diagnostic and prognostic parts of the health monitoring [1]. They will be expected to provide sufficient information on the presence of defects, possible initiation of cracks and model the growth of damages including cracks to

B. Dattaguru (✉)
Institute of Aerospace Engineering and Management, Jain University, Bangalore, India
e-mail: datgurb@gmail.com

B. Dattaguru
TechMahindra, Bangalore, India

K. J. Vinoy et al. (eds.), *Micro and Smart Devices and Systems*,
Springer Tracts in Mechanical Engineering, DOI: 10.1007/978-81-322-1913-2_22,

help estimate the remaining life and residual strength of structural components [2]. When an aircraft comes for inspection, operators will look for information on diagnosis of damages or cracks in the structure. Damages if any may be left unattended only if they are minor. Major damages are those which are likely to significantly affect the remaining life of the structure and they are repaired. The operators would like to know the extent of remaining life of the structure before repair and also evaluate the repair efficiency by estimating the life to failure after the repair. For such evaluation it is necessary to study damage (or crack) growth rate and this is the essential aspect to be dealt with in prognosis. If the problem has non-linearity (stresses vary non-linearly with load) then it should be recognized and the integrations should be carried out appropriately.

Large scale primary components in aerospace structures are often made in parts and assembled using joints. Fastener (bolted, rivetted) joints are the most commonly used types of joints in metallic structures over the years. These are detachable type of joints and provide ease of assembly and disassembly for the purposes of operational events such as inspection, maintenance and repairs. Adhesively bonded joints have become popular after the advent of composite structures for several high technology applications and in wing type of structures where one looks for large area bonding. These are semi-permanent joints whereas welded and spot welded joints used extensively in automotive vehicles are permanent joints. This chapter deals with fastener and bonded joints which are mostly used in aerospace structures.

Lug joints are the commonly known fastener joints. The pin-hole combination in these joints could be interference, push or clearance fit. Load transfer through these fastener joints could result in partial contact at the pin-hole interface and leading to moving boundary value problems. The changing contact/separation regions around the pin-hole interface makes the stress variation in the joints to vary non-linearly with increasing load transfer [3]. Further higher extent of interference could cause material yielding resulting in a second type of non-linearity [4].

Similarly in adhesively bonded joints the eccentricity of load path results in geometric non-linearity and further, the adhesives could be beyond its yield point resulting in material non-linearity [5]. Both these types of joints are to be analyzed in states of two-way non-linearity and these are the critical aspects to be dealt with in prognostic analysis on crack growth life and residual strength of the components.

Non-linear finite element analysis is conducted using an inverse method of analysis in fastener joints to study the load-contact behaviour. The point of crack initiation is identified as the location of maximum tensile hoop stress on the pin-hole interface and the cracks are grown from there to estimate the remaining life [6, 7]. Further, the case of yielding around the lug joint is considered due to the interference in the pin-hole fit and the crack growth phenomenon is studied through the inelastic region [8]. In case of adhesively bonded joints, experimental studies were conducted on the the de-bond growth in single lap joints. The finite element analysis combines both material and geometric non-linearity [9].

The methods, procedures and numerical studies presented in this chapter are of direct relavance to the prognostics strategy in critical aerospace components.

When defects are diagnoized the information needed will be the remaining life in the presence of the identified defect and the residual strength. Depending on the extent of damage, the area containing the defect could be repaired and the remaining life is to be recalculated in the repaired condition [10]. Joints are one of the primary locations where structural health monitoring exercise need be carried out. The results presented in this chapter are a small part of a larger effort to provide relevant information at the inevitable structural joints in aerospace structures. This chapter presents the effect of non-linearities on the stress distributions in typical joints in both fastener and bonded joints.

2 Fastener Joints

Lug joints are typical fastener joints which will be discussed in the present chapter. The configuration considered is a lap joint between two plate/panel type of two dimensional components. They transfer load through a round pin and a typical multi-pin joint configuration in Fig. 1. It is well known that at each joint in the top plate certain amount of load is transferred through the pin bearing to the bottom plate and the remaining load is bypassed in the top plate. Assuming that the joints are at a distance apart, it is sufficient to consider for analysis a pin in a hole bearing against the hole boundary. Figure 1 also shows a finite element mesh for this types of joints.

The pin hole fit could be interference, push or clearance fit. The pin of diameter $2a(1 + \lambda)$ is introduced into a hole of diameter $2a$. The joint is an interference fit if λ is positive, clearance if λ is negative and push or neat fit if λ is zero. The pin transfers load P and this load transfer creates partial contact situation. The partial contact/separation configurations are shown in Fig. 2. Consider the interference fit in Fig. 2. At no load transfer ($P = 0$) there is all round contact at the pin-hole interface and there are compressive radial stresses at all points on the interface. During load transfer the compressive stresses increase at point A and they decrease at point B. At certain load level the point B will reach zero stress and the pin-hole interface will separate. Further loading will cause the separation to spread symmetrically about the point B. However the extent of separation will reach an asymptotic limit beyond which it can not separate. Similarly in case of clearance fit the pin hole interface has no compressive contact to start with at zero load. At an infinitesimally small load transfer the pin-hole interface will come into contact at point A and for further loading the contact spreads symmetrically about A. Here too the extent of contact would reach an asymptotic limit and contact can not go beyond a limiting value. The limiting value is same for both interference and clearance and that configuration is the extent of contact/separation for push fit.

The problem of interference and clearance fits are non-linear with load transfer due to changing contact/separation whereas the case of push fit is a linear problem and the stresses and deformations are linear with load transfer. The partial contact/separation configurations for all the three fits are shown in Fig. 2. The figure also

Fig. 1 Two-dimensional view with FE mesh: three fastener lap joint

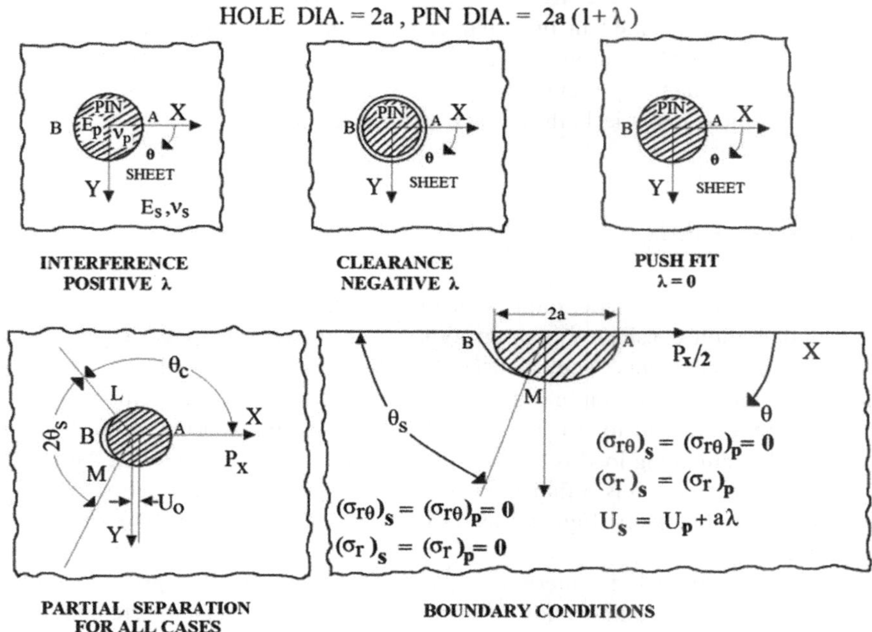

Fig. 2 Partial contact configurations and boundary conditions for different fits with pin load

shows the boundary conditions to be imposed in the regions of separation and contact. Here U, V are the radial and tangential displacements and the notation for stresses is standard. Suffix with p refer to the pin values and suffix with s refers to the sheet. These fall into the moving boundary value problems.

The non-linear problem of these joints is analysed using an inverse method of analysis [11]. This will not be described here and it is well covered in the literature referred. The finite element mesh used for the analysis is made of four node quadrilateral elements. The hoop stress variation with the level of load transfer is shown in Fig. 3 qualitatively for the case of interference fit joint. The non-linear variation of the stresses is seen with the load level. At zero load transfer there is initial hoop stresses due to the introduction of the interference pin. As the load transfer level is increased the actresses vary linearly with applied load till there is pin-hole separation. The non-linearity becomes perceptible after this load level. Finally when the separation reaches an asymptotic value the contact/separation extents do not change and the variation becomes nearly linear. Obviously the variation of hoop stress is high in the post separation region where as in the full

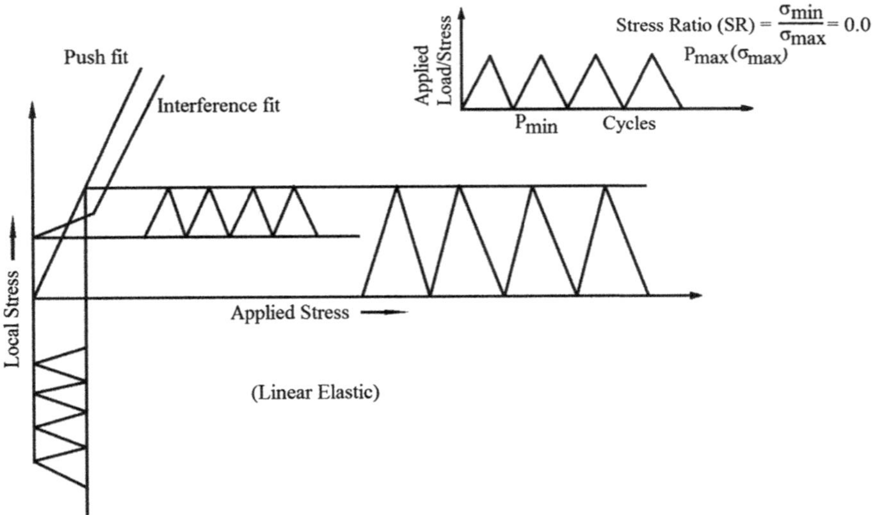

Fig. 3 Effect of interference in joints shown to improve the fatigue life of the joints

contact region before the inception of separation the variation of the hoops stress with load level is very low. Shown also in Fig. 3 is the variation of hoop stress for push fit which is a linear problem. For a cyclic applied stress variation on the joint, the local stress variation is shown. The cyclic local stresses at the pin hole interface undergo lesser alternating component for the case of interference joints, whereas the same is much larger for the case of push fit. This is the reason, an interference joint provides higher fatigue life. There is an increase of mean stress which marginally off-sets the advantage due to the decrease in alternating stress. There were instances reported in literature that the mean stress can also be decreased by cold working of the holes by introducing a large size mandrel into the hole and removing it [4]. This will produces plastically stretched material around the hole boundary causing compressive residual hoop stresses decreasing the mean stress levels.

Figure 4 shows the variation of stress intensity factor variation for a particular crack length. This refers to the lug joint shown as an inset in Fig. 4. The Stress Intensity factor is estimated using Modified Crack Closure Integral [12]. Obviously when the crack length grows the stress intensity factor at the crack becomes larger. Also the slope of the stress intensity factor variation with load level is also larger beyond the region of separation. For a given fatigue cyclic loading the ΔK (the change in SIF between the maximum and minimum load in the cycle) is smaller for interference joint when the full contact is maintained and at load levels before the separation. The crack growth obviously is slower in case of interference joint in the full contact region.

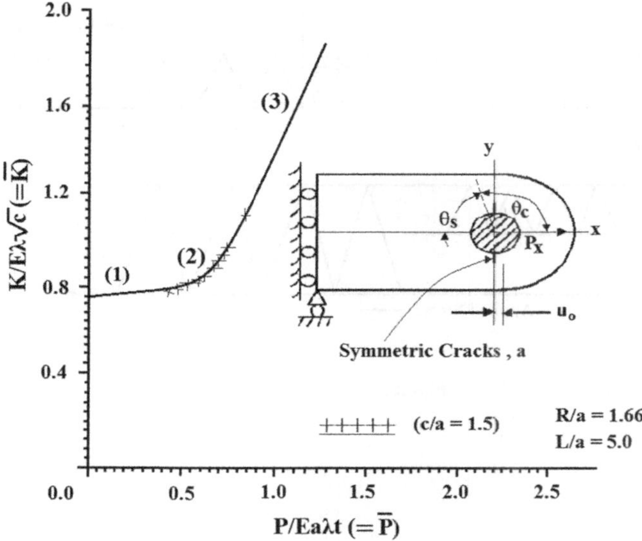

Fig. 4 The stress intensity factor variation with applied load level for various crack lengths in an interference fit joint

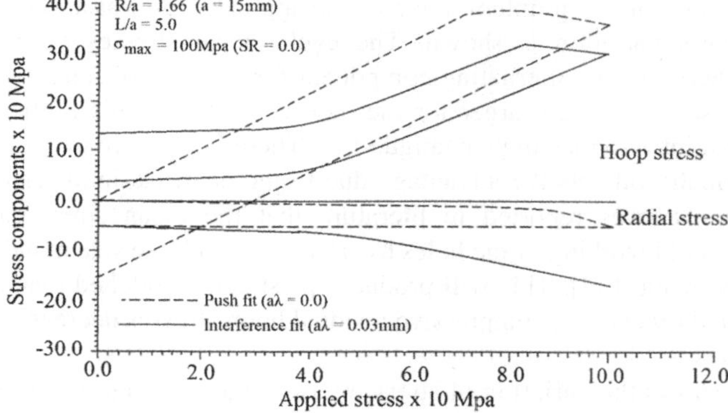

Fig. 5 Variation of radial and hoop stress with applied stress beyond the yield point of the material

The introduction of interference fit of larger interference value could cause plastic deformation around the hole. This can be checked with von-Mises stress criteria form the radial and hoop stresses caused by the interference. In this case the hoop and radial stress distribution within and beyond the load level where plastic deformation [13], is shown in Fig. 5. This is for the case of elastic-perfectly

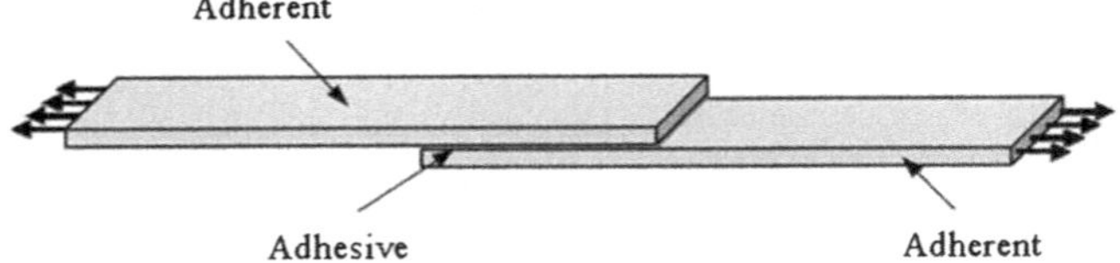

Fig. 6 Typical adhesively bonded lap joint

plastic non-linear stress-strain curve for the lug material. The stress distribution during loading would follow constant value beyond the yield point. Later, unloading will follow a different path and this will significantly affect the use of this information for prognostic crack growth estimations. The procedure follows like the case of strain life estimation. Here the position of the cycle has to be mapped on to the loading or unloading curve and the effect of each cycle is to be evaluated depending on the position where it is mapped on to this curve.

3 Adhesively Bonded Joints

As observed before, the adhesive bonding has become popular with the advent of composites as preferred materials for aerospace structures. Laminated composites have become the most preferred for aircraft wing type structures. Extensive experimentation was done in the past to understand the damage tolerance behavior of these joints [5] and it was also the topic of considerable interest in computational field.

The configuration on which current analysis is carried out is shown in Fig. 6. The adherends could be of both adherends made of metallic and/or composite materials are used in the study. The adhesive considered is Redux—319A epoxy. Fatigue de-bond growth experiments were carried out with specimen designed on ASTM guidelines. A typical result showing the variation of rate of de-bond growth with de-bond length is shown in Fig. 7. A power law variation is expected and the fit shows a variation similar to Paris law [14] as

$$\frac{\mathrm{d}a}{\mathrm{d}N} = C\Delta G_{1\max}^{m}$$

Where $m = 5.845$, $c = 2.2$ E-15 and G is the mode I Strain Energy Release Rate. The prognosis analysis is to estimate the crack growth depending on the G variation based on non-linear analysis.

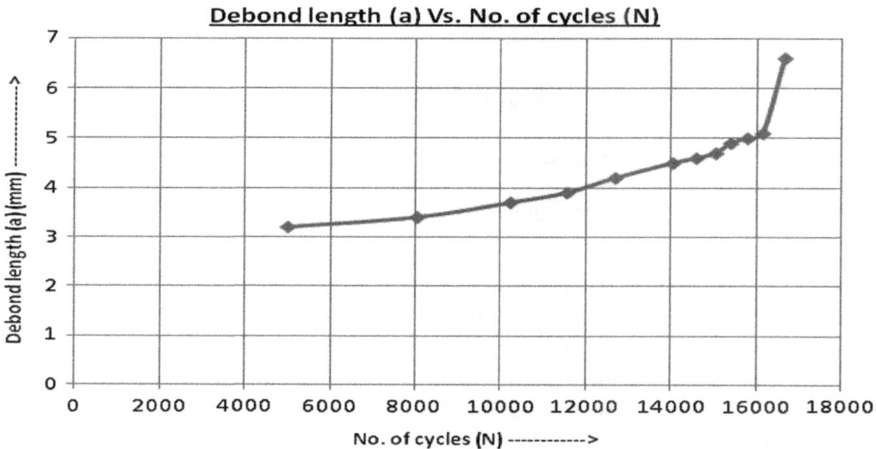

Fig. 7 The growth of de-bond in adhesively bonded lap joint

4 Concluding Remarks

Prognostics analysis is a major part of Structural Health Monitoring system. After the diagnostic part detects possible damage in any structure, one needs to know the remaining life and the remaining strength with the damage. The damage may be left unattended if it is minor or repaired if it is major, the operator would like to know certain approximate extent of the remaining life expectancy of the component. In structural systems one should be aware that joints are major locations where SHM should play an important role in both diagnostic and prognostic parts. The stresses with load often vary non-linearly and the remaining life should account for this. Fasteners which are commonly used in aerospace structures have non-linearity due to contact stress and material non-linearity. The adhesively bonded joints exhibit geometric non-linearity as well as material non-linearity in the adhesive. These issues have been discussed with typical numerical results.

Acknowledgement The author likes to pay special tribute to Dr. V.K. Aatre (currently Visiting Professor, IISc and formerly Scientific Advisor to Defence Minister, Govt. of India) who is responsible for the author's entry into the field of Smart Structures and SHM. The author's association with the National programs (NPSM and NPMASS) which Dr. Aatre initiated, helped the author to take forward his contributions on Structural Mechanics to Flight Safety and SHM systems. The author thanks many of his doctoral students who contributed to most of the work presented in this chapter. Thanks are due to Drs. P.D. Mangalgiri, V.A. Eshwar, S.P. Ghosh, K. Satishkumar, K. Badarinarayana and P.K. Sahoo, and also Dr. P.R. Arora who primarily contributed along with the author in the work on fatigue.

References

1. Spencer BF Jr, Ruiz-Sandoval ME, Kurata N (2004) Smart sensing technology for structural health monitoring. In: Proceedings of 13th world conference on earth quake engineering, Vancouver, Canada
2. Goranson UG (1993) Damage tolerance: facts and fiction. 14th Plantema memorial lecture, presented at the 17th ICAF symposium, Stockholm, Sweeden
3. Morgetson J, Morland LW (1970) Separation of smooth circular inclusion from elastic and visco-elastic plates subjected to uni-axial tension. J Mech Solids 18:295–309
4. Arora PR, Dattaguru B, Subramanya Hande HS (1992) Estimation of elastic-plastic boundary around cold-worked holes. J Test Eval (JTEVA) 20(5):369–375
5. Dattaguru B, Everett RA Jr, Whitcomb JD, Johnson WS (1984) Geometric non-linear analysis of adhesively bonded joints. J Eng Mater Technol (ASME) 106:59–65
6. Schijve J, Hoeymakers ANW (1979) Fatigue crack growth in lugs. Eng Mater Struct 1:185–201
7. Hsu TM (1981) Analysis of cracks at attachment lugs. J Aircr 18(9):755–760
8. Satish Kumar K, Dattaguru B, Ramamurthy TS (1996) Analysis of a cracked lug loaded by an interference fit pin. Int J Mech Sci 38(8–9):967–969
9. Sahoo PK, Dattaguru B, Manjunath CM, Murthy CRL (2013) Stress prediction methods for adhesively bonded lap joints between composite-composite/metal adherends. In: Advances in modelling and design of adhesively bonded systems. Scrivener Publishing, Wiley, Hoboken, New Jersey, USA
10. Dattaguru B (2012) Structural integrity and damage tolerance strategy for joints in aerospace structures, (Plenary lecture). In: Proceedings of ICCMS 2012, Hyderabad
11. Mangalgiri PD, Dattaguru B, Rao AK (1984) Finite element analysis of mechanically fastened joints. Nucl Eng Des 78:303–311
12. Rybicki EF, Kanninen MF (1977) A finite element calculation of stress intensity factors by a modified crack closure integral. Eng Fract Mech 9:931–938
13. Dattaguru B (2013) Effect of non-linear behaviour of joints and on the damage tolerance analysis in aerospace structures. Proc Indian Natl Sci Acad (Special Issue—Part A) 79(4):553–562
14. Anderson TL (2005) Fracture mechanics: fundamentals and applications, 3rd edn. CRC Press, Boca Raton, Florida, USA

Part V
Systems and Applications

Smart e-Textile-Based Nanosensors for Cardiac Monitoring with Smart Phone and Wireless Mobile Platform

Prashanth Kumar, Pratyush Rai, Sechang Oh, Robert E. Harbaugh and Vijay K. Varadan

Abstract Wireless monitoring of Cardiac activity for diagnostic purposes as well as rehabilitative applications has gained significant traction over the past decade. The integration of the nanotextile-based sensors in regular clothing can make the health monitoring system completely unobtrusive and "invisible" for everyday use. A significant research effort has been underway to bring these technologies out of the lab and encourage their use in clinical practice. In this paper, we review the promising efforts made in this direction and address some of the remaining impediments to the wide adoption of these technologies. Based on the existing literature, we conclude that the clinical adoption of these systems may require a trade-off between existing diagnostic techniques for Cardiovascular Diseases (CVDs) and techniques that can be reproducibly and robustly implemented on textiles.We focus specifically on the measurement of Cardiac Biopotentials (CBP) using smart nanotextile garments and wireless systems.

Keywords Textile nanosensor · Smart textile · Wireless · Bioelectromagnetism · ECG · EEG · EOG · EMG · Cardiovascular disorder · Neurological disorder

---

P. Kumar · P. Rai · S. Oh · V. K. Varadan (✉)
Department of Electrical Engineering, University of Arkansas, Fayetteville, AR, USA
e-mail: vjvesm@uark.edu

V. K. Varadan
Department of Biomedical Engineering, University of Arkansas, Fayetteville, AR, USA

R. E. Harbaugh · V. K. Varadan
Department of Neurosurgery, Penn State University Medical School, Hershey, PA, USA

V. K. Varadan
Global Institute of Nanotechnology in Engineering and Medicine, Fayetteville, AR, USA

K. J. Vinoy et al. (eds.), *Micro and Smart Devices and Systems*, Springer Tracts in Mechanical Engineering, DOI: 10.1007/978-81-322-1913-2_23,

1 Introduction

The leading causes of mortality across the globe are Cardiovascular Diseases (CVDs) and Strokes and they are projected to increase the mortality to 23.3 million by 2030 [1]. On the one hand, a widely accepted strategy to minimize risk of CVDs is to follow a healthy lifestyle of adequate physical activity, healthy balanced diet, and refraining from excessive alcohol consumption, and smoking tobacco. On the other hand, if an individual has an occult CVD that remains asymptomatic, no medical attention is afforded to the individual until these diseases manifest clinically as angina or shortness of breath. The latter case often leads to mortality that could have been prevented with appropriate intervention and treatments if the disease was identified earlier. Over the past decade, with concomitant advancements in portable and wireless electronics, preliminary cardiac diagnostic data like Electrocardiograph (ECG), Vectorcardiography (VCG), and Blood Pressure can be measured noninvasively with minimal to no effort or training on the part of the patient.

The essential components of the smart textile cardiac monitoring system are the same as conventional biomedical monitoring equipment—the sensor used to measure the biomedical signal, the signal acquisition and conditioning hardware, the wired or wireless transmission of the acquired data, and the final presentation of the data for medical personnel. In the case of an e-textiles system, the sensors are made of textile materials. Figure 1 shows an illustration of the components and their integration toward an end-to-end cardiac monitoring system.

Nanomaterials and nanocomposite-based components of smart textile system have been developed and commercialized previously for consumer applications. Cloth keyboard developed by Softswitch incorporate micro/nanocapsules of conductive material in insulator matrix made of paraffin as tunneling effect switches. By application of pressure, the material's resistance drops from mega ohms to less than one ohm [2]. Graphene coating of natural and synthetic fibers have been demonstrated by [3] to show improved mechanical strength, fire-retardant and a sturdy platform for developing textile-based electronics.

The research challenges, specific to e-textile systems in cardiac care, not only arise from the functional design of the system, but also in the ergonomic and reliable design decisions to be made without compromising the quality of signals acquired, the ease of use, and an intuitive user interface. These factors play a vital role in the acceptance of these technologies for the average user. At this juncture, it is important to note that the same factors govern the type of information architecture used, in terms of wireless communication protocols and trade-offs in the amount of data processing done on-board, i.e., on the smart textile garment, and processing to be done at the data logging device (smart device or PC) or a remote server where the data is uploaded. In this chapter, the schools of thought behind smart textile systems design are visited both from the textile nanosensor perspective as well as the wireless communication perspective. Examples are

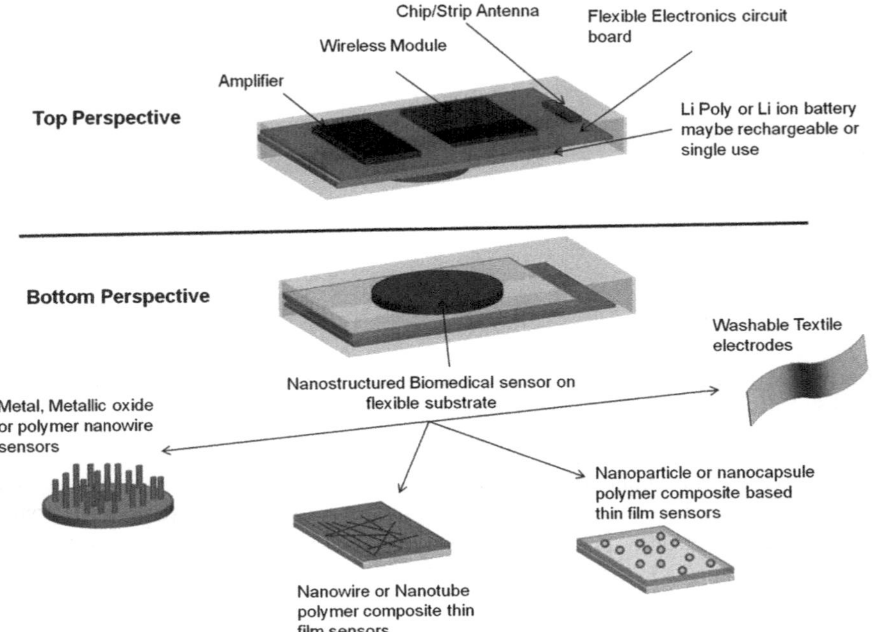

Fig. 1 An illustration of the integration of sensors, electronics and wireless communication

provided for these approaches as wearable healthcare textiles systems for cardiac monitoring e-bro for men and e-bra for women. In conclusion, the challenges and future directions for healthcare smart textiles are summarized.

2 Cardiac Biopotential Measurements: ECG Versus VCG

The cardiac electromotive activity is modeled as a dipole vector, which is instantaneously generated at each moment of the depolarization phase or cardiac muscle contraction phase of every cardiac cycle [4]. The summation of all dipole vectors created at every instant is the resultant electromotive force generated by cardiac activity, called the cardiac vector.

There are two types of CBP measurements used to observe this cardiac vector, namely, ECG and VCG. ECG is a temporal depiction of the cardiac vector as observed from various electrodes placed on the frontal surface of the human body. On the other hand, VCG is a spatiotemporal depiction of the actual cardiac vector in three dimensions. VCG traces the tip of the cardiac vector in 3D as time elapses and provides an additional perspective through the back electrode (*M* in Fig. 2) that is not available in the 12 lead ECG.

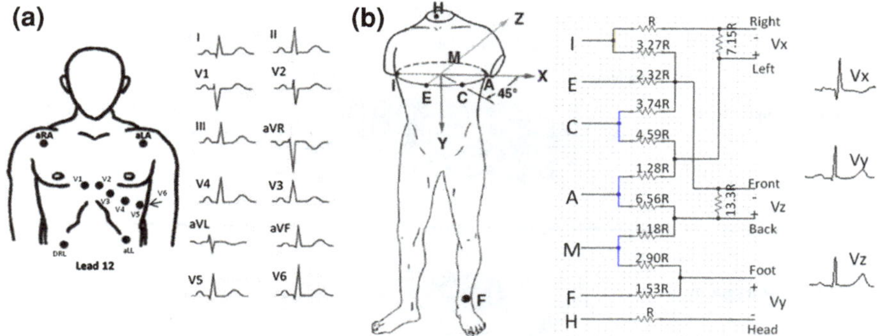

Fig. 2 **a** 12 Lead ECG electrode locations and depictions of typical signals **b** Frank XYZ [5] Lead system electrode placement for VCG

The two measurements principally differ in terms of electrode placements. For ECG, a standard 12-lead electrode placement is used for diagnosis purposes. This involves 10 electrodes and 12 signals are acquired between combinations of electrodes. For VCG, the Frank XYZ [5] lead system is the most widely used electrode placement setup.

ECG's 12 lead system has linearly dependent leads i.e. some of the lead signals can be obtained as algebraic sums of other lead signals. The three leads of the VCG are mutually orthogonal and linearly independent. For a more exhaustive treatment of both ECG and VCG in terms of electrode placements and developmental history, the reader is referred to [6]. Figure 2 shows the electrode locations for ECG and Frank XYZ VCG.

2.1 Diagnostics: ECG Versus VCG

In ECG-based diagnostic tests, parameters like lead polarity, duration of ECG waves, and wave amplitudes are used as criteria to diagnose diseases. A comprehensive list of diagnostic parameters and associated diseases can be found in [7].

In the case of VCG, parameters like maximum magnitude of cardiac vector, Volume of the QRS loop, area, perimeter, ratio of area and perimeter, major axis, minor axis, ratio of major to minor axis, planar area, etc. are used as diagnostic criteria [8–11].

VCG has been shown to be clinically superior to ECG in certain conditions like detection of atrial enlargements, clearing suspicion of electrically inactive areas in the heart, correlation to left ventricular mass, detecting acute myocardial infarction, etc. An exhaustive list of cases where VCG may be superior to ECGElectrocardiograph has been compiled by Riera et al. [12].

2.2 Sensors for CBP Measurement

The function performed by the sensor in the measurement of CBP is the transduction, at the skin level, of ionic currents associated with the stimulation of cardiac muscle contraction, into electronic potential variations that can be amplified and visualized. There are two classification criteria for CBP electrodes. *First*, based on whether a conductive gel is used to lower contact impedance between skin and electrode—wet and dry electrodes. *Second*, based on whether a conductive layer or an insulating layer of the electrode is in contact with the skin—capacitive or resistive electrodes. Capacitive electrodes can acquire CBP through thin layers of insulating fabric [13], while resistive electrodes require direct contact with the skin. For a comprehensive review of the fundamentals of textile-based CBP measurements, from individual cardiac myocyte activity to the evaluation of electrode properties for CBP, the reader is directed to the review article by Xu et al. [14].

From the perspective of the electrodes used, in case of wet electrodes the skin electrode impedance is fairly low (5–10 KΩ) as compared to the contact impedance for dry electrodes (may vary from a few hundred KΩ to a few hundred MΩ depending on moisture content of the skin). Therefore, choice of materials to reduce the skin electrode impedance in textile electrodes improves the Signal to Noise performance of the system. Nanomaterials and nanostructures have a high surface area to volume ratio as compared to planar electrode structures. One way to reduce skin-electrode impedance is to increase the surface area of the electrode in contact with the skin through the use of nanomaterials and nanostructures.

3 Nanomaterial SensorFabrication

There are two broad approaches to the fabrication of garments with ECG sensor electrodes—(1) through functionalization or integration of finished garments with sensor elements (2) introduction of smart materials during the garment fabrication process. The former may involve the integration of finished electrodes into finished garments by simply stitching the electrodes at the appropriate locations on the garment or using deposition techniques to transfer the functional materials at the appropriate locations. The latter approach entails the use of textile fabrication techniques to form woven or nonwoven fabrics with the inclusion of functional materials.

3.1 Prefabricated Electrodes Stitched into Finished Garments

The Several flexible and rigid materials have been fabricated and evaluated for use as electrodes. Among resistive electrodes—Flexible Polydimethylsiloxane (PDMS) [15], CNT array electrodes named ENOBIO [16], Carbon Nanotube

(CNT)/PDMS nanocomposites [17], Flexible polymeric dry electrodes [18], and vertically aligned metallic nanowires [19].

Among Capacitive Electrodes—Ti/TiN electrodes [20], IrO-coated electrodes [21] MEMS spiked electrodes [22]. A comprehensive review of these contact and noncontact dry electrodes has been presented in [23]. Smart textile implementations can be achieved by stitching these electrodes onto finished garments.

3.2 Textile Electrodes

Textile electrodes are essentially conductive fabrics made by the inclusion of conductive metallic filaments. Different properties and textures of various fabrics can be obtained by varying design parameters at four essential steps—(1) the material used to form the smallest filament (2) Process used for the conversion of the filaments into yarn (3) Whether the yarn is knitted or weaved to form the fabric (4) The coatings used to functionalize the finished fabric.

There are two strategies followed to make fabrics conductive—(a) Inclusion of thin conductive filaments in the yarns used to make the fabric and (b) Coating the finished fabric with conductive materials by various coating techniques.

The first strategy has been extensively researched by [24] with the use of stainless steel. Single Walled Carbon Nanotubes (SWCNTs) have also been adsorbed in cotton to impart conductivity to the yarn as well as improve its mechanical properties for smart textile applications [25]. In the second approach: sputtering, screen printing, electro spinning, carbonizing, and evaporation deposition techniques have been explored for coating of yarns before weaving them into fabric. It is concluded by [26] that fabric with coated conductive layers offer higher conductivity, but the washability is lower because of cracking of the conductive layer as the number of wash cycles increases.

Fabrication of Nanostructures on textile fabrics or forming nanoscale filamentous structures on fabric, rather than incorporating nanomaterials into fabrics is referred to as *Nanotextiles*. This is a relatively new research area that has gained traction in the last few years. In Oh et al. have evaluated electrospun Silver plated Polyvinylidene fluoride (PVDF) nanofibers as potential long-term dry biopotential recording electrodes [27]. Rai et al. [28] have formed vertically aligned nanostructures using a fabric flocking technique. As shown in Fig. 3, the nano Flock electrodes show skin-electrode impedance lower by an order of magnitude in the low frequency range, as compared to a conventional wet Ag/AgCl electrode.

3.3 Functionalization of Textiles: Nanocomposite Inks

An alternative approach to fabricating textile sensors is to use screen printing to print functional inks on finished fabrics to give them sensing properties. Screen

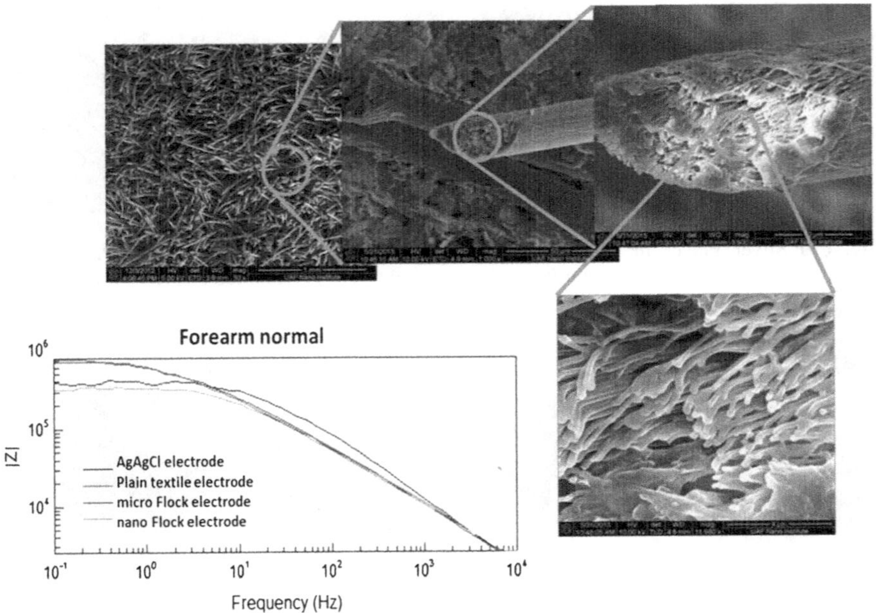

Fig. 3 SEM images showing the HeterostructuredNanoelectrodes on textile and the corresponding impedance spectroscopy performed on the forearm of a normal subject

printing is an attractive manufacturing technique due to its high throughput and compatibility to conventional textile manufacturing techniques.

In case of ECG smart textiles, this involves the formulation of conductive inks or printable capacitive structures on fabric. Conductive traces and transmission lines have been printed on nonwoven textile surfaces using Polymer Thick Film (PTF) technologies [29].

Stretchable conductive inks on textiles have been demonstrated through conductive traces using Silver flakes in Polyurethane-based binder by Araki et al. [30]. poly (3,4-ethylenedioxythiophene) poly(styrenesulfonate) (PEDOT:PSS) films have been successfully deposited on Rubber latex substrates using ink-jet printing by Romaguera et al. [31].

Multi-Wall Carbon Nanotubes (MWCNTs) and polyaniline nanoparticles (PANP) core shell-based nanocomposite conductive inks were synthesized and successfully screen printed on woven cotton by Rai et al. [32]. Conductive traces drawn from inks with Silver flakes in an acrylic binder have been used to make connections between ECG electrodes and flexible Printed Circuit Boards (PCBs) [33] (Fig. 4).

A similar system for a single lead ECG has been implemented on a women's brassiere called the e-bra system [34]. In addition to conductive materials for resistive type ECG electrodes, a flexible capacitive type ECG electrode structure

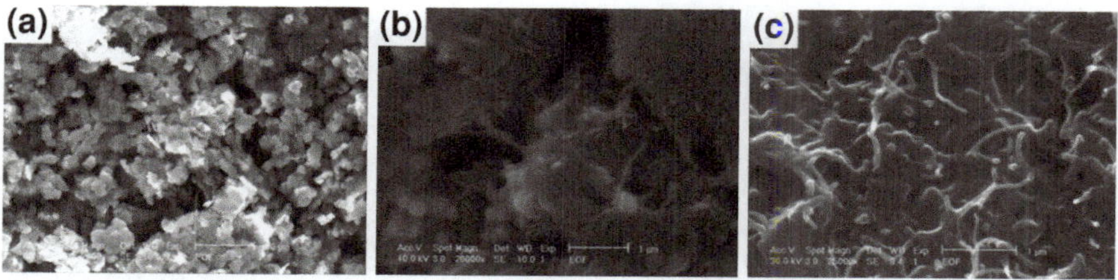

Fig. 4 **a** Silver Flakes in Modified Acrylic Paste (4:1 v/v) **b** CNTs in 70 % PANP 30 % Modified Acrylic Paste (2:1 v/v) **c** Polyaniline coated CNTs in Modified Acrylic Paste (2:1 v/v)

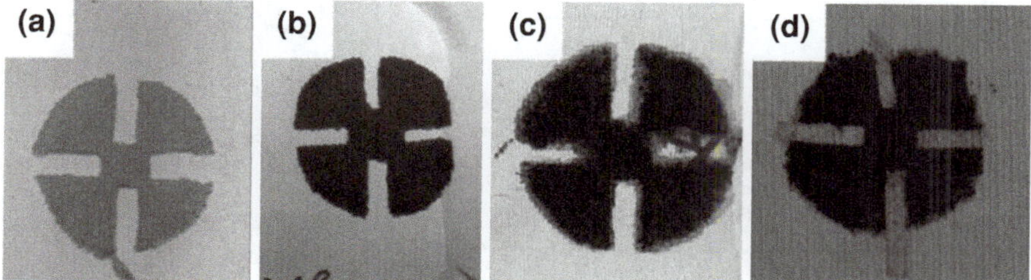

Fig. 5 Sheet resistance test structures for conductive inks on woven cotton **a** Silver flakes composite **b** CNT Composite **c** CNT and PANP **d** Polyaniline coated CNT

consisting of conductive CNT-acrylic nanocomposite inks sandwiched between two layers of acrylic inks has been fabricated and tested for ECG acquisition by Kumar et al. [35] (Fig. 5).

4 Data Acquisition

Cardiac biopotential signals measured by the electrodes are in the order of a few hundred microvolts to a few millivolts depending on electrode size and position [36]. These signals need to be amplified appropriately in order to be able to digitize and visualize them. An amplifier is used to enhance the signal strength. Key attributes of amplifier design were large common mode rejection ratio (CMRR), small input offset voltage, and low power consumption. The amplifier is designed in multiple stages. The instrumentation amplifier provides the first high impedance stage with a gain of 10 to avoid amplifier's saturation due to impedance mismatch. An instrumentation amplifier is needed because the input impedance needs to be sufficiently higher than the skin-electrode contact impedance to ensure that the largest fraction of ECG signal appears as input to the instrumentation amplifier circuit.

Analog filters with a pass band between 0.2 Hz and 100 Hz make sure the mismatch is not seen by the later high gain stages. Two more stages of

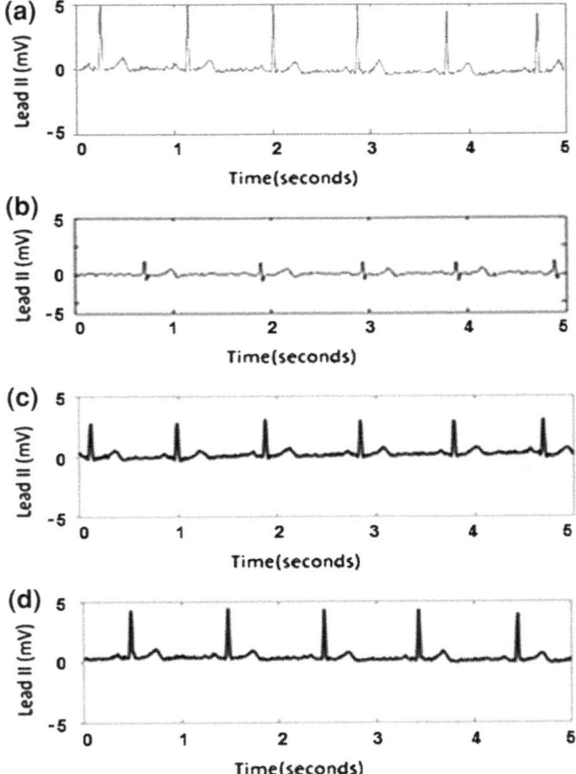

Fig. 6 Lead II ECG signals acquired using (**a**) conventional Ag/AgCl wet electrodes (**b**) dry CNT acrylic composite capacitive electrodes [35] (**c**) dry planar conductive textile electrode Silver coated Nylon [33] (**d**) a nanoflock electrode

non-inverting amplifiers are used to further improve the signal quality. The amplified analog signal is then digitized by the Analog to Digital Converter (ADC) on the microprocessor at a fixed sampling rate of 200 Hz. The microprocessor then communicates with a wireless module using the Universal Asynchronous Receiver/Transmitter (UART) interface. The ECG data may then be transmitted wirelessly via a wireless module using any of the standard wireless communication protocols like Bluetooth^{TM}, ZigBee, etc.

ECG Lead 2 signals acquired from four different electrode types are shown in Fig. 6. The variation in amplitudes is due to differences in electrode sizes and contact impedances [36]. It can be observed that the ECG waveform characteristics P, QRS, and T waves are similar.

3 orthogonal signals obtained with Frank XYZ system are plotted in Fig. 7. The corresponding spatiotemporal plot is also plotted for a single cardiac cycle showing the prominent QRS loop. In addition, the best fit plane for the QRS loop, indicated by the black square mesh, obtained using orthogonal regression, is also shown.

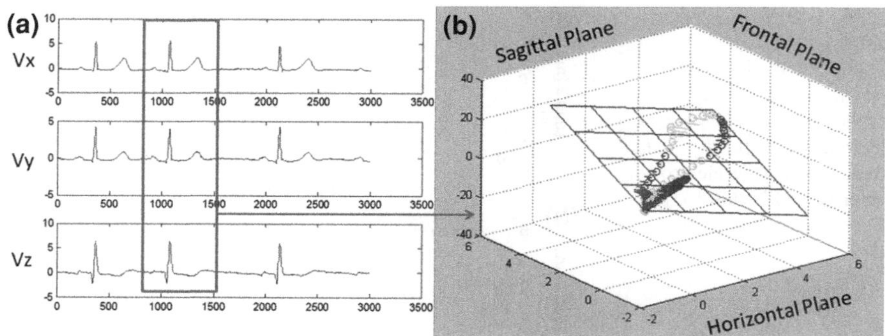

Fig. 7 **a** Shows the three orthogonal Frank XYZ leads from a healthy subject **b** Shows the VCG loop associated with a single cardiac cycle marked in *red*

5 Wireless Communication Protocols and Architectures

As mentioned earlier, the data acquired from the smart textiles system needs to be wirelessly transmitted to a smart device or a PC at the patient's home or to the hospital where the data can be interpreted and diagnostically relevant parameters can be extracted. The type of wireless communication protocol used depends on the distance between the transmitting smart textile system and the receiving station.

In one embodiment, short range wireless protocols covering a maximum of 10 m like ZigBee or Bluetooth™ can be used to transfer the data to a PC in the vicinity of the patient or user [33] as shown in Fig. 8. The data from the e-vest or e-bra can also be sent to a smartphone via Bluetooth™ [37] as shown in Fig. 9. These devices can then transfer the data to the hospital through their internet connections. A depiction of this architecture is shown in Fig. 8. In case of health monitoring outdoors, as is the case of health monitoring for athletes, a high power ZigBee radio can be used to extend the range to 350 m with line of sight communication as demonstrated by Kumar et al. [38].

Alternatively, the data from the smart textile system can be directly transmitted to the hospital through the mobile network using GPRS or 3G/4G [39]. This functionality is vital to the success of efforts to provide cardiac monitoring systems to under privileged residents of remote villages. In this implementation there will be no need for a smart device with internet connectivity. A mobile clinic with such wireless capabilities will be of significant benefit and such a mobile clinic is being setup by Vijay Varadan at Kotagiri Mission Fellowship (KMF) Hospital, Kotagiri, Tamilnadu, India.

Readers with interest in a comprehensive review on all the presently explored communication architectures and wide scale deployments are directed to a review article by Custodio et al. [40].

The schematic in Fig. 10 depicts the connectivity paths from sensors to cloud storage and 24X7 health monitoring as the ultimate objective of all wireless wearable healthcare systems.

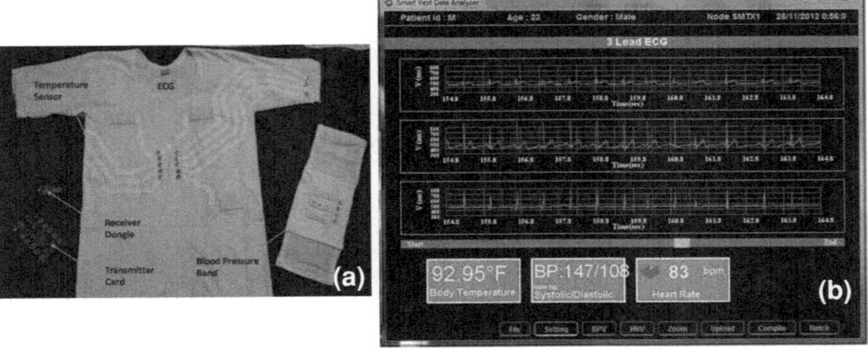

Fig. 8 e-vest with conductive textile electrodes for 3 lead ECG, printed traces to connect sensors and flexible electronics, Photoplethysmography band for Pulse Transit time based blood pressure estimation, wireless communication through ZigBee and real-time data display [33]

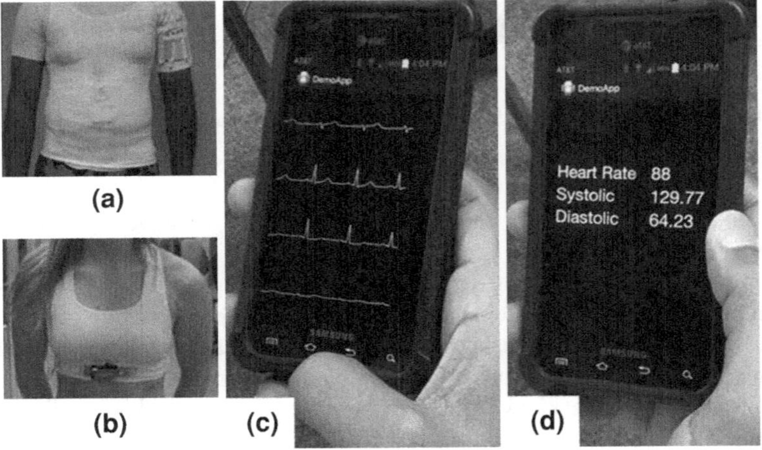

Fig. 9 **a** e-bro (**b**) e-bra (**c**) 3 Lead ECG and Pulse signal Real-time Plot (**d**) Heart rate and Blood Pressure estimates

6 Challenges and Future Directions

Despite the widely published research efforts demonstrating the ability to integrate healthcare monitoring sensors into textiles and effectively acquire biomedical signals, these systems have not gained appreciable acceptance in clinical practice.

The challenges are the following; first, the placement of electrodes for the measurement of ECG is vital to acquire a diagnostically accurate signal. Wrong placements can lead to misdiagnoses, which will be highly detrimental to the patient. Therefore, it is imperative that each smart garment is custom designed

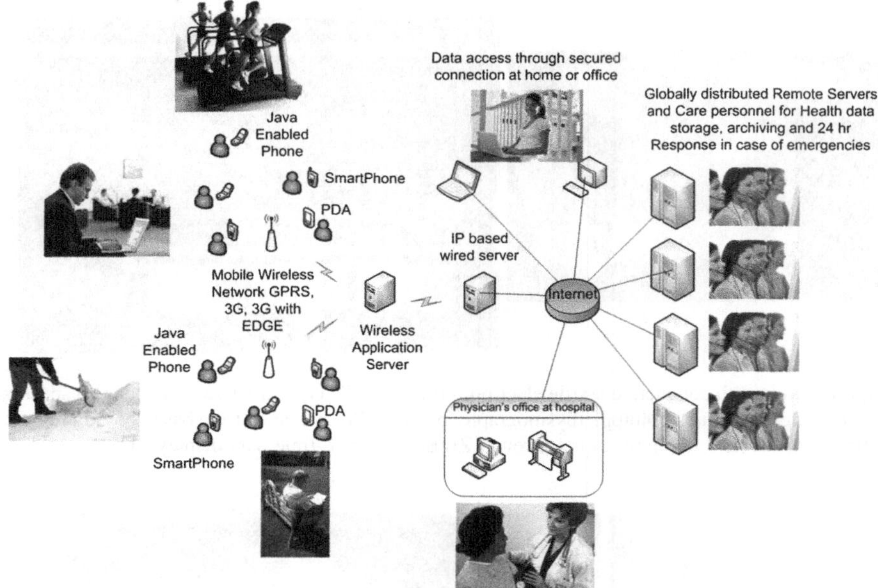

Fig. 10 Schematic for the use of Mobile devices as relay stations and establishing Cloud connectivity

based on the patient's body shape and dimensions. Consequently, the cost of implementing smart garment solution may in fact be significantly higher than conventional holter monitoring systems. A potential solution to this problem is to determine new electrode placement positions along with sufficiently accurate and robust transformations to translate these signals into diagnostically standard lead systems. The EASI lead placement system is one such promising approach [41], but the transformations are known to have lower accuracy for obese patients. This challenge may be overcome by research into designing and evaluating a new set of ECG based diagnostic criteria more suitable for uncontrolled and ambulatory environments, such as non-amplitude dependent criteria and criteria that are dependent on variability rather than on absolute measures like heart rate variability.

Second, the use of dry electrodes in smart garments is a highly advantageous prospect as they are less likely to cause skin irritation and can be washed and reused. However, dry electrodes in smart garments are more susceptible to motion artifacts caused by relative motion of skin and the electrodes. Compression garments or harnesses and belts have been used to alleviate motion artifacts, but these approaches may cause discomfort to the patients when worn over long periods of time. In addition to that, the nanotextile can be augmented with motion artifact rectification algorithm at the software side. Alternatively, design of garments with compression only in the regions where electrodes are present, thereby reducing patient discomfort, is a promising approach that needs further research.

Acknowledgments One of the authors, Vijay K. Varadan would like to extend his gratitude to Dr. Vasudev K. Aatre for his constant and continuing support, encouragement, and mentorship during the course of his research work in acoustics, electromagnetics, smart materials and systems, and health care. This research was conducted with the endowment for smart textiles for health care, from Global Institute for Nanotechnology in Engineering and Medicine Inc., 700 Research Center Blvd., Fayetteville, AR 72701.

References

1. Mathers CD, Loncar D (2006) Projections of global mortality and burden of disease from 2002 to 2030. PLoS Med 3(11):e442
2. Sastry URB (2004) An overview of worldwide developments in smart textile. Tech Text Int 13(4):31–35
3. Liu Y (2008) Functionalization of cotton with carbon nanotubes. J Mater Chem 18:3454–3460
4. Helm RA (1955) Theory of vectorcardiography: a review of fundamental concepts. Am Heart J 49:135–159
5. Frank E (1956) An accurate, clinically practical system for spatial vectorcardiography. Circulation 13:737–749
6. Malmivuo J, Plonsey R (1995) Bioelectromagnetism: principles and applications of bioelectric and biomagnetic fields. Oxford University Press, USA
7. Brosche TAM (2010) EKG handbook. Jones and Bartlett, Sudbury, M.A., Sect. 10, pp 87–98
8. Correa R, Arini PD, Valentinuzzi ME et al (2013) Novel set of vectorcardiographic parameters for the identification of ischemic patients. Med Eng Phys 35(1):16–22
9. Yang H (2011) Multiscale recurrence quantification analysis of spatial cardiac vectorcardiogram signals. IEEE Trans Biomed Eng 58(2):339,347
10. Huang C, Li-Wei K, Lu S et al (2011) A vectorcardiogram-based classification system for the detection of Myocardial infarction. In: Annual international conference of the IEEE Engineering in Medicine and Biology Society, EMBC, 2011, pp 973,976, 30 Aug–3 Sept 2011
11. Correa R, Laciar E, Arini P et al (2010) Analysis of QRS loop in the vectorcardiogram of patients with Chagas' disease. In: 2010 Annual international conference of the IEEE Engineering in Medicine and Biology Society (EMBC), 2561,2564, 31 Aug–4 Sept 2010
12. Riera ARP, Uchida AH, Filho CF (2007) Significance of vectorcardiogram in the cardiological diagnosis of the 21st Century. Clin Cardiol 30(7):319–323
13. Chi YM, Cauwenberghs G (2010) Wireless non-contact EEG/ECG electrodes for body sensor networks. In: 2010 International conference on body sensor networks (BSN), pp 297,301, 7–9 June 2010
14. Xu PJ, Zhang H, Tao XM (2008) Textile-structured electrodes for electrocardiogram. Text Prog 40:183–213
15. Fernandes MS, Lee KS, Ram et al (2010) Flexible PDMS -based dry electrodes for electro-optic acquisition of ECG signals in wearable devices. In: 2010 Annual international conference of the IEEE Engineering in Medicine and Biology Society (EMBC), pp 3503,3506, 31 Aug–4 Sept 2010
16. Ruffini G, Dunne S, Farres E et al (2007) ENOBIO dry electrophysiology electrode; first human trial plus wireless electrode system. In: 29th Annual international conference of the IEEE Engineering in Medicine and Biology Society, 2007. EMBS 2007, pp 6689,6693, 22–26 Aug 2007
17. Jung H, Moon J, Baek D et al (2012) CNT/PDMS composite flexible dry electrodesfor long-term ECG monitoring. IEEE Trans Biomed Eng 59(5):1472,1479

18. Baek J, An J, Choi J et al (2008) Flexible polymeric dry electrodes for the long-term monitoring of ECG. Sens Actuators A 143(2):423–429
19. Varadan VK, Oh S, Kwon H, Hankins P (2010) Wireless point-of-care diagnosis for sleep disorder with dry nanowire electrodes. J Nanotechnol Eng Med 1(3):031012-031012-11
20. Fiedler P, Griebel S, Fonseca C et al (2012) Novel Ti/TiN Dry Electrodes and Ag/AgCl: A Direct Comparison in Multichannel EEG. In: 5th European conference of the international federation for medical and biological engineering, V. 37 IFMBE proceedings, pp 1011–1014
21. Dias NS, Carmo JP, Ferreira da Silva A et al (2010) New dry electrodes based on iridium oxide (IrO) for non-invasive biopotential recordings and stimulation. Sens Actuators A: Phys 164(1–2):28–34
22. Chiou J, Ko L, Lin C et al (2006) Using novel MEMS EEG sensors in detecting drowsiness application. In: IEEE Biomedical circuits and systems conference, 2006. BioCAS 2006, pp 33, 36, 29 Nov–1 Dec 2006
23. Chi YM, Tzyy-Ping J, Cauwenberghs G (2010) Dry-contact and noncontact biopotential electrodes: methodological review. IEEE Rev Biomed Eng 3:106,119
24. Mestrovic MA, Helmer RJN, Kyratzis L et al (2007) Preliminary study of dry knitted fabric electrodes for physiological monitoring. 3rd International conference on intelligent sensors, sensor networks and information, 2007. ISSNIP 2007, pp 601–606, 3–6 Dec 2007
25. Shim BS, Chen W, Doty C, Xu C, Kotov NA (2008) Smart electronic yarns and wearable fabrics for human biomonitoring made by carbon nanotube coating with polyelectrolytes. Nano Lett (12):4151–4157
26. Rattfalt L, Chedid M, Hult P et al (2007) Electrical properties of textile electrodes. 29th Annual international conference of the IEEE engineering in medicine and biology society, 2007. EMBS 2007, pp 5735–5738, 22–26 Aug 2007
27. Oh TI, Yoon S, Kim TE et al (2013) Nanofiber web textile dry electrodes for long-term biopotential recording. IEEE Trans Biomed Circuits Syst 7(2):204, 211
28. Rai P, Oh S, Shyamkumar P et al (2014) Nano- bio- textile sensors with mobile wireless platform for wearable health monitoring of neurological and cardiovascular disorders. J Electrochem Soc 161(2):B3116–B3150
29. Karaguzel B, Merritt CR, Kang T et al (2008) Utility of nonwovens in the production of integrated electrical circuits via printing conductive inks. J Text Inst 99(1):37–45
30. Araki T, Nogi M, Suganuma K et al (2011) Printable and stretchable conductive wirings comprising silver flakes and elastomers. IEEE Electron Device Lett 32(10):1424,1426 (Mathers CD, Loncar D (2006) Projections of global mortality and burden of disease from 2002 to 2030. PLoS Med 3(11):e442)
31. Romaguera VS, Madec MB, Yeates SG (2009) Inkjet printing of conductive polymers for smart textiles and flexible electronics. In: Materials and devices for flexible and stretchable electronics materials research society symposium proceedings, vol 1192, pp 26–31
32. Rai P, Lee J, Mathur GN, Varadan VK (2012) Carbon nanotubes polymer nanoparticle inks for healthcare textile. In: Proceedings of the SPIE 8548, Nanosystems in Engineering and Medicine, 854822, 24Oct 2012
33. Rai P, Kumar PS, Oh S, Kwon H, Mathur GN, Varadan VK, Agarwal MP (2012) Smart healthcare textile sensor system for unhindered-pervasive health monitoring. In: Proceedings of the SPIE 8344, Nanosensors, biosensors, and info-tech sensors and systems 2012, 83440E, 26 April 2012
34. Varadan VK, Kumar PS, Oh S et al (2011) e-bra with nanosensors for real time cardiac health monitoring and smartphone communication. J Nanotechnol Eng Med 2(2):021011-021011-7
35. Kumar PS, Rai P, Oh S et al (2012) Nanocomposite electrodes for smartphone enabled healthcare garments: e-bra and smart vest. In: Proceedings of the SPIE 8548, Nanosystems in engineering and medicine, 85481O, 24 Oct 2012
36. Geddes LA, Baker LE (1996) The relationship between input impedance and electrode area in recording the ECG. Med Biol Eng 4(5):439–450
37. Varadan VK, Kumar PS, Oh S et al (2011) e-Nanoflex sensor system: smartphone-based roaming health monitor. J Nanotechnol Eng Med 2(1):011016-011016-11

38. Kumar PS, Oh S, Kwon H, Rai P, Varadan VK (2013) Smart real-time cardiac diagnostic sensor systems for football players and soldiers under intense physical training. In: Proceedings of the SPIE 8691, nanosensors, biosensors, and info-tech sensors and systems 2013, 869108, 9 April 2013
39. Kwon H, Oh S, Kumar PS, Varadan VK (2012) E-Bra system for women ECG measurement with GPRS communication, nanosensor, and motion artifact remove algorithm. In: Proceedings of the SPIE 8548, nanosystems in engineering and medicine, 85482 N, 24 Oct 2012
40. Custodio V, Herrera FJ, Lopez G et al (2012) A review on architectures and communications technologies for wearable health-monitoring systems. Sensors 12:13907-13946
41. Field DQ, Feldman CL, Horacek BM (2002) Improved EASI coefficients: Their derivation, values, and performance. J Electrocardiol 35(4):23–33

Polymer-Based Micro/Nano Cantilever Electro-Mechanical Sensor Systems for Bio/Chemical Sensing Applications

Rajul S. Patkar, Manoj Kandpal, Neena Gilda, Prasenjit Ray and V. Ramgopal Rao

Abstract In this chapter, we present the status of polymer cantilever sensor platforms for biochemical sensing and energy harvesting applications. We introduce a novel process flow for polymer microstructure fabrication called the SPARE MEMS, which involves **S**pinning of sacrificial/structural layers, **P**atterning, **A**nchor formation and the final **RE**lease of the device stack along with the anchor from the substrate. In this process the wafer is spared and is reusable. An organic/thin film FET embedded cantilever devices (CantiFETs) have been demonstrated using this process in order to reduce the noise levels and to achieve high deflection sensitivities. We have also used the SPARE MEMS process to fabricate a variety of other piezoresistive polymer cantilever devices with the highest reported deflection sensitivity (>100 parts-per-million/nm) to surface stress. Electronic circuit design approaches for the detection of ΔR down to sub parts-per-million level of resolution for piezoresistive cantilevers are also discussed. Using various surface coatings, development of sensor systems and sensor nodes for the detection of nitro-based explosive compounds, cardiac proteins, and environmental sensors are demonstrated. For powering the sensors, a novel piezoelectric polymer composite platform has been proposed.

Keywords Polymer MEMS cantilever · Canti-FET · SPARE MEMS process · Energy harvesting · Piezoresistive · Piezoelectric · Surface functionalization · Signal conditioning techniques

---

R. S. Patkar · M. Kandpal · N. Gilda · P. Ray · V. R. Rao (✉)
Department of Electrical Engineering, Indian Institute of Technology Bombay, Mumbai, India
e-mail: rrao@ee.iitb.ac.in

K. J. Vinoy et al. (eds.), *Micro and Smart Devices and Systems*, Springer Tracts in Mechanical Engineering, DOI: 10.1007/978-81-322-1913-2_24,

1 Introduction

Polymer-based sensing devices offer advantages in terms of simple processing, low manufacturing cost (cheaper material), biocompatibility, and higher sensitivity because of low Young's modulus. Polymers like SU-8 and PMMA are available in different viscosities which aids in the fabrication of thin film structures to structures with a high aspect ratio. Also in some cases the receptor molecules can be functionalized directly onto the surface of SU-8 cantilevers for variety of sensing applications avoiding additional deposition of a layer for immobilization purpose. Because of their low cost, label free detection, and high sensitivity, these sensors have a huge potential in applications like explosive sensing, health-care diagnostics, mass spectrometry, etc. Low cost polymer cantilever-based sensors are an ideal choice of sensors where disposable sensors are often required. The various polymers which are used in MEMS application are polyimide [1, 2], parylene [3], polycarbonate [4], photoresist [5–9], and similar materials. New class of materials with desirable physical properties such as piezoelectric, piezoresistive, and magnetic can be formulated by embedding various nanoparticles into the flexible polymer matrix [7, 9–11].

2 Polymers for MEMS

SU-8, a negative tone photo resist has been favored as a structural material for MEMS application because of its mechanical stability, resistance to chemicals during wet etching, and direct surface immobilization. SU-8 is a high contrast, negative tone, epoxy-based, high tensile, near UV photoresist which has been developed and patented by IBM. It was formerly developed as a high resolution mask for semiconductor applications but subsequently used for applications in MEMS and microfluidics. It is shown to be a very good biocompatible material and hence can be used as an implant material, biomolecular encapsulation, and in bio-MEMS [12].

SU-8 is highly optically transparent in the UV region (350–400 nm) and hence is suitable for fabrication of relatively thick structures with nearly vertical sidewalls. It has maximum absorption at 365 nm wavelength. When exposed with suitable light near UV radiation, cross-linking process begins. This process of polymerization is carried out in two steps. In the first step strong photoacid (triarylsulfonium salts) is formed during the exposure process and subsequently thermal cycle of post exposure bake(PEB) process causes cationic chain growth by ring opening polymerization of the epoxide groups. The cross-linking of long molecular chains of SU-8 results in solidification of the material. Cured films of SU-8 are highly resistant to chemicals and radiation and shows very low levels of outgassing in vacuum [13]. They also have very good mechanical and thermal stability and hence very suitable for MEMS application.

The current series of SU-8 resist, i.e., SU-8 2,000 and SU-8 3,000 uses cyclopentanone as the primary solvent. This new formulation can give better adhesion on some substrates. A single coat of SU-8 can give film thickness ranging from 0.5 μm to >200 μm by using resist with different viscosities. The other physical properties of SU-8 like lower Young's modulus, makes it ideal for sensor application. Biocompatibility of SU-8 can be further improved by various surface treatments like oxygen plasma treatment, chemical treatment with acid and base, ethanolamine and grafting of the surface with polyethylene glycol [10].

Processing of SU-8 resist involves steps like spin coat, soft bake, exposure, PEB, and finally development of the resist. 1-Methoxy-2-propanol acetate is mainly used as a SU-8 developer. In some cases where SU-8 is part of the device, a hard bake is advisable to further cross-link the polymer.

The use of Parylene [poly (*p*-xylylene)] as a structural material in MEMS devices has been a topic of interest for many researchers for its excellent mechanical and electrical properties. One of the derivatives of Parylene—Parylene C is a biocompatible polymer which is used in sensing and drug delivery application. Thin films of parylene can be formed by deposition process. Coatings of parylene are completely conformal, free of pinholes, and have a uniform thickness and can be etched with plasma etching process. Because of its compatibility with the microfabrication processes, it can be easily integrated with the MEMS technology. Parylene coatings provide a good electrical insulation, passivation, and moisture penetration. Hence, it can also be used as a passivation layer for many polymer-based devices, which are not very stable in the presence of moisture [14].

3 Electro-Mechanical Sensing Techniques for Cantilevers

The cantilever sensing principle is based on the accurate detection of surface stress or the deflection in real-time. The transduction techniques which are commonly employed are optical and electro-mechanical. The various transduction schemes which can be grouped under electro-mechanical type are piezoresistive, piezoelectric, capacitive, tunneling, and thermoelectricity. By using electro-mechanical transduction, one can get rid of bulky optical components, which limits the portability of the device. Integrating an electro-mechanical transduction system as part of the sensor has an advantage that a readout system for the sensor can be a part of the chip.

3.1 Piezoresistive

The word piezo is derived from the Greek word *piezein,* meaning "to squeeze". A piezoresistive material when subjected to strain shows a change in resistance. One of the popular classes of MEMS devices is a piezoresitive microcantilever.

Apart from the advantages like integrated readout, it is also useful wherein the sensing is done in an opaque liquid medium. Piezoresistive transduction technique is implemented in a cantilever platform by embedding a piezo layer within the cantilever. This is done during the fabrication process of the cantilever. The drawback of such a process is that the fabrication of a composite beam becomes more complex. The other drawback of piezoresistive sensor is Joule heating effect. The power dissipation in the resistor generates heat and causes thermal drift. This drift may result in inaccurate measurements. Part of this problem can be taken care by having a reference cantilever during measurement. Also in some cases, heating of surface can be used as an advantage to maintain polymerase chain reactions [15]. It can also be used for regeneration of surface in some cases. There are numerous materials which have been studied for piezoresistive effect and one can theoretically use any one of these provided that they are compatible with fabrication process. One such material is doped silicon. Piezoresistivity of a material is determined by gauge factor which is a material property. Gauge factor of a material is defined as relative change in resistance as a function of strain. This change in resistance can be attributed to geometrical change and change in resistivity due to the movement of atoms in the material because of strain. Recently materials like CNTs, graphene, carbon black, silicon nano wires have shown promising results. Gauge factor of graphene is reported to be as high as 1.8×10^4 [16]. Materials with high gauge factor are desirable for sensing application where piezoresistor is used as a transducer to convert mechanical strain into change in resistance.

3.2 Piezoelectric

Piezoelectric materials are the special class of materials, which accumulate an electric charge when they are mechanically strained or vice versa. Some of the thin film piezoelectric material which are used in MEMS fabrication are lead zirconium titanate (PZT-Pb(Zr, Ti)O_3), $BaTiO_3$, ZnO, and AlN. Recently, with significant development in the high performance piezoelectric nanomaterial and the limitation of traditional inorganic ceramics opened up a new area of research in the field of polymer-based piezoelectric materials [17]. These new class of materials were formulated by embedding various nanoparticles into the flexible polymer matrix and designed for various physical properties. One can also integrate this material into devices by sputtering or sol-gel process. The use of polymer piezoelectric materials for energy harvesting applications, by utilizing the ambient available vibrations are of recent interest for researchers due to their low cost, flexibility of design, and improved physical properties. Piezoelectric transduction technique offers similar advantages as piezoresistive technique in terms of portability. No external optical set up or laser alignment is required. The primary drawback of this technique is that the thicker layer of piezoelectric material is required to generate an adequate output signal. This adds to the stiffness of cantilever and reduces the sensitivity of the device. However, these devices can be used as self-powered sensors. More about piezoelectric and power generation using piezoelectric cantilevers will be discussed in the later sections.

3.3 OFET/TFT Integrated Cantilever Structures (CantiFET)

It was observed that piezoresistive layers comprising of SU-8 composites/CNTs have variability issues due to the nonuniform particle dispersion. This problem can be addressed by a novel transduction technique in which a strain sensitive transistor was integrated in the cantilever [18–20]. The principle behind this technique was that strain generated due to cantilever deflection modulates the mobility of channel and hence change in drain current of the transistor. This method offered advantages like improved sensitivity, reduction of noise, and as a switching matrix to choose from array of sensors. The various designs based on the above principle are MOSFET-embedded silicon nitride, microcantilever with gold as a sensing layer, organic FET embedded in a polymer microcantilever (CantiFET), and thin film transistor integrated in a cantilever of SU-8 polymer.

4 Fabrication and Characterization of Polymer Cantilevers

Although conventional microcantilevers were fabricated with silicon and its compounds, mechanically stable polymer material SU-8 has been favored as a structural material. The reason being lower Young's modulus of this material makes them highly sensitive to surface stress-based sensing devices. SU-8 cantilevers with piezoresistive readout with different strain sensing layers such as gold [21], polysilicon [6], MWCNT [22], polymer composite SU-8/CB [7], etc., have been reported in the literature. Similarly, work has been done on cantilevers with piezoelectric readout for sensing/energy harvesting [9] applications.

This process starts with a clean substrate. A fast etching sacrificial layer of adequate thickness is formed on the substrate either by deposition or by spinning. This layer will be subsequently etched at the end to release the devices. In the next step, a polymer layer (SU-8 2002) is spun on the sacrificial layer which forms the structural layer. This layer was patterned by prebaking, UV exposure using a mask, post baked and developed. Next a transducer layer of a piezoresistive/piezoelectric material is formed on the structural layer by deposition/spinning. Next metal contacts are made to the piezo layer using evaporation or metal sputter. To encapsulate piezo layer, an encapsulation layer of SU-8 2002 is spun and patterned on to piezo layer. One should take care while deciding the thickness of each layer of cantilever process that strain sensing layer doesn't fall in the neutral axis of the

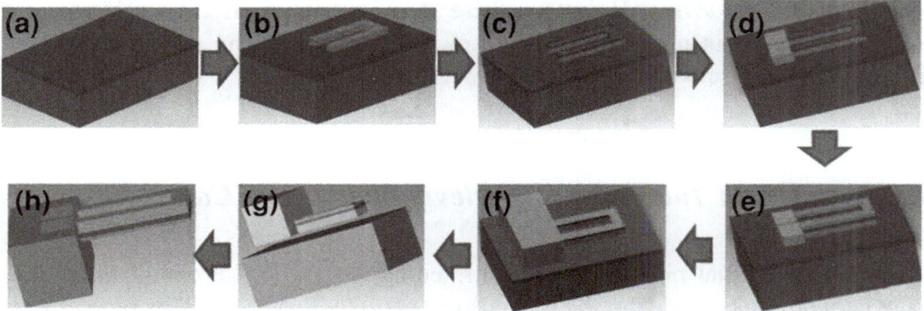

Fig. 1 Fabrication sequence of polymer cantilever with SPARE MEMS process **a** substrate with sacrificial layer **b** patterned SU-8 layer **c** piezo layer **d** chrome/gold contact pads **e** encapsulation layer of SU-8 **f** thick SU-8 anchor **h** released cantilever die

device. To get the maximum compressive/tensile strain in the piezo layer, this layer should be far away from the neutral axis. The anchor or the base was formed by SU-8 2100 resist which was highly viscous. It was spun at the speed which resulted in thickness of more than 100 μm. Resist was exposed and cured to form the base of the cantilever. The last step was to release the cantilever die from the substrate. This is achieved by completely etching away the sacrificial layer. The cantilever die lifts off from the substrate and the substrate is reusable.

The process therefore involves **S**pinning of sacrificial/structural layers, **P**atterning, **A**nchor formation, and final **RE**lease of the stack along with the anchor from the substrate. This will be referred to as SPARE MEMS process in this work. Fabrication sequence of SPARE MEMS process is shown in Fig. 1. Here in the process the wafer is **SPARE**d and is reusable. With the similar process, one can encapsulate any piezoresistive polymer layer in stack of SU-8 polymer layer.

4.1 SU-8 Polymer Cantilevers with Different Piezoresistive Layers by SPARE MEMS Process

SU-8 cantilevers with gold as strain sensing layer are not very sensitive because of lower gauge factor of gold. This can be improved by using a material with higher gauge factor. The advantage of polymer and polycrystalline composite is that one can have higher deflection sensitivity because of lower Young's modulus of polymer layer and higher gauge factor of polycrystalline film that gives higher change in relative resistance on deflection.

Kale et al. [6] fabricated and characterized polymer cantilever with HWCVD encapsulated polysilicon piezoresistor. Deposition of polysilicon on a polymer layer is tricky because of the temperatures involved. This is resolved by using low temperature HWCVD process for deposition. The cross-section view of the cantilever as shown in Fig 1c, shows different layers of the cantilever. The bottom layer which

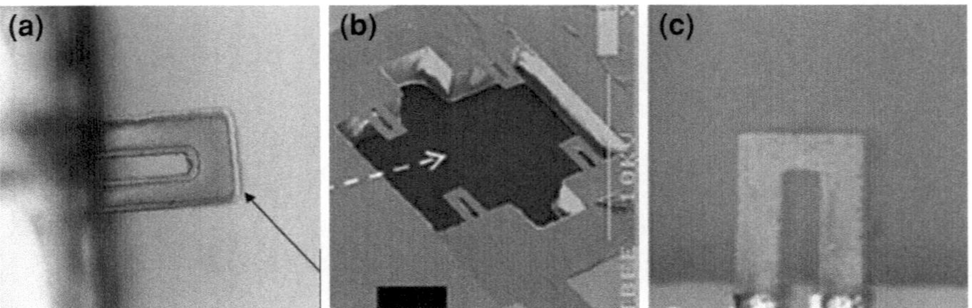

Fig. 2 Images of released cantilever die **a** SU-8/poly/SU-8 **b** SU-8/CB/SU-8 **c** SU-8/ZnO nanorods/SU-8

is also the structural layer and the top most layer which functions both as an immobilization layer and encapsulating layer are made from polymeric layer. The middle layer is p-type polycrystalline silicon, which acts as a piezoresistive layer is deposited by low temperature HWCVD process by using gases silane and diborane. The anchor layer is made of SU-8 100. Sacrificial layer silicon oxide was etched using buffered hydrofluoric acid and when the oxide was completely dissolved, the die lifted off from the substrate. Optical image of released die is shown in Fig. 2a.

The SU-8/poly/SU-8 cantilevers are less sensitive compared to all polymer cantilevers because of higher Young's modulus of polysilicon which makes the cantilever stiff. This can be improved by integrating the composite of SU-8 with carbon black (CB) as a strain sensing layer. This was reported in [16] where a SU-8 cantilever with thickness of 7 μm having a gauge factor in the range of 15–20 was reported. On the same lines Seena et.al improved the performance of these devices by reducing the thickness of cantilever and improving the dispersion of CB in SU-8 [7]. The polymer nanocomposite of a high-structured carbon black (CB), was prepared by ultrasonic mixing of CB in SU-8 and Nanothinner (Microchem) in the ratio 1:1. The fabrication of SU-8/composite/SU-8 cantilever is based on SPARE MEMS process. Figure 2b shows the SEM image of fabricated device. Herein the piezo layer is a polymer composite. A cantilever device with structural layers of SU-8 and MWCNT as a piezoresistive layer was also demonstrated [22].

A polymer cantilever with ZnO nanorods as a piezoresistive layer was demonstrated [8]. The deflection sensitivity of ZnO nanowire-based polymer cantilever is high because of the lower Young's modulus of polymer and the higher strain sensitivity of ZnO nanowires. In this process a layer of vertically grown ZnO nanowires are encapsulated in polymer. To grow ZnO nanowires on a structural SU-8 layer, a seed layer of ZnO was sputtered, patterned, and etched. ZnO nanowires were vertically grown by low temperature hydrothermal process. To grow ZnO nanowires by hydrothermal process, an equimolar (30 mM) aqueous solution has been prepared with zinc nitrate hexahydrate ($Zn(NO_3)_2{\cdot}6H_2O$) and methenamine ($C_6H_{12}N_4$). De-ionized (DI) water of resistance 18.2 MΩ-cm was used in the process. Sample was kept in a solution for 6 h at 95 °C in a hot air

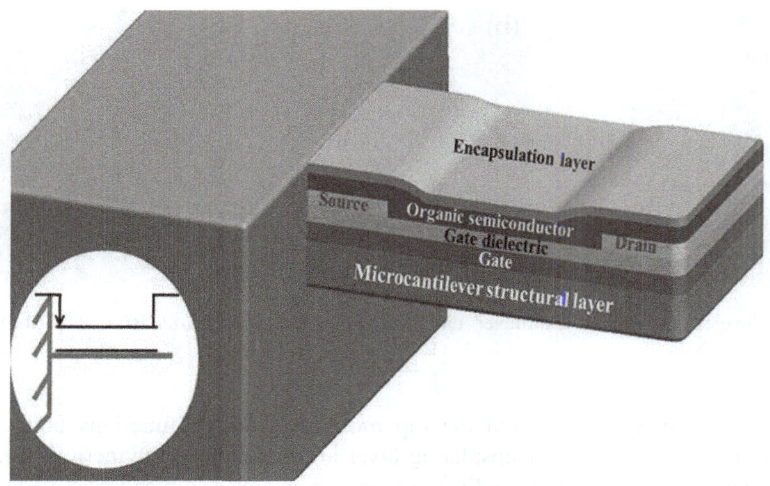

Fig. 3 Concept of an cantiFET device. Reprinted with permission from [19] ©2012 IEEE

oven. Sample was thoroughly cleaned with DI water to remove residual salt and then dried. Later an encapsulation layer and anchor layer are formed and sacrificial oxide was etched to release the die. The fabrication process is similar to SPARE MEMS process. Optical image of released die is shown in Fig. 2c.

The grown ZnO nanowires are approximately of length 2.5 microns and XRD have shown a strong (002) X-ray diffraction pattern (34.4°). Mode of conduction through a vertically grown ZnO nanowire film is mainly because of thermo ionic diffusion [24, 25].

The fabrication of cantiFET and TFT-based microcantilever is based on SPARE MEMS process. In fabrication of cantiFET [19], source, drain and gate electrodes were made of chrome/gold, gate dielectric, structural layers, and anchor were made of SU-8 and pentacene was thermally evaporated on released structures. To improve the stability of these devices, a very thin layer of silicon nitride was deposited by low temperature HWCVD process. For cantilevers embedded with TFT [20], aluminum-doped zinc oxide (AZO) of thickness 50 nm was deposited on the released structure by sputtering. The schematic of a cantiFET device is shown in Figs. 3 and 4 shows the released structures of cantiFET and TFT-based cantilever.

4.2 Device Characterization

The mechanical properties like resonance frequency and spring constant play an important role in sensing application. Laser Doppler Vibrometer is one such system which can be used to measure the resonant frequency. The laser beam from

Fig. 4 Released cantiFET and TFT integrated cantilever device **a** Photographs of released arrays of organic cantiFETs. [19] **b** SEM micrograph of fabricated organic cantiFET device [19] **c** Bottom and top enlarged view of cantilever portion of the cantiFET from SEM. Reprinted with permission from [19] ©2012 IEEE **d** Optical image of top and bottom view of AZO deposited TFT based cantilever. Reprinted with permission from [20] ©2013 AIP Publishing LLC

Table 1 Comparison of Electro-Mechanical characteristics of cantilever devices

Characteristics	Device			
	SU-8/Poly/ SU-8 [6]	SU-8/ CB/ SU-8 [7]	SU-8/ZnO nanorods/SU-8 [8]	TFT based μ cantilever [20]
Spring constant (Nm^{-1})	0.25	0.4	1.16	0.9
Resonant frequency (kHz)	39	22	30.4	50
Deflection sensitivity (ppm-nm^{-1})	Not reported	1.1	128	116

the LDV was directed to the surface of the cantilever which was actuated by a piezo buffer. Motion of the cantilever caused the shift in Doppler frequency and the amplitude and the frequency of vibration was extracted from this shift. The spring constant of the cantilever was measured by nanoindenter (Hysitron Triboscope) using beam bending technique. The tip of the indenter was placed at the apex of the cantilever and the load versus displacement curve was obtained. The slope of the curve gives the stiffness of the cantilever. Atomic force microscope can also be used to measure these characteristics [6]. Piezoresistive performance of the device can be confirmed by measuring the deflection sensitivity of the device. This can be measured by deflecting the tip of the cantilever by known amount (few nms or μms) and measure the current and voltage simultaneously. Slope of the curve between the normalized value of change in resistance as a function of deflection gives the deflection sensitivity. Devices which are discussed earlier were characterized and their properties are given in the Table 1.

5 Harvesting Energy with Low Frequency Mechanical Vibration with SU-8/ZnO Piezoelectric Polymer MEMS Cantilever Systems

Powering the sensors in a sensor array is an important problem. Energy scavenging by utilizing the available ambient vibrations with the piezoelectric cantilevers offers a good solution to this problem. When the cantilever gets deformed, the asymmetric stress developed in the nanocomposite piezoelectric layer due to the mechanical excitation by ambient vibrations generates the electric charge. This charge can be extracted and stored in the capacitor upon rectification. Although, the electric power generated with this energy system is low but is sufficient enough for powering the micro devices and wireless sensor networks for numerous applications. The schematic for Cantilever-based energy harvesting system is shown in the Fig. 5a.

In literature researchers postulated different configurations of piezoelectric nanocomposite-based upon the connectivity of piezoelectric nano filler inside the polymer matrix [26]. Two common nanofiller configuration inside the matrix are reported in literature due to their ease of fabrication. It can be broadly classified into 0–3 or 3–3 types. The first configuration of type 0–3 represents the nanofiller that are randomly distributed inside the polymer matrix whereas in the second configuration 1–3 the nanofiller are connected in one dimension. The schematic representations for these two configurations are as shown in Fig. 5b, c. To fabricate a polymer-based piezoelectric device with composite of SU-8/ZnO as piezoelectric layer, the major challenge was to make this composite photo patternable. For this ZnO nanoparticles of dimensions less than 100 nm were incorporated into SU-8 matrix. Suspensions of 5, 10, 15, 25 wt% of ZnO nanoparticles were prepared by dispersing the ZnO nanoparticles in solution of SU-8 and thinner in the ratio 1:1. Thin films were made on low resistivity silicon wafer by spinning and exposed at various doses and characterized for their photopatternability and piezoelectricity [9]. The SEM micrograph of dispersed ZnO in SU-8 is shown in Fig. 5d.

The studies showed that photo patterning is possible up to 15 wt%. The loss of photo patternability behavior of the polymer beyond certain threshold of ZnO wt% concentration was due to the wide bandgap nature of ZnO. Once the concentration goes beyond 15 wt%, major fraction of the light was absorbed by ZnO nanoparticles. This created hindrance in the cross-linking, hence SU-8 loses its photo patternabilty [27, 28]. Experimental results are shown in Fig. 6a, b.

The piezoelectric material can be characterized by measuring its piezoelectric coefficient. This method can be classified into two categories, direct method and indirect method. Direct method includes normal loads, cantilever method, single beam optical interferometry, double beam interferometry, and piezo force microscopy (PFM). Indirect method includes bulk acoustic wave, surface acoustic wave, and composite resonance [29]. The cantilever and PFM techniques are commonly used for estimation of d33 coefficient [30, 31]. In order to probe nano mechanical expansion of embedded ZnO nanoparticles, PFM technique was used. PFM response was recorded by applying alternate current voltage of modulated

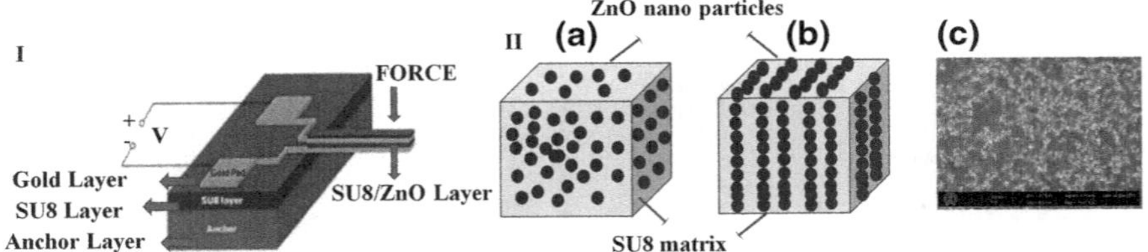

Fig. 5 **i** Conceptual schematic for piezoelectric cantilever energy harvester **ii** Connectivity models for nanofiller configurations inside piezoelectric polymer *a* 0–3 connectivity *b* 3–3 connectivity *c* scanning electron microscopic picture for SU8/ZnO nanocomposite thin films with 15 wt% ZnO filler concentration

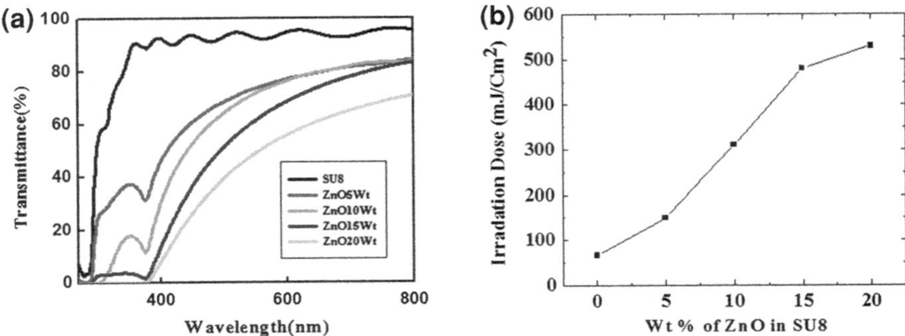

Fig. 6 **a** Transmission characteristic. Reprinted with permission from [9] ©2012 AIP Publishing LLC **b** Optimized dose for patterning thin films of SU8/ZnO nanocomposite with different ZnO wt% concentrations [9]

frequency of 14.2 kHz. Schematic for PFM measurement and results are as shown in Fig. 7a. Figure 7b shows a linear relationship between piezoresponse amplitude and piezoresponse drive voltages which varied in the range of 0–12 V. The measured effective piezoelectric coefficient was recorded in the range of 12–23 pm/V which is higher as compared to bulk zinc oxide [31, 32]. The high piezoelectricity at several points is due to the smaller size and morphology of the ZnO nanofiller. A recently reported theoretical and experimental study showed that as the surface to volume ratio goes higher, the atomic polarization of surface atoms changes and it significantly improves the piezoelectricity [32, 33].

Fabrication of SU-8/ZnO Piezoelectric cantilever-based energy harvester was done by SPARE MEMS process as discussed earlier. The structural layers are made from SU-8 2002 and the piezoelectric layer is made of SU-8/ZnO nanocomposite of 15 wt%. Device were anchored and released from the substrate. Optical and SEM image of piezoelectric cantilever is shown in Fig. 8i.

The fabricated device was then tested for its output power generation to the mechanical vibrations of tuned shaker. The experimental arrangement and schematic representation for the same are shown in Fig. 8ii(a). An oscilloscope was

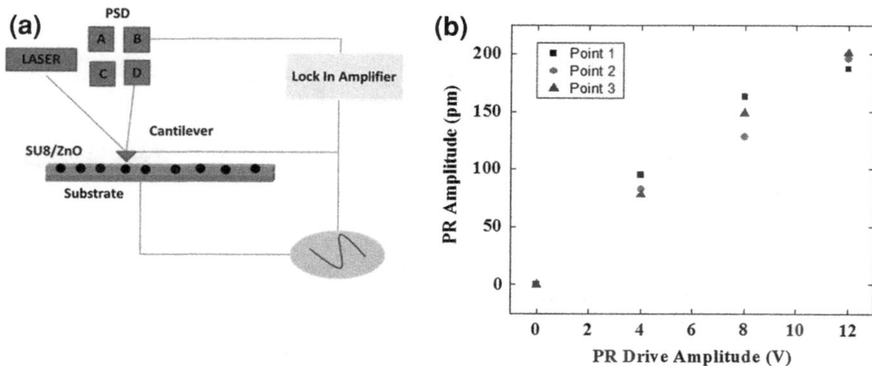

Fig. 7 **a** The experimental schematic used for piezoforce microscopy experiment. **b** Piezoelectric measurement of ZnO nanoparticles measured at three different points of surface. Reprinted with permission from [9] ©2012 AIP Publishing LLC

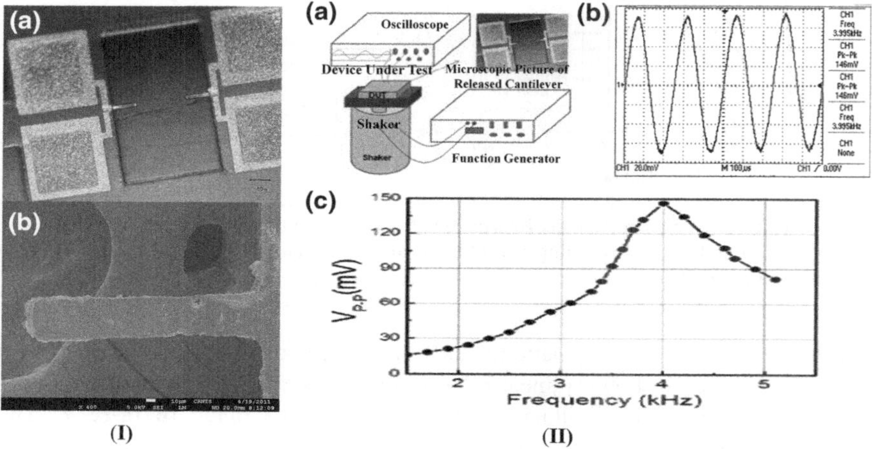

Fig. 8 **i** Microscopic image and scanning electron micrograph of released die are as shown in *a* and *b* respectively. Reprinted with permission from [9] ©2012 AIP Publishing LLC. **ii** *a* Schematic for experimental setup used for shaker experiments for mechanical excitation of cantilever *b* Peak to peak measured output voltages with respect to various tuned mechanical excitation frequency *c* oscilloscope output corresponding to 4.3 kHz. Reprinted with permission from [11] ©2013 IEEE

used for monitoring the output signal amplitude of device under test. The graph of electrical output voltages as a function of shaker excitation frequency are as shown in Fig. 8ii(c). Device showed 8 mV output signal at zero vibration which can be considered as noise. Maximum AC peak to peak ($V_{peak-peak}$) output of 143 mV was observed at excitation frequency of 4.3 kHz and then it decreased. This frequency is very close to the theoretical value of resonant frequency of fabricated device which is 4.2 kHz. The maximum output voltage was observed at 4.3 kHz due to the resonance of cantilever.

The device characterizations for practical applications were demonstrated by evaluating the power with 100 kΩ load resistance (R) and was estimated to be 0.025 μW using the Eq. 1 [11].

$$\text{Power} = \left[\frac{V_{peak-peak}}{2} \Big/ \sqrt{2}\right]^2 \Big/ R \quad (1)$$

6 Surface Functionalization of Receptor Molecules for Cantilever Sensor Applications

In previous sections, we have briefly discussed the fabrication aspects of a polymer-based microcantilever sensor. The next logical step to use these devices as sensors for various applications [34–39] like explosive sensing, cardiac diagnostics, CO sensing, bacterial detection, drug discovery, humidity, etc., is to make these surfaces receptive to the target molecules. This can be done by asymmetrically immobilizing the surface of these devices with receptor molecules. Biomolecules on polymer surface can be immobilized by adsorption, encapsulation, entrapment, and covalent binding. Immobilization by covalent binding is a preferred method so as to minimize nonspecific adsorption, enhance stability, and improve the biomolecule activity. This can be realized by modifying the surface of the polymer to have one of the functional groups like SH, NH_2, CHO, etc. These functional groups can bind to active biomolecules [40].

Often antibody-antigen binding is used for detection of specific proteins using cantilevers. Antibody immobilization can be performed on polymerized SU-8 surface by grafting amine groups on SU-8 surface. This was achieved by treating the surface of SU-8 by sulfochromic acid followed by silanization [40]. One can also get amine group on SU-8 surface without wet chemistry by implanting amine groups using pyrolytic dissociation of ammonia in a hotwire CVD setup [41]. Next step was to attach a cross-linker which will attach antibody to the amine group. This was done by treating the sample with Glutaraldehyde. Further, we attach antibodies by incubating the sample in human Immunoglobulin HIgG (0.5 ml/ml in phosphate buffer saline (PBS)) suspension. Then the sample was rinsed in phosphate buffer saline solution to get rid of loosely bound antibodies.

For explosive (TNT) detection, 4-MBA is used as a receptor molecule as this forms a hydrogen bond with nitroaromatic explosive molecules (TNT). To get a monolayer of 4-MBA on one side of the cantilever surface, gold was sputtered on the selected side as 4-MBA forms a stable monolayer with gold based on thiol chemistry. Monolayer of 4-MBA was formed on gold-coated cantilevers by incubating them for 24 h in a 6 mMol solution of 4-MBA in ethanol and cleaned in ethanol to get rid of the residues of 4-MBA [7].

Similarly, the (Fe(III)[T(4,5(OCH_3)$_2$ P)P]Cl) which is an iron porphyrin was used to detect CO gas. 1 mg of iron porphyrin was dissolved in 20 ml of IPA and the solution was drop-coated on SU-8 piezoresistive cantilever using a microdispenser. Experiments were performed to show that (Fe(III)[T(4,5(OCH_3)$_2$ P)P]Cl) has strong selectivity toward CO gas [39].

7 Signal Conditioning Techniques for Cantilever Sensors

The change in the resistance (ΔR) of the piezoresistive sensor can be in the order of sub parts-per-million. This necessitates signal conditioning circuits to be insensitive to drift, temperature changes, humidity changes, and other noise interferences. One of the solutions to this is to use differential measurement technique. This technique uses the reference sensor with similar mechanical properties but without selective material layer to sense the target molecules. The simultaneous measurement reduces the above-mentioned problems from measurement to some extent. Most popular technique used for a differential measurement is the Wheatstone's bridge arrangement with a sensor in one arm [42]. This simplest arrangement provides the differential output voltage proportional to the change in sensor resistance (ΔR). Sensitivity of this technique depends on the bridge excitation voltage and hence cannot be increased infinitely, considering power performance of the circuit. Furthermore, thermoelectric voltages and stray noise components are still present in the measurements that limit the performance drastically. Other methods based on Wheatstone's bridge sensing include, constant current excitation [43], resistance to frequency conversion [44], lock-in-amplification [42], etc. The sensitivity in above listed configurations is still limited by the factors like excitation voltage, sensor base resistance, and noise. The bidirectional Current Excitation Method (CEM) [45], explained in next section, promises better sensitivity over low voltage and hence low power performance. Figure 9a, b show different Wheatstone's bridge configurations commonly used for piezoresistive sensor applications. In Current Excitation Method (CEM), the two arms of the half bridge configuration are actuated by two identical, bidirectional, and temperature invariant current sources as shown in Fig. 9c.

7.1 Current Excitation Method and Implementation

In traditional Wheatstone's bridge configuration the differential output voltage (V_o) can be given as

$$V_o \approx V_A \times \left(\frac{\Delta R}{4R}\right) \tag{2}$$

This says that the sensitivity of the bridge is a function of excitation voltage (V_A) and base resistance of the sensor (R). Hence, ultrasensitive measurements

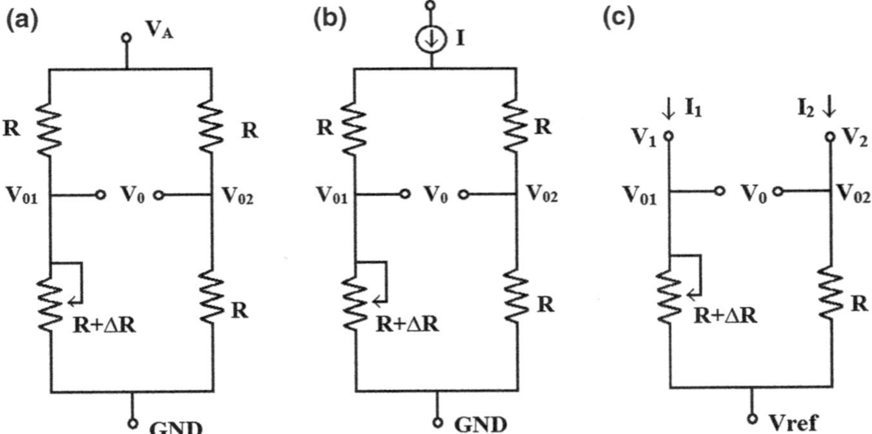

Fig. 9 Different Wheatstone's bridge configurations **a** Voltage excited **b** DC current excited **c** Bidirectional current excited half bridge. Reprinted with permission from [45] ©2013 IEEE

may necessitate very large excitation voltage which further leads to high power dissipation. Also, the effects like thermoelectric offsets, non-ohmic contacts, and different thermal drifts for different sensors become dominant here as change in resistance ΔR is a direct function of sensor base resistance R. Many groups have tried AC voltage excitation of bridge to get rid of thermoelectric offset and thermal drift but this method still needs higher excitation voltage for better sensitivity. AC current excitation of a bridge can accommodate these problems. Considering Fig. 9c, when there is change in resistance (ΔR), the half bridge output voltage (V_o) can be expressed as [45],

$$V_o = [(I \times R) - I \times (R + \Delta R)] - [I \times (R + \Delta R) - (I \times R)] = 2(I \times \Delta R) \quad (3)$$

Considering both the arms are excited with identical current sources $I_1 = I_2 = I$ and the base resistance of the bridge is R. Equation (3) explains that the bridge output voltage depends only on change in resistance (ΔR) and excitation current (I), eliminating the need of higher excitation voltage (V_A), and base resistance (R) dependence. One can choose wide range of base resistances as the output voltage V_o is independent of base resistance R from Eq. (3). To understand a low voltage excitation aspect of CEM, large signal voltage levels can be considered here. From Eqs. (2) and (3), excitation voltages required for two different configurations are $V_A/4$ (for Wheatstone's bridge configuration) and $V_{\text{ref}} + 2IR$ (for AC current excitation configuration) for the same sensitivity (V_{ref} can be as low as 0 V here). Hence, $V_A = 8IR$ that gives factor of 8 reduction in excitation voltage for CEM. Figure 10 gives an idea about the complete implementation of current excitation method [45, 46]. The bidirectional current source module comprises of two identical floating current sources connected in diode bridge configuration. The bridge is excited with a square wave to sink/source the current with respect to the reference voltage

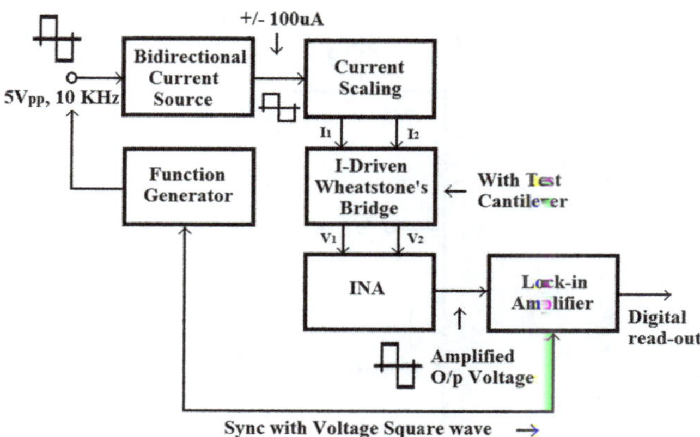

Fig. 10 Block diagram of the current excitation method. Reprinted with permission from [45] ©2013 IEEE

generated at load end as shown in Fig. 11. The diodes in the opposite arms of the bridge are ON at a time and accordingly current direction at load changes. Bidirectional current is further applied to current scaler circuit. Current scaler scales down the current magnitude and so the voltage drop across the sensor doesn't exceed indefinitely. One can control the gain of the sensor system by varying current magnitude. This keeps signal levels within the input voltage range of instrumentation amplifier (INA) next stage. Also, it guarantees that INA doesn't get saturated due to large change in resistance (ΔR) of the sensor. It can be noted that current scaler is basically a current divider followed by a very low noise operational amplifier (op-amp). The scaled down bidirectional current is allowed to flow through a sensor arm and reference arm of the half bridge configuration.

The differential output of half bridge configuration is fed to an ultralow noise, programmable gain INA inputs. Single-ended INA output is further fed to Lock-in Amplifier stage. Lock-in amplifier measures the amplitude of signal at particular frequency, i.e., the switching frequency of bidirectional current source in this case. Figure 12a shows the actual implementation of the proposed method on a single PCB board. Ultrasensitive Polymer composite SU-8 piezoresistive cantilevers are used as explosives detector sensors in E-Nose system. The selective material layer embedded on top of these cantilevers has affinity toward explosives molecules. The binding chemistry at the surface allows the cantilever bend further changing its resistance by ΔR. Hand held explosive detector system (E-Nose) is implemented to sense parts-per-billion level changes in the base resistance of the cantilever [47]. It uses current excitation method (CEM) with ADC and microcontroller interface. The cantilever structure is mounted on PCB and covered with Teflon cap to avoid any false positive detection. 9 V battery operated micro-diaphragm DC pump is used for purging purpose. Based on sensor readings, detection threshold is set with the help of microcontroller. Also, LCD and buzzer

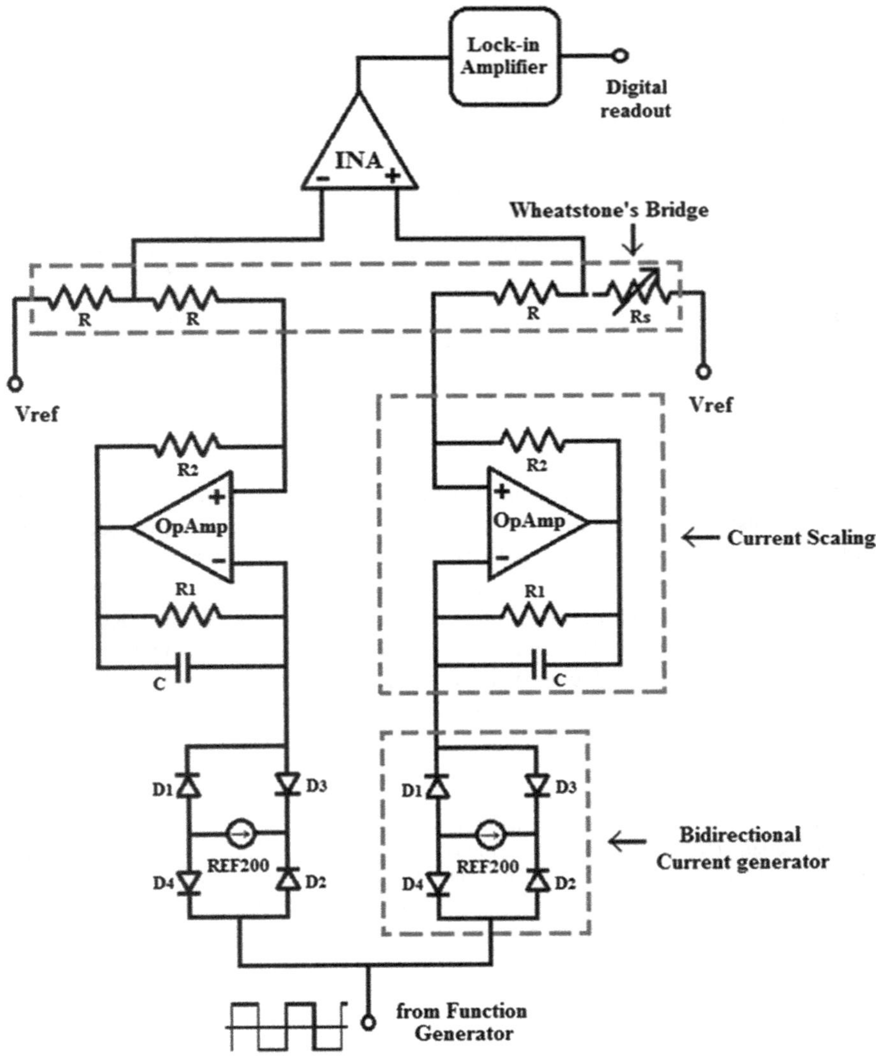

Fig. 11 Operational amplifiers and current sources based circuit implementation of CEM Reprinted with permission from [45] ©2013 IEEE

are configured in the system with the help of microcontroller. Figure 12b shows actual measurement results with the implemented system. This system is used to detect traces amount of TNT present in a gaseous form in the atmosphere. TNT vapor generator was used for controlled flow of TNT molecules. TNT exposure was turned ON and system responded in couple of seconds. As soon as TNT exposure was turned OFF, response came back to its original level.

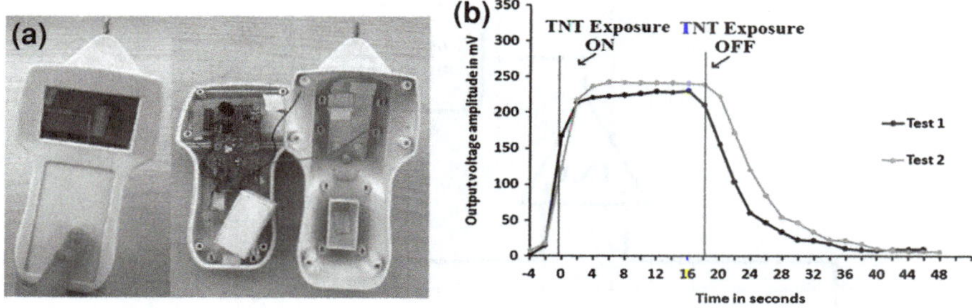

Fig. 12 **a** System level implementation of explosives detecter system (e-nose) [47] **b** Actual piezoresistive microcantilever response. Reprinted with permission from [45] ©2013 IEEE

8 Conclusions

This chapter introduces the polymer MEMS devices for a variety of sensing and energy harvesting applications. A novel process flow has been discussed in this work, which is referred to as the SPARE MEMS process. This process allows the substrate to be reused. It has been shown that, by utilizing novel transduction techniques with polymer microcantilevers, fabricated using the SPARE MEMS technology, it is possible to achieve a record high deflection sensitivity (>100 parts-per-million change in electrical parameters) for a nanometer of deflection. Using a novel current sensing approach, we have achieved circuit sensitivity down to 0.3 ppm for ΔR detection for piezoresistive cantilever sensors.

Acknowledgments Authors wish to acknowledge the financial support for the project 'A PoC system development for the Cardiac Diagnostics' sponsored by the "National Programme on Smart Materials (NPSM)" and the "National Programme on Micro and Smart Systems (NPMASS)", Govt. of India for the work presented in this chapter. Department of Electronics and Information Technology (DeitY), Govt.of India is also gratefully acknowledged for the Center of Excellence in Nanoelectronics facilities at IIT Bombay.

References

1. Ibbotson RH, Dunn RJ, Djakov V et al (2008) Polyimide microcantilever surface stress sensor using low-cost, rapidly-interchangeable, spring-loaded microprobe connections. Microelectron Eng 85:1314–1317
2. Huanga S, Zhanga X (2005) Application of polyimide sacrificial layers for the manufacturing of uncooled double-cantilever microbolometer. Presented at the MRS Fall Meeting, Boston, MA, 2005
3. Chen C-L, Lopez E, Makaram P et al (2007) Fabrication and evaluation of carbon nanotube-parylene functional composite films. Transducers and eurosensors' the 14th international conference on solid-state sensors, actuators and microsystems, Lyon, France, 10–14 June 2007
4. Shiraishia N, Ikehara T, Dao DV et al (2013) Fabrication and testing of polymer cantilevers for VOC sensors. Sens Actuators A 202:233–239

5. Kshirsagar A, Duttagupta SP, Gangal SA (2012) Optimization of poly(methyl methacrylate) as sacrificial layer for application in low temperature MEMS. 1st international symposium on physics and technology of sensors (ISPTS), pp 114–117
6. Kale NS, Nag S, Pinto R et al (2009) Fabrication and characterization of a polymeric microcantilever with an encapsulated hotwire CVD polysilicon piezoresistor. J Microelectromech Syst 18(1):79–87
7. Seena V, Fernandes A, Pant Prita et al (2011) Polymer nanocomposite nanomechanical cantilever sensors: material characterization, device development and application in explosive vapour detection. Nanotechnology 22:295501
8. Ray P, Rao VR (2013) ZnO nanowire embedded strain sensing cantilever: a new ultra-sensitive technology platform. J Microelectromech Syst 22(5):995–997
9. Kandpal M, Sharan C, Poddar P et al (2012) Photopatternable nano-composite (SU-8/ZnO) thin films for piezo-electric applications. Appl Phys Lett 101:104102–104105
10. Sutter M, Ergeneman O, Zurcher J et al (2011) A photopatternable superparamagnetic nanocomposite: material characterization and fabrication of microstructures. Sens Actuators B 156:433–443
11. Prashanthi K, Naresh M, Seena V et al (2012) A novel photoplastic piezoelectric nanocomposite for MEMS applications. J MEMS Lett 21:259–261
12. Nemani KV, Moodie KL, Brennick JB et al (2013) In vitro and in vivo evaluation of SU-8 biocompatibility. Mater Sci Eng C 33:4453–4459
13. Boisen A, Dohn S, Keller SS et al (2011) Cantilever like micromechanical sensors. Rep Prog Phys 74:036101
14. http://www.parylene.com
15. Fletcher PC (2006) Piezoresistive geometry for maximizing microcantilever array sensitivity. Master of engineering thesis
16. Hosseinzadegan H, Todd C, Lal A et al Graphene has ultra-high gauge factor. The SonicMEMS Laboratory, School of Electrical and Computer Engineering Cornell University, Ithaca, NY, USA
17. Safari A (1994) Development of piezoelectric composites for transducers. J Phys III Fr 4:1129–1149
18. Shekhawat G, Tark S-H, Dravid VP (2006) MOSFET-embedded microcantilevers for measuring deflection in biomolecular sensors. Science 311(5767):1592–1595
19. Seena V, Nigam A, Pant P et al (2012) Organic cantiFET: a nanomechanical polymer cantilever sensor with integrated OFET. J Microelectromech Syst 21(2):294–301
20. Ray P, Rao VR (2013) Al-doped ZnO thin-film transistor embedded micro-cantilever as a piezoresistive sensor. Appl Phys Lett 102:064101
21. Rasmussen PA, Thaysen J, Hansen O et al (2003) Optimised cantilever biosensor with piezoresistive read-out. Ultramicroscopy 97:371–376
22. Ray P, Seena V, Khare RA et al (2010) MRS proceedings, vol 1299. Cambridge University Press, Boston, Massachusetts
23. Huang MH, Wu Y, Feick H et al (2001) Catalytic growth of zinc oxide nanowires by vapor transport. Adv Mater 13(2):113
24. Donald AN (2011) Semiconductor physics and device. McGraw-Hill, New York, p 320
25. Nagata T, Ahmet P, Yoo YZ et al (2006) Schottky metal library for ZNO-based UV photodiode fabricated by the combinatorial ion beam-assisted deposition. Appl Surf Sci 252(7):2503–2506
26. Newnham RE, Skinner DP, Cross LE (1978) Connectivity and piezeoelectric-pyroelectric composites. Mater Res Bull 13:525–536
27. Teh WH, Durig U, Drechsler U et al (2005) Effect of low numerical-aperture femtosecond two-photon absorption on (SU-8) resist for ultrahigh-aspect-ratio micro stereo lithography. J Appl Phys 97:054907–054911
28. Ge J, Zeng X, Tao X, Li X et al (2010) Preparation and characterization of PS-PMMA/ZnO nanocomposite films with novel properties of high transparency and UV-shielding capacity. J Appl Polym Sci 118:1507–1512

29. Liu JM, Pan B, Chan HLW et al (2002) Piezoelectric coefficient measurement of piezoelectric thin films: an overview. Mater Chem Phys 75:12–18
30. Kalinin V, Rar A, Jesse S (2006) A decade of piezoresponse force microscopy: progress, challenges, and opportunities. IEEE Trans Ultrason Ferroelectr Freq Control 53:2226–2252
31. Zhao MH, Wang ZL, Mao SX (2004) Piezoelectric characterization of individual zinc oxide nanobelt probed by piezoresponse force microscope. Nano Lett 4:587–590
32. Agrawal R, Espinosa DH (2011) Giant piezoelectric size effects in zinc oxide and gallium nitride nanowires: a first principles investigation. Nano Lett 11:786–790
33. Dai S, Gharbi M, Sharma P, Park HS (2011) Surface piezoelectricity: size effects in nanostructures and the emergence of piezoelectricity in non-piezoelectric materials. J Appl Phys 110:104305–104307
34. Senesac L, Thundat TG (2008) Nanosensors for explosive detection. Mater Today 11(3):28–36
35. Longo G, Alonso-Sarduy L, Rio LM et al (2013) Rapid detection of bacterial resistance to antibiotics using AFM cantilevers as nanomechanical sensors. Nat Nanotechnol 8(7):522–526
36. Kosaka PM, Tamayo J, Ruz JJ et al (2013) Tackling reproducibility in microcantilever biosensors: a statistical approach for sensitive and specific end-point detection of immunoreactions. Analyst 138:863–872
37. Lang HP, Baller MK, Berger R et al (1999) An artificial nose based on a micromechanical cantilever array. Anal Chim Acta 393:59–65
38. Sen X, Mutharasan R (2009) Cantilever biosensors in drug discovery. Inf Healthc 4(12):1237–1251
39. Reddy CVB, Khaderbad MA, Gandhi S et al (2012) Piezoresistive SU-8 cantilever with Fe(III)porphyrin coating for CO sensing. IEEE Trans Nanotechnol 11(4):701–706
40. Joshi M, Pinto R, Rao VR, Mukherji S (2007) Silanization and antibody immobilization on SU-8. Appl Surf Sci 25(6):3127–3132
41. Joshi M, Kale N, Lal R, Rao VR, Mukherji S (2007) A novel dry method for surface modification of SU-8 for immobilization of biomolecules in bio-MEMS. Biosens Bioelectron 22:2429–2435
42. Nag S, Kale NS, Rao V et al (2009) An ultra-sensitive delta R/R measurement system for biochemical sensors using piezo resistive micro-cantilevers. Conf Proc IEEE Eng Med Biol Soc 2009:3794–3797
43. Sangtong S, Thanachayanont A (2007) Low-voltage CMOS instrumentation amplifier for piezoresistive transducer. ETRI J 29(1):70–78
44. Ferrari V, Ghisla A, Vajna ZK et al (2007) ASIC front-end interface with frequency and duty cycle output for resistive-bridge sensors. Elsevier Sci Direct Sens Actuators A 138:112–119
45. Gilda NA, Nag S, Patil S et al (2013) Current excitation method for ΔR measurement in piezo-resistive sensors with a 0.3-ppm resolution. IEEE Trans Instrum Meas 61(3):767–774
46. Gilda NA, Patil S, Seena V et al (2011) Piezoresistive 6-MNA coated microcantilevers with signal conditioning circuits for electronic nose. ASSCC 2011, pp 121–124
47. Gilda NA, Surya S, Joshi S et al (2011) A low-cost, ultra-sensitive hand-held system for explosive detection using piezo-resistive micro-cantilevers. ISOCC 2011, pp 325–328

Smart Materials Technology for Aerospace Applications

S. Gopalakrishnan

Abstract In this article, we present some very interesting aerospace applications related to smart materials. These applications are presented in the form of three case studies. These studies relate to the control of vibration in an aircraft component, de-icing of aircraft surfaces, and the active control of broadband excitations. Although there are many different smart materials available commercially, in this article only those technologies that are based on piezoceramic materials are used in solving the above-mentioned applications. Some current and future technologies that can be developed using smart materials are discussed toward the end of the article.

Keywords Piezoelectric materials · Laminated composites · Active vibration control · De-icing · Active spectral element · Active wave control

1 Introduction

Smart materials are those that respond to two or more different environmental variables [1–3]. Some environmental variables that these materials respond to include force, strain, temperature, pressure, electric fields, magnetic fields, voltage, etc. For example, piezoelectric materials produce electric field when the material is loaded. On the other hand, when the same material is subjected to an electric field, the material is strained. Thus, in such material, energy conversion takes place, and in the case of piezoelectric material, mechanical energy is converted to electrical energy and vice versa. This conversion process can be used in a variety of applications especially in the domain of aerospace. This article will provide some glimpse of the use of these materials in aerospace application.

S. Gopalakrishnan (✉)
Department of Aerospace Engineering, Indian Institute of Science,
Bangalore 560 012, India
e-mail: krishnan@aero.iisc.ernet.in

K. J. Vinoy et al. (eds.), *Micro and Smart Devices and Systems*,
Springer Tracts in Mechanical Engineering, DOI: 10.1007/978-81-322-1913-2_25,

Since a smart material can respond to two or more environmental variables, it will have as many constitutive relations as the number of environmental variables that the material responds to [1]. For example, piezoelectric material responds to both strain and the electric field, and hence it has two constitutive relations in terms of these environmental variables and they are highly coupled. One of the constitutive relations is called the *sensing law* and these are normally employed in sensor applications, while the second constitutive relation is called the *actuation law.*

Some common problems experienced by aircraft, which smart materials can solve, are the following: (i) Engine vibrations, (ii) flow separation due to fluid turbulence, (iii) flutter, (iv) high cabin noise levels, and (v) heavy ice formation on wing and control surfaces in extremely cold climates. In addition to solving the problems, smart material technologies are used in some novel aircraft applications to enhance their performance and operability. Some such applications are the wing morphing and flapping wing technologies [4, 5]. Such technologies are currently prevalent in unmanned air vehicles, and are actively pursued in many countries for military and civil applications.

A smart material system has three basic components, namely the host structure, a sensing element, and an actuating element. The sensing element can sense any environmental variables and if the values of these exceed the threshold values, the actuating element is triggered to inject the necessary energy to control the values of the sensed environmental variable. This additional energy is provided by the smart material actuator. This injection process has to be precisely controlled and the smart material system will thus require a robust and dependable control unit. Note that due to sensing and actuation constitutive laws, a smart material can act both as a sensor and an actuator.

This article is organized as follows. First, the constitutive laws for piezoelectric and piezofiber composite (PFC) materials are outlined. This will be followed by the presentation of three case studies, relating to active vibration control, de-icing of the aircraft wing, and the active broadband wave control of a multimodal excitation. The chapter ends with some concluding remarks and some of the advances made in this area of technology.

2 Constitutive Laws for Piezoelectric Materials

Piezoelectric material is available in many forms, namely crystals, polymers, and ceramics. Of all these forms, the ceramic form of piezoelectric materials is extensively used in actuator applications. The polymer form of this material is called polyvinyl di-fluoride (PVDF). This form has low actuator authority and is hence used mostly used as a sensor. All these forms have similar constitutive laws. However, the constitutive law of PFC is quite different and, hence, they are given separately in this section.

All forms of piezoelectric materials have the constitutive laws in the following form

$$\{\sigma\}_{3x1} = [C]^{(E)}{}_{3x3}\{\varepsilon\}_{3x1} - [e]_{3x2}\{E\}_{2x1} \tag{1}$$

$$\{D\}_{2x1} = [e]_{2x3}{}^{T}\{\varepsilon\}_{3x1} + [m]^{(\sigma)}{}_{2x2}\{E\}_{2x1}. \tag{2}$$

The first constitutive law is called the actuation law, while the second is called the sensing law. Here $\{\sigma\}$ is the stress vector, $\{\varepsilon\}$ is the strain vector, $[e]$ is the matrix of piezoelectric coefficients of size, $\{E\}^T = \{ V_x/t \quad V_y/t \}$ is the applied field in two coordinate directions, where V_x and V_y are the applied voltages in the two coordinate directions, and t is the thickness parameter. $[\mu]$ is the permittivity matrix, measured at constant stress and $\{D\}$ is the vector of electric displacement in two coordinate directions. $[C]$ is the mechanical constitutive matrix measured at a constant electric field. Equation (1) can also be written in the form

$$\{\varepsilon\} = [S]\{\sigma\} + [d]\{E\}. \tag{3}$$

In the expression above, $[S]$ is the compliance matrix, which is the inverse of the mechanical material matrix $[C]$ and $[d] = [C]^{-1}[e]$ is the electromechanical coupling matrix, where the elements of this matrix are direction dependent.

The first part of Eq. (1) represents the stresses developed due to a mechanical load, while the second part of the same equation gives the stresses due to voltage input. From Eqs. (1) and (2), it is clear that the structure will be stressed due to the application of an electric field even in the absence of a mechanical load. Alternatively, when the mechanical structure is loaded, it generates an electric field. In other words, the constitutive law above demonstrates electromechanical coupling, which is exploited for a variety of structural applications.

The reversibility between strain and voltage makes piezoelectric materials ideal for both sensing and actuation. Among the many forms of piezoelectric materials, PVDF has a very low electromechanical coupling coefficient and hence cannot be used for actuation. However, it is a very good sensor material. PZT and PFC, on the other hand, have very high coupling coefficients and, hence, higher actuator authority. They are extensively for multimode wave control.

Next, we will discuss the constitutive model of PFC. PFC is those wherein the active piezofibers are integrated into composites with interdigital electrodes as shown in Fig. 1.

Obtaining its constitutive model is far more complex and is done by considering a representative volume element (RVE) and the modified rule of mixtures. Complete details are given in [6, 7] and only the final form of the constitutive model is reported here, and is given by

$$\begin{Bmatrix} \sigma_1 \\ \sigma_2 \\ \sigma_3 \\ D_3 \end{Bmatrix} = \begin{bmatrix} c_{11}^E & c_{12}^E & c_{13}^E & -e_{31} \\ c_{12}^E & c_{22}^E & c_{23}^E & -e_{32} \\ c_{13}^E & c_{23}^E & c_{33}^E & -e_{33} \\ e_{31} & e_{32} & e_{33} & \varepsilon_{33}^s \end{bmatrix} \begin{Bmatrix} \varepsilon_1 \\ \varepsilon_2 \\ \varepsilon_3 \\ E_3 \end{Bmatrix}. \tag{4}$$

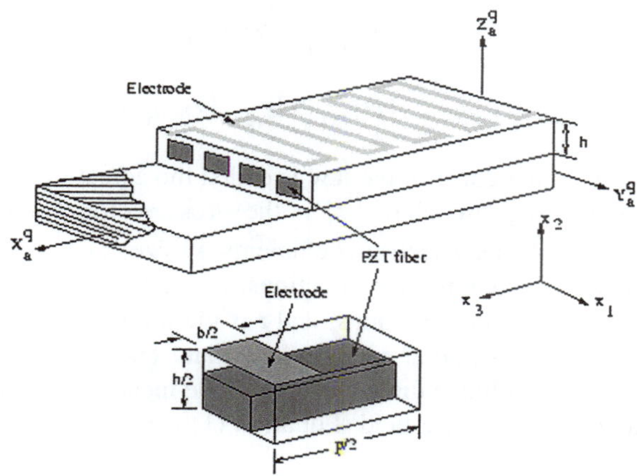

Fig. 1 Laminated composite with integrated piezofibers and electrodes

Equation (4) is the expanded form of Eqs. (1) and (2) of the piezoelectric material and all the notations have the same meaning. Inserting appropriate boundary conditions along 1 and 2 directions, the effective 1-D actuation law can be written as

$$\sigma_3 = c_{33}^{eff} \varepsilon_3 - e_{33}^{eff} E_3. \tag{5}$$

In Eq. (5), the expressions for c_{33}^{eff} and e_{33}^{eff} are quite long and complex and it is given in [6, 7]. The constitutive law given by Eq. (5) can be used for actuation along the axial direction (or direction 3 shown in Fig. 1).

3 Aerospace Applications Using Smart Materials

In this section we present three different aerospace applications using smart materials technology. The first is active vibration control in a thin-walled box beam structure, the second is the de-icing application in an aircraft wing, and the last example is broadband multimodel wave control using a PFC actuator. The first and last examples are the work done in the Department of Aerospace engineering at the Indian Institute of Science, while the second case study is the outcome of a project done under National Program on Micro and Smart Systems (NPMASS), Government of India.

3.1 Active Vibration Control of Thin-Walled Composite Structures

Thin-walled constructions are extensively used in aircraft structural components due to their inherent advantages of high stiffness and low weight. The analysis of these structures is more complicated than that of conventional structures. There are many secondary effects such as warping, ovaling, etc., which are very important in their behavior. In addition, thin-walled structures in aircrafts give coupled motions. For example, a bending motion will give rise to torsional motion or an axial load will give rise to a bending motion. Reference [8] gives a complete overview of the behavior of thin-walled structures in aerospace and civil structures. The analysis of such structures requires a good mathematical model to obtain the plant matrix, which is necessary for designing the controller. A finite element model can be used for this purpose, and such a model for thin-walled structures is reported in [1, 9, 10]. In this section, we will use this FE model and outline the design procedure for the controller in performing vibration control. The output obtained from the designed controller is verified experimentally, as well as numerically through FE simulations.

A glass-epoxy composite box beam with two bimorph surface-mounted ***SP5H*** PZT patches is used for an experimental study of vibration suppression of transverse bending modes. The PZT patches act as actuators, through which both the exciting electrical load and control signals are applied. Sensing is done through ***Structcel 330A*** (PCB Electronics, USA) accelerometer with a sensitivity of $20.4\,\mathrm{mV/m^2/s}$. The material properties and dimensions of the box beam and PZT patches are given in Table 1. The cantilever beam has a ply layup of $[0_3/90_2/0_3]$on all four sides. The schematic diagram of the beam is shown in Fig. 2. The vibration in the box beam is induced through a single patch of a PZT actuator using an oscillator. The generated vibration is sensed through an accelerometer (***Structcel 330A***, PCB Electronics USA) with sensitivity 200 mV/g, which is then fed to the A/D converter after signal conditioning. The PI controller is implemented in the DSP-based vibration control card. The card is designed around the **ADSP-2181 DSP**. The sensed voltage is proportional to the acceleration of the ADSP controller, which outputs the calculated control voltage through D/A converter channels. The control voltage is amplified by a high voltage amplifier and then fed into the PZT actuators for bending actuation to suppress the transverse vibration.

The DSP-based vibration controller card is developed for implementing the digital control system. The card is designed around **ADSP-2181 DSP**. This card also contains one ADC with an eight channel inbuilt multiplexer, eight channel serial DAC, UART, and Boot Flash. Eight analog sensor signals from the input to card and eight analog channels are the outputs. The **ADSP-2181 DSP** from analog devices has been chosen to implement vibration control laws and to acquire sensor inputs. This DSP features high-speed processing, single cycled instruction execution, and multifunction instructions. The signal conditioner analog sensor inputs are connected to their respective ADCs. Prior to ADC sampling, all sensor inputs are filtered by a second-order Butterworth low pass filter of 2 kHz cutoff frequency. Analog outputs (DAC outputs) are buffered and brought to a connector. One asynchronous serial channel is provided for downloading **ADSP-2181** programs and debugging. Eight general purpose inputs/outputs of DSP are also

Table 1 Material properties of glass-epoxy composites and an SP5H actuator

Material properties: glass fiber composite
$E_{11} = 23.69$(GPa), $E_{22} = 7.63$(GPa), $G_{12} = G_{13} = G_{23} = 3.37$(GPa), $\nu_{12} = 0.26$, $\rho = 1985$ kg/m^3
Dimensions (in meters): length (L) = 0.655, width (2b) = 0.057, depth (2h) = 0.019, ply thickness $t_p = 0.00025$ (m)
Material properties: SP5H PZT actuator:
E (Young's modulus) = 62.0(GPa), ν (Poisson's ratio) = 0.31, thickness (m) = 0.0005, density $\rho = 7400$ kg/m^3
Piezoelectric coefficients: $d_{31} = d_{32}$(m/V) $= -166 \times 10^{-12}$, d_{33}(m/V) $= 300 \times 10^{-12}$

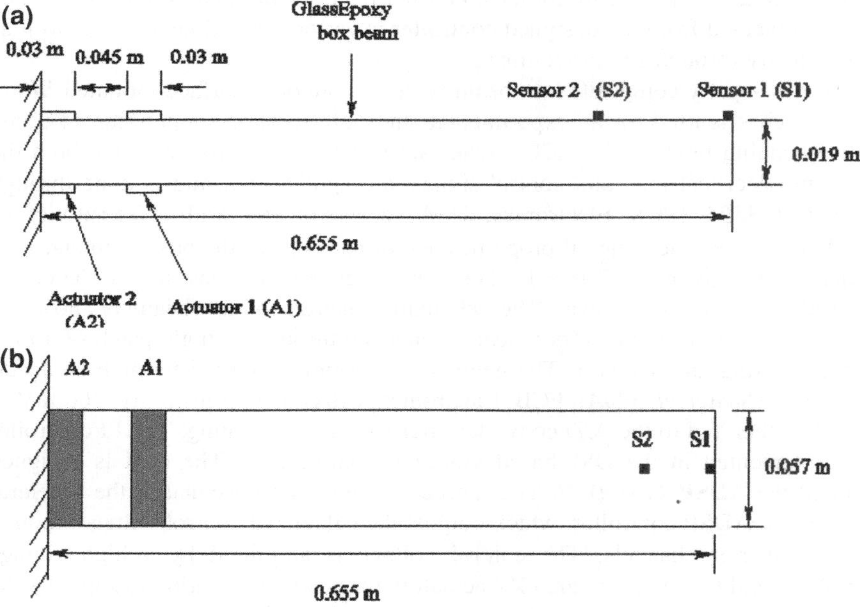

Fig. 2 Test specimen with actuators and sensors: **a** side view and **b** top view

brought out to a berg stick connector. The DSP-based vibration controller card specifications are given in Tables 2 and 3 for the digital interface, analog inputs and analog outputs, respectively.

The design methodology for the PI controller is explained in [1]. Here, we implement a PI controller to reduce the vibration in a thin-walled box beam. First, the closed-loop responses using the PI controller are obtained experimentally using an SISO and a single output two input system. The test specimen is excited in a transverse bending mode by applying single frequency sinusoidal voltage through PZT actuators. The experimental results are correlated with the results simulated via state space modeling.

Here, the mass, and stiffness matrix for controller design is generated using the FE model described in [1, 9, and 10]. Vibration suppression is performed

Table 2 Digital interface specifications of DSP vibration controller card

DSP ADSP 2181 KS-160 specification	
Operating frequency	40 MHz
Memory	Data 16 × 32 (on chip), program 16 × 24 (on chip), boot 64 × 8 (external FP ROM)
Serial channels	One RS 232 port
External ADC	AD 7891, 12 bit, 0–5 V
Digital inputs and outputs	8 ports
Power supply	5 V @ 1 A
Form factor	5.3 in × 3.5 in

Table 3 Analog channel specifications for analog inputs and outputs (output specification in brackets)

No: of channels	15 (8 buffered)
Voltage range	−5 V to +5 V (0 V to +5 V)
Frequency	0.5 Hz–2 KHz
Signal conditioner	Second-order Butterworth filter, cutoff frequency at 2 KHz
Resolution	12 bit (12 bit)
Interface to DSP	Parallel

experimentally by a single and two PI controllers. In both the cases, tip acceleration is measured and the feedback is given as proportional gain K_P times tip acceleration $\ddot{x}_{tip}$ (this is measured by the accelerometer sensor S_1 (see Fig. 2) and integral gain $\boldsymbol{K_I}$ times velocity $\dot{x}_{tip}$ obtained by integrating $\ddot{x}_{tip}$. In the first case, using a single PI, a sinusoidal voltage having an amplitude of 95 V and frequency 34.5 Hz is applied through actuator 1 ($\boldsymbol{A_1}$) (see Fig. 2b) and the control signal is fed back through actuator 2 ($\boldsymbol{A_2}$). In Fig. 3a the experimental open-loop tip acceleration and the controlled tip acceleration with $K_P = 1.8$ and $K_I = 0.54$ is shown. These K_P and K_I values used in the experiment are obtained through tuning using the Ziegler–Nichols rule [11]. The uncontrolled response has amplitude of 1.88 V, which is reduced to 0.97 V in the controlled response. In the figure above 800 sample points with a sampling interval of 0.00025 s are shown. Figure 3b shows the corresponding simulated results with K_P and K_I, the same as that in experiment. This figure shows the steady-state responses between 3.8 and 4.0 s and the time step for numerical simulations is 0.0001 s. The simulated open- and closed-loop amplitudes are 1.93 and 0.99 V, respectively, giving a net suppression of 48.7 %.

3.2 De-icing of an Aircraft Wing

This project was performed at the Department of Aerospace Engineering as a part of the NPMASS project.

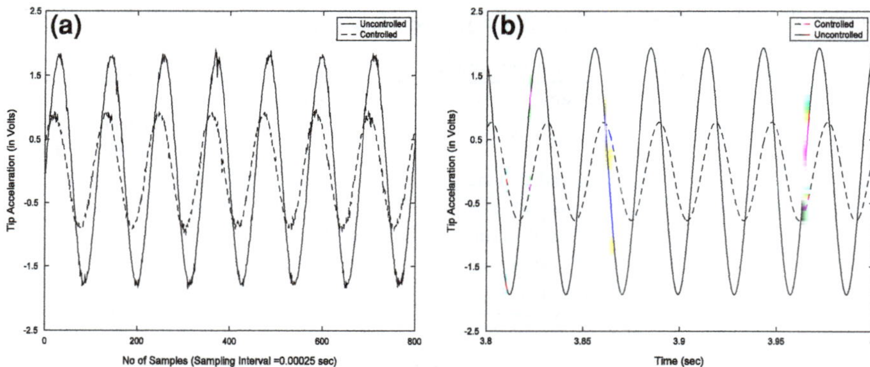

Fig. 3 Uncontrolled and controlled responses of the box beam using a single PI controller **a** Experimental and **b** Theoretical

Ice accretion on the surface of an aircraft wing is a major concern in the aircraft industry as it affects aircraft performance by degrading its ability to generate lift. Ice accretes near the leading edge of an aircraft wing as it is a stagnation point. Ice deposition changes the leading edge profile of the aerofoil and severely affects the lift-generating ability of the wing. Ice formation takes place either on the ground or in flight. In-flight ice formation is a significant problem in propeller-driven aircraft since they cruise at altitudes at which ice- bearing clouds occur. The presence of ice on airplane surfaces prevents the flow of air, which increases drag and reduces lift. Ice on wings is especially dangerous during take-off, when a sheet of ice of the thickness of a compact disk can reduce lift by 25 % or more. Ice accumulated on the tail of an aircraft (a spot often out of the pilot's sight) can throw off a plane's balance and force the craft to pitch downward, a phenomenon known as a tail stall.

Following are some of the techniques used in de-icing to prevent and remove ice from aircraft: (a) Electro-thermal de-icing, (b) fluid de-icing/anti-icing systems, (c) pneumatic de-icing systems, (d) electro impulse de-icing systems, and (f) electro-vibratory de-icing systems. Each of them has their own advantages and disadvantages. In this article, we will show how the electro-vibratory approach (EVA) can be employed to de-ice the system.

The EVA uses vibration or mechanical disturbance to dislodge ice. The basic idea is to vibrate the structure at its resonance frequencies at sufficiently high g-forces to debond the ice. The power requirement for this technique is quite small and the only drawback that is perceived is the fatigue loading of the structure. However, if the structure is subjected to excitation only for short intervals of time, then this problem may not occur at all. An alternate approach to de-ice using vibration is by subjecting the ice–substrate interface to shear horizontal (SH) disturbances. These disturbances are generated using a piezoceramic actuator, and the interface bonds are broken when the shear stresses due to the SH disturbances exceed the shear strength of the interface. These results are encouraging, and the benefits in terms of low power consumption of the order of 10 W [12] and low weight make this quite an attractive candidate for de-icing/anti-icing applications on aircraft surfaces. The proposed de-icing principle is based on the generation of debond shear strength using PZT patches. The structure when vibrating at its natural frequency, generate high stresses. The stresses vary with different natural frequencies due to the associated mode shapes. The modes where shear stresses on the surface of the aircraft wing leading edge are higher can be utilized to break the adhesive bond.

An aircraft wing model conforming to NACA0020 is fabricated with commercial aluminum. The leading edge module is fabricated in such a way that the profile is a part of the wing and can be fixed in the wing. In order to match the exact condition of the leading edges, the *T section* and *L sections* are considered for the supporting of the leading edge skin. The aircraft wing is analyzed for its global and local vibration characteristics using finite element analysis (ANSYS) and correlated with experimental investigations. The finite element analysis results are shown in Table 4.

A wing model is fixed to the I-section as shown in Fig. 4a and tested for its natural frequencies. An instrumented impulse hammer (PCB) is used along with a B&K 4344 accelerometer and a B&K 2635 charge amplifier. The bending frequency of the wing was found to be 9 Hz and the twist frequency was 11.25 Hz. These low frequency global modes will not significantly contribute in the de-icing process. The leading edge modules in between the cross ribs have local shell modes and help in generating the required debond stress from the actuation of PZTs bonded underneath the leading edge. The analytical and experimental investigation has confirmed the large number of local modes that help to de-ice the leading edge of the wing. The airfoil wing leading edge model shown in Fig. 4b is placed in the ice chamber. Water was sprayed on the model using a sprayer keeping the nozzle of the sprayer directed horizontal to the edge of the model at −14 °C. The temperature on the model surface rises. The model with water sprayed on it was allowed to cool. Once the temperature of the model surface reached −14 °C, the piezoactuators are energized. A sine-wave generator signal fed the signal to an amplifier that amplified the sine wave to a peak value of 200 V. The frequency of the sine wave is varied across the resonance frequencies. De-icing of the model did not occur during the initial frequencies.

When the model was excited at its resonance frequency of about 1–2 kHz (Sinesweep), the ice layer accreted on the model surface started de-bonding. The thickness of the ice layer deposited on the air foil leading edge typically has a higher thickness at the stagnation point as shown in Fig. 4a. Figure 5 shows the process of de-icing at different times.

In conclusion, the high strain zones for a particular mode can be identified to create high interfacial stresses so that de-icing can be successfully implemented. The piezoceramic (PZT) patches can be effectively used to de-ice the leading edge of the aircraft wings. The space available in the leading edge inner surface can be used to bond the PZT patches.

Table 4 Comparison of experimental and theoretical natural frequencies of a wing

Mode	Natural frequency–FEM in Hz	Natural frequency–experiments in Hz	Remarks
1	7.2	9	Bending
2	10.6	11.25	Twisting
3	10.8		Twisting

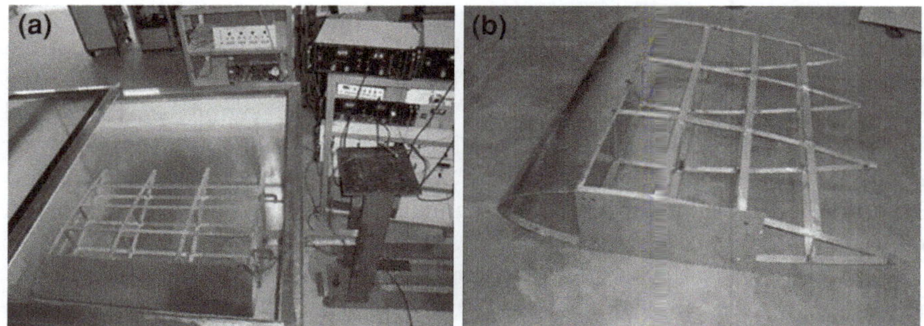

Fig. 4 **a** Experimental setup for wing de-icing and **b** Airfoil section of the wing

3.3 Broadband Active Wave Control Using PFC Actuators

The concept of broadband distributed active control in flexible structures has evolved in recent times. Tremendous technological success in the field of micro-electro-mechanical systems (MEMS) has laid the path toward the implementation of such concepts, especially structures made of multifunctional multiphase composites (see, e.g. [13, 14, and 15]). Currently, a number of designs are available, which use different forms of PZT ceramic fibers and conventional matrices. Following the performance standardization of these active composites as distributed actuators, the main task remains on how to use these for better control performance, particularly in transverse actuation. Some of the physical insight into the macroscopic behavior of these PFC actuators has been studied [16]. In this article, we consider PFC with an interdigitated surface electrode as a distributed actuator element, and a broadband vibration sensor capable of measuring far-field as well as near-field. The computational model used is an active spectral element model (ASEM) [7], which accounts for axial–flexural coupling due to out-of-plane actuation efforts and unsymmetric mechanical stiffness and inertia across the beam thickness. Here, it is important to note that most of the available studies on local wave control using distributed actuation deal with collocated sensor–actuator configurations, where the actuator has to cancel the disturbance arising only due to near-field disturbances. The lack of information regarding approaching waves from distant boundaries may increase the level of actuation input required at the local control point. This necessitates the placement of the actuators upstream from

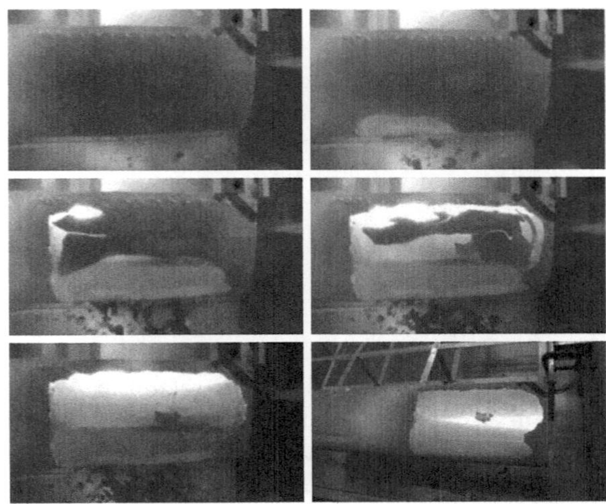

Fig. 5 De-icing process at different times

the local control point to minimize the approaching disturbances well in advance. In addition, direct feedback from the control point is found to be more suitable when a large number of higher modes are to be controlled. The fundamental behavior of this noncollocated sensor–actuator configuration for distributed active low authority control resembles that of the disturbance propagation in structural network. The proposed ASEM is based on a general finite element formulation in frequency domain, and provides remarkably faster and efficient computation [6].

The formulation of the ASEM model is given in [6, 7]. This model uses the constitutive model given in Sect. 2. Here, a 1-m-long unsymmetric composite cantilever beam with a unit impulse load at its tip is considered. The beam is considered to be 2 cm thick and made of AS/3501-6 graphite-epoxy with ply stacking sequence $[0_5/90_5]$. The effective control of multiple spectral peaks in the frequency response of transverse tip displacement is chosen as the local control objectivity. This procedure, if satisfied, will also guarantee the stability of the closed-loop system since all the resonant modes, and hence the poles will be suppressed with no modal truncation error in spectral computation. We consider the location of a PFC actuator, bonded on the top surface of the beam as shown in Fig. 6.

It is important to note that the stress waves that travel from the tip toward the fixed end of the cantilever are of larger amplitude near the fixed end. Also, for local control at the tip, it is absolutely necessary to suppress the scattered waves from the fixed end. Due to this reason, the actuator (of length L_a) can be placed adjacent to the fixed end. The sensor is assumed at x_s, which is considered near the tip for direct feedback to the actuator. The total configuration is shown in Fig. 6 In this study, we consider only velocity feedback. However, specific applications of acceleration and displacement feedback can be performed to augment the amount of closed-loop damping. A nondimensional scalar feedback gain g_r is derived from the actual gain γ_r to perform parametric studies. These two are related as

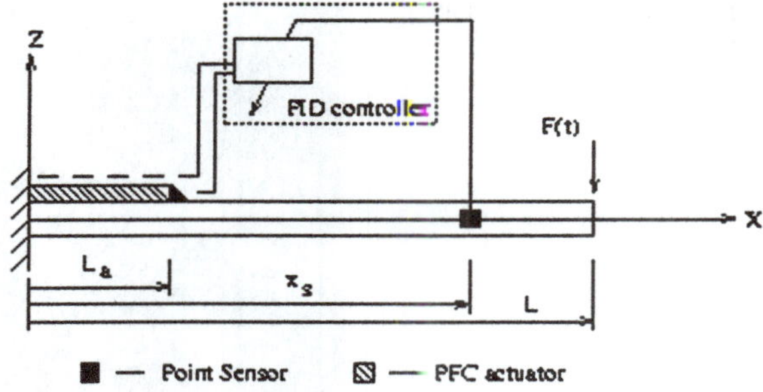

Fig. 6 Composite cantilever beam with surface-bonded PFC actuator

$g_r = (c_0 \alpha \beta)\gamma_r / V_0$. Here, V_O is the reference AC voltage and c_O is the speed of sound in air. A near-optimal performance can be achieved in the numerical simulation by an iterative scheme with varying gain g_r and sensor–actuator location (L_a, x_s). One such closed-loop performance corresponding to $L_a = 0.25$ m, $x_s = 1.0$ m (at the tip) and $g_r = 3.4 \times 10^6$ is shown in Fig. 7. From the figure, it is seen that the configuration is able to suppress almost all the resonant modes. Location of all the zeroes, except those at 10 kHz, remained unaffected showing minimal pole-zero interaction.

However, Fig. 7 does not aid in identifying the nature of change in the amplitude level over the whole frequency range due to parametric variation. To study this particular aspect in the example considered above, we first assume that the feedback gain is optimal and not sensitive to very small variations in the other parameters such as L_a and x_s. Next, L_a is slowly varied from 0.15 m to 0.35 m corresponding to velocity feedback from various sensor locations x_s moving away from the cantilever tip (Fig. 6). A total change in the amplitude level for the closed-loop system over the frequency range of 20 kHz is evaluated as

$$\hat{\prod} = \sum^{N/2} 20 \left[\log_{10} \left| \hat{w}(\omega_n)^2_{\text{open}} \right| - \log_{10} \left| \hat{w}(\omega_n)^2_{\text{close}} \right| \right]. \tag{6}$$

In Fig. 8, the variation in $\hat{\prod}$ is shown by the shades over a two- dimensional solution space comprising various combinations of L_a (length of PFC actuation) and x_s (sensor location). This plot confirms with the results in Fig. 7 in a sense that one optimum solution exists at ($x_s/L = 1.0$, $L_a/L = 0.25$) and yields a total suppression of nearly 6.025 dB. Figure 8 also predicts that one better solution could exist at ($x_s/L = 1.0$, $L_a/L = 0.15$) with a lower actuation length. It is also clear from the plot that a sensor placed before 0.96 m never performs effectively over the significant resonant modes, whereas nearly collocated configuration before 0.8 m may not produce stable performance.

Hence, in this study, the concept of active fiber composites for distributed actuation is employed. Accurate and computationally efficient simulation studies on the control of broadband wave are performed based on an active spectral element model (ASEM). The model developed can easily handle any combination

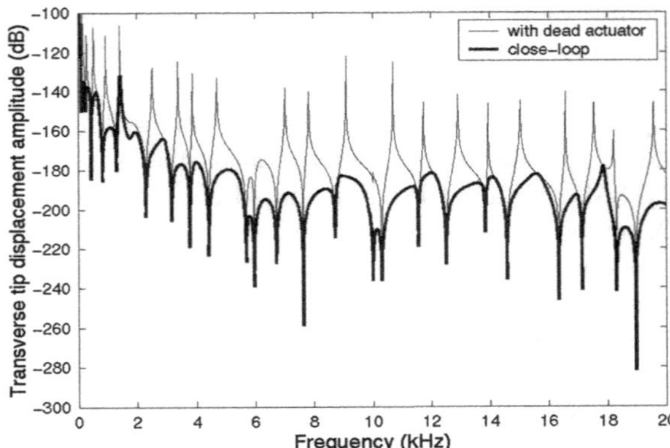

Fig. 7 Closed-loop performance of surface-bonded PFC

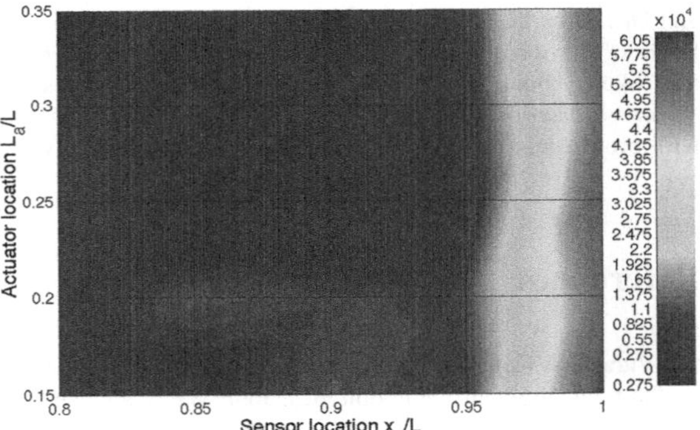

Fig. 8 Performance of various sensor–actuator configurations

of sensor inputs into the compensator with constant feedback gains. The generalized computational model is used for a very few iterations to converge into an optimal solution for a partially known set of control parameters.

4 Conclusions

In this article, three case studies on the use of smart materials for solving some of the problems associated with aircraft applications were presented. The applications of these materials in fixed wing and rotary wing aircrafts are plenty. In this article, we showed the use of only one of the smart materials, that is piezoelectric materials and its derivatives. The use of other materials such as magnetostrictive materials (TEFENOL-D), electrostrictive material (lead manganese niobate–lead titanate (PMN-PT)), and shape memory alloys, are extensively reported in the literature for aerospace applications. In addition to the problems solved in this article, many challenging problems such as wing morphing, flapping wing aircraft, trailing edge flap actuations for exterior noise reduction, interior cabin noise reduction, flutter suppression, and engine mount vibration reduction have been addressed and solved using smart materials by many researchers the world over.

India contributes significantly to this technology. Thanks to the funding provided by national programs such as the National Program on Smart Materials (NPSM), National Program on Micro and Smart Systems (NPMASS), and the Development Initiative for Smart Aircraft Systems (DISMAS), many of the problems identified in the previous paragraph, have been solved by Indian researchers using a variety of smart actuators. In addition to solving application oriented research, these national programs also funded actuator development programs. Due to these efforts, India is in a position to make indigenous stack actuators of different ranges of force and displacement that are comparable to the best commercially available actuators, for varied aerospace applications.

Acknowledgments The author would like to thank his former graduate students Dr Mira Mitra and Dr D Roy Mahapatra for providing the data for the first and third case studies in this article. Some of the work reported here was part of their respective doctoral theses. The author would also like to thank Dr S.B. Kandagal, the Principal Investigator of the NPMASS project titled *Electro-Vibratory deicing of airfoils with Electrostrictive and Piezoceramic actuators* (PARC 3 with project number 3.2) for providing the data and the write up for the de-icing case study.

References

1. Varadan VK, Gopalakrishnan S, Vinoy KJ (2005) Smart material systems and MEMS: design and development methodologies. John Wiley and Sons, United Kingdom
2. Culshaw B (1996) Smart structures and materials. Artech House, United Kingdom
3. Gandhi MV, Thompson BS (1992) Smart materials and structures. Kluwer Academic Publishers, Boston
4. Platzer MF, Jones KD, Young J, Lai JCS (2008) Flapping wing aerodynamics: progress and challenges. AIAA J 46(9):2136–2149
5. Jha AK, Kudva JN (2004) Morphing aircraft concepts, classifications and challenges. In: Proceedings of SPIE 5388, smart structures and materials 2004: industrial and commercial applications of smart structures, 213, San Diego, USA
6. Gopalakrishnan S, Chakraborty A, Mahapatra DR (2007) Spectral finite elements. Springer , United Kingdom
7. Mahapatra DR, Gopalakrishnan S (2000) An active spectral element model for PID feedback control of wave propagation in composite beams . In: Proceedings of VETOMAC I, October 25-27, Bangalore, India
8. Megson, THG(1974) Linear analysis of thin walled elastic structures. Surray University Press, United Kingdom
9. Mira Mitra (2003) Active vibration suppression of composite thin walled structures: M.Sc (Engg) Thesis, Indian Institute of Science, Bangalore, India
10. Mitra M, Gopalakrishnan S, Bhat MS (2004) Vibration control in acomposite box beam with Piezoelectric actuators. Smart Mater Struct 13:676–690
11. Ziegler JG, Nichols NB (1942) Optimum settings for automatic controllers. ASME Trans 64:759–768

12. Kandagal SB, Venkatraman K (2005) Piezo-actuated vibratory deicing of a flat plate. In: Proceedings of 46th AIAA/ASME/ASCE/AHS/ASC structures, structural dynamics and materials conference, Austin, Texas, 18–21 April 2005
13. Vaidya UK (1999) Integrated and multi-functional thick section and sandwich composite materials and structures. In: Proceedings of ISSS-SPIE, International Conference on smart materials structures and systems, Bangalore, India, pp 311–318
14. French J, Weitz J, Luke R et al (1997) Production of continuous Piezoelectric ceramic fibers for smart materials and active control devices. In: Proceedings of SPIE, 3044
15. Bent AA (1997) Active fiber composites for structural actuation. Ph.D. Thesis, Massachusetts Institute of Technology
16. Hagood NW, Kindel R, Ghandi K et al (1993) Improving transverse actuation of Piezoceramics using integrated surface electrodes. In: Proceedings of SPIE 1917–25

Electronic Circuits for Piezoelectric Resonant Sensors

M. Umapathy, G. Uma and K. Suresh

Abstract Resonant sensors have established themselves in the market place, usually in applications that require high resolution, for measuring physical and chemical phenomena. The electronics for collocated and self-sensing piezoelectric resonators are described as they are important components in resonant sensor design. The applicability of the closed-loop electronics and self-sensing electronics to the cantilever-based piezo actuated/sensed resonant sensor for measuring the thickness and DC current is presented.

Keywords Resonant sensors · Piezoelectric · Self-sensing actuation · Cantilever beam · Mass sensor · Closed-loop electronics

1 Introduction

In the past few years, resonant sensing has become a very popular method for measuring physical and chemical phenomena and it has established itself in the marketplace, usually in applications that require high accuracy. This is especially true in the field of micro electro mechanical systems (MEMS) sensors, since the electronics required for bringing a system to a state of oscillation can be fairly simple and go well with integration. Another advantage of resonant sensors is that the measurement principle is usually based on the mechanical properties of materials rather than the electronic properties, and these can be shown to offer sensors with good stable performance, resolution, reliability, and response time [1]. In resonant sensors, the frequency at resonance of a mechanical structure is changed by a physical or chemical stimulus to be measured, in such a way that the value of

M. Umapathy (✉) · G. Uma · K. Suresh
Department of Instrumentation and Control Engineering, National Institute of Technology, Tiruchirappalli 620015, India
e-mail: umapathy@nitt.edu

K. J. Vinoy et al. (eds.), *Micro and Smart Devices and Systems*,
Springer Tracts in Mechanical Engineering, DOI: 10.1007/978-81-322-1913-2_26,

the measurand can be derived by the shift in the resonant frequency of the structure. The principle can be applied to several measurands, the amplitude and phase of the output at resonance can be used for detecting the change in resonance frequency [2]. Frequency output resonant sensors typically offer high sensitivity and stability, potentially leading to accuracy and resolution. In addition the quasi-digital nature of the frequency signal eases interfacing to readout and processing circuitry and improves noise immunity. The resonance sensor can be used for sensing any parameter which changes either the mass or the stiffness of the resonating structure.

Cantilevers are the simplest resonating structures that can be easily machined and mass-produced on the meso- and micro-scale. Measuring the resonant frequency change of a cantilever first requires them to be excited by a perturbing driving force so that the resonant frequency can be realized. Second, the resulting motion must be both measured and gathered for post-processing by a readout technique. Among the various excitation and detection techniques used in resonant sensors in recent years, the use of piezoelectric excitation and detection to measure physical, chemical, and biological quantities have increased apart from their application in structural vibration measurement and control [3]. Moreover, the use of a single piezoelectric element for sensing and actuation, i.e., a self-sensing piezoelectric actuator for measurement and control, has gained importance as it minimizes sensor/actuation signal conditioning electronics, weight and cost as compared to non-collocated and collocated sensed, actuated systems.

A self-sensing actuator is a new concept for intelligent systems, where a piece of a piezoelectric element simultaneously acts as both a sensor and an actuator [4]. In self-sensing, more emphasis is given to the electronics required for separating the sensed signal from the actuation signal as it is a crucial element in achieving the required performance. In addition, it is advantageous to use self-sensing piezoelectric actuators in resonant sensor design, which requires the bonding or deposition of piezoelectric material on one side of the resonating mechanical structure. At present, self-sensing actuators have found many applications in active vibration control, structure health monitoring, system parameters identification, and control algorithm optimization [5]. Self-sensing actuator systems are lighter, minimize the sensor/actuation signal conditioning electronics, and are less expensive than non-collocated and collocated sensed, actuated systems.

2 Theory of Operation of Resonant Sensor

A resonant sensor is a device with an element vibrating at resonance, which changes its resonant frequency, i.e., mechanical resonance frequency as a function of a physical or chemical parameter. The vibrating structure can be a cantilever, diaphragm, wire, etc. The conversion from the measurand to the variation in resonance frequency of the vibrating element can be accomplished by, with a change in the stress, mass, or shape of the resonator [6]. A resonant sensor consists of a vibrating structure, excitation source, and detection element with appropriate

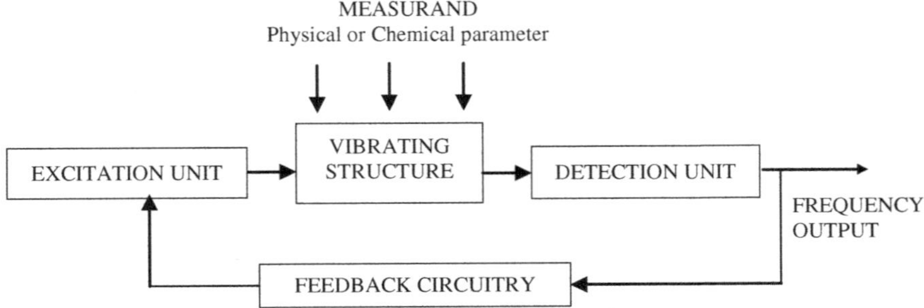

Fig. 1 Block diagram of resonant sensor

electronics in a closed loop to keep the resonant structure at resonance in spite of variations in the measurand as shown in Fig. 1. The excitation and detection technique used in resonant sensors are electrostatic excitation and capacitive detection, dielectric excitation and detection, piezoelectric excitation and detection, resistive heating excitation and piezoresistive detection, optical heating and detection, and magnetic excitation and detection [7, 8].

3 Electronics for Resonant Sensor

3.1 Closed-Loop Electronics for a Piezoelectric Resonance Sensor

An electronic circuitry which maintains the resonator at the desired resonance when a measurand acts on it as shown in Fig. 2. A uniform cantilever beam has two identical piezoelectric patches bonded onto the upper and lower surfaces of a beam over the region r_1 to r_2. The piezo on the upper surface acts as an actuator and the one on the lower surface acts as a sensor. In closed-loop electronics, an operational amplifier is used and the sensing piezo is connected in series with an RC network. With high gain for the amplifier, the output of the circuit is connected to the actuating piezoelectric patch. When the system is switched ON, the piezoelectric sensor responds to the noise and produces a small voltage which is used by the oscillator to buildup a steady-state oscillating condition. By designing the value of the feedback resistance R_f and the RC network for providing the required gain and phase shift, respectively, the measurement system is made to oscillate at its resonant frequency. The closed-loop resonant circuit tracks the change in resonance frequency and vibrates the beam with the new resonance frequency whenever the measurand on the cantilever beam changes. The shift in resonant frequency is the measure of an unknown input present on the cantilever beam. The transfer function between the sensor output voltage $V_c(t)$obtained due to the voltage applied to the actuating piezoelectric patch $V_a(t)$ is given as [9, 10],

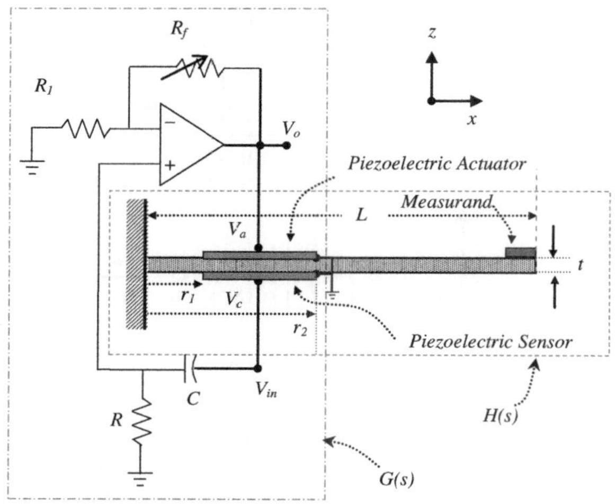

Fig. 2 Piezoelectric laminated cantilever beam with closed-loop electronics

$$H(s) = \frac{V_c(s)}{V_a(s)} = \sum_{i=1}^{\infty} \frac{-C_s C_a[\varphi_i'(r_1) - \varphi_i'(r_2)]^2}{\rho A L^3(s^2 + 2\zeta_i\omega_i s + \omega_i^2)} \tag{1}$$

where $\omega_i = \lambda_i^2\sqrt{EI/\rho A}$, E—is the modulus of elasticity of the beam material, I—is the moment of inertia of the beam, ρ—is mass density of the beam material, A is the cross-sectional area of the beam, L—is the length of the beam, $\lambda_i L$—is weighted mode frequencies, C_s and C_a are the constants which depends on the geometry of the system, ω_i, ζ_i, and $\varphi_i(r)$ are the ith mode frequency, damping ratio, and shape function of the cantilever beam, respectively. Since the thickness of the piezoelectric patch is small compared with the cantilever beam thickness, we assume that EI and ρA are uniform over the length of the beam.

The feedback electronics *G(s)* in Fig. 2 consists of a non-inverting amplifier in series with an RC lead network; its transfer function is given as,

$$G(s) = \frac{V_o}{V_{\text{in}}} = k\left(\frac{RCs}{1 + RCs}\right) \tag{2}$$

where $k = 1 + (R_f/R_1)$ is the gain of the operational amplifier. As the Barkhausen criteria for continuous oscillation at ω is [11],

$$|G(j\omega)| \times |H(j\omega)| = 1 \tag{3.a}$$

$$\angle G(j\omega) + \angle H(j\omega) = \varphi_{RC} + \varphi_{\text{beam}} = 0^{\circ} \tag{3.b}$$

where $G(j\omega)$ and $H(j\omega)$ are the sinusoidal transfer function of the feedback electronics and the piezolaminated cantilever beam, respectively, with ω as the continuous oscillation frequency. Here, the piezoelectric laminated cantilever beam in Fig. 2 is considered a forward path element $H(s)$ and the RC lead network with op-amp is considered a feedback path element $G(s)$. The magnitude $|H(j\omega)|$and phase $\angle H(j\omega)$ of the piezolaminated cantilever beam obtained from Eq. (1) and the magnitude $|G(j\omega)|$and phase $\angle G(j\omega)$ of the RC lead network obtained from Eq. (2) are,

$$|H(j\omega)| = \sum_{i=1}^{\infty} \frac{-C_s C_a [\varphi_i'(r_1) - \varphi_i'(r_2)]^2}{\rho A L^3} \times \frac{1}{\sqrt{(-\omega^2 + \omega_i^2)^2 + 4\zeta_i^2 \omega^2 \omega_i^2}} \tag{4}$$

$$\angle H(j\omega) = \varphi_{\text{beam}} = -\tan^{-1}\left(\frac{2\xi_i \omega \omega_i}{-\omega^2 + \omega_i^2}\right) \tag{5}$$

$$|G(j\omega)| = \frac{kRC\omega}{\sqrt{1 + R^2 C^2 \omega^2}} \tag{6}$$

$$\angle G(j\omega) = \varphi_{RC} = \tan^{-1}\left(\frac{1}{\omega RC}\right) \tag{7}$$

When the measurand is added, the free oscillating frequency of the piezolaminated cantilever beam drives the RC network to a new phase angle. The phase of the RC network cannot be altered on its own. To meet the phase angle criteria, the frequency of oscillation of the beam slightly shifts from the resonance frequency and catches up the sustained oscillation at a frequency close to the resonant frequency with the closed-loop electronics. The phase angle of the beam at the new oscillating frequency can be derived from the phase angle of the RC network at the free oscillation frequency using the phase condition of the Barkhausen criteria. From the compensating phase angle of the beam, the new oscillating frequency can be obtained. By substituting Eqs. (5) and (7) in Eq. (3.b), the continuous oscillation frequency ω is obtained as [12],

$$\omega = \sqrt{\frac{\omega_i^2}{1 + 2\xi_i \omega_i RC}} \tag{8}$$

This is illustrated in Fig. 3 that with the mass m_1, the open-loop resonance frequency of the piezolaminated beam is ω_{m1}, but, in a closed loop, the phase condition is satisfied only at the frequency of ω_1 and hence the resonator oscillates with this frequency, i.e., closed-loop resonance frequency ω_1 instead of ω_{m1}. The amplitude and the phase of the RC network and the beam at sustained oscillation is A_1,φ_1, $-A_2$, and$-\varphi_1$, respectively. Similarly with masses m_2 and m_3, the phase conditions are satisfied only at ω_2 and ω_3 and hence the resonator oscillates at these closed-loop frequencies instead of oscillating at ω_{m2} and ω_{m3}, and the corresponding amplitudes and phases are shown in Fig. 3.

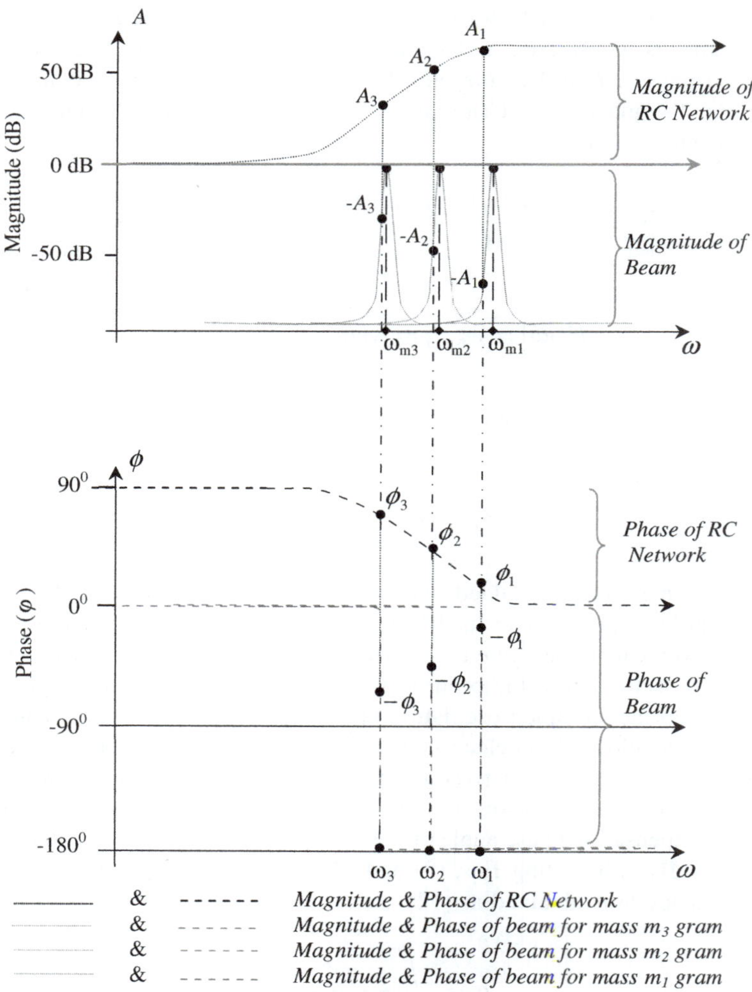

Fig. 3 Frequency response representation of measurement system with closed-loop electronics

3.2 Self-Sensing Electronics for Piezoelectric Resonance Sensor

The self-sensing actuator electronics used for the piezoelectric resonant sensor is shown in Fig. 4. The piezoelectric patch is surface bonded on the top surface of the cantilever structure. The rectifier converts the full sine wave into a half wave, the voltage follower is used to isolate the points P_1 and P_2, and the Schmitt trigger is a threshold circuit, which converts the sine wave into a square wave. A MOSFET—metal oxide semiconductor field effect transistor is used as a high efficient

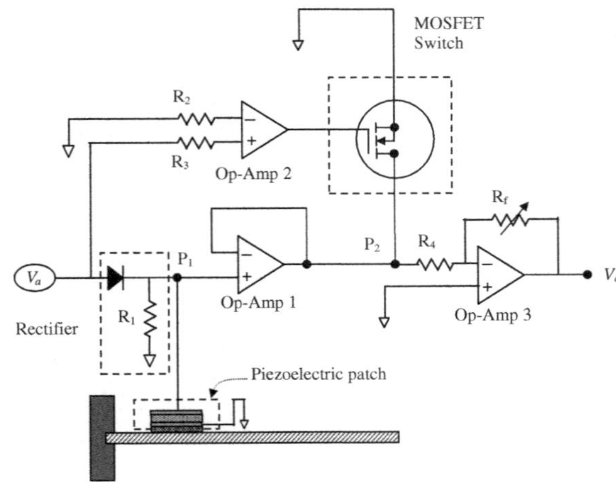

Fig. 4 Cantilever beam with piezoelectric self-sensing electronics

switching device. The signal applied to the actuator and the signal produced from the sensor are given to the drain of the MOSFET, and the source of the MOSFET is grounded. The control input to the MOSFET is given to the gate from the Schmitt trigger output which switches the MOSFET to the ON and OFF state [13].

A sinusoidal signal from function generator (V_a) is rectified using a half-wave rectifier, and the positive half cycle is used to excite the piezoelectric patch which, in turn, vibrates the beam. The same sinusoidal signal is used to switch the MOSFET switch ON and OFF through an Schmitt trigger, which is used to separate the actuating and sensing signal, available at point P_2 in the circuit. During the negative half cycle of the excitation signal, the piezoelectric patch acts as a sensor and produces an output voltage proportional to the strain induced in the beam. Due to this, the actuating signal and sensed signal are made available at the point P_2 from P_1 through a voltage follower. The sensor signal is separated by a MOSFET switch, operated by an Schmitt trigger circuit. During the positive half cycle of the excitation signal, the MOSFET switch is ON and hence the actuating signal at P_2 will be routed to the ground. During the negative half cycle, the MOSFET switch is OFF; the sensed output signal is amplified by the op-amp 3 and its output V_0 is proportional to the strain. The output of a typical cantilever oscillating at its first resonance with this self-sensing electronics is shown in Fig. 5.

3.3 Impedance-Based Self-Sensing Electronics

An impedance-based piezoelectric self-sensing actuation is shown in Fig. 6. The piezoelectric patch is surface bonded on the top surface of the cantilever beam. A resistor R is connected in a series with a piezoelectric patch to form a voltage

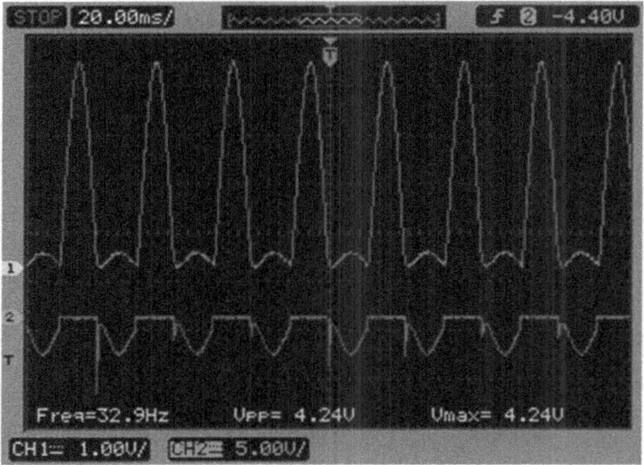

Fig. 5 Output of self-sensing electronics

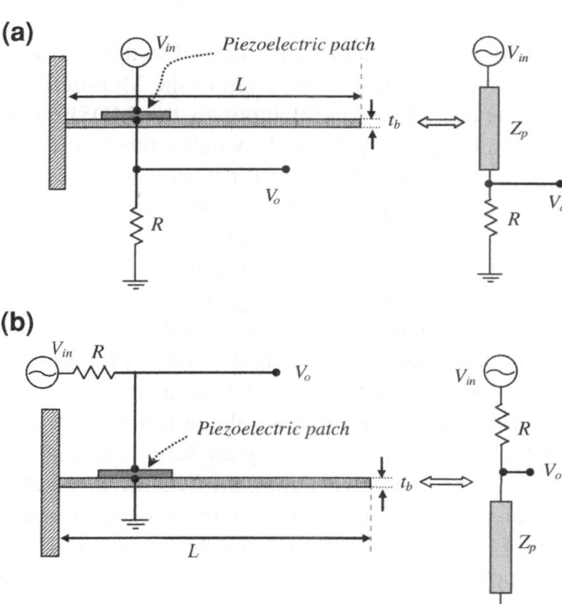

Fig. 6 Piezoelectric impedance-based self-sensing electronics. **a** Configuration A and **b** configuration B

divider network. A sinusoidal signal V_{in} is applied as an input to the voltage divider circuit, the variation in the output voltage V_o is due to the variation in impedance of the piezoelectric patch Z_p around the system natural frequency [14, 15]. Two different self-sensing electronic configurations A and B are shown in Fig. 6. In configuration A, the voltage is measured across the resistor R, which has the characteristics of a high pass filter. The voltage across the piezoelectric patch is measured in configuration B, which has the characteristics of a low pass filter.

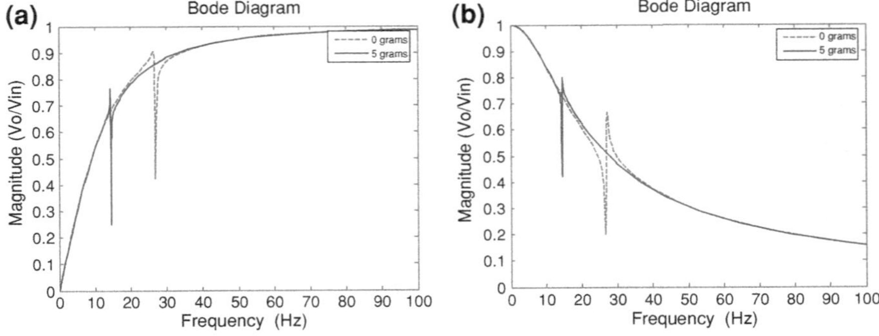

Fig. 7 Frequency responses of impedance-based self-sensing electronics. **a** Configuration A and **b** configuration B

Considering the electrical equivalent of the cantilever beam with the proposed self-sensing electronics, the transfer function between the output voltage (V_o) and input voltage (V_{in}) for configuration A is

$$\frac{V_o(s)}{V_{in}(s)} = \frac{L_1C_1CRs^3 + RR_1CC_1s^2 + R(C_1 + C)s}{L_1C_1CRs^3 + (RR_1CC_1 + L_1C_1)s^2 + (R(C_1 + C) + R_1C_1)s + 1} \tag{9}$$

for configuration B is

$$\frac{V_o(s)}{V_{in}(s)} = \frac{L_1C_1s^2 + R_1C_1s + 1}{L_1C_1CRs^3 + (RR_1CC_1 + L_1C_1)s^2 + (R(C_1 + C) + R_1C_1)s + 1} \tag{10}$$

Since configuration A acts like a high pass filter, the output magnitude increases as the frequency increases, with the appearance of a peak at the natural frequency of the cantilever beam. Similarly, as configuration B acts like a low pass filter, the output magnitude decreases as the frequency increases, with the appearance of a peak at the natural frequency of the cantilever, Fig. 7.

4 Resonance-Based Thickness Measurement

Thickness is the smallest of the three dimensions that define an object. All three dimensions are of major importance in manufacturing industries. Thickness measurement and control are of particular importance in manufacturing industries that are involved in producing sheeted, webbed, and extruded end productions. Examples include sheets of metal, plastic, paper, veneer, and plate glass. Thickness is also important in the production of various films and coatings or plated materials. There are various methods of measuring the thickness of sheets such as inductive, capacitive, ultrasonic, and nuclear radiation [16, 17]. A piezoelectric

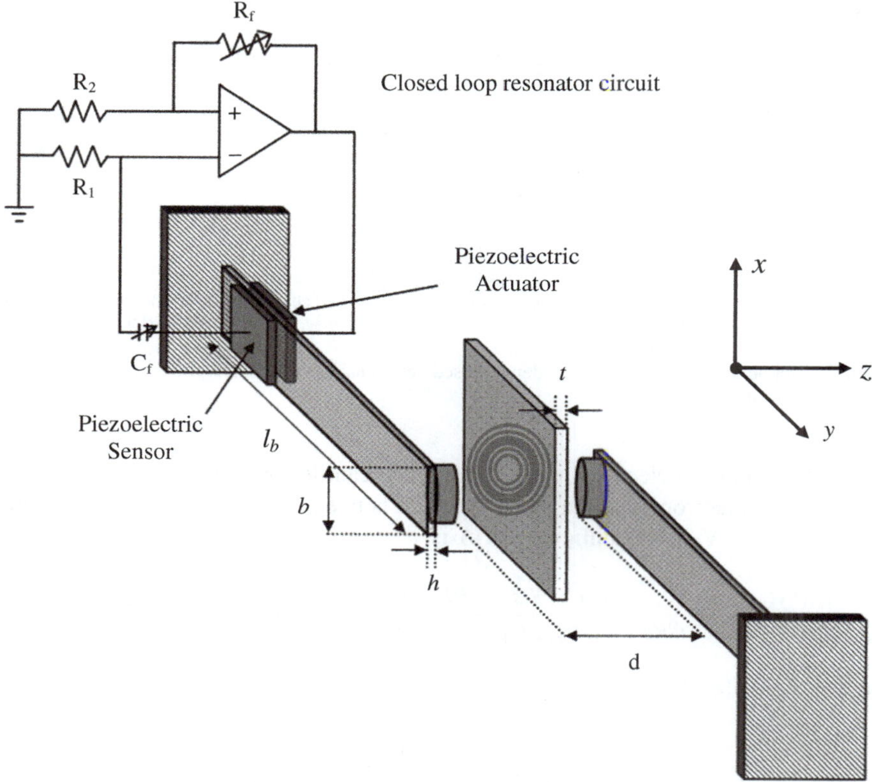

Fig. 8 Resonance-based thickness measurement system

resonant-based thickness measurement system for a non-magnetic conducting sheet is shown in Fig. 8. The closed-loop electronics described in Sect. 3.1 is used in this measurement system.

The measurement system consists of two identical aluminum cantilever beams with permanent magnet mounted at the free end. The cantilever beams are arranged to have a gap of "d" in the horizontal z-plane, in such a way that the permanent magnets at its tip are fixed face to face as shown in Fig. 8. The magnetic force from the magnet induces additional stiffness on the vibrating element which, in turn, alters the resonance frequency of the measurement system. Due to the presence of the magnetic force between the cantilevers, the effective stiffness of the beam increases if the magnetic force is repulsive and decreases if the magnetic force is attractive. This is because the effective displacement of the cantilever beam for an external force (eddy current force) will be less or high due to the presence of the repulsive or attractive magnetic force (which depends on d). Hence, the effective stiffness will be higher for the repulsive force and lower for the attractive force.

Fig. 9 Photograph of the thickness measurement system

One of the cantilever beams is made to vibrate by exciting it with a piezoelectric actuator, which causes the other beam to synchronize with it. This produces an oscillating magnetic field between the free ends of the beams. When a conducting sheet of thickness "t" is placed in the gap "d", an eddy current is induced on it, which alters the stiffness of the spring introduced by the oscillating magnetic force between the tips of the cantilever beams. The eddy current generated in the conducting sheet depends on its permeability (μ), resistivity (ρ), and thickness (t). Keeping μ and ρ constant, the change in resonant frequency ω_t of the measurement system can be related to the thickness of the sheet as [18],

$$\omega_t = \left(\left(K_{\text{sys}} - k2\pi\sigma t \int_0^{r_c} y B_y^2(y, d)\, \mathrm{d}y \right) \Big/ m_{\text{sys}} \right)^{1/2} \tag{11}$$

where K_{sys} and m_{sys} are the effective stiffness and mass of the measurement system, respectively, σ is the conductivity of the sheet, t is the thickness, B_y is the magnetic field in the y-direction, and ω_t is the change in natural frequency of the measurement system with respect to the thickness of the sheet.

The closed-loop resonant circuit tracks the change in resonance frequency and vibrates the beam with the new resonance frequency whenever the thickness of the sheet changes. The shift in resonant frequency is a measure of thickness of the conducting sheet. A photograph of the thickness measurement system is shown in Fig. 9, and the variation in the resonant frequency with thickness for the aluminum and the copper suet is shown in Fig. 10.

5 Resonance-Based Current Measurement

Information regarding current flow is required in a wide variety of electrical and electronics applications. Each application has different performance requirements in terms of cost, isolation, precision, bandwidth, measurement range, or size, and

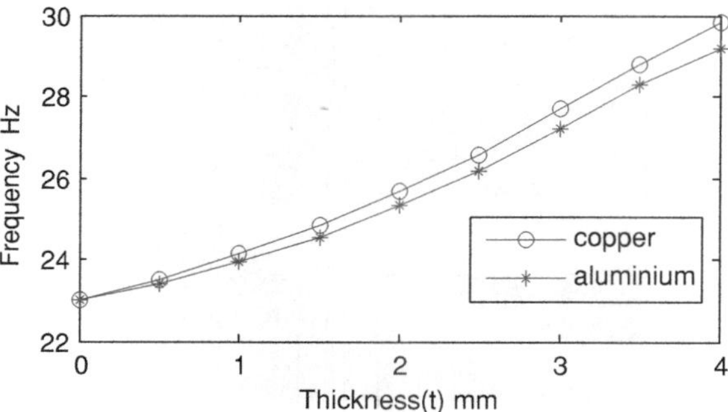

Fig. 10 Variation in natural frequency with thickness

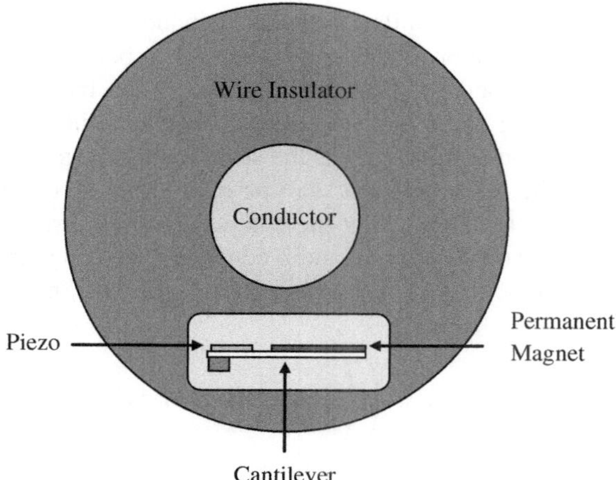

Fig. 11 Resonance-based current sensor

many different current measurement methods have been developed to satisfy these demands [19–21]. One such method is to use a resonant sensor to measure DC. A cantilever-based resonant sensor with a tip magnet is placed in close proximity to the DC-carrying conductor. The force between the permanent magnet and the current-carrying conductor induces an additional stiffness (positive for a repulsive force and negative for an attractive force) on the structure, and, hence, the resonant frequency of the structure changes. The $K_{eff} = K_{current} + K_{beam}$, where $K_{current}$ is the additional stiffness on the beam produced by a magnetic force (F), due to the current (I)-carrying conductor. The closed-loop electronics described in Sect. 3.1 is used to adapt to the change in the resonance frequency and this makes the structure vibrate at its new resonance frequency [22]. This change in resonant frequency is the measure of the current through the conductor. The sensor placed in the current-carrying conductor is shown in Fig. 11, and the typical input–output characteristics are shown in Fig. 12.

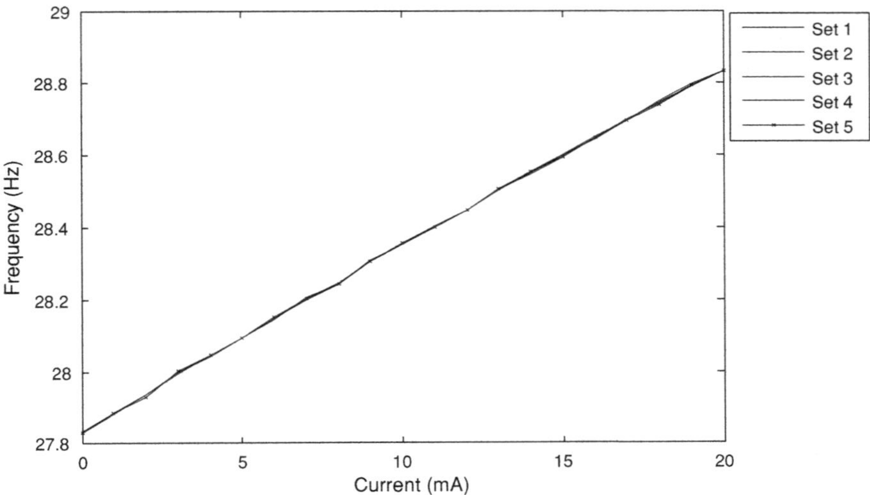

Fig. 12 Input–output characteristics of current sensor

6 Conclusion

The closed-loop electronics and two self-sensing electronics for a piezoelectric excited resonance sensor have been discussed. As applications, the measurement of thickness and a DC sensor using a resonant principle with the proposed electronics are described. The resonant sensor can be designed for various parameters which change the mass or stiffness of the resonating structure on both the meso- and the micro-scale.

Acknowledgments The authors would like to thank Dr. V.K.Aatre and his team for their continuous support and encouragement to work in the area of smart materials and MEMS. They would also like to thank the team members of NPSM and NPMASS for sanctioning the project: "Establishment of new national MEMS design centers (Project No: 2:10)".

References

1. Stemme G (1991) Resonant silicon sensors. J Micromech Microeng 1(2):113–125
2. Yan T, Jones BE, Rakowski RT, Tudor MJ, Beeby SP, White NM (2004) Design and fabrication of thick-film PZT-metallic triple beam resonators. Sens Actuators, A 115:401–407
3. Uma G, Umapathy M, Suneel Kumar K, Suresh K, Maria Josephine A (2009) Microcontroller based split mass resonant sensor for absolute and differential sensing. J Smart Struct Syst 5(3)
4. Dosch JJ, Inman DJ, Garcia E (1992) A self-sensing piezoelectric actuator for collocated control. J Intell Mater Syst Struct 3:166–185
5. Simmers GE, Hodgkins JR, Mascarenas DD, Park G, Sohn H (2004) Improved piezoelectric self-sensing actuation. J Intell Mater Syst Struct 15:941–953

6. Bentley JP (2005) Principles of measurement systems, 4th edn. Pearson Education Limited, England
7. Suresh K, Uma G, Santhosh BVMP, Kumar U, Varun Kumar M, Umapathy (2011) Piezoelectric based resonant mass sensor using phase measurement. Measurement 44(2):320–325
8. UmaG , Umapathy M, Maria Josephine A, Aishwarya S, Suresh K (2008) Design of microcontroller based resonant sensor with piezoelectric excitation and detection. J Instrum Sci Technol 36(4)
9. Reza Moheimani SO (2000) Experimental verification of the corrected transfer function of a piezoelectric laminate beam. IEEE Trans Control Syst Technol 8(4):525–530
10. Steidel RF (1998) An introduction to mechanical vibration, 3rd edn. Wiley, New York (ISBN-0470-82091-1)
11. Franco S (2002) Design with operational amplifiers and analog integrated circuits. Tata McGraw-Hill, 3rd edn. Wiley, New York (ISBN 0-07-053044-0)
12. Suresh K, Uma G, Umapathy M (2013) New observation on automatic phase adjustment of smart cantilever resonator. J Vib Control. doi:10.1177/1077546313497244 published on July 31
13. Suresh K, Uma G, Umapathy M (2012) A new self sensing electronics for piezoelectric resonance sensors. J Intell Mater Syst Struct 23(5):587–593
14. Park G, Inman DJ (2007) Structural health monitoring using piezoelectric impedance measurements. Philos Trans R So A 365:373–392
15. Suresh K, Uma G, Umapathy M (2012) Design of resonance based mass sensor using self sensing piezoelectric actuator. Smart Mater Struct 21:025015
16. Challa VR, Prasad MG, Shi Y, Fisher FT (2008) A vibration energy harvesting device with bidirectional resonance frequency tunability. Smart Mater Struct 17:015035
17. Sodano HA, Bae J-S, Inman DJ (2006) Improved concept and model of Eddy current damper. Trans ASME 294:128
18. Suresh K, Uma G, Umapathy M (2011) A new resonance based method for the measurement of non magnetic conducting sheet thickness. IEEE Trans Instrum Meas 60(12):3892–3897
19. Leland ES, Wright PK, White RM (2009) A MEMS AC current sensor for residential and commercial electricity end—use monitoring. J Micromech Microeng 19:094018
20. Santhosh Kumar BVMP, Suresh K, Varun Kumar U, Uma G, Umapathy M (2010) Design and simulation of resonance based DC current sensor. Int J Interact Multiscale Mech 3(3)
21. Ziegler S, Woodward RC, Ho-Ching Iu H (2009) Current sensing techniques: a review. IEEE Sens J 9(4):354–376
22. Santhosh Kumar BVMP, Suresh K, Varun Kumar U, Uma G, Umapathy M (2011) Resonance based DC current sensor. Measurement 45(3):369–374

A Universal Energy Harvesting Scheme for Operating Low-Power Wireless Sensor Nodes Using Multiple Energy Resources

Design of Rectennas and Power Management Electronics

K. J. Vinoy and T. V. Prabhakar

Abstract Low-power wireless and sensor technologies are fast proliferating everyday life. Medical and structural implants are common examples of devices based on these technologies. A new term "Internet of Things" has been coined to encompass many such sensors and wireless nodes. One of the critical concerns for their deployment is the source of energy, especially in operational scenarios where wall power is not available. Batteries run out of energy in due course. Solar or other alternatives are not always dependable. Combining various means of energy harvesting schemes assumes significance in this context. Low energy density radiations such as ambient RF signals from various broadcast and cellular towers have been found to be a convenient and widespread source of energy. Incorporating RF harvesting circuits into such a universal energy harvesting platform also enables intentional wireless power transfer to energize the device using an RF transmitter. This chapter explains electronic circuits required for a universal energy harvesting platform to capture, store, and efficiently utilize RF energy at different power levels in combination with other sources of ambient energy such as the Solar (for high energy). For demonstration, a low power radio and the required power management circuit have been integrated with this platform.

Keywords Energy harvesting · Wireless power transfer · Internet of things · Rectena · Tuned rectifier

K. J. Vinoy (✉)
Department of Electrical Communication Engineering, Indian Institute of Science, Bangalore, India
e-mail: kjvinoy@ieee.org

T. V. Prabhakar
Department of Electronic Systems Engineering, Indian Institute of Science, Bangalore, India

K. J. Vinoy et al. (eds.), *Micro and Smart Devices and Systems*,
Springer Tracts in Mechanical Engineering, DOI: 10.1007/978-81-322-1913-2_27,

1 Introduction

With the availability of low-power wireless and sensor technologies [1], many medical and structural implants using these technologies are getting popular by the day. A new term "Internet of Things" refers to many such sensors with wireless nodes [2]. These devices are expected to interact with each other through unique addressing schemes, and to cooperate with their neighbours to reach common goals. Many of these may operate intermittently, with only an internal control circuit keeping track of the overall performance continuously. One of the critical concerns in their deployment is meeting their energy requirements. Wall power may not be dependable in many operational scenarios of such devices, such as in remote locations. Batteries, on the other hand, run out of energy in due course and may require recharging or replacement. Means for energy harvesting from multiple sources assumes significance in this context.

Energy sources typically preferred in practical harvesting solutions have high energy density. However, there are niche situations where such forms of energy are not at all available. For example, light energy although abundant outdoors or in a well-lit room, cannot be expected outdoors at night or inside a closed chamber. Storing sufficient energy to sail through such situations may add cost as batteries get expensive and inefficient. Low energy density radiations such as ambient RF signals due to various broadcast and cellular towers have been found to be convenient and widespread means of energy [3, 4]. Thermo electric devices (TED), on the other hand, can scavenge energy from thermal gradients naturally available in many deployment situations. Recent efforts in [5], addresses combination of energy harvested from vibration, thermal and solar sources.

This chapter explains electronic circuits developed by our groups to harvest, store and efficiently utilize RF energy at two different power levels, in combination with other sources of harvesting energy such as the solar (for high energy). We therefore combine RF. Solar and thermal harvesting with a universal energy harvesting platform (UEHP). It may be noted that incorporating RF harvesting circuits into a UEHP also enables intentional wireless power transfer to energize the device using an RF transmitter [6]. Yet, unlike classical experiments involving high powers, significant RF power is not available at the receiving units. One of the theoretical reasons for this is the free space loss factor. In contrast, high transmission efficiency has been achieved in inductive power transfer, e.g., between two coils [7].

One of the major concerns in the low power RF energy transfer is the low efficiency of rectification and power management electronics. It is clear that the efficiency is a function of the load one uses. A high efficiency scheme designed for high load impedances, fails when realistic active loads such as CMOS circuits are connected to its output. Hence, a low power radio is included in our UEHP to investigate its performance. The power management circuit has also been integrated with this platform.

2 System Description

The primary focus of this work was to demonstrate the operation of a low energy wireless node using energy harvested using a UEHP that can combine various energy sources. Available energy sources include mechanical actions such as water flows, wind, vibration, or flutter; or various forms of ambient energy such as temperature gradient, RF radiations, or solar; or intentional transfer such as in wireless power transfer.

Harvested energy converted into the electrical domain may be either in AC or DC form. Examples of AC inputs are vibration and RF sources. These are rectified and stored. Examples of DC energy include solar and thermoelectric generators. Commercial energy harvesters are available from several vendors. These could be compared based on power per unit area/volume [8] or power per unit weight [9]. In specific situations such as in aircraft, where the overall weight is important, the thermoelectric generator provides more power density compared to other sources, including solar [8]. Although poor in terms of overall weight, vibration harvesters also are promising with a possibility to provide sufficient power to drive embedded communication devices when electric charges from the vibration harvester are collected over a sufficiently long duration (15–20 min). For instance, the V21BL piezo wafer from Mide Volture [10] provides 2.66 mW (maximum) power when subjected to mechanical vibrations tuned to 40 Hz with a tip mass of 4.89 g. In general, since the output voltage generated from a piezo-based vibration harvester is higher than operating voltages of most embedded systems, step down converters may be required. However, in this work, our chapter focuses on step up (or boost) converters.

Table 1 summarizes power output measurements conducted for various harvesting sources. A solar panel produces 30 mW of power under a sufficiently bright light. The energy generation in Thermo energy generator (TEG) is based on the principle of Seebeck effect. This is a phenomenon that generates a voltage when a temperature differential is applied across two dissimilar metals such as tellurium and bismuth. The maximum power obtained is 6 mW at a temperature differential of 25 °C. A micro hydrogenerator has a small impeller which rotates when water flows, and is capable of producing a maximum power of 1 watt. Linear motion harvesters work on the principle of electromagnetic induction to generate the energy. It is found that a peak power of 300 mW is generated from one simple rocker switch action.

In this work, an UEHP is designed based on the energy requirements for sensors and wireless electronics. However, energy requirements for sensing components vary widely. Some of these (e.g., thermistor, passive infra red (PIR) sensors) may require very low or no energy while there are others (e.g., accelerometer, humidity sensor) that require substantial energy, especially to operate their integrated electronics to achieve the required sensing characteristics. Hence, the energy requirements for wireless communication alone are budgeted for in the design of the UEHP presented.

Table 1 Characterization of some of the energy harvesting sources

Energy source	Size (mm)	Output Power (mW)
Solar	95 × 65	30 @ 20,000 Lux
TEG	2.5 × 2.5	6 @ $\Delta T = 25$ °C
Hydro	84.5 × 64.5 × 81	1,000 @ flow > 1.5 L/min
Linear motion	33.3 × 22.0 × 10.8	300/actuation

Table 2 A comparison of energy requirements for some commercially available low power radio chips [3]

	Jennic JN5148	TI-CC430	BLE	Zarlink ZL70250
Active mode current at 16 MHz [mA]	6	4	6.7	3.2
Deep sleep current [nA]	100	1,000	400	20
Transmission current [mA] @ Tx-power [dBm]	15 @ 2.5	18 @ 0	36 @ 2	2 @ −10
Transmit frequency	2.4 GHz	2.4 GHz	2.4 GHz	868 MHz
Wakeup time [ms]	1	3	0.12	0.16
Energy consumption for a transmission cycle of 2 ms [μJ]	183	300	196	32
Power supply voltage [V]	2.2–3.6	1.8–3.6	2–3.6	1.2–1.8

Power characteristics of some commercially available wireless transceivers are compared in Table 2. Although options are available at multiple frequencies, 2.4 GHz ISM bands are typically preferred. As noted from Table 2, an output voltage of 2.2–3.6 V at about 15–18 mA would be required to drive a low power wireless radio such as Jennic JN 5148. The energy consumption is computed for one transmission cycle (Sleep → Active → Transmission → Sleep). It is clear that some energy sources such as solar may meet this requirement instantaneously. On the other hand, to meet this requirement using the harvested energy at low incident power, we require a two-level accumulate-and-use topology. Energy efficient power gating and leakage management circuits are developed around a boost converter to enable this operation. A high-level block diagram for the UEHP is shown in Fig. 1. The four inputs used in this scheme cover different categories listed in Table 3.

In the present scheme, commercial sensors were used for solar and TED. But RF inputs require antennas followed by appropriate rectification circuits developed in-house. The first block of an RF energy harvesting system is usually called a rectenna [6, 11] consisting of a rectifier circuit and an antenna that can operate as the receiving unit that converts RF to DC. Since the efficiency of this part affects the overall performance of the harvesting scheme, the design of RF components is discussed in the following section.

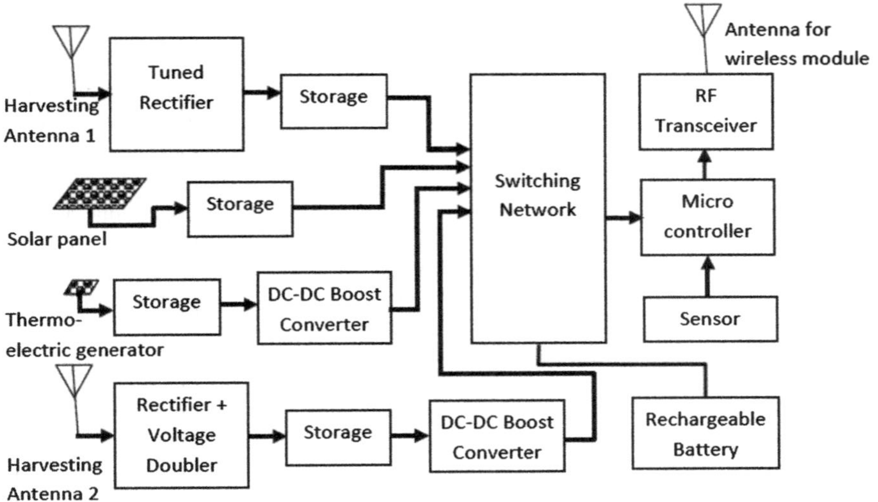

Fig. 1 Block diagram of universal energy harvesting platform

Table 3 Schemes incorporated in UEHP for harvesting different categories of energy

Power level	AC/DC	Example	Harvesting scheme
Low (<−10 dBm)	AC	Ambient RF	Tuned rectifier + DC–DC boost converter + storage
Low	DC	TED	DC–DC boost converter + storage
High	DC	Photovoltaic (Solar)	Storage
High (>−10 dBm)	AC	Mobile radiations, intentional RF transmission	Voltage doubler-rectifier + storage

3 Design of RF Rectennas

The first block of RF energy harvesting systems is a rectenna consisting of a rectifier circuit and an antenna that can operate as the receiving unit to convert the RF to DC. The word "rectenna" became popular since the wireless power transmission experiment of Brown in the 1960s [12] for a system consisting of an antenna, an input filter, a rectifying circuit (a balanced bridge or a single diode) with an output filter and a matched resistive load [11]. However, it may be noted that rectennas used in an energy harvesting system are somewhat different from wireless power transmission as the power levels associated with practically available RF radiations are significantly low. Therefore, several modifications have been attempted to operate at these low power levels.

3.1 Antennas

Antennas that receive RF radiations were developed in-house. In order to maximize the performance, these antennas are required to have a high efficiency, high gain, and radiation patterns in appropriate directions. Photographs of some antennas that operate around 900 MHz ISM bands are shown in Fig. 2. Their performance parameters are compared in Table 4. The biquad antenna has the highest gain and is used in our RF harvesting experiments reported here. The compact planar inverted-F antenna (PIFA) is primarily designed for the cellular uplink, which also overlaps with an ISM band in many parts of the world. The suspended patch antenna covers both these bands comfortably.

3.2 RF Rectifier Circuits

RF radiations are received using antennas with standard port impedance. At low incident power levels, the incident voltage at the rectifier diode is low and hence to overcome the drop at the diode, a high Q tuning circuit with appropriate voltage magnification is required. On the other hand, at high incident RF power levels, such tuning circuit may not be necessary; yet a voltage doubler circuit is used.

3.2.1 Tuned Rectifier

This configuration assumes that a high gain antenna with a terminal impedance of 50 Ω is used for capturing the radiations. For an input RF power of −18 dBm, the maximum DC voltage level possible is about 30 mV, which would be insufficient to operate any active device. Hence, a matching circuit is required between the antenna and the rectifying diode. The quality factor of this matching section enhances the level of the voltage, so that the impact of the drop in the diode (AVAGO HSMS2852) can be minimized and reasonably high voltage is available at the output. However, realizable Q would restrict the output voltage to several 100s of mV. Even this is not sufficient to drive commercially available boost converters, and hence a supercapacitor may be used to accumulate the harvested energy.

The matching circuit is designed using Agilent ADS assuming that the tuned rectifier is driving the supercapacitor (not shown) at the rectified output side of the circuit shown in Fig. 3. Transmission line stub and the series inductor are designed to get the desired quality factor for the tuned rectifier. The circuit parameters are designed for a center frequency of 945 MHz.

The performance of the tuned rectifier output is evaluated by investigating the RF characteristics at its input as well as for RF–DC energy conversion characteristics under various conditions. The voltage across the supercapacitor is observed over a period of 20 min and the final output for different frequencies and power levels are compared in Table 5. For a given input, the maximum output is

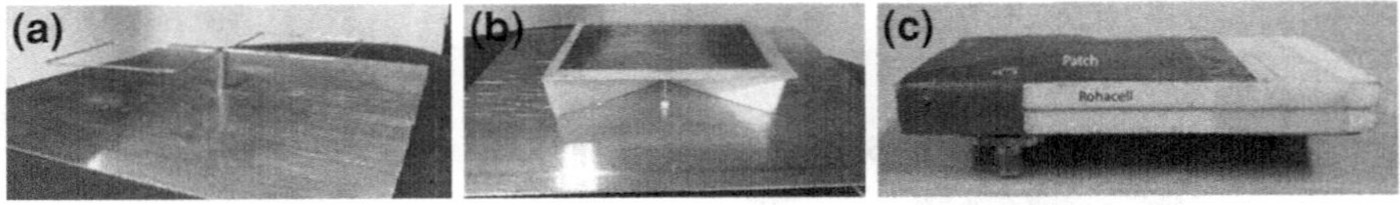

Fig. 2 Photographs of antennas designed for RF energy harvesting. **a** Biquad antenna. **b** Suspended patch antenna. **c** Planar inverted F antenna

Table 4 Comparison of performance of antennas developed for RF energy harvesting

Antenna	Size (overall) [mm^2]	Measured gain [dBi]	Operational frequencies (S11 < −10 dB) [MHz]
Biquad [13]	220 × 110 (250 × 250)	10.48	904–967
Suspended patch [14]	110 × 85 (250 × 250)	7.2	910–1,320
Planar inverted F (PIFA) [15]	61 × 45 (85 × 41)	1.5	907–937

Fig. 3 Design of the impedance matching components for the tuned rectifier

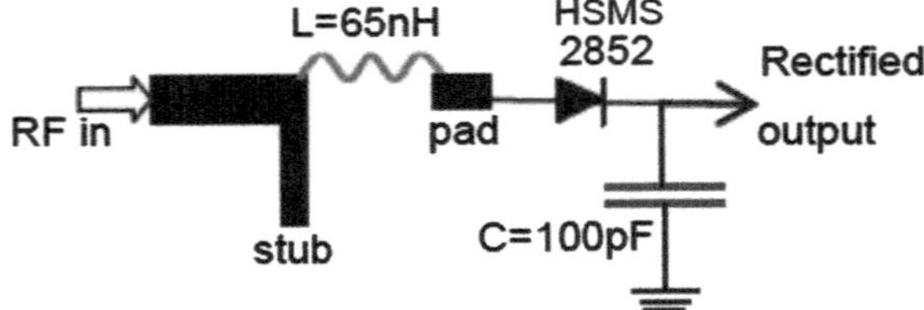

Table 5 Measured response of the tuned rectifier for different frequency and power of input

P_in→ Freq. ↓	−10 dBm	−13 dBm	−16 dBm	−20 dBm	−25 dBm
930 MHz	917 mV	664 mV	469 mV	281 mV	131 mV
945 MHz	1,016 mV	736 mV	515 mV	300 mV	132 mV
955 MHz	1,038 mV	747 mV	513 mV	289 mV	122 mV
960 MHz	1,032 mV	736 mV	499 mV	276 mV	114 mV
Peak efficiency (%)	51	47	39	33	20

observed across the supercapacitor at 955 MHz. The performance of the rectifier degrades at other frequencies due to the fact that this is a tuned circuit.

3.2.2 Voltage Doubler-Rectifier

At high RF power levels (one example is the uplink RF power from a mobile phone, captured in close range), high instantaneous power is received for a very short duration. Hence, HSMS 2822 diodes with a higher power rating are used. High voltages can be developed across the storage capacitor using the doubler network shown in Fig. 4. This relaxes the need of a boost converter, which is

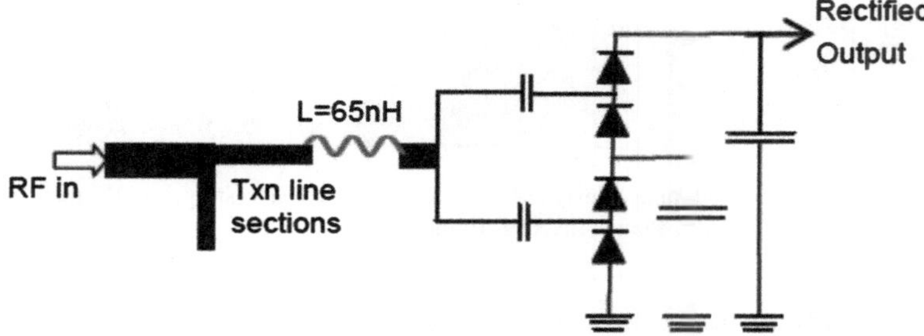

Fig. 4 Voltage doubler-rectifier circuit for high RF power. In this circuit, HSMS 2822 diodes are used. All capacitors shown are of 100 pF

Table 6 Measured behavior of the voltage doubler rectifier circuit

Power level (dBm)	Charging time (ms)	Efficiency (%)
0	40	64.11
−2	55	63.77
−3	67.5	63.5
−5	90	63
−7	230	59.89
−10	370	56.78
−12	500	53.89
−15	900	45
−18	2,000	20.56

typically inefficient for intermittent operation. The equivalent circuit of the diode is used to design the matching network for this stage. Similar to the previous case, transmission line sections are used for the tuned rectifier. The efficiency of this circuit (Table 6) is above 60 % for RF input power exceeding −5 dBm.

4 Power Management Circuits

Since energy harvesting sources provide variable output power, it is often necessary to operate these at an operating point corresponding to the maximum power availability point. Maximum power point tracking (MPPT) prevents the harvester electronics from draining energy so that the source output is always kept above a certain voltage. This is particularly true for PV panels that do not work well when their output voltage falls below a certain value. In such cases, the idea is to take less power from the source and keep them efficient. A simple MPPT circuit makes a open circuit source voltage measurement and draws power from the source until the source's voltage falls to 1/2 of its open circuit voltage. Such a circuit typically comprises of a current source and a comparator. The LTC 3105 is one such boost converter with maximum power point control. [16]. However, it is easy to see that,

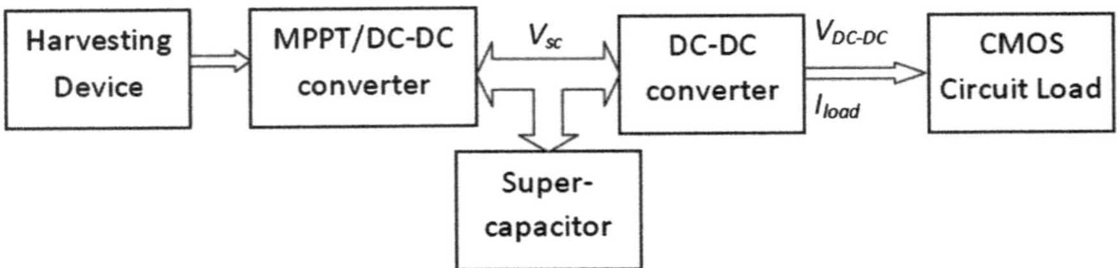

Fig. 5 Block diagram of an energy harvesting scheme to power a CMOS circuit as the load

in order to perform MPPT, the electronic circuits require power. Therefore, the harvesting electronics perform power conditioning and store a small amount of energy to meet its own requirements. Once this power is available, the excess power can be conditioned and stored in batteries or supercapacitors. In addition to providing impedance matching, this circuit rectifies the voltage whenever necessary, and delivers the power to a storage device. MPPT algorithms are required to ensure that the source impedance variations due to the varying input power is matched to the load impedance to ensure transfer of the maximum power from the source to the storage device.

The output of harvesting electronics is shown in Fig. 5 as V_{sc}. The output DC–DC converter is a voltage regulator that feeds power to the load and is shown as V_{DC-DC}. Due to efficiency issues in harvesting electronics with DC–DC converters, a critical input voltage is required to start the DC–DC converter for perennial functioning of the load. In the present circuit, this tipping point voltage is 1.8 V and hence if the input drops below this voltage the DC–DC converter shuts down.

Figure 6 shows the voltages across the DC–DC converter and the supercapacitor as well as the current drawn (I_{load} the load current) by a CMOS circuit in a low input power scenario. The initial start-up current requirement is approximately 23 mA. Each time the output DC–DC converter switches on, the load attempts to draw I_{load} as indicated by the voltage waveform. The voltage builds up across supercapacitor (the storage device), and as soon as this crosses the minimum input voltage for the DC–DC converter (1.8 V for TPS63031 from Texas Instruments), the output switches on (3.3 V), and loads the input harvester. This behavior occurs when the input power is low, leading to wastage of energy. This is a typical problem that requires adequate attention from system designers. One of the approaches to address this problem is to incorporate a power gating logic. A simple comparator and a MOSFET-based circuit is used to ensure that sufficient charge is present in the supercapacitor before turning on the load. This simple measure ensures that there is no loading effect on the supercapacitor by the electronics and hence allowing it to charge sufficiently above the electronics turn on voltage, before the charge can be made available to it for its operation.

The current waveform in boost converter is discontinuous for low input power, and hence the boost converter cannot be operated in a continuous conduction mode. The operation of boost converter is said to be continuous when the input voltage to the boost converter stays above a threshold to enable uninterrupted energy availability.

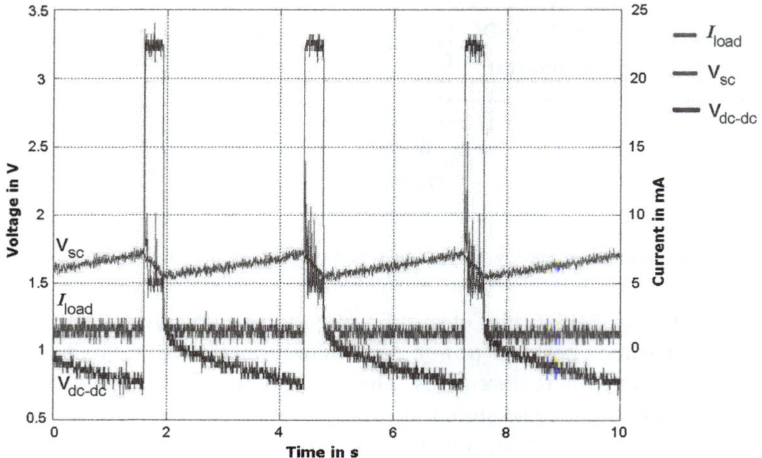

Fig. 6 Current and voltage waveforms: harvester–load interaction

The boost converter can be designed either in a closed loop or an open loop configuration. In closed loop, the constant output voltage is maintained by a feedback path by controlling the duty cycle of operation. On the other hand, there is no feedback in open loop operation, and the output voltage is maintained constant by putting a voltage source at the output. In this case, the boost converter transfers energy from low voltage source to a high voltage source. However, the duty cycle of operation is to be kept fixed. A rechargeable battery serves as a voltage source at the output. The converter parameters and the duty cycle are designed to match the discontinuous operation. The open loop boost converter topology eliminates the necessity of feedback path thereby reducing the power loss. Energy from the ambient sources is accumulated on a capacitor and when sufficient energy is available, the boost converter is switched on to transfer that energy to the rechargeable battery. The charging time of the capacitor is relatively higher than the operating time of boost converter and hence the boost converter is almost isolated from the transducer during its operation. Therefore, there is no need to track the MPP of the transducer, which avoids resistance emulation.

The circuit in Fig. 7 shows a boost converter circuit. When the MOSFET (M1) is ON, energy in the capacitor is transferred to the inductor and when it is OFF, this energy is transferred to the battery. A diode is used to detect the inductor zero-crossings and to ensure unidirectional power flow from source to the battery.

The switching operation of converter is controlled using an oscillator. The duty cycle of the oscillator is fixed at 50 % for discontinuous conduction mode. The minimum voltage across the input capacitor for good efficiency is 0.2 V. The inductor obtained by simulations is 150 uH for the best efficiency. The ripple on input capacitor (ΔV) is 15–30 %, the switching period of oscillator (T) is 40 μs, with on-period ($t1$) is 20 μs, and input capacitor of 100 μF.

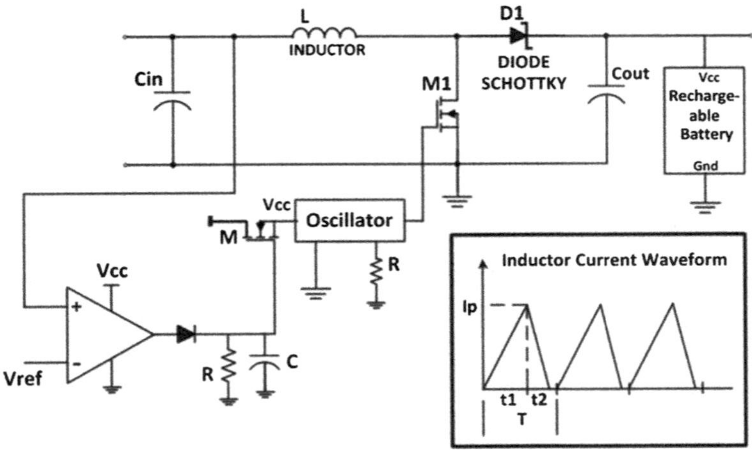

Fig. 7 Circuit diagram of the boost converter for low energy harvesting

The comparator is needed as the voltage across the capacitor has to be monitored and when it reaches 0.2 V, the energy is transferred to a rechargeable battery at the output at 4.1 V at the output. An operational amplifier-based comparator is designed for this purpose. Whenever the voltage on the input capacitor reaches 0.2 V, the comparator output becomes V_{cc} and it is stored on the capacitor, C. The RC circuit is designed in such a way that it will switch on the MOSFET M for the required duration to control the ripple.

The power for the oscillator is drawn from the battery. The comparator has to be always on as it needs to compare the voltage on the capacitor with a fixed value of reference which is 0.2 V. This comparator is also powered from rechargeable battery. Therefore, the energy requirements for the comparator and oscillator affects the energy transferred the output and hence the efficiency. The overall efficiency of the boost converter can be calculated using:

$$\text{Efficiency} = \left[E_{\text{out}} - P_{\text{osc}} * t_{\text{dis}} - P_{\text{comp}} * (t_{\text{ch}} + t_{\text{dis}})\right] / E_{\text{in}}$$

where E_{out} and E_{in} are the energy drawn by the load and at the input; P_{osc} and P_{load} are the power requirements of the oscillator and the load; t_{ch} and t_{dis} are the charging and discharging time.

5 Characterization of UEHP Using a Wireless Module

In Sect. 2, the block diagram of the universal energy harvester design is described. There are several applications to use scavenged energy for indoor environments. For instance, a pyroelectric effect-based passive infrared (PIR) sensor can be used

Table 7 Performance of a photovoltaic panel used for indoor applications

Indoor intensity (Lux)	Power at battery (μW)
100	12
200	20
500	30
1,000	45

for detecting human motion and several gestures to activate home appliances. For example, a TV can be switched or door access can be controlled based on gestures. One major advantage of PIR sensor is the power consumption for sensing. An always-on PIR sensor consumes typically 3–5 μW of power. Thus, this sensor is amenable for energy harvesting technologies. We used a small PV panel (5 × 2.5 cm) that harvests energy from indoor lighting systems. The maximum output power is 60 μW. Our measurements indicate that indoor light intensity may be 100–1,000 Lux and the corresponding output at the battery may range from 12 to 45 μW (Table 7). Similar characterizations have also been conducted for Thermo electric generator (TEG) as well as RF inputs.

5.1 Power Management Policies for Wireless Communication

While PIR-based sensing is amenable for energy harvesting, the energy consumed is limited to sensing operation. Whereas, actuation of a load device means that the command signal from the embedded system will require communication from the transmitter to the receiver. Since indoor environments can contribute to channel impairments such as fading, shadowing and path loss, unless the transmitted power is sufficiently high, a data packet is bound to suffer from outages. Therefore, one has to design good algorithms and policies for power management. For example, the algorithm might consider several acknowledgement (ACK) and negative acknowledgements (NACK) packets gathered over time to decide the transmission power. One such happy situation is when no outage occurs due to the channel being in a good state, although the transmission power chosen was low due to a low harvested energy.

5.2 Results and Discussion

Keeping in mind the requirement for communication between the sensor node and a device to be actuated in an indoor application, we conducted extensive measurements using the three input power sources to gain insight on the frequency of packet transmission with near-zero outage probability. Table 8 shows the results of our measurement. When energized using a solar panel, the scheme is able to

Table 8 Packet transmission frequency for three energy harvesting sources

Solar		RF		TEG	
Light intensity (Lux)	Duty cycle of operation (s)	Power level (dBm)	Duty cycle of operation (s)	Temperature differential (°C)	Duty cycle of operation (s)
1,000	7	0	3	55	9
300	11	−5	6	45	13
200	20	−7	20	35	240
100	42	−10	50	–	–
–	–	−12	240	–	–

operate the wireless transceiver and transmit a data packet successfully every 20 s under low lighting conditions (200 lux). An incident RF power of −7 dBm (~0.2 mW) also ensures a similar rate of data transmission. An appropriately oriented 20 mW source with a high gain antenna (~10 dB) can reach this RF power at a low gain rectenna (e.g., using PIFA) at 1 m distance. Current wireless emission norms allow much higher radiations at 900 MHz ISM bands and hence the scheme proposed here can rely on wireless power transfer as backup in case other harvesting schemes fail.

6 Conclusions

In this work, we have identified some of the key problems with the power harvesting and management in an energy harvesting system. In some cases, even though the input power is sufficient to drive an embedded system, the power train efficiency is an important parameter to retain the operation of the boost converter. We have described the design of an universal energy harvested module with the objective of ensuring high efficiency, using an open loop DC–DC converter that is capable of achieving up to 60 % efficiency at low incident power levels. Proper impedance matching, efficiency in power conversion, MPPT, and low leakage storage are some of the considerations in its design.

Since packet transmission is one of the key energy consuming tasks, energy management policies are required. The wireless module integrated with this UEHP is able to transmit a packet with 14 bytes of data once every 20 s, for a received RF power of −7 dBm. As this can be easily ensured using a low power transmitter in an indoor environment, this offers the possibility of employing wireless power transfer to tide over difficult sensor operation periods such as at night.

Acknowledgments The authors would like to extend their gratitude to Dr. Vasudev K. Aatre for his constant encouragement and continuing support. They also thank their colleagues Gaurav Singh, Rahul Ponna, Aditya Mitra, Chaitanya, and Uday Sainy for their contributions to this work.

References

1. Akyildiz IF, Su W, Sankarasubramaniam Y, Cayirci E (2002) Wireless sensor networks: a survey. Comput Netw 38(4):393–422
2. Atzori L, Iera A, Morabitoc G (2010) The internet of things: a survey. Comput Netw 54:2787–2805
3. Singh G, Ponnaganti R, Prabhakar TV, Vinoy KJ (2013) A tuned rectifier for RF energy harvesting from ambient radiations. Int J Electron Commun 67(7):564–569
4. Popovic Z, Falkenstein EA, Costinett D, Zane R (2013) Low-power far-field wireless powering for wireless sensors. Proc IEEE 101(6):1397
5. Bandyopadhyay S, Chandrakasan AP (2012) Platform architecture for solar, thermal, and vibration energy combining with MPPT and single inductor. IEEE J Solid-State Circuits 47(9):2199–2215
6. Brown WC (1984) The history of power transmission by radio waves. IEEE Trans Microw Theory Tech MTT-32:1230–1242
7. Kurs A, Karalis A, Moffatt R, Joannopoulos JD, Fisher P, Soljačić M (2007) Wireless power transfer via strongly coupled magnetic resonances. Science 317(5834):83–86
8. Calhoun BH, Daly DC, Verma N, Finchelstein DF, Wentzloff DD, Wang A, Cho S-H, Chandrakasan AP (2005) Design considerations for ultra-low energy wireless microsensor nodes. IEEE Trans Comput 54(6):727–740
9. Becker T, Kluge M, Schalk J, Otterpohl T, Hilleringmann U (2008) Power management for thermal energy harvesting in aircrafts. In: Proceedings of IEEE Sensors Conference
10. http://www.mide.com/pdfs/Volture_Datasheet_001.pdf
11. Choi SH, Song KD, Glen GC, Woodall C (2002) Rectenna performances for smart membrane actuators. In: SPIE conference proceedings smart electronics and MEMS, vol 4700. 18–21 March 2002, pp 213–221
12. Brown WC, George RH (1964) Rectification of microwave power. IEEE Spectr 1(10):92–97
13. Straw RD (ed) (2007) ARRL antenna handbook. ARRL, Newington (Chapter 12)
14. Kasabegoudar VG, Vinoy KJ (2010) Coplanar capacitively coupled probe fed microstrip antennas for wideband applications. IEEE Trans Antennas Propagat 58(10):3131–3138
15. Muniganti H, Mannangi V, Vinoy KJ, Bommer JP, Marston SE (2013) Immersible antenna for RF energy harvesting. In: IEEE applied electromagnetics conference AEMC 2014, Bhubaneshwar, 18–20 Dec 2013
16. http://cds.linear.com/docs/en/datasheet/3105fa.pdf

RF MEMS True-Time-Delay Phase Shifter

Shiban K. Koul and Sukomal Dey

Abstract A radio frequency microelectromechanical system (RF MEMS)-based true-time-delay (TTD) phase shifter is one of the key components in a modern electronically steerable phased array for satellite communication, radar systems, and high precision instrumentation. The phase shifter controls the signal phase in order to steer the direction of the beam [3]. RF microelectromechanical systems (MEMS) technology provides a superior performance in terms of low loss, low power consumption, and excellent linearity compared to other technologies. MEMS-based digital phase shifters provide a large phase shift and low sensitivity to electrical noise with a high tuning ratio compared to analog versions. This chapter describes different types of TTD phase shifters utilizing MEMS switches and MEMS varactors. A gold-based surface micromachining process is used to develop different kinds of MEMS phase shifters on alumina ($\varepsilon_r = 9.8$) substrates. All phase shifters are implemented using coplanar waveguide (CPW) transmission lines and actuated by electrostatic actuations. These include the analog-type distributed MEMS transmission line (DMTL) phase shifter using push-pull actuation, a 5-bit DMTL phase shifter using MEMS bridge and a fixed capacitor, 5-bit switched line phase shifter using DC-contact MEMS switches and a 2-bit and 5-bit phase shifter using MEMS SP4T and SPDT switches. This chapter includes details on the design, development, and characterization of MEMS phase shifters. Furthermore, all experimental results are validated with a circuit analysis and full-wave EM simulation.

Keywords Phase shifter · Push–pull actuation · Phase error · True time delay · Switched line · Distributed MEMS transmission line

S. K. Koul (✉) · S. Dey
Centre for Applied Research in Electronics, Indian Institute of Technology Delhi, Hauz Khas, New Delhi, India
e-mail: shiban_koul@hotmail.com

K. J. Vinoy et al. (eds.), *Micro and Smart Devices and Systems*,
Springer Tracts in Mechanical Engineering, DOI: 10.1007/978-81-322-1913-2_28,

1 Introduction

A low loss and miniature microwave phase shifter is a critical component of a transmit/receive (T/R) module in passive electronically scanned arrays (ESAs) used widely in military and commercial applications [5]. TTD phase shifters are designed using switched-line or distributed loaded-line configurations. In comparison with switched line, a distributed loaded line phase shifter gives wide band performance, good aspect ratios and works well at higher frequencies. However, a practical digital DMTL phase shifter would require a large number of movable bridges and accordingly the length of the phase shifter becomes larger than the switched line phase shifter [7].

In general, DMTL phase shifters are implemented on a coplanar waveguide (CPW) line which is periodically loaded with MEMS bridges. When a single analog control bias voltage is applied to the center conductor, the bridges are pulled closer to the center conductor, which, in turn, increases the loading capacitance on the distributed line, besides varying the propagation characteristics and decreasing the phase velocity of the DMTL. The resulting change in the phase velocity of the DMTL produces the TTD phase shifts. On the other hand, the switched-line configuration is the most straightforward approach where the RF signal flows between two transmissions lines of different lengths using two single-pole double-throw (SPDT) switches.

2 Design of Push–Pull-Type DMTL Phase Shifter

The model for a push–pull-type DMTL phase shifter consists of a CPW transmission line and torsional MEMS actuators, as shown in Fig. 1a. The mobile plate has a thickness of 1.25 μm, and it has 2.5 μm of an air gap from a bottom fixed electrode of thickness 1 μm. All relevant structural parameters of the push–pull bridge are listed in Table 1 and a 3D view of a push–pull bridge is shown in Fig. 2. The pull-in voltage (V_{pi}) is quasi-static in nature, and it gives a lower boundary on the pull-in voltage. The electromechanical modeling of the push–pull bridge can be found in [1].

After the successful release of the push–pull bridge, the nonuniform distribution of gap height was captured by SEM and shown in Fig. 1b–d. The unit cell of the phase shifter has three operating states: push, up, and pull state. In the up state, the model comprises a transmission line of length s, with a capacitor to the ground due to a shunt bridge (C_{up}) and impedance Z_{up}. In the push state, the central part of the beam is lifted upward under V_{push}, leading to increased loaded impedance (Z_{lpush}) with low capacitance (C_{push}). In the pull state, the bridge moves down within the pull-in limit, up to a gap height of 0.75 μm from the zero bias state, leading to decreased loaded impedance (Z_{lpull}) with high capacitance (C_{pull}) [1].

The impedance for each state of the unit cell is given by [1]

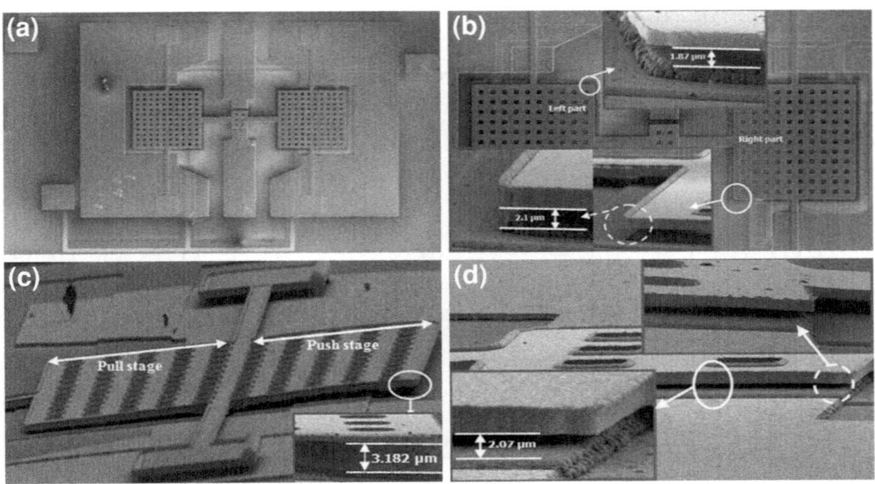

Fig. 1 **a** Unit cell of TTD phase shifter, **b** initial deformation of *left* and *right* part of the mobile plate over the pull electrodes (*inset* shows the deformation), **c** initial deformation of push stage mobile electrode (*inset* shows the deformation), and **d** initial deformation of the middle part of the bridge (*inset* shows the deformation). Copyright/used with permission of/courtesy of Institute of Physics and IOP publishing limited

Table 1 Designed structural parameters of the micromachined push–pull actuator

Parameter	Value (μm)
Fixed electrode length, l_{fe}	90
Fixed electrode width, w_{fe}	230
Mobile electrode length, l_{me}	200
Mobile electrode width, w_{me}	230
Lever length, l_l	85
Lever width, w_l	20
Length of the torsional spring, l_{ts}	190
Width of torsional spring, w_{ts}	15
Thickness of torsional spring, t_{ts}	1.25
Thickness of lever, t_t	1.25

Fig. 2 3D view of push–pull actuator. Copyright/used with permission of/courtesy of Institute of Physics and IOP publishing limited

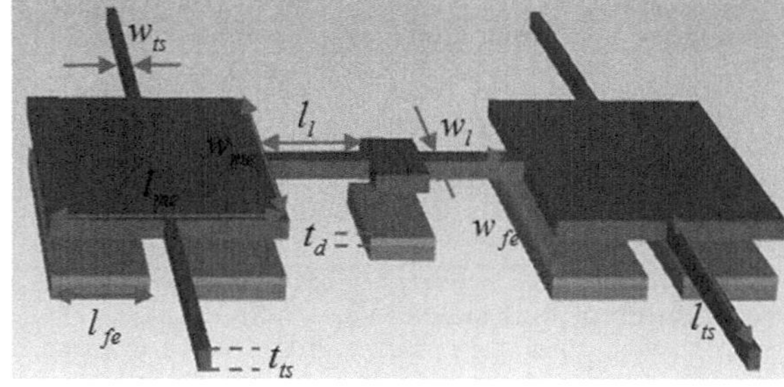

$$Z_{up} = \sqrt{\frac{sL_t}{sC_t + C_{up}}},\ Z_{push} = \sqrt{\frac{sL_t}{sC_t + C_{push}}},\ Z_{pull} = \sqrt{\frac{sL_t}{sC_t + C_{pull}}} \tag{1}$$

where L_t and C_t are the per unit length inductance and capacitance. These are expressed as

$$C_t = \frac{\sqrt{\varepsilon_{eff}}}{cZ_0},\ L_t = C_t Z_0^2 \tag{2}$$

in which ε_{eff} is the effective dielectric constant of the transmission line, Z_0 is the unloaded line impedance, and c is the free space velocity.

All functional capacitances like C_{push}, C_{up}, and C_{pull} can be obtained by

$$C_{push} = \frac{\varepsilon_0 w_{cb} W_1}{g_1 + \frac{t_d}{\varepsilon_r}} + C_f,\ C_{up} = \frac{\varepsilon_0 w_{cb} W_1}{g_0 + \frac{t_d}{\varepsilon_r}} + C_f,\ C_{pull} = \frac{\varepsilon_0 w_{cb} W_1}{g_2 + \frac{t_d}{\varepsilon_r}} + C_f \tag{3}$$

where w_{cb} is the width of the contact beam and C_f is the fringing capacitance, g_1 is the maximum travel range when the beam is lifted upward under V_{push} and g_2 is the effective gap from the top or zero bias state, within the pull-in range (considered to be 0.75 μm in this work) under V_{pull}.

The TTD phase shifter has a cutoff frequency called the Bragg frequency (f_B) [4] near the point where almost total reflection occurs and impedance becomes zero with no power transfer. For this reason, the Bragg frequency should be considered carefully to determine the upper operational frequency limit.

The Bragg frequency for the unit cell can be expressed as [5]

$$f_B = \left[\pi\sqrt{sL_t(sC_t + C_b)}\right]^{-1} \tag{4}$$

The length of the unit cell or spacing between two MEMS bridges (s) can be obtained from (5), as given in [5]

$$s = c\left(\pi f_B \sqrt{\varepsilon_{eff}}\right)^{-1} \tag{5}$$

All functional bridge capacitances (C_{push}, C_{up}, and C_{pull}) can also be found from the length of the unit cell (s), loaded impedances (Z_{push}, Z_{up}, and Z_{pull}), and line inductance and capacitance (L_t and C_t).

$$C_{push} = s\left[\frac{L_t}{Z_{push}^2} - C_t\right],\ C_{up} = s\left[\frac{L_t}{Z_{up}^2} - C_t\right],\ C_{pull} = s\left[\frac{L_t}{Z_{pull}^2} - C_t\right] \tag{6}$$

Here, C_{up} is the capacitance at zero bias state, whereas other two capacitances are at different bias states (V_{push} and V_{pull}).

Using (1)–(5), the phase constants in each state (β_{push}, β_{pull}) and the net phase shift ($\Delta\varphi$) can be expressed as [1]

$$\beta_{\text{push}} = \frac{360}{2\pi} s\omega \left[\sqrt{L_t C_t \left(1 + \frac{C_{\text{push}}}{C_t} \right)} \right]$$

$$\beta_{\text{pull}} = \frac{360}{2\pi} s\omega \left[\sqrt{L_t C_t \left(1 + \frac{C_{\text{pull}}}{C_t} \right)} \right]$$

$$\Delta\phi = \beta_{\text{push}} - \beta_{\text{pull}} = \frac{360}{2\pi} s\omega \left[\sqrt{L_t C_t \left(1 + \frac{C_{\text{push}}}{C_t} \right)} \right] - \sqrt{L_t C_t \left(1 + \frac{C_{\text{pull}}}{C_t} \right)} \tag{7}$$

In the push state, the center beam is lifted upward so the phase constant (β_{push}) decreases due to smaller loading capacitance (C_{push}), leading to high loaded impedance (Z_{push}). The center beam is deflected downward in the pull state, leading to a higher phase constant (β_{pull}) due to higher loading capacitance (C_{pull}) with low loaded impedance (Z_{pull}). Therefore, the difference in the phase constants ($\Delta\varphi$) between the two states increases and introduces a differential phase shift.

In order to maintain acceptable matching over a wide band, it is always adjustable not to overload the transmission line with an excessively large MEMS capacitance. In this circuit, a sawtooth-shaped center conductor is used at the place where the MEMS bridge is built to reduce the overlapping area leading to the reduction of down state impedance. The complete layout of the DMTL analog phase shifter is shown in Fig. 3. The complete area of the phase shifter is 8.5 mm^2. The electromechanical modeling of the push–pull bridge is carried out in a Coventor ware, saber platform and it is validated with measured responses up to a reasonable extent [1].

A CPW line is used as a base transmission line with 70 nH/cm inductance and 14.17pF/cm capacitance, giving an unloaded characteristic impedance of 70 Ω, as shown in the equivalent circuit model of Fig. 4a. The line has a 100-μm wide center conductor with 140 μm gaps on either side and is printed on the alumina substrate. This line is loaded with one MEMS bridge with a line length of 780 μm to make a unit cell which is the fundamental building block of a complete distributed cell. An extra 200 μm line of 50 Ω (width = 100 μm, gap = 45 μm) has been kept on either side of the unit cell for carrying out the RF measurements, as shown in Fig. 4a. The MEMS bridge loads this line with 0.205 pF/cm capacitance (C_{up}) along with 28.38pH bridge inductance and 0.37 Ω resistance that is in a series with C_{up} at a zero bias condition. The center conductor is narrowed (saw-shaped) where the MEMS bridge is loaded, and then an inductance (4.02nH) is introduced on the transmission line on either side of the variable capacitance (C_r) as shown in Fig. 4a [1].

HP8510C network analyzer and calibrated using the short-open-load-through (SOLT) on-wafer standards. The measurement results of the unit cell for three

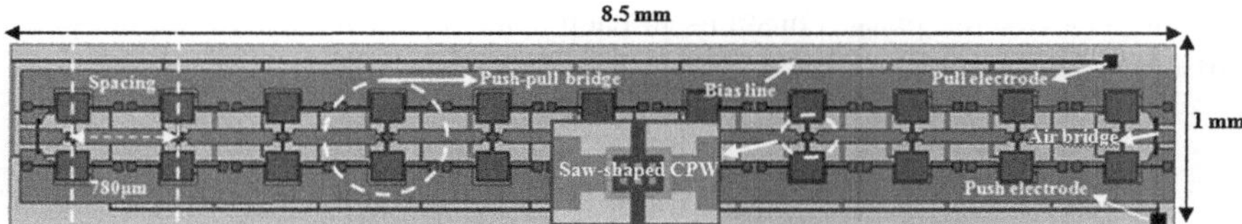

Fig. 3 Layout of complete TTD phase shifter, *inset* shows the saw-shaped CPW. Copyright/used with permission of/courtesy of Institute of Physics and IOP publishing limited

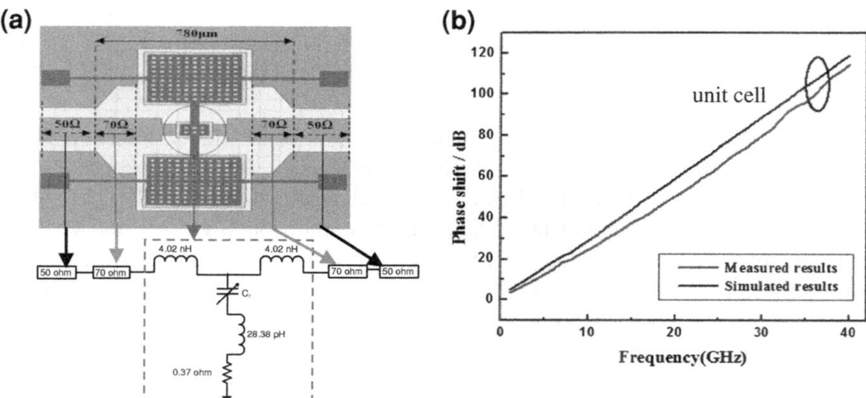

Fig. 4 Schematic top view of the unit cell phase shifter with equivalent circuit model: *red circle* indicates the *red* block in circuit model and *blue arrows* indicates the *blue* blocks. **b** Measured versus simulated phase shift per dB performance of the unit cell. Copyright/used with permission of/courtesy of Institute of Physics and IOP publishing limited

different states have been validated using FEM-based simulation through HFSS and are shown in Fig. 5. The reflection loss (S_{11}) is better than 12 dB and the worst-case transmission loss (S_{21}) of 0.28 dB has been obtained over a 1–40-GHz frequency band from a unit cell. The discrepancy between measured and simulation results in S-parameters is attributed to the overall height (g_0) nonuniformities in the MEMS bridges that lead to an asymmetric distribution of loaded line capacitances. A measured phase shift of 32.12° is obtained from a unit cell. The figure-of-merit in degrees per decibels for the unit cell phase shifter is approximately 114.64°/dB at 40 GHz which is found to be within 5.4 % tolerance compared with the simulation results (Fig. 4b).

The CPW transmission line is loaded with 11 MEMS bridges with a spacing of 780 μm to form a distributed loaded transmission line of 8.35 mm length. A return loss of better than 11.5 dB is achieved up to 40 GHz as shown in Fig. 6a. A maximum insertion loss of 3.75 dB is noticed up to 20 GHz. Furthermore, it increases to 5.7 dB at 40 GHz in the pull states as shown in Fig. 6b. The agreement between the measured and simulated loss is found to be within a 15 %

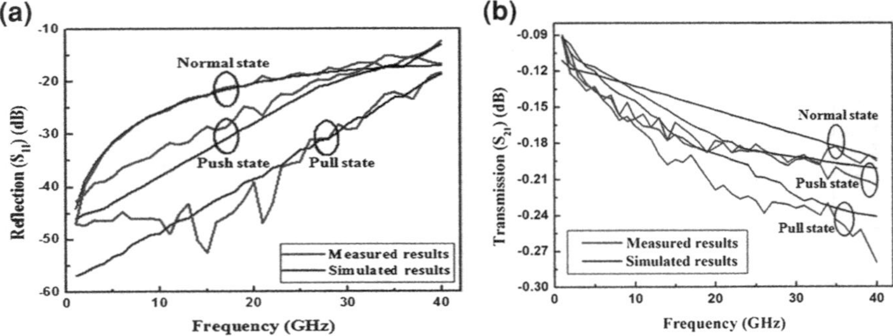

Fig. 5 Measured versus simulated loss for unit cell **a** reflection and **b** transmission. Copyright/used with permission of/courtesy of Institute of Physics and IOP publishing limited

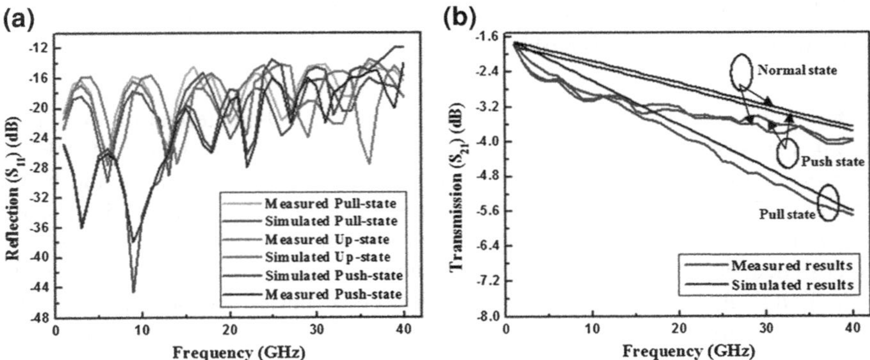

Fig. 6 Measured versus simulated loss for distributed cell **a** reflection and **b** transmission. Copyright/used with permission of/courtesy of Institute of Physics and IOP publishing limited

tolerance limit due to an asymmetric distribution of a gap profile throughout the TTD. Typical height nonuniformities in the bridges are approximately 0.43–0.68 μm. Furthermore, the increase in the transmission loss (S_{21}) of DMTL compared to the unit cell loss (0.28 dB) is due to signal leakage via the Cr bias lines. The 36 % from 34.3 kΩ to 21.7 kΩ reduction of bias resistance has been found from the distributed cell to the unit cell. The capacitance variation in the pull state is higher than the other two states (normal and push) and loss is also high, as depicted in Fig. 6b. A phase shift of 317.15°/cm has been obtained at 40 GHz with 0 to 4.1 volt actuation bias in the pull state as shown in Fig. 7a. In the push state, a 44.1°/cm phase shift has been achieved with reference to the normal state at 40 GHz with an 8.1 volt actuation bias as shown in Fig. 7b. A continuous phase shift of 0–360° has been obtained from the fabricated TTD device from the push to pull state.

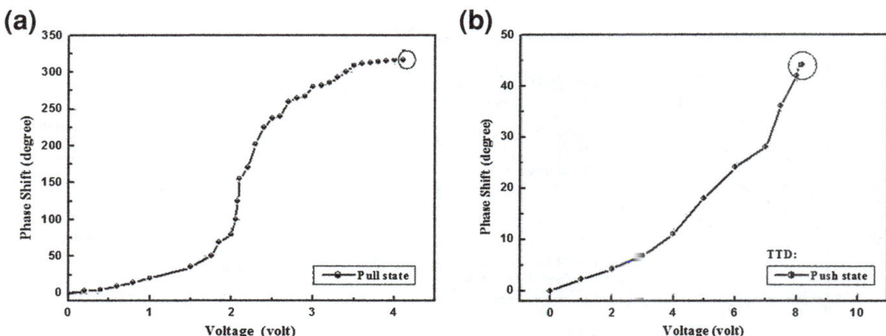

Fig. 7 Measured variation of phase shift with applied bias at **a** pull and **b** push state. Copyright/used with permission of/courtesy of Institute of Physics and IOP publishing limited

3 MEMS 5-Bit Switched Line Phase Shifter Using DC-Contact Series Switches

In this section, the detailed design, development, and characterization of a 5-bit digital switched-line phase shifter with an in-line metal–metal contact MEMS series switch utilizing a CPW-based transmission line is discussed. A gold-based surface micromachining process has been used to develop the complete 5-bit phase shifter. Initially, different primary phase bit sections (11.25°/22.5°/45°/90°/180°) are designed, experimentally investigated and then combined to build the complete 5-bit digital phase shifter. A systematic analytical design methodology on two different topology-based 5-bit switched line phase shifters is presented in this section.

All individual primary bits are fabricated separately to get the desired phase response and these are then fitted to realize the complete 5-bit phase shifter. The design is based on SPDT switches where two MEMS switches are placed in a back-to-back configuration as shown in Fig. 8f. Thus, each bit uses only four MEMS switches. As the insertion loss of the phase shifter is mainly caused by conductor and switch losses, lower switches will surely result in low insertion losses. MEMS switches are employed to route the input RF signal through the appropriate length of matched transmission lines. The phase shift is given by (8)

$$\Delta\phi = \frac{2\pi(l_d - l_r)}{\lambda} = \frac{2\pi f\sqrt{\varepsilon_{\text{eff}}}(l_d - l_r)}{c} \tag{8}$$

where l_d and l_r are the lengths of the delay line and reference line, respectively. c and f are the velocity of light and the operating frequency, respectively. The microscopic images of individual and different primary bits of the 5-bit phase shifter are shown in Fig. 8a–e. The individual-bits have been measured using

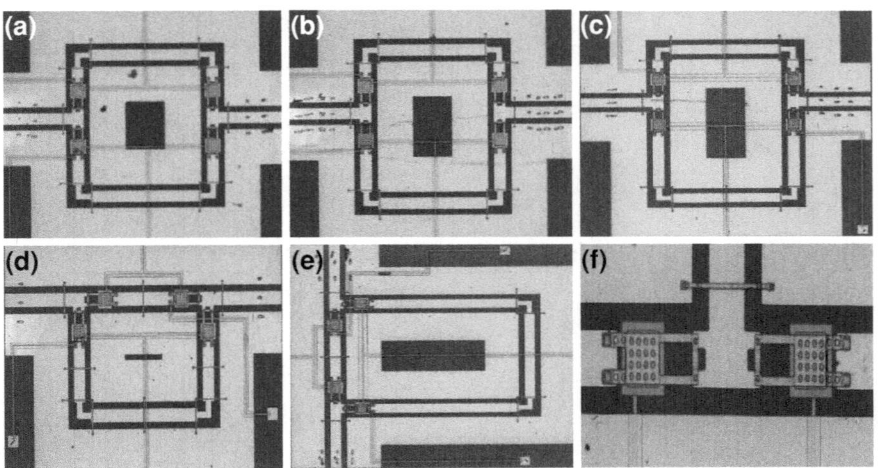

Fig. 8 Microscopic images of five primary individual switched-line phase shifter bits utilizing MEMS DC-contact switch; **a** 11.25°, **b** 22.5°, **c** 45°, **d** 90°, **e** 180°, and **f** show an image of two back-to-back MEMS SPST switches. Copyright/used with permission of/courtesy of Institute of Physics and IOP publishing limited

Agilent PNA series E8361C VNA utilizing Cascade DC probes and validated using simulation, as shown in Fig. 9.

The lengths of the reference line for 11.25°, 22.5°, 45°, 90°, and 180° phase bits are 1675, 1675, 1675, 570, and 570 μm, respectively. Similarly, the length of the delay line for 11.25°, 22.5°, 45°, 90°, and 180° phase bits are 1925, 2150, 2620, 2480, and 4320 μm, respectively. The maximum return loss of better than 16.5 dB and the worst-case insertion loss of 1.05 dB have been obtained from individual phase bits over 13 to 17.25 GHz as shown in Fig. 9a–e. An almost desired phase shift is achieved from individual phase bits with +0.79° of a phase error which is also validated through simulation as shown in Fig. 9f.

In order to achieve a compact design with a minimum phase error, corners of the individual switched-line phase shifter bits are realized using 90° CPW bends as shown in Fig. 10a. The equivalent circuit of individual phase shifter bits is shown in Fig. 10b. The dimensions of the bends are optimized using FEM-based simulation in HFSS to achieve small transmission distortion caused by intra-coupling in the line. For CPW ground discontinuities, ground planes are connected by air-bridges at each discontinuity to short out the parasitic slotline modes that are easily exited. The maximum error of 12 % has been observed between the measured and simulated loss performance. This can be attributed to nonuniform overall height (g_O) and also nonuniformities in the in-line MEMS switches. Furthermore, signal leakage through a TiW bias line also contributes to deviations between simulation and the measured results. Finally, all individual phase bits are cascaded together in a proper arrangement to build a complete 5-bit phase shifter [2]. The SEM image of the complete phase shifter is shown in Fig. 11. The reference and delay paths

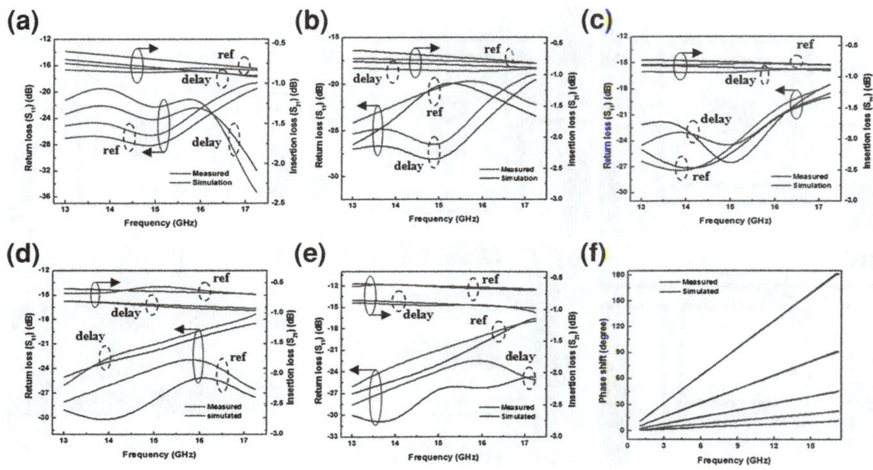

Fig. 9 Measured versus simulation results of individual-bits of the phase shifter. **a** 11.25°, **b** 22.5°, **c** 45°, **d** 90°, **e** 180°, and **f** phase shift versus frequency characteristics of all five individual phase bits; reference and delay path loss are shown by name “ref” and “delay”, respectively. Copyright/used with permission of/courtesy of Institute of Physics and IOP publishing limited

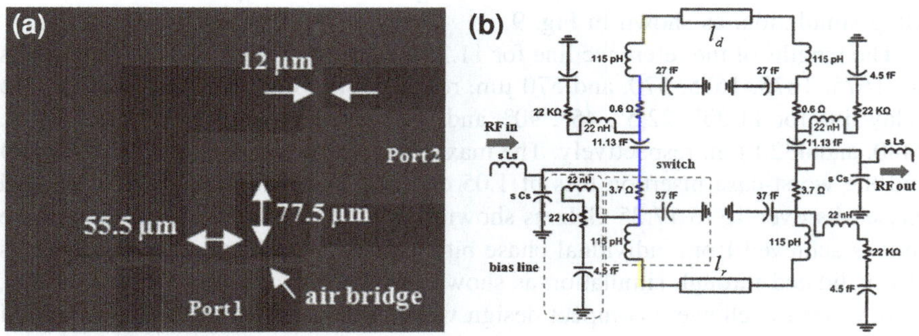

Fig. 10 **a** SEM image of 90° CPW bend at the corner of the phase shifter and **b** equivalent circuit of individual switched-line phase shifter bits. Copyright/used with permission of/courtesy of Institute of Physics and IOP publishing limited

are controlled by dedicated bias pads on either side of the phase shifter. The total area of the phase shifter is $11.4 \times 4\ \text{mm}^2$. The length and width of the reference and delay arms are similar for the individual phase bits. Finite-ground CPW lines with uneven ground plane widths may cause a higher loss and noise of localized points in CPW circuits [2]. The CPW ground plane width (300 μm) is three times that of the t-line width (100 μm). CPW-type transmission lines are comprised of three separate conductors, and they support two independent quasi-TEM modes. For symmetric CPWs, these modes are typically called the CPW or the even mode

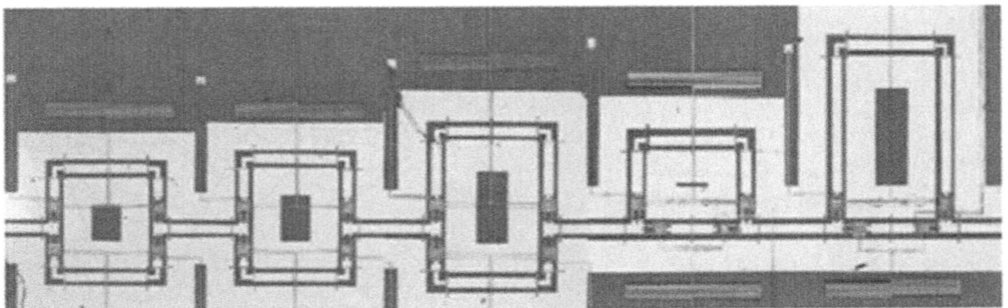

Fig. 11 Microscopic image of complete 5-bit phase shifter (PS1) using SW1 switches. The area of the phase shifter is 11.4 × 4 mm^2. Copyright/used with permission of/courtesy of Institute of Physics and IOP publishing limited

and the coupled slotline or the odd mode. Either of these two modes can propagate along the transmission line independently if they are excited, and they are coupled to each other at discontinuities. In general, the coupled slotline mode is excited in the CPW circuits if there is a discontinuity or an asymmetry in the transmission line. It is for this reason that air-bridges are used to equalize the voltages on the two ground planes of the CPW lines at the discontinuities and also at periodic intervals along the uniform CPW line [2].

The connecting line length (shown in Fig. 12a) between two individual phase bits is also very important in the CPW-based switched-line phase shifter. When an electromagnetic wave propagates from the output port of one-bit to the input-port of the next individual-bit, different modes (CPW mode, coupled slotline mode) are coupled with each other if the line length is not optimized and this can lead to off-path resonance at the operating band of frequency. To overcome this resonant behavior, the connecting line length (l_c) needs to be optimized using the FEM-based simulator HFSS. In this work, l_c has been chosen to be 1400 μm that is λ_g /5, where λ_g is the guide wavelength [2].

The complete phase shifter has been characterized on Agilent PNA series E8361C VNA using cascade DC probes and calibrated with on-wafer short-open-load-through (SOLT) standards. The average measured return loss of better than 13 dB and the worst-case insertion loss of 5.3 dB have been obtained from 32 states over 13 to 17.25 GHz with an actuation voltage of 26 Volt from a complete 5-bit phase shifter as shown in Fig. 12b–c. The measured results have been validated through FEM-based simulation in HFSS and a circuit model in ADS. The same equivalent circuit model (shown in Fig. 10b) is used to build the complete 5-bit phase shifter using different delay and reference path lengths and MEMS switches. The off-path resonance due to CPW discontinuity and asymmetric gap profiles (g_0) between ten switches (at a time) results in an 18 % deviation between the measured and average simulated insertion loss. Furthermore, signal leakage through the TiW bias lines is also responsible for the deviation between the two results. The measured phase response of 32 states in the 5-bit phase shifter is shown in Fig. 12d. The measured maximum phase shift per decibel or figure-of-

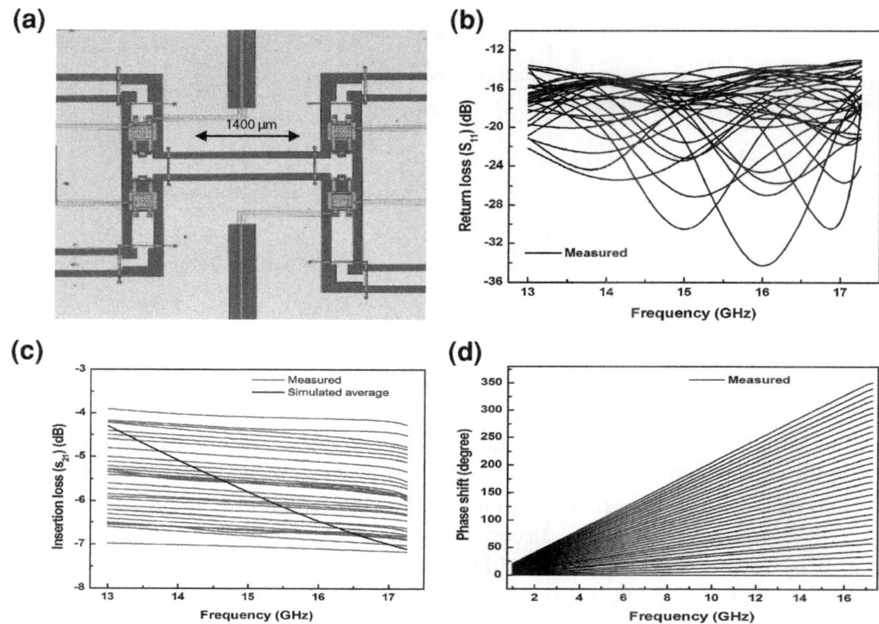

Fig. 12 **a** Microscopic image of the connecting section of two individual phase bits. Measured S-parameter response of complete 5-bit switched line phase shifter, **b** return loss, **c** insertion loss, and **d** phase shift versus frequency response. Copyright/used with permission of/courtesy of Institute of Physics and IOP publishing limited

merit (FOM) performance of 48.9°/dB is obtained from the phase shifter at 17.25 GHz. A maximum phase error ($\Delta\phi_E$) of $-0.7°$ to $+3.1°$ is achieved from the complete 5-bit phase shifter [2].

The average loss, phase accuracy, and area of the switched-line phase shifter are improved using a different switch configuration [4], while keeping the same dimensions of the reference and delay arm lengths. The SEM image of the phase shifter is shown in Fig. 13. In this type of phase shifter, two reference and delay arms are sharing one CPW ground plane which leads to the reduction of 25 % of the chip area compared to the other model (Fig. 11). The total area of the phase shifter is 9×4 mm^2 [2].

The phase shifter exhibits a measured return loss of better than 14 dB and the worst-case insertion loss of 5.48 dB over 13 to 17.25 GHz. The maximum differential time delay of 56.36 ps is obtained at 17.25 GHz. In addition, a maximum phase error of $-1.12°$ to $+1.35°$ is achieved from this phase shifter.

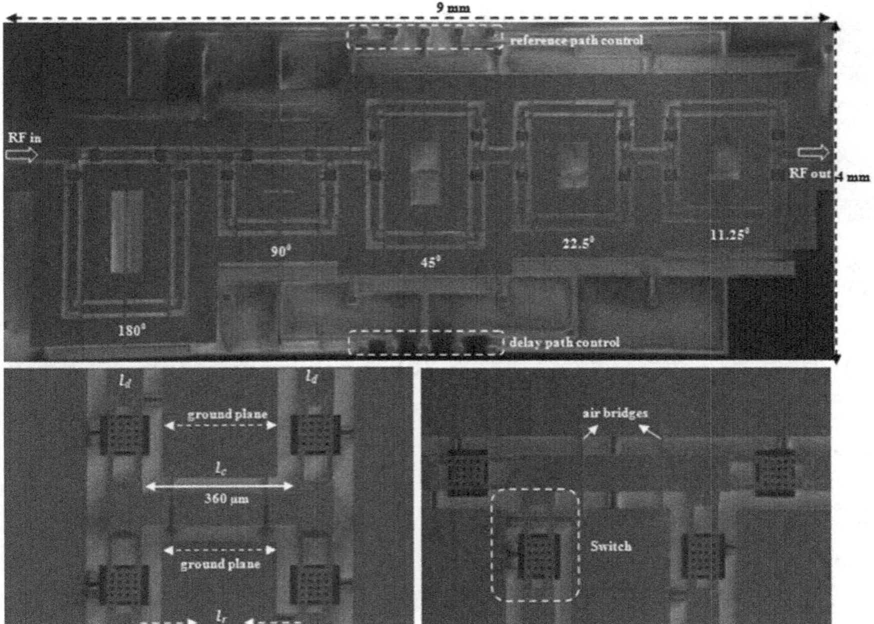

Fig. 13 SEM images of PS2-type 5-bit phase shifter using SW2 switch. The area of the phase shifter is 9 × 4 mm^2. Copyright/used with permission of/courtesy of Institute of Physics and IOP publishing limited

4 MEMS 2-bit and 5-bit TTD phase shifters Using SP4T and SPDT Switches

A Ku-band 2-bit phase shifter is designed and fabricated using an SP4T switch discussed in [4], using a 25mil (635 µm) alumina substrate. Four delay lines and two SP4T switches are connected to the input and output t-line.

The design aim is to achieve an optimum performance within 13–18 GHz with a low insertion loss and compact size. The microscopic images of two different 2-bit phase shifters are shown in Fig. 14a–b. These are fine-bit (11.25°/22.5°/33.75°) and coarse-bit sections (45°/90°/135°) of a 5-bit phase shifter operating at 17.25 GHz. In order to obtain low loss and compact design, the reference line is optimized to be 429 µm (electrical length: 21° @ 17.25 GHz) for both sections. The three delay lines of the individual section contain a section equal to the reference line plus an additional delay line as per the desired phase shift. Hence, the length of the fine-bit delay lines for 11.25°, 22.5°, and 33.75° are 690, 945, and 1180 µm, respectively. The coarse-bit contains 1560, 2664, and 3662 µm length delay lines for 45°, 90°, and 135° phase states, respectively. In order to achieve a compact design with minimum phase error, all delay lines are routed with 90°

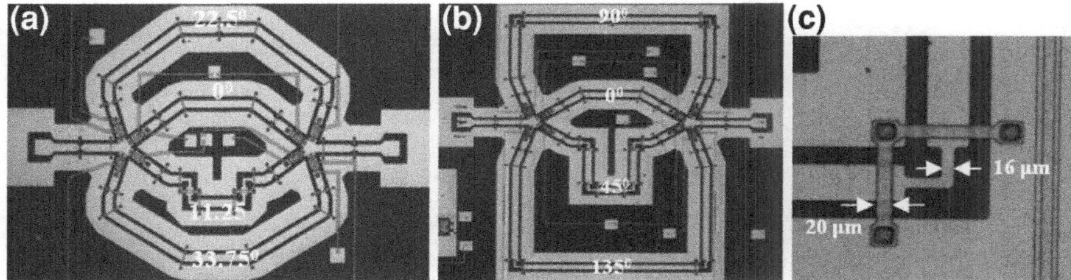

Fig. 14 Microscopic images of **a** fine-bit and **b** coarse-bit section of 2-bit phase shifter **c** 90° CPW bend at the corner

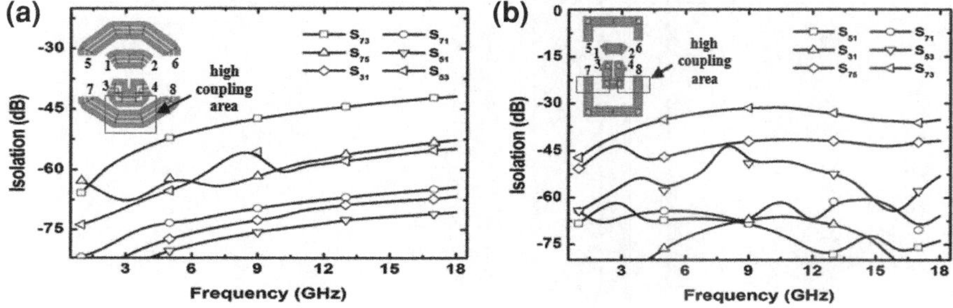

Fig. 15 Simulated coupling among the delay lines: **a** fine-bit section and **b** coarse-bit section

CPW bends at each corner, as shown in Fig. 14c. The dimensions of the 90° CPW bends are optimized using a full-wave simulator HFSS to achieve small transmission distortions caused by intra-coupling in the line. Inductive bends are used in all the delay paths to overcome the excitation of coupled slotline modes at CPW discontinuities [6].

To check the effect of coupling between various delay lines, an eight-port simulation was done using HFSS. The isolation is better than 32 dB in all the cases up to Ku-band. As 11.25°/22.5° and 45°/135° lines are closely packed and parallel to each other, coupling is 6–9 dB more (S_{73}) than other combination of lines as depicted in Fig. 15. High impedance (30–50 KΩ) bias lines are made with 12 μm width TiW and covered with SiO_2 (0.7 μm). These can be routed easily underneath the t-line and this improves the simplicity of the device with a negligible effect on the insertion loss of the complete phase shifter.

The design is verified using HFSS simulation with a return loss better than 28 dB and an insertion loss less than 0.5 dB. No off-path resonance is observed in the band of interest. A sanity check of two complete 2-bit phase shifters was carried out in ADS with the bend (90° bends and inductive bends) data taken from HFSS and combined with a measured SP4T response for completeness. As a result, an average return loss of better than 23.5 dB and maximum insertion loss of 0.78 dB have been obtained over a 13–18-GHz band as shown in Fig. 16a–f.

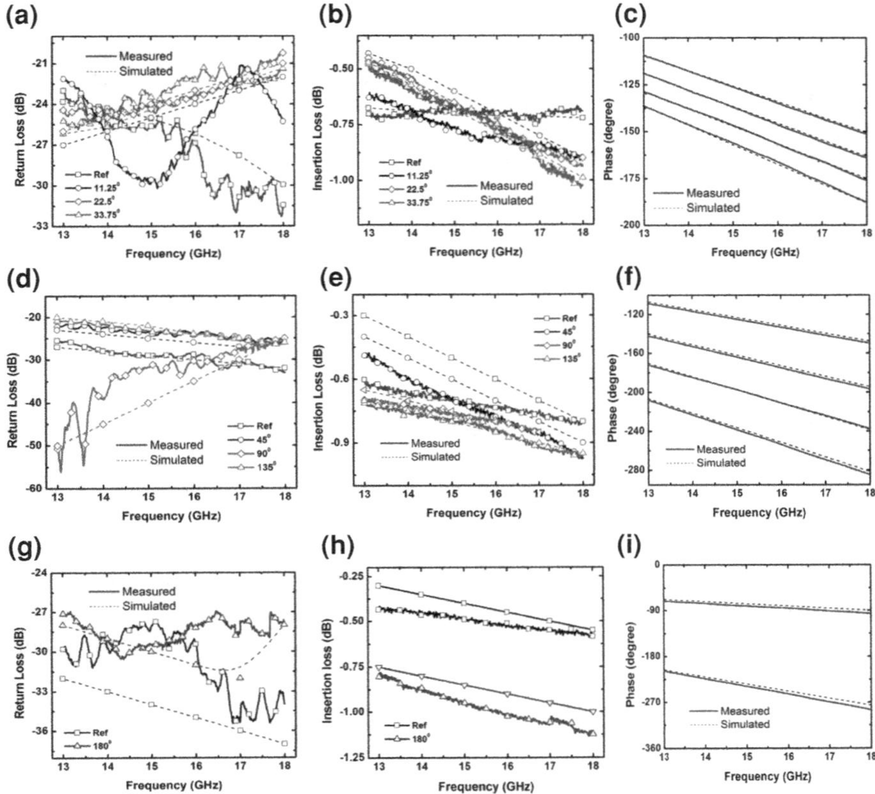

Fig. 16 Measured versus simulated S-parameters of individual phase shifters: **a** return loss, **b** insertion loss, and **c** phase of fine-bit section; **d** return loss, **e** insertion loss, and **f** phase of coarse-bit section; **g** return loss, **h** insertion loss, and **i** phase shift of the 1-bit section

The overall size of the fine-bit section is 3.84 mm^2 (2.21 × 1.74 mm^2) (Fig. 14a) and the coarse-bit section is 6.45 mm^2 (2.34 × 2.76 mm^2) (Fig. 14b). The actuation voltage applied at each state of the phase shifter is 53–70 V that is quite high compared to the single switch actuation voltage (43–53 V). This is due to the added capacitance between the delay line and ground plane and the non-uniform overall height (g_O) distribution of dc-contact switches after the release process. An optical profilometer shows 0.4 μm to 0.57 μm variation in the tip deflection of the switch along the line which leads to 3.97–5.6 fF variation in C_{off} (at zero bias) and 2.8–3.7 Ω variation in R_c (applied bias) through out different delay lines. The measured return loss of better than 21 dB and less than 0.82 dB of insertion loss are obtained between 13 and 18 GHz from the "fine-bit" section, as shown in Fig. 16a–b. The coarse-bit section gives a return loss better than 24 dB from 16 to 18 GHz and 20 dB within 13–15 GHz. The minimum and maximum measured losses are 0.65 dB (0°) and 0.92 dB (135^0) respectively over the band of

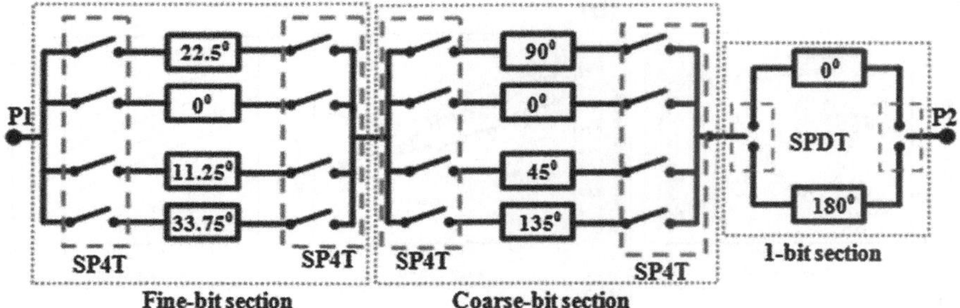

Fig. 17 Schematic diagram of the 5-bit TTD MEMS switched-line phase shifter based on SP4T and SPDT switches

interest. The different delay line lengths cause the variation of insertion loss at different phase states. Figure 16c, f show the measured versus simulated phase response of the fine-bit and coarse-bit sections. A maximum phase error of 0.62° is obtained experimentally at 17.25 GHz. No off-path resonance was observed in the measured insertion loss over the band of interest.

A 1-bit phase shifter is also designed, fabricated, and tested which provides a 180° phase shift at 17.25 GHz. A measured return loss of better than 27 dB is obtained from both states within 13–18 GHz as shown in Fig. 16g. A minimum insertion loss of 0.53 dB and a maximum insertion loss of 1.1 dB are achieved from reference and delay states, respectively, over the band (Fig. 17h). A maximum phase error of 0.47° at 17.25 GHz is obtained from the 1-bit section (Fig. 16i).

4.1 5-Bit MEMS Phase Shifter

Different topologies of a 5-bit phase shifter can be designed and optimized using four SP4T and two SPDT switches with different delay and reference line combinations. Figure 17 shows a possible schematic of a 5-bit TTD phase shifter based on SP4T and SPDT switches.

Compared with the conventional TTD phase shifter, in which a minimum of 10 switches are actuated at a time, the present design requires only six switches to be actuated for 5-bit operation. The fine- and coarse-bit sections of the 5-bit phase shifter have already been discussed earlier. Now, all these sections are cascaded together to build a complete 5-bit phase shifter. The microscopic image of the final phase shifter is shown in Fig. 18. This phase shifter occupies an area of 16.4 mm^2 (5.17 × 3.19 mm).

To overcome off-path resonance, connecting line lengths (L_1, L_2) between each section are optimized using full-wave simulation within an 8–18-GHz frequency band. Connecting line length between fine- to coarse-bit (L_1) and coarse- to 1-bit

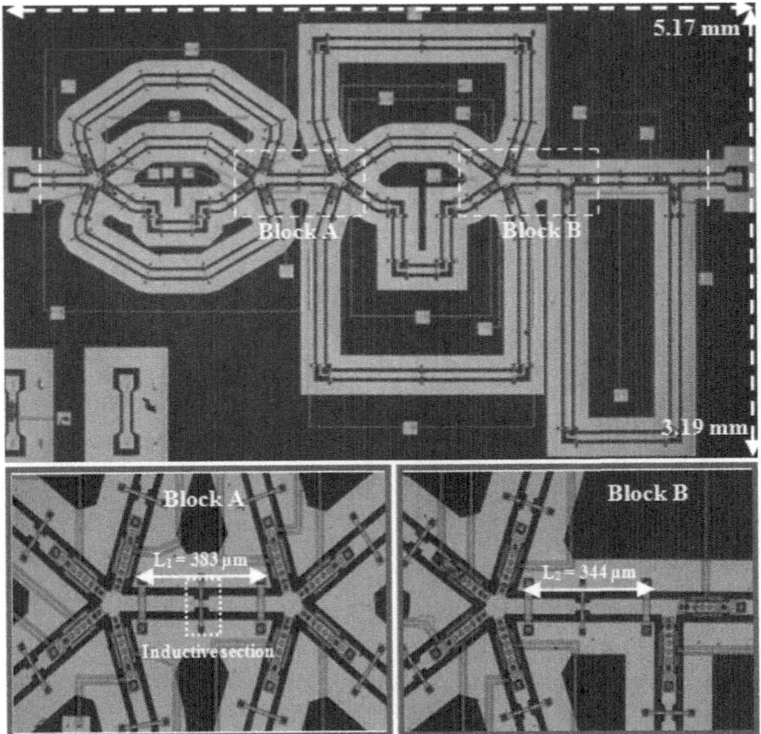

Fig. 18 Microscopic image of the fabricated Ku-band 5-bit TTD phase shifter

sections (L_2) are optimized to be 383 and 344 μm, respectively. In addition, a small inductive section is introduced at the middle of the connecting lines L_1 and L_2. As a result, the effect of off-path resonance is completely eliminated with an almost flat response in insertion loss in all the 32 phase states over the band. The complete 5-bit phase shifter (Fig. 18) is simulated in HFSS and the response is also verified in ADS for completeness. The simulated results of the optimized phase shifter are depicted in Fig. 19a. A simulated average return loss of 22 dB and an insertion loss of 3.3 dB are obtained from 13 to 18 GHz with a maximum phase error of 0.38° at 17.25 GHz.

A measured return loss of better than 20 dB and an insertion loss less than 3.8 dB are achieved experimentally from all the states over the entire frequency band of 13–18 GHz, as shown in Fig. 19b–c. A maximum phase error of 1.14° is obtained at a 258.75° phase state compared to other phase states. Phase shift versus frequency response of the primary phase bits is depicted in Fig. 19d.

The actuation voltage at each section is different as each path consists of six MEMS switches which have undergone nonuniform tip deformation. Hence, it leads to the variation of contact resistance (R_c) throughout the line. It is observed

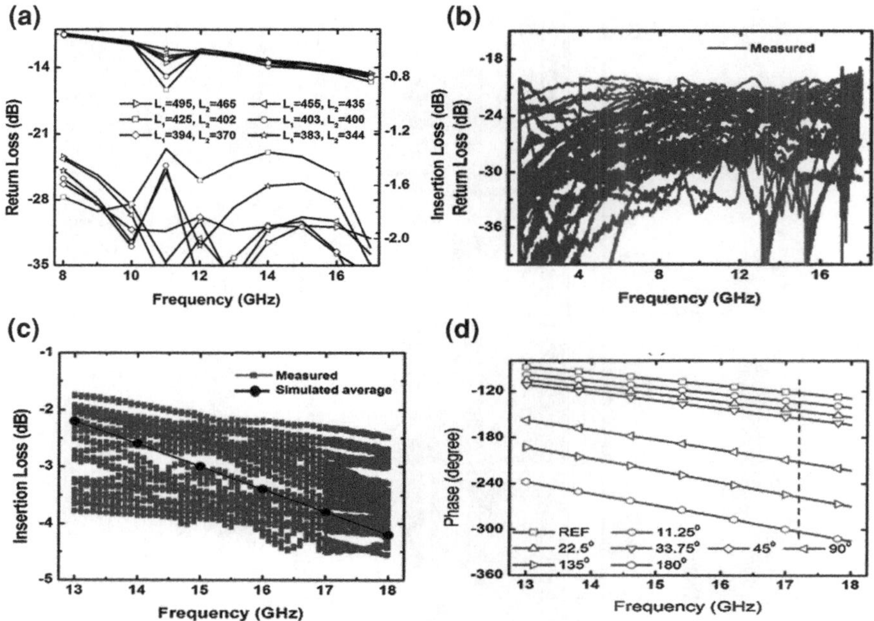

Fig. 19 **a** Effects of insertion and return loss under different connecting lengths. Measured S-parameter response of 5-bit phase shifter **b** return loss from 0 to 18 GHz, **c** insertion loss, and **d** phase versus frequency response of primary phase bit of 5-bit phase shifter within 13–18 GHz

that R_c is maximum (4.7 Ω) in the 258.75° state and minimum (2.4 Ω) at the reference state. Nonuniform tip deflection can be improved with a thicker gold beam (>3 μm) and testing under a hermetic environment can improve the performance of the overall phase shifter.

5 Other Phase Shifter Designs

The Centre for Applied Research in Electronics (CARE), Indian Institute of Technology, Delhi, is currently working on a very small MEMS TTD phase shifter. The expected area could be 11–13 mm^2 (including all bias lines and pads) at 17 GHz without compromising the phase shifter performance. Furthermore, the development of frequency reconfigurable 5-bit MEMS phase shifters at 13.4 and 17.2 GHz is also planned as future work.

6 Conclusions

This chapter presents a detailed design methodology of DMTL and switched-line phase shifters. A push–pull-type analog DMTL phase shifter gives 0°–60° phase shift variation with an 8-V actuation bias at 40 GHz. A conventional switched-line phase shifter using 20 MEMS switches is presented and the measured response is validated with simulations up to a reasonable extent at 17 GHz operating frequency. Two different kinds of switched line phase shifters are discussed. Furthermore, MEMS SP4T- and SPDT-based 2-bit and 5-bit phase shifters show an excellent performance over the band of interest. The advantages of this new configuration over any available MEMS 5-bit phase shifters are highlighted.

Acknowledgments The authors would like to express their profound gratitude to the National Program on Micro and Smart Systems (NPMASS), Govt. of India for setting up a MEMS design lab and RF characterization facilities at CARE, Indian Institute of Technology, Delhi, India. The work reported here was carried out under the project titled "Design and Development of MEMS phase shifter and SPDT switches." Authors also would like to acknowledge Institute of Physics and IOP publishing limited for granting the permission of using few figures from the Journal of Micromachining and Microengineering.

References

1. Dey S, Koul SK (2012) Design and development of a surface micro-machined push-pull-type true-time-delay phase shifter on an alumina substrate for Ka-band T/R module application. J Micromech Microeng 22:125006–125025
2. Dey S, Koul SK (2014) Design and development of a CPW-based 5-bit switched-line phase shifter using inline metal contact MEMS series switches for 17.25 GHz transmit/receive module application. J Micromech Microeng 24:1–24
3. Koul SK, Bhat B (1991) Microwave and millimeter wave phase shifter-Vol. II. Artech House, Norwood, MA
4. Koul SK, Dey S (2014) RF MEMS Single-pole- multi- throw switching circuits. This volume
5. Rebeiz GM (2003) RF MEMS theory, design, and technology. Wiley, Hoboken, NJ
6. Simons RN (2001) Coplanar waveguide circuits, components, and systems. Wiley, New York
7. Zhang WM, Hsia RP, Liang C, Song G, Domier CW, Luhmann NC Jr (1996) Novel low-loss delay line for broadband phased antenna array applications. IEEE Microw Guided Wave Lett 6:395–397

MEMS Sensors for Underwater Applications

V. Natarajan, M. Kathiresan, K. A. Thomas, Rajeev R. Ashokan, G. Suresh, E. Varadarajan and Shiny Nair

Abstract Microelectromechanical system (MEMS)-based sensors for marine environment help to realize new systems that bring enhanced levels of perception, control, and performance to sonar systems and sensors related to marine environments. Processing, assembly, packaging, testing, and manufacturing methods are all highly dictated by the intended application of MEMS devices; hence, these disciplines are being honed up to meet the demands with new materials and performance requirements across a wide spectrum of underwater applications. Five basic parameters are measured in the ocean to define its physical state: temperature, salinity, pressure, density, and velocity of sound. These can be obtained using a pressure sensor, temperature detector, and a conductivity sensor. Biologically inspired MEMS shear stress sensors comprising a piezoresistive floating element offer the potential to make flow measurements in fluid with unprecedented sensitivity, and spatial and temporal resolution. In order to get finer resolution of underwater objects in turbid waters, it is imperative to work at MHz frequencies. Different types of transducers such as CMUT, PMUT, and Helmholtz resonator are also realized by MEMS fabrication and are readily scalable in size. In addition, multiplexing, pulsing, and pre-amplifying electronics can be easily integrated on the same chip with the transducers or on a separate chip via flip-chip bonding. This allows for 1D and 2D arrays of elements to be easily steered electronically. Thus, fabrication of a large number of transducers with built-in pre-amplifiers required in a planar array configuration is possible with MEMS-based technology.

Keywords MOSFET acoustic sensors · CTD sensors · MEMS shear stress sensors · Biaxial MEMS accelerometer

V. Natarajan (✉) · M. Kathiresan · K. A. Thomas · R. R. Ashokan · G. Suresh · E. Varadarajan · S. Nair
Naval Physical and Oceanographic Laboratory, Thrikkakara, Kochi 682021, India
e-mail: natarajan_vinay@yahoo.com

K. J. Vinoy et al. (eds.), *Micro and Smart Devices and Systems*,
Springer Tracts in Mechanical Engineering, DOI: 10.1007/978-81-322-1913-2_29,

1 Introduction

Microelectromechanical systems (MEMS) is the integration of mechanical elements, sensors, actuators, and electronics on a common substrate through the utilization of micro fabrication technology. MEMS devices have led to the development of products that sense, think, act, communicate, self-power, and navigate. It has great potential in the area of marine environment—hydrophones, tilt sensors, Conductivity, Temperature, and Depth (CTD) sensors, flow sensors, micro-fluidics, bio-MEMS, micro-power and security/defence applications. Integrating MEMS devices is more complex than manufacturing them.

The key to rapid research and development of MEMS is due to both technology and fabrication advances through innovative modeling, design, fabrication, and characterization and test tools that have evolved over the last few years. MEMS-based sensors for marine environment help to realize new systems that bring enhanced levels of perception, control, and performance to sonar systems and sensors related to marine environments. MEMS sensors find applications where size, weight, and power must be reduced simultaneously with an increase in functionality and low cost achieved by batch fabrication and high precision in their manufacture.

Processing, assembly, packaging, testing, and manufacturing method are all highly dictated by the intended application of MEMS devices; hence, these disciplines are being honed to meet the demands with new materials and performance requirements across a very wide spectrum of applications.

Piezoelectric materials form the basis of advanced MEMS sensors and energy harvesting techniques that are gaining prominence in industrial, medical, and in underwater applications. Researchers are laying emphasis to perfect a lead–zirconate–titanate (PZT) deposition process compatible with mass production and integrate it into established MEMS processes for the above applications.

Five basic parameters are measured in the ocean to define its physical state. These are temperature, salinity, pressure, density, and velocity of sound. These can be obtained using (a) pressure sensor to monitor depth, (b) platinum resistance temperature detector, and (c) conductivity sensor utilizing platinum electrodes or inductive type cells; the last two parameters are derived from CTD data. Apart from the sensor it is of equal importance to design a signal conditioning circuit to convert the responses of various sensors into useful form

MEMS shear stress sensors offer the potential to make flow measurements in fluid with unprecedented sensitivity, along with spatial, and temporal resolution. Most MEMS shear stress sensors have been developed for measurements in air and utilize indirect methods. The development of micro machined, distributed flow sensors based on a biological inspiration, the fish lateral line sensors, is the technology of the day. It comprises piezoresistive floating element shear stress sensors for direct dynamic measurement.

Acoustic sensors can be scalar or vector, meaning they can measure only the magnitude of acoustic pressure at a point or can measure any of the acoustic vectors, i.e., displacement, velocity, or acceleration in terms of its magnitude and direction. Scalar sensors have omnidirectional pattern and large arrays of scalar sensors are used for direction of arrival (DOA) estimation. Vector sensors with their inherent directional pattern of "figure of eight" offer greater advantage in terms of their smaller dimensions and selective noise rejection capabilities.

In order to get finer resolution of the underwater objects in turbid waters, it is imperative to work at MHz frequencies. A large number of transducers are required in a planar array configuration. This is possible with MEMS-based technology, which has a built-in pre-amplifier in close proximity to the sensor. Different types of transducers are being practiced: CMUT, PMUT, and Helmholtz resonator. MUTs are realized by batch fabrication techniques employed by the semiconductor industry and are readily scalable in size. In addition, multiplexing, pulsing, and pre-amplifying electronics can be easily integrated on the same chip with transducers or on a separate chip via flip chip bonding. This allows for 1D and 2D arrays of elements to be easily steered electronically.

2 Types of Sensors

2.1 Acoustic Sensor

A novel MEMS design for an underwater acoustic sensor, hydrophone, has been attempted. A hydrophone is a device that picks up the underwater acoustic energy and converts it into electrical energy. Generally a hydrophone is used to measure the intensities at points within acoustic fields. For this purpose, it is desirable that the hydrophone is relatively small compared to the wavelength, so that the effect of the hydrophone upon the field is negligible.

Many Lead Zirconate Titanate-polymer (PZT) composite transducers with various connectivity have been developed, which consist of small PZT elements embedded in a low-density polymer matrix. These composites have lower acoustic impedance than conventional PZT, leading to an improved bandwidth [1]. One of the major problems associated with these piezoelectric composite materials, however, is the high fabrication cost.

The piezoelectric PVDF, poly vinylidene fluoride (PVDF) film has been found to be especially suited for underwater acoustic field measurements because of its low acoustic impedance and lateral coupling coefficient. With the development of silicon IC technology, it is desired to integrate the mechanical elements with microelectronic devices on a common silicon substrate to fabricate low cost, highly sensitive and miniaturized sensors. In this structure Fig. 1a, a sheet of PVDF film is bonded to the extended gate of a MOSFET, resulting in a structure named as piezoelectric-oxide-semiconductor field-effect transistor (POSFET) [2]. Since the POSFET structure combines the active electronics with the sensing element on the same chip, it has been widely used and studied as an essential

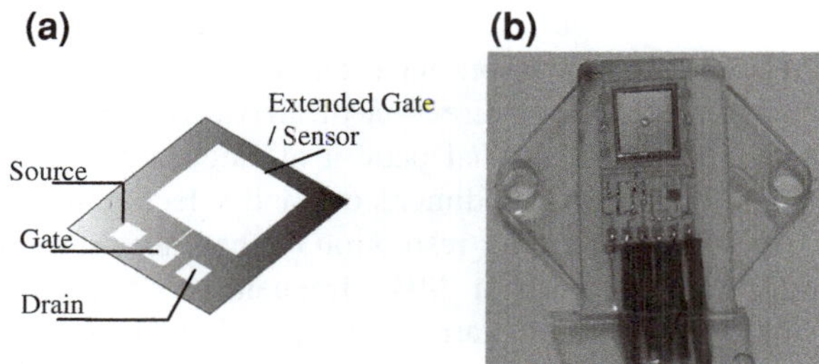

Fig. 1 **a** Schematic of MOSFET-Acoustic sensor. **b** Fabricated MOSFET-Acoustic senor with pre-amplifier and rabbit ears for mounting

structure for piezoelectric or pyroelectric sensors [3, 4]. However, there is one disadvantage with this structure, which is the capacitance existing between the extended gate electrode and the semiconducting silicon substrate, with the insulating silicon dioxide layer acting as a dielectric material. Due to the limitations of IC technology, the silicon dioxide layer is thinner than the sensor. This capacitance shunts a large part of the piezoelectric signal generated by the sensor and results in a significant reduction of the device sensitivity.

Some methods have been developed to minimize the parasitic capacitance using high resistivity polysilicon or sapphire substrate to replace the silicon under the lower electrode. Using this technique, the effective sensitivity is predicted to be increased by a factor of five or more. An integrated transducer structure with the extended-gate electrode of the MOSFET padded up with a polyimide dielectric layer is reported and a sensitivity improvement of over 13 dB was achieved [5]. Some have reported the use of SU-8 instead of polyimide below the piezoelectric sensor due to its compatibility with photolithography processes [6, 7]. The schematic and fabricated device is shown in Fig. 1a, b.

2.2 Conductivity, Temperature, and Depth Sensors

In SONAR, sound is used for the detection and classification of underwater bodies. The SONAR performance depends on the sound speed profile in water which varies with location and seasons. Sound profile in a region in ocean is highly influenced by the depth, temperature, and salinity (conductivity) of seawater.

Micro CTDs are being attempted to be manufactured using MEMS polymer technology [8]. The crux of the problem with pressure sensors for underwater applications is their packaging. Temperature sensors have been developed. The conductivity sensor is the most difficult of them all. The fluid has to pass through the micro channels of the sensor and flushed at a steady rate so as to record CTD parameters. Apart from the sensor it is of equal importance to design a signal conditioning circuit to convert the responses of various sensors into useful form.

2.2.1 Conductivity Sensor

Conductivity is an intrinsic property of seawater from which salinity and density can be derived. From the measure of conductance, conductivity can be calculated after taking into consideration the "cell constant" that reflects the ratio of length and cross-sectional area of the sampled water volume in which the electrical current actually flows. The flow of current through electrolytic conductors (liquid) is accomplished by the movement of electric charges (positive and negative ions) when the liquid is under the influence of an electrical field. The conductance of a liquid can be defined by its electrical properties—the ratio of current to voltage between any two points within the liquid. As the two points move closer together or further apart, this value changes.

Conductivity is measured by two design approaches: electrodes or transformers (inductive). The electrode method has four electrodes of sufficiently low resistance Fig. 2a. The transformer (inductive) method uses a transformer to couple a known voltage to the water and detect the resulting current flow with a second transformer core. The electrode method is simpler and accurate. The measurement can be carried out using a constant current source and measuring the voltage. The conductance Y is measured by reading the voltage drop across the sensing electrodes in the presence of a constant current flowing through it using Ohm's Law. This is multiplied by the cell constant to obtain the conductivity. The temperature compensation to be applied. Maintenance of stable cell geometry is the limiting design challenge and is crucial for high accuracy when considering the effects of seawater such as coatings due to mineral depositions and water pollutants due to oil slick, bio fouling, industrial wastes, and marine growth. Anti-fouling coatings can be given near the entry/exit ports of the electrode cell to enhance the life of the conductivity cell. A design to overcome this effect has been developed and several sensors have been fabricated and used at sea [9].

2.2.2 Temperature Sensor

The electrical resistance of a conductor varies according to its temperature and this forms the basis of resistance thermometry. The effect is most commonly exhibited as an increase in resistance with increasing temperature, a positive temperature coefficient of resistance. When utilizing this effect for temperature measurement, a large value of temperature coefficient is ideal; however, its stability over the short and long term is equally vital. The relationship between the temperature and the electrical resistance is usually nonlinear. For the measurement of temperature, a thin film of platinum provides an extremely stable and sensitive thermometer. Temperature is measured indirectly by reading the voltage drop across the sensing resistor in the presence of a constant current flowing through it using Ohm's Law. For accurate measurement of resistance, a Wheatstone bridge circuit is incorporated.

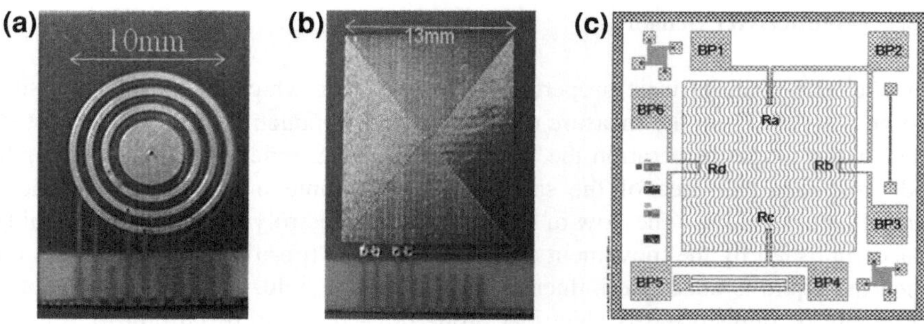

Fig. 2 **a** Conductivity. **b** Temperature and **c** Pressure sensor

Platinum film resistor on a ceramic/polymer substrate is an easy approach Fig. 2b. Thin film sensors have fast thermal response and their small thermal mass minimizes intrusion in the media being tested. The resistance thermometer sensor is protected from the environment by a thin film of thermally conductive oxide. The shown device is on a flexible substrate and several variants have been produced and used [10].

2.2.3 Pressure Sensor

Silicon micro machined pressure sensors are the most mature commercial sensors technology available today. The pressure sensors used for measurement of high pressures are based on piezoresistive sensing technique [11].

The basic structure of a piezoresistive pressure sensor consists of four sensing elements in a Wheatstone bridge configuration to measure stress within a thin, crystalline silicon membrane. The stress is a direct consequence of the membrane deflecting in response to an applied pressure differential across the front and backsides of the sensor.

The sensors' sensitivity can be improved by finding the optimum membrane shape and resistor configuration using finite element analysis. The bridge is made of four piezoresistors located on the four edges of the sensor membrane, close to the edges where the stress is the maximum when vertical pressure is applied to the center of the membrane Fig. 2c. Two of the resistors are positioned parallel to the direction of the stress, and their resistance increases with pressure. The other two resistors are oriented perpendicular to the direction of the stress and their resistance decreases with pressure. In the absence of applied pressure, the bridge is balanced and the output is zero. On application of pressure, there is a change in the resistor R_a, R_b, R_c and R_d values, resulting in a bridge output voltage proportional to the input pressure. For the poly-silicon resistor across the diaphragm, the resistance change caused by pressure due to change in the dimensions of the resistors can be expressed as

$$\Delta R/R = (1 + 2\upsilon)\Delta l/l + \Delta r/r \tag{1}$$

where ΔR and R are the change in resistance and original resistance of the resistor υ is the Poisson's ratio of the material, Δl is the change in length of the resistor due to pressure; l is the original length of the resistor, $\Delta \rho$ is the change in density, and ρ is the original density.

The fabricated devices have been packaged for underwater operations with data logging electronics and used up to 1,000 m depth in the marine environment.

2.3 Shear Stress Sensor

2.3.1 Introduction

Micro Electromechanical Systems shear stress sensors offer the potential to make flow measurements in fluid with unprecedented sensitivity, and spatial and temporal resolution [12, 13]. Most MEMS shear stress sensors have been developed for measurements in air [14] and utilize indirect methods. Substantial work on thermal-based sensors (hot wire/film anemometry) has been reported [15], but these devices require a priori knowledge of flow profiles, in situ calibration under identical conditions, and are limited by heat transfer in water.

These sensors are designed to study the effect of hydrodynamics and surface roughness on flow profiles and mass transfer. Arrays of these sensors allow the first direct measurement of shear stress profiles under unsteady wave-driven flow over a coral reef canopy (natural rough surfaces), as well as in oscillatory flowing cell-culture, and cardiovascular mock ups. Robust underwater shear stress sensors are required for measurements with fine spatial, ~100 μm–1 mm, and temporal (110 kHz) resolution, as well as sensitivity over the range of 0.01–100 Pa. These sensor arrays provide an exciting platform to explore factors affecting wall shear stress, such as roughness, as well as spatial variation along and across the flow. These sensors will allow detection of flow reversals in turbulent flow and normal force due to flow separation.

2.3.2 Design

The floating element sensor concept consists of a plate element suspended by four tethers, as shown in Fig. 3. The uniqueness of this design is in its transduction scheme which uses sidewall-implanted piezoresistors to measure lateral force (and shear stress), along with traditional top-implanted piezoresistors to detect normal forces. Piezoresistors are placed at the root of each tether from which shear stress can be inferred. Each sensor measures normal and lateral forces simultaneously. The orientations of the piezoresistors are chosen such that two of them are sensitive to lateral (along the flow direction), while the other two are sensitive to out-of-plane deflections.

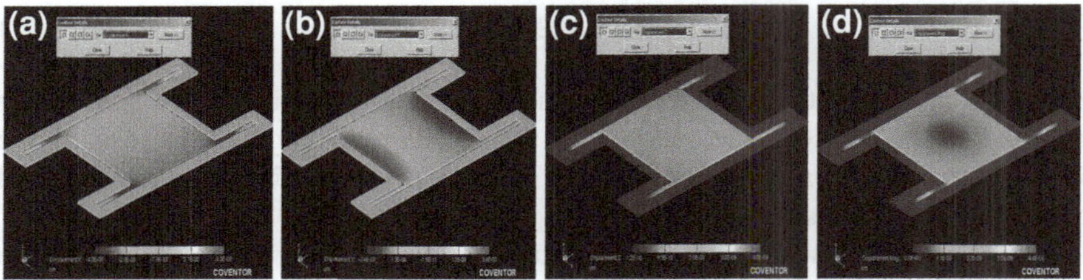

Fig. 3 Displacements due to flow along **a** x, **b** y, **c** z, directions and **d** magnitude of displacement

2.3.3 Simulation

Each tether is modeled as a fixed-guided beam (fixed at one end to the substrate and guided at the other end by a quarter of the plate element). As fluid flows on top of the sensor, it exerts shear stress on the top surface (plate element and tethers), causing the tethers to bend. The bending moments due to the resultant fluid forces and the stress at the root of the tether where the piezoresistor is located can be calculated and then the change in the resistance of the piezoresistor due to applied stress is determined. The piezoresistors are oriented along the <110> direction of (100) p-Silicon, which gives the maximum value for πl ($\sim 71 \times 10^{-12}$ cm^2/dyne).

2.3.4 Results

Figure 3 shows the simulated responses of the sensor for a hydrostatic force of 10 μN along different directions of fluid flow. From the detailed simulation studies of various dimensions [16], the sensor with tether length 250 μm shows high displacement, Fig. 3d, of the plate element (diaphragm); hence finalized for fabrication.

2.4 Acoustic Vector Sensor

2.4.1 Introduction

Acoustic sensors can be scalar or vector, meaning they can measure only the magnitude of acoustic pressure at a point or can measure any of the acoustic vectors, i.e., displacement, velocity or acceleration in terms of its magnitude and direction. Scalar sensors have omnidirectional pattern and large arrays of scalar sensors to be used for DOA estimation [17, 18]. Vector sensors with their inherent directional pattern of "figure of eight" offer greater advantage in terms of their smaller dimensions and selective noise rejection capabilities.

An acoustic vector sensor based on a piezoresistive MEMS accelerometer is designed. Piezoresistive accelerometers have a simple structure, batch-fabrication potential, a dc response, simple readout circuits, high reliability, and low cost but suffer from dependence of temperature. Many acoustic vector sensor designs based on MEMS accelerometers have been proposed but only few of them cater to a wide bandwidth requirement. A high sensitivity bionic structure of 702 Hz resonant frequency has been reported in [19]. This MEMS accelerometer based on a four-beam microstructure and cylinder is geometrically modified to meet high bandwidth specification. Analytical and FEM studies prove that an eight-beam mass structure has higher sensitivity and stiffness than a quad beam structure. Combining the merits of an eight-beam mass structure and a hollow cylinder gives a novel MEMS vector sensor.

2.4.2 Design

A novel design is based on a biaxial MEMS accelerometer. The two commonly used configurations of micro machined accelerometers are the cantilever supported mass and multiple supported mass. Cantilever configurations have low lateral sensitivity, whereas multiple supported configurations have greater lateral sensitivity and hence can be used as biaxial sensor [20]. Various multiple supported configurations (Table 1) have been compared for their sensitivities, resonant frequencies, and cross-axe sensitivities to determine the optimum design for high bandwidth and sensitivity [21]. Configurations (a) and (b) have beams supporting the seismic mass along the four sides. Configurations (a) has one beam on each side at its middle, while config. (b) has them near the corners, and config. (c) has two beams near the corners on each opposite sides but no beams on the other two opposite sides, and config. (d) has two beams near the corners on all the four sides of the seismic mass. The stiffness constants Kx, Ky and Kz and corresponding resonance frequencies of each configuration are calculated. K_{Θ_x} and K_{Θ_y} are the rotational stiffness along x and y-axes and from these, sensitivities and cross-axes sensitivities of the configurations are derived [22]. It is evident from Table 1 that the sensitivities of the third and fourth configurations are the maximum in the two lateral directions and the fourth configuration is twice as stiff compared to all other structures, as it has twice as many beams. The length, width, and thickness of the beam are 400, 120, and 50 μm; width and thickness of the central block are 600 and 50 μm and height and radius of cylinder are 4,000 and 100 μm Fig. 4a.

Hence the fourth multiple beam configuration, i.e., an eight-beam mass structure, is found to be the optimum beam design due to its high resonance frequency and equal lateral sensitivities. The sensitivity of the structure, however, decreases further with increase in resonant frequency.

The relationship of sensitivity and resonant frequency with the stiffness constant and the effective mass is given in the following equations:

Table 1 Comparison of different configurations

Configuration	Mass (kg)	K_z (N/m)	f_r (MHz)	K_{Θ_x}/K_z	K_{Θ_y}/K_z
(a)	$6.43e^{-8}$	158,440	1.59	0.021	0.021
(b)	$6.43e^{-8}$	158,440	1.59	0.041	0.041
(c)	$6.43e^{-8}$	158,440	1.59	0.028	0.028
(d)	$8.67e^{-8}$	316,875	2.21	0.028	0.028

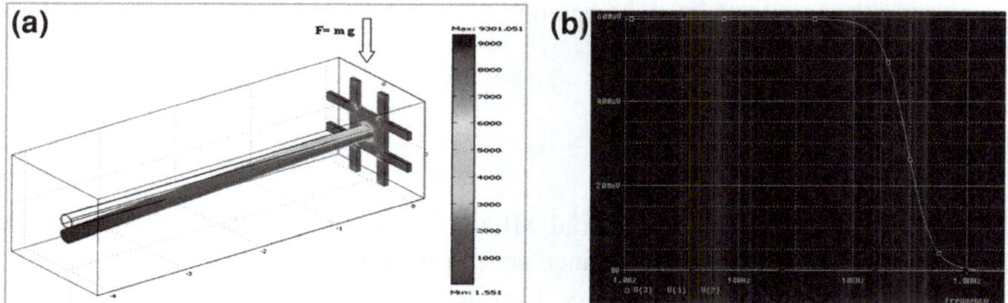

Fig. 4 **a** Eight-beam structure with hollow cylinder. **b** Amplified sensitivity

$$\text{Sensitivity} \propto (\text{displacement}/\text{acceleration}) \propto M_{\text{eff}}/K, \tag{2}$$

$$2\pi f_r = \sqrt{(K/M_{\text{eff}})} \tag{3}$$

where f_r is the natural frequency or resonant frequency, K is the stiffness constant, and M_{eff} is the effective mass.

Hence, to design a structure with optimum sensitivity and high bandwidth various novel structures have been introduced. The dependence of natural frequency of the quad beam central cylinder micro accelerometer on the device geometry is analyzed by simulating the analytical equation using Matlab. The results led to finalization of the dimensions of the novel structure which satisfies the high bandwidth requirement of 15 kHz (Table 2).

FEM studies of central cylinder with different materials show that high strength and low density seem to be the desirable material properties and carbon fiber reinforced plastics and silicon are the most promising and it is observed that the resonant frequency increases as the rigidity of the structure increases. In order to achieve high resonant frequency of the structure the required Young's modulus is extremely high.

From the above discussions it emerges that silicon and carbon reinforced fiber prove to be promising materials for the cylinder since they give comparatively closer values of resonant frequency. The thickness of the hollow silicon cylinder and its influence on the resonant frequency study concluded that a cylinder wall thickness of 10 μm gives the required bandwidth of 15 kHz. Thus, a novel biaxial MEMS accelerometer-based vector sensor is designed with multiple supported

Table 2 Comparison of three different beam structures

Characteristics	Quad beam without cylinder	Quad beam with hollow cylinder	Eight beam with hollow cylinder
First eigen frequency, *fr*	1.57 MHz	15.67 kHz	16.32 kHz
Position of piezoresistors	10 μm from either ends	20 and 10 μm from the mass and frame end	10 and 5 μm from the mass and frame end
Maximum stress developed	2617 Pa	7,859 Pa	9,301 Pa
Acc. sensitivity	0.00015 μV/g	0.008 μV/g	0.0054 mV/g
Cross-axis sensitivity	0.009 nV/g	0.038 nV/g	0.0405 nV/g
Amplified sensitivity, gain = 100	16.5 nV/g	0.09 mV/g	0.6 mV/g

beams and a central square mass on which is attached a hollow silicon/CFRP cylinder as shown in Fig. 4a. Wheatstone bridge is used for the measurement of resistance change of piezoresistors on sensors [23], since a bridge circuit increases the sensitivity. The eight-beam structure requires 16 piezoresistors, two on each beam ends. The change in the resistances under *X*, *Y*, and *Z* accelerations and the bridge configuration has been studied. The property of piezoresistivity is used as the transduction mechanism. Boron doped silicon material with an unstressed resistivity of $180e^{-6}$ Ωm and piezoresistive coefficients of $\pi_{11} = 6.6e^{-11}$ (1/Pa), $\pi_{12} = -1.11e^{-11}$ (1/Pa) and $\pi_{44} = 143.6e^{-11}$ (1/Pa) are used.

2.4.3 Simulation

The simple quad beam structure, the quad beam structure with hollow cylinder, and the eight-beam structure with hollow central cylinder are simulated using FEM modeling software COMSOL with piezoresistive elements of dimension 50 × 50 × 2.5 μm. The frame ends of the beam are fixed and rest of the structure is free. The structure is meshed using tetrahedral elements and eigen frequency analysis of eight beam with hollow cylinder structure is studied. For an applied acceleration of 1 g, maximum linear stress region is determined to locate the position of piezoresistors to be implanted. Circuit connections are set up using SPICE circuit editor in COMSOL Multiphysics and the amplified output is shown in Fig. 4b.

2.4.4 Results

The results obtained from the FEM studies for the three different configurations are summed in Table 2. The criteria are the bandwidth, maximum stress developed, position of maximum stress developed, sensitivity, and cross-axis sensitivity. Based on the results, a hollow cylinder of 10 μm wall thickness was finalized.

Wheatstone bridge circuit is designed so as to operate at maximum imbalance and to cancel out lateral sensitivities. The sensitivity of the sensor is further improved by amplification and common mode noise rejection capabilities of an instrumentation amplifier with a gain of 100.

A novel MEMS acoustic vector sensor with high bandwidth, high sensitivity, and low cross axis sensitivity and high linearity for underwater applications has been finalized for fabrication [22].

2.5 Imaging Sensor

2.5.1 Introduction

Visibility is very poor to detect objects in the littoral waters, which is a significant requirement of underwater acoustic imaging sensors. An acoustic-based system can provide a solution for underwater imaging. It has a clear advantage over optical methods in situations where the water is murky and in the detection of buried objects. In order to get finer resolution of the objects, it is imperative to work at MHz frequencies and also a very large number of transducers are required in a planar array configuration. This is possible with MEMS-based technology, which has built-in pre-amplifier in close proximity to the sensor. Different types of transducers are being practiced: CMUT, PMUT, and Helmholtz resonator.

Capacitive micro machined ultrasonic transducers have become very popular for more than a decade in non-destructive evaluation, underwater acoustic imaging, medical imaging, etc. With the advent of MEMS and silicon micromachining techniques, it is possible to make capacitors with submicron gaps where electric fields of over 10^8 V/m can be sustained [24–26]. The merit of such a high electric field is that it results in transducers where the electromechanical coupling coefficient can get close to unity, and thus be very competitive with the best piezoelectric material in terms of bandwidth, dynamic range, and sensitivity. A CMUT cell is an electroded membrane suspended over a highly doped silicon substrate (Fig. 5a). The receiver output is a function of device capacitance, change in capacitance and DC bias. During reception when an ultrasound signal hits the surface of the DC biased membrane, a current is generated due to the change in capacitance. The use of CMUT for underwater immersion applications is largely limited compared to air and medical applications due to the special package requirements.

2.5.2 Design and Simulation

CMUT is designed by FEM and optimized for acoustic imaging at 1 MHz for operating at a depth of 10 m underwater. In underwater, the CMUT membrane is deflected by the hydrostatic pressure as a function of depth and applied DC bias. In

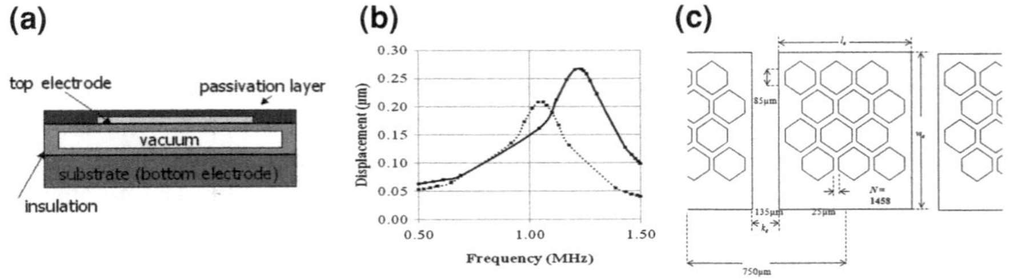

Fig. 5 **a** CMUT. **b** Fluid–structure interaction. **c** Array dimensions for 1 MHz

addition, the sensor has to sense the acoustic pressure (dynamic). For the same applied pressure of 0.21 MPa, the peak displacements of square, circular, and hexagonal membranes were 0.5, 0.703, and 0.685 μm respectively. In view of complexity involved in fabricating circular membranes, hexagon is selected as the suitable geometry as its response can be approximated to that of a circular membrane. In addition, hexagon offers a better packing density compared to a circle [27].

2.5.3 Results and Discussion

The model incorporates the damping effect imposed on the membrane by the surrounding fluid medium, DC bias, geometrical, and material parameters of membrane and gap height for their effect on the receiver performance. The fluid–structure interaction study shows that the peak membrane displacement is decreased by 22 % due to fluid loading and the membrane resonant frequency decreased from 1.23 to 1.06 MHz, Fig. 5b [28]. The hexagonal silicon membrane with 85 μm edge length and 1.5 μm thickness is selected as the optimized membrane for 1 MHz application. Simulations on hexagonal models show that the output change in capacitance increases with radius and thickness. An increase in gap height results in weakening the response of CMUT. Simulations revealed that silicon membrane offered a larger deflection of 0.173 μm compared to that of silicon nitride membrane with a deflection of 0.072 μm. The resonant frequency, collapse voltage, and input capacitance of the unit cell are 1.135 MHz, 51.64 V, and 0.186 pF, respectively. In order to obtain the output characteristics of unit cell, an acoustic pressure varying up to 0.01 MPa is applied on the CMUT unit cell, in addition to DC bias (90 % collapse voltage) and hydrostatic loading of 0.21 MPa. A change in capacitance of 9.59 fF is observed for 10 kPa acoustic pressure, which is very small. This necessitates the design of an array combining many unit cells, in which all cells of an element work in unison.

A CMUT array represents a periodic arrangement of elements; each element composed of many unit cells. Each cell in an element works in parallel to contribute to the element response. Since the cells are connected in a parallel configuration, the capacitance change in an element will be the sum of the changes in capacitance of each of the cells comprised in it. As a result, effective resistance seen by the succeeding electronic circuitry will be equal to 1/*N*th of the equivalent resistance of the unit cell where *N* denotes the number of cells in an element. The number of cells per element was selected such that the total effective resistance of the combination of cells match with the input impedance of the consecutive electronic circuitry in order to achieve maximum power transfer. The number of cells per element *N* in the array is calculated to be 1,458. In order to incorporate the maximum number of cells and at the same time allowing appropriate element kerf, it necessitates including three unit cells in a row. The proposed 1D array design consists of 128 elements, each of dimensions 0. 615 × 74.17 mm. The finalized simulated design detail of array is shown in Fig. 5c [28].

3 Conclusions

Various types of underwater MEMS sensors and devices such as CTD sensors, variants of acoustic sensors both for detection and acoustic imaging for oceanographic applications have been developed at the Naval Physical and Oceanographic Laboratory. Some of the challenges faced during these developments are the low leakage current, 10^{-12} A, at the gate of the MOSFET for acoustic sensor, which could be overcome by optimizing the fabrication process parameters, packaging of pressure sensors, and CTD for underwater deployment. The response of the flexible conductivity and temperature sensors is better than the conventional sensors being used due to the large area thin film. The other sensors have been designed but not yet fabricated.

Acknowledgments The authors thank the Director, NPOL for the encouragement and permission to publish this work, and his colleagues for discussions on sensors.

References

1. Hayward G, Bennett J, Hamilton R (1995) A theoretical study on the influence of some constituent material properties on the behaviour of 1–3 connectivity composite transducer. J Acoust Soc Am 98(4):2187–2196
2. Swartz RG, Plummer JD (1979) Integrated silicon-PVF2 acoustic transducer array. IEEE Trans Electron Devices ED-26(12):1921–1931
3. Fiorillo AS, Spiegel JV, Bloomfield PE, Esmail-Xandi D (1990) A P(VDF-TrFE)—based integrated ultrasonic transducer. Sens Actuators, A21–A23:719–725
4. Uma G, Umapathy M, Sumy J, Natarajan V, Kathiresan M (2007) Design and simulation of PVDF-MOSFET based MEMS hydrophone. J Instrum Sci Technol 35(3)329–339
5. Zheng XR, Lai PT, Liu BY, Li B, Cheng YC (1997) An integrated PVDF ultrasonic sensor with improved sensitivity using polyimide. Sens Actuators A 63:147–152

6. Zhu B, Varadan VK (2002) Integrated MOSFET-based hydrophone device for underwater applications. In: Proceedings of SPIE on smart structures and materials 2002: smart electronics, MEMS, and nanotechnology, vol 4700, pp 101–110
7. Gopikrishna M, Natarajan V, Kathiresan M (2005) Proceedings of international conference on MEMS and semiconductor nanotechnology, vol TM1.7. IIT, Kharagpur, pp 12–13
8. Fries D, Steimle G, Natarajan S, Vanova S, Broadbent H, Weller T () Maskless lithographic PCB/laminate MEMS for a salinity sensing system. University of South Florida, USA
9. Rajeev RA, Thomas KA, Natarajan V (2009) Design, fabrication and evaluation of 'zero external field' conductivity cell for CTD. Proceedings of SYMPOL 2009, pp. 128–132
10. Thomas KA, Rajeev RA, Natarajan V (2012) Low-cost flexible micro conductivity and temperature sensor for oceanography applications. In: 5th ISSS National Conference on MEMS, Smart Structures & Systems (2012), pp.17–22
11. Clark SK, Wise KD (1979) Pressure sensitivity in anisotropically etched thin-diaphragm pressure sensors. IEEE Trans Electron Devices 26(12):1887–1896
12. Barlian AA, Park SJ, Mukundan V, Pruitt BL (2005) Design and characterization of microfabricated piezoresistive floating element-based shear stress sensors. In: Proceedings of IMECE 2005, pp 1–6
13. Barlian AA, Narain R, Li JT, Quance CE, Ho AC, Mukundan V, Pruitt BL () Piezoresistive MEMS underwater Shear Stress Sensors. In: MEMS 2006, Turkey, 22–26 Jan 2006, pp 626–629
14. Horowitz S et al (2004) A wafer-bonded, floating element shear-stress sensor using a geometric moire optical transduction technique. In: Solid-state sensor, actuator and microsystems workshop, USA, 2004
15. Saunvit P, Yingchen Y, Douglas LJ, Jonathan E, Chang L (2006) Multisensor processing algorithms for underwater dipole localization and tracking using MEMS artificial lateral-line sensors. EURASIP J Appl Signal Process 2006(Article ID 76593):1–8
16. Diana ZM, Natarajan V, Elizabeth R (2009) Design and simulation of piezoresistive flow sensor. In: SENNET 2009, international conference on sensors and related networks, VITU, India
17. Shipps JC, Abraham BM (2004) The use of vector sensors for underwater port and waterway security. In: Sensors for industry conference, New Orleans, USA, 27–29, pp 41–44, Jan 2004
18. Charles HS, John LB (2007) Transducers and arrays for underwater sound. Springer, Berlin
19. Chenyang X, Shang C, Wendong Z, Binzhen Z, Guojun Z, Hui Q (2007) Design fabrication and preliminary characterisation of a novel MEMS bionic vector hydrophone. Microelectron J 38:1021–1026
20. Amarsinghe R, Dao DV, Toriyama T, Sugiyama S (2005) Design and fabrication of miniaturized six-degree of freedom piezoresistive accelerometer. In: MEMS 2005 conference, pp 351–354, 2005
21. van Kampen RP, Wolffenbuttel RF (1998) Modelling the mechanical behaviour of bulk-micromachined silicon accelerometers. Sens Actuators A 64:137–150
22. Roshna BR, Natarajan V () A novel MEMS vector sensor. In: ISSS-NC6, 6–7 Sep 2013, Pune, India
23. Chenyang X, Shang C, Hui Q, Wendong Z, Jijun X, Binzhen Z, Guojun Z (2008) Development of a novel two axis piezoresistive micro accelerometer based on silicon. Sensor Lett 6:1–10
24. Ladabaum X, Jin HT, Atalar A, Khuri-Yakub BT (1998) Surface micromachined capacitive ultrasonic transducers. IEEE Trans Ultrason Ferroelectr Freq Control 45:678–690
25. Omer O, Ergun AS, Cheng CH, Johnson JA, Karaman M, Khuri-Yakub BT (2002). Underwater acoustic imaging using capacitive micromachined ultrasonic transducer arrays. In: IEEE OCEANS'02, vol 4, pp 2354–2360
26. Oralkan O, Ergun AS, Johnson JA, Karaman M, Demirci U, Kaviani K, Lee TH, Khuri-Yakub BT (2002) Capacitive micromachined ultrasonic transducers: next generation arrays for acoustic imaging. IEEE Trans Ultrason Ferroelectr Freq Control 49:1596–1610

27. Anil A, Ramgopal, Maheshkumar, Pant BD, Dwivedi VK, Chandrashekhar, Babar A, Rudrapratap, George PJ (2008) Fabrication of capacitive micromachined ultrasonic transducer using wafer bonding technique. Sens Transduc J 93(6):15–20
28. Suresh G, Natarajan V, Srijith K, Raghavan S (2013) Design and modelling of 1 MHz CMUT for underwater applications. In: Acoustics 2013, New Delhi, India

Author Index

K. J. Vinoy et al. (eds.), *Micro and Smart Devices and Systems*,
Springer Tracts in Mechanical Engineering, DOI: 10.1007/978-81-322-1913-2,

Subject Index

K. J. Vinoy et al. (eds.), *Micro and Smart Devices and Systems*, Springer Tracts in Mechanical Engineering, DOI: 10.1007/978-81-322-1913-2,

T

V

W

X

Y

Z

ˋSIA information can be obtained at www.ICGtesting.com
ˋed in the USA
ˊ01*1104150614
ˋ5LV00010B/416/P

9 788132 219125